Selected Titles in This Series

64 **G. Ferreyra, R. Gardner, H. Hermes, and H. Sussmann, Editors,** Differential geometry and control (University of Colorado, Boulder, 1997)

63 **Alejandro Adem, Jon Carlson, Stewart Priddy, and Peter Webb, Editors,** Group representations: Cohomology, group actions and topology (University of Washington, Seattle, 1996)

62 **János Kollár, Robert Lazarsfeld, and David R. Morrison, Editors,** Algebraic geometry—Santa Cruz 1995 (University of California, Santa Cruz, July 1995)

61 **T. N. Bailey and A. W. Knapp, Editors,** Representation theory and automorphic forms (International Centre for Mathematical Sciences, Edinburgh, Scotland, March 1996)

60 **David Jerison, I. M. Singer, and Daniel W. Stroock, Editors,** The legacy of Norbert Wiener: A centennial symposium (Massachusetts Institute of Technology, Cambridge, October 1994)

59 **William Arveson, Thomas Branson, and Irving Segal, Editors,** Quantization, nonlinear partial differential equations, and operator algebra (Massachusetts Institute of Technology, Cambridge, June 1994)

58 **Bill Jacob and Alex Rosenberg, Editors,** K-theory and algebraic geometry: Connections with quadratic forms and division algebras (University of California, Santa Barbara, July 1992)

57 **Michael C. Cranston and Mark A. Pinsky, Editors,** Stochastic analysis (Cornell University, Ithaca, July 1993)

56 **William J. Haboush and Brian J. Parshall, Editors,** Algebraic groups and their generalizations (Pennsylvania State University, University Park, July 1991)

55 **Uwe Jannsen, Steven L. Kleiman, and Jean-Pierre Serre, Editors,** Motives (University of Washington, Seattle, July/August 1991)

54 **Robert Greene and S. T. Yau, Editors,** Differential geometry (University of California, Los Angeles, July 1990)

53 **James A. Carlson, C. Herbert Clemens, and David R. Morrison, Editors,** Complex geometry and Lie theory (Sundance, Utah, May 1989)

52 **Eric Bedford, John P. D'Angelo, Robert E. Greene, and Steven G. Krantz, Editors,** Several complex variables and complex geometry (University of California, Santa Cruz, July 1989)

51 **William B. Arveson and Ronald G. Douglas, Editors,** Operator theory/operator algebras and applications (University of New Hampshire, July 1988)

50 **James Glimm, John Impagliazzo, and Isadore Singer, Editors,** The legacy of John von Neumann (Hofstra University, Hempstead, New York, May/June 1988)

49 **Robert C. Gunning and Leon Ehrenpreis, Editors,** Theta functions – Bowdoin 1987 (Bowdoin College, Brunswick, Maine, July 1987)

48 **R. O. Wells, Jr., Editor,** The mathematical heritage of Hermann Weyl (Duke University, Durham, May 1987)

47 **Paul Fong, Editor,** The Arcata conference on representations of finite groups (Humboldt State University, Arcata, California, July 1986)

46 **Spencer J. Bloch, Editor,** Algebraic geometry – Bowdoin 1985 (Bowdoin College, Brunswick, Maine, July 1985)

45 **Felix E. Browder, Editor,** Nonlinear functional analysis and its applications (University of California, Berkeley, July 1983)

44 **William K. Allard and Frederick J. Almgren, Jr., Editors,** Geometric measure theory and the calculus of variations (Humboldt State University, Arcata, California, July/August 1984)

43 **François Trèves, Editor,** Pseudodifferential operators and applications (University of Notre Dame, Notre Dame, Indiana, April 1984)

(Continued in the back of this publication)

(Photograph courtesy of Harolyn Gardner)

Robert Brown Gardner
February 27, 1939–May 5, 1998

Proceedings of Symposia in
PURE MATHEMATICS

Volume 64

Differential Geometry and Control

Summer Research Institute on
Differential Geometry and Control
June 29–July 19, 1997
University of Colorado, Boulder

G. Ferreyra
R. Gardner
H. Hermes
H. Sussmann
Editors

American Mathematical Society
Providence, Rhode Island

PROCEEDINGS OF A SUMMER RESEARCH INSTITUTE ON DIFFERENTIAL GEOMETRY AND CONTROL HELD AT THE UNIVERSITY OF COLORADO, BOULDER JUNE 29–JULY 19, 1997

with support from the
National Science Foundation, Grant DMS-9713413

Any opinions, findings, and conclusions or recommendations expressed in this material are those of the authors and do not necessarily reflect the view of the National Science Foundation.

1991 *Mathematics Subject Classification.* Primary 49–XX, 53–XX, 93–XX; Secondary 22–XX, 60–XX, 35–XX.

Library of Congress Cataloging-in-Publication Data

Summer Research Institute on Differential Geometry and Control (1997 : University of Colorado, Boulder)

Differential geometry and control : Summer Research Institute on Differential Geometry and Control, June 29–July 19, 1997, University of Colorado, Boulder / G. Ferreyra... [et al.], editors.

p. cm. — (Proceedings of symposia in pure mathematics, ISSN 0082-0717 ; v. 64)

Includes bibliographical references.

ISBN 0-8218-0887-7 (alk. paper)

1. Control theory—Congresses. 2. Geometry, Differential—Congresses. 3. Exterior differential systems—Congresses. I. Ferreyra, Guillermo Segundo, 1953– . II. Title. III. Series.

QA402.3.S887 1997

629.8′312—dc21 98-38711

CIP

∞ The paper used in this book is acid-free and falls within the guidelines
established to ensure permanence and durability.
Visit the AMS home page at URL: http://www.ams.org/

10 9 8 7 6 5 4 3 2 1 04 03 02 01 00 99

Contents

Preface

The 1997 American Mathematical Society Summer Research Institute on Differential Geometry and Control was held in June-July 1997 at the University of Colorado in Boulder. It was an exciting and scientifically rewarding summer institute with the participation of more than 90 mathematicians and mathematically-oriented engineers from Bulgaria, Chile, Colombia, France, Germany, Israel, Italy, Mexico, Poland, Portugal, Russia, Switzerland and the United States.

The organizer of the conference was Hector J. Sussmann with Henry Hermes as co-organizer. Guillermo S. Ferreyra and the late Robert B. Gardner were additional members of the organizing committee.

This edited volume represents a cross section of the presentations made at the conference. The papers included here were refereed to the highest standards and contain complete results, as well as interesting open problems, in the area of Geometric Control Theory. The reader interested in this topic will find here an excellent source of current challenging research problems and results. The objective of this collection of papers is to provide a coherent global perspective of recent developments and important open problems in the area of Geometric Control Theory.

Control theory is a very large and active area of contemporary science and engineering. In the United States, most work in control takes place in industry, and academic activity in this field occurs primarily in departments of engineering, especially electrical, mechanical and aerospace. The main professional society of control theorists and engineers is the IEEE Control Systems Society, which has about 10,000 members, i.e. a size comparable to about half that of the entire American Mathematical Society. There is also significant activity by mathematicians, working in engineering or mathematics departments. Within the Society for Industrial and Applied Mathematics (SIAM) there is an activity group in control with about 300 members, which sponsors a triennial conference with about 500 participants. There are many large control conferences, such as the IEEE Conference on Decision and Control, held every year in December, where about 1,000 papers are presented each year. As counterpart to the wide breadth of the large conferences held yearly in control theory, the 1997 NSF/AMS Summer Research Institute focused mainly on the differential geometric aspects of control theory. Participants exchanged ideas and presented recent work to colleagues and graduate students.

The field of "Differential-geometric nonlinear control theory" was born in the 1960's, with the pioneering work of R. Hermann, H. Hermes, and C. Lobry, who were the first to adopt the perspective that a control system should be regarded primarily as a collection of vector fields. Starting from this vector-field point of view it is reasonable to expect that the natural algebraic structure of the space of vector fields —given by the linear operations and the Lie bracket— ought to play a significant role in the study of control-theoretic problems, and the work of Hermann, Hermes and Lobry provided strong support to this idea, in the form of concrete applications to control problems. Since then, the field has grown and expanded in several directions. The differential-geometric point of view —and in particular the systematic search for properties that are invariant under the appropriate groups of nonlinear transformations, and the representation of control systems as systems of vector fields on manifolds— has led to the development of new classification theories for nonlinear control systems, and the discovery of new results on feedback stabilization and optimal control. Applications have been devised to numerous problems such as satellite and helicopter control, robot trajectory planning, and optimal drug therapy. In the opposite direction, methods from optimal control have had an impact on the understanding of differential-geometric questions such as the problem of the structure of sub-Riemannian minimizers and various questions about Lie semigroups. The interaction between geometry and control theory involves the use of various differential-geometric tools, in particular Lie algebras of vector fields, free Lie algebras, nilpotent Lie algebras, Lie algebra filtrations, Lie groups and Lie semigroups, symplectic geometry, vector bundles, distributions (in the differential-geometric sense of "subbundles of the tangent bundle") and their dual objects (i.e. codistributions), exterior differential systems, Cartan prolongations and equivalence, infinite-dimensional manifolds ("diffieties") of map jets, as well as tools from "analytic geometry" (real-analytic, semianalytic and subanalytic sets, stratifications and desingularization), from control theory (the "Maximum Principle" of Optimal Control and its generalizations, theorems on local and global controllability) and from dynamical systems (Lyapunov techniques, stable and unstable manifolds, center manifolds). Most of these topics are covered in this volume.

With sorrow we learned last month of the untimely death of our friend and great colleague Robert Gardner. Upon our invitation, George Wilkens has written an article for this proceedings in tribute to Robert Gardner's lasting contributions to mathematics.

We gratefully acknowledge the financial support of the National Science Foundation, and of the University of Colorado in Boulder. We would like to thank Jim Maxwell (American Mathematical Society) for his help with the administration of the NSF grant and Wayne Drady (American Mathematical Society) for the on-site organization of the Summer Institute and associated social events. We also acknowledge the superb efforts of the many anonymous referees that were involved with the papers submitted to this publication. Finally, we wish to thank Christine Thivierge of the American Mathematical Society, for her excellent support in the editorial and prepublishing process for this volume.

Guillermo Ferreyra

Henry Hermes

Hector J. Sussmann

Proceedings of Symposia in Pure Mathematics
Volume **64**, 1999

Remembering the Mathematics of Robert Brown Gardner

George R. Wilkens

1. Mathematical Interests

Robby Gardner, one of the organizers of this Summer Research Institute, passed away on May 5, 1998[1]. Robby led an extremely active mathematical life producing more than fifty papers, three books, eighteen masters students and thirteen Ph.D students. He seemed to spend a sizeable fraction of every year traveling to conferences and meetings around the world. I was continually amazed by the number of mathematicians who had met Robby. He was one of those personalities that everyone, regardless of field, knew.

Robby was part of a long heritage of great differential geometers, tracing a direct line from Élie Cartan, through S.S. Chern, to himself. Robby's own mathematical interests clearly reflect this lineage. Even a casual observer of Robby's work could not help noticing the central role of exterior differential systems. Nor could anyone fail to notice Robby's enthusiasm for Cartan's work and his vigorous efforts to bring this work to a much larger audience. Robby certainly succeeded. Numerous "heirs to Cartan" dot the current mathematical landscape, and many of them have Robby as an ancestor on their family tree.

Robby's mathematical interests covered a number of areas. A list of these areas includes the following:

- geometric partial differential equations
- Cartan's equivalence method
- Cartan's method of moving frames
- exterior differential systems

I want to thank Robert Bryant, Niky Kamran and William Shadwick for their comments and support.

[1]To honor his contributions to graduate education, the UNC Mathematics Department created the Robert Brown Gardner Memorial Fund devoted to supporting graduate student activities. Contributions may be sent to Ms. Janet Farrell, Robert Brown Gardner Fund, Department of Mathematics, CB#3250, University of North Carolina, Chapel Hill, NC 27599.

- submanifolds of Euclidean space
- integral geometry
- calculus of variations
- exterior algebra over commutative rings
- invariant theory
- control theory
- affine geometry
- Lie groups, Lie algebras and symmetries
- the geometry of shape in medical imaging.

I believe it is fair to say that Robby worked near the junction of geometry, algebra and analysis.

Before getting started, I want to issue a few apologies to the reader. Due to the limited time available to prepare this article, I will only be able to discuss some of the above topics. Further, time did not allow me to adequately refresh my memory of Robby's earlier work so my coverage will not be very balanced. Finally, I decided not to indicate which work is joint within the article, leaving that instead to the nearly complete bibliography. I hope this will not detract from the overall spirit of the article, and that the reader will forgive any shortcomings.

2. Exterior Differential Systems

Exterior differential systems (EDS) permeate nearly all of Robby's work. Many of his papers exemplify skillful and artistic applications of EDS. One of his very earliest papers [**Gar67**] deals with differential systems in their own right. Specifically this paper discusses the invariants of one of the most commonly occurring differential systems, Pfaffian systems.

In this paper, Robby lays out tools he will use his entire life. The first invariant of a Pfaffian system is its Cauchy characteristics. This is an involutive distribution, and the Pfaffian system is locally generated by 1-forms involving only first integrals of the Cauchy system and their differentials. Moreover, a maximal independent set of first integrals of the Cauchy system gives the minimum number of variables required for any set of generators of the Pfaffian system. This number of variables, which is also the co-dimension of the Cauchy characteristic system, is called the class of the Pfaffian system.

In addition to the Cauchy characteristics, there is also the derived flag of a Pfaffian system. The length of the flag is called the derived length, and the co-dimensions of this flag of subspaces give further integer invariants, called the type numbers. Another basic integer invariant is the Engel rank. These integer invariants are not arbitrary, and Robby establishes a number of inequalities they must satisfy.

The real meat of the paper is contained in the last two sections. Robby shows how to intrinsically define a symmetric conformal tensor on Pfaffian systems with even co-dimension. In the case of co-dimension 4, the conformal tensor is bilinear. The rank, index and any isotropic subspaces of this tensor are immediately seen to be invariants of the Pfaffian system. Robby applies these results to a single second order pde in the plane and shows that its classification as elliptic, parabolic, or hyperbolic corresponds exactly to the classification of the conformal tensor. Since this tensor is contact invariant, an immediate consequence is that the type of the pde is contact invariant. Robby goes on to show how the characteristic curves belonging to a solution of a hyperbolic or parabolic pde are completely determined by the isotropic subspaces of the conformal tensor.

In the early 1980s, Robby began participating in a new comprehensive treatment of the theory of exterior differential systems. This wonderful book [**BCG$^+$91**] is now the standard reference for the subject. A tremendous amount of effort went into its production, and I remember a period of about ten years where it was regularly referenced in papers as "EDS, to appear." With ten chapters and nearly five hundred pages, the book covers a wide range of topics and strikes an appropriate balance between theory and application. Robby, of course, was a master at applying EDS and any reader wanting to supplement [**BCG$^+$91**] with additional applications may consult the reference section of this article.

3. Submanifolds and Integral Geometry

Robby loved submanifold theory and integral geometry. He wrote a number of papers on these subjects including [**Gar69c, Gar70b, Gar70c, Gar71, Gar72b, Gar75b, Gar77b**].

In [**Gar70b**], Robby shows how the Dirichlet integral formula can be used to generate numerous integral formulas in differential geometry. In addition to showing how this technique leads to several established results, Robby proves a new rigidity theorem for Euclidean spheres.

Robby goes to town in [**Gar77b**] by digging into the study of arbitrary co-dimension submanifolds of Euclidean space. As one might guess from the title, *New viewpoints in the geometry of submanifolds of* $\mathbf{R}^N$, the new viewpoints are Robby's own and they are wonderfully original. Robby develops the basic invariants in a way that is both geometrically and algebraically natural. He defines new invariants, gives new meaning to old ones, and develops new methods for studying submanifolds.

4. Invariant Theory

While this section is very short, it would be very remiss not to mention one of Robby's most scholarly works [**Gar80**]. Interestingly, it is a review of a book on invariant theory. The book is only reviewed in the last paragraph while the rest of the article is a review of invariant theory itself.

Robby brings the reader through a historical development of the subject in a mere ten pages. He traces this development from the naive beginnings with Cayley and Sylvester to the post-Hilbert period of then present 1980. While, at first glance, this attention to invariant theory might seem out of place in the work of a differential geometer, Robby understood its central importance in the early development of differential geometry and, indeed, how important the interplay between the two subjects was. However, Robby's interest was not just scholarly. He certainly viewed invariant theory as the ultimate technique in tensor analysis and used it wherever possible.

5. Geometric PDE[2]

Robby's papers contain beautiful developments of EDS concepts applied to the geometric study of PDEs. In [**Gar69a**] and [**Gar69b**], Robby defined and studied a rather broad notion of a Cauchy problem for Pfaffian systems. He was notably able to prove an existence theorem for the Cauchy problem in the C^∞ category for Pfaffian systems admitting what Robby called k-stable vector fields. These are vector fields which essentially stabilize the Pfaffian system after k iterations of the Lie derivative. They can thus be thought of as natural generalizations of Cauchy characteristics. He then defined a powerful general notion of characteristic and focal systems for Pfaffian systems endowed with distinguished invariant subsystems, such as the maximal isotropic subsystems associated to an intrinsic conformal symmetric bilinear form. The fundamental examples are the Monge and Friedrichs characteristics for the Pfaffian systems associated to scalar second-order pdes or first-order determined systems in the plane. This enabled him to prove in [**Gar69b**] and [**Gar77a**] several elegant existence theorems and finite escape-time theorems for the smooth Cauchy problem for non-linear parabolic equations. Paper [**Gar79**] gives a very elegant explanation in EDS terms of a notoriously difficult memoir of Goursat on Bäcklund and Laplace transformations for scalar second-order hyperbolic pdes in the plane. Papers [**GK93**], [**GK95**] and [**GK97**] contain further applications of this body of ideas to second-order hyperbolic pdes and first-order determined systems in two independent variables.

6. Control Theory

Robby began his work in control theory in the early 1980s. He had a long interest in this vital area of twentieth-century research, especially in the work by the very active group in geometric control. Problems related to the classification of orbits of control systems with respect to various transformation groups held a special interest for Robby. He was, after all, one of the few experts at that time in a technique ideally suited to study such problems.

In a series of four talks given at a conference on geometric control theory at Michigan Tech, Robby introduced exterior differential systems and Cartan's equivalence method to a new audience. The corresponding paper [**Gar83**] showcases Robby's original perspectives on differential geometry, and I think it is an important mile marker in his career. On reexamining

[2] I am extremely grateful to Niky Kamran for contributing this section.

this paper, I can not help but feel the connection between his early papers (e.g. [**Gar67**], [**Gar68**], [**Gar69b**]) and his later books on the method of equivalence [**Gar89**] and on exterior differential systems [**BCG$^+$91**].

Robby began to study the orbits of control systems under the action of the pseudo-group of (static) feedback transformations. One of the principal tools of this investigation was Cartan's equivalence method. A series of papers describing normal forms, new integration methods, geometric isomorphisms, relationships with classical geometries, and a universal bundle for control systems issued forth.

Much of the early work is summarized very nicely in the paper [**Gar91**]. This paper begins by describing the representation of the feedback equivalence problem as a G-structure equivalence problem. Following that is a discussion of the G-structure equivalence associated to constrained variational problems of the sort found in Phillip Griffiths' book on calculus of variations. One immediately observes that these G-structures are formally identical which suggests a geometric isomorphism between the two problems. Some low dimensional examples are examined in detail. In addition to identifying systems with control-linear and linear normal forms, systems with special geometric structures are also identified. Examples include systems with intrinsically attached pseudo-Riemannian metrics and contact structures.

Robby generated tremendous interest in his applications of the equivalence method to control theory. Interest was so great that in 1988 Robby was invited to be the principal speaker at a joint AMS-SIAM CBMS regional conference at Texas Tech. Robby gave a series of ten lectures on Cartan's method of equivalence. Robby covered the general theory and presented numerous applications to differential geometry and control theory. Robby's book [**Gar89**] was the first to clearly exhibit the algorithmic nature of the equivalence method. His book remains the best one-stop source for general theory, examples and computational technique.

A particularly important piece of work emanated from studying the feedback orbits of Brunovsky normal forms. While the problem of identifying which non-linear systems are feedback equivalent to a Brunovsky normal form was already solved, some interesting new results were obtained. For instance in [**GSW89a**] it was shown that a Brunovsky normal form is characterized by its symmetry subgroup. Following this was a series of two papers [**GS91a**], [**GS92**] showing how the symmetries give rise to a new algorithm for finding a particular feedback transformation to linearize the original non-linear system. This algorithm, called the GS-algorithm, was a substantial improvement over then existing methods. The GS-algorithm identifies a set of Frobenius systems, minimal in dimension and number, to be integrated. A maximal independent set of solutions to these systems forms part of a linearizing coordinate system. The remaining coordinates are then determined by routine differentiation and linear algebra. The efficiency of the GS-algorithm led to its adoption by NASA as part of an automatic flight controller.

A beautiful normal form theorem lies at the heart of GS-algorithm. The spirit of this new theorem is very similar to other normal form theorems in exterior differential systems, such as the Engel, Goursat and Bryant normal forms (see [**Gar83**] or [**BCG$^+$91**] for details on these normal forms.) I believe that this result inspired some of the work by Shankar Sastry, Richard Murray and their colleagues on applications of exterior differential systems.

The paper on control structures [**BG95**] provided an intrinsic and global representation of control systems in terms of an affine co-distribution in the co-tangent space of a manifold. The co-distribution must satisfy some reasonable necessary conditions. The interesting properties of these structures are that they naturally determine a manifold of states, their presentation in local coordinates has the expected form, and the given co-distribution is realized as the pull-back of a canonical co-distribution on the punctured tangent bundle to the state manifold. This last property represents the universal bundle construction belonging to control structures. One of the main technical advantages of control structures is that tools like the method of equivalence can be directly and easily applied.

The work on control structures developed from the interpretation of a control system as a subbundle of the tangent bundle to state space, and the interpretation of a feedback transformation as the canonical extension of a state space diffeomorphism to the tangent bundle. From these interpretations, Robby observed that a control system embeds a submanifold in each tangent space and that a feedback transformation induces a linear action on each of these submanifolds. He reasoned that an important first step in understanding the geometry of the subbundle is to understand the geometry in each fiber. The next step is to piece the fiber geometries together in a way that exhibits their evolution over state space.

The fiber geometry can be described as the study of submanifolds of a finite dimensional vector space under the action of the full general linear group. This turns out to be a subject in classical differential geometry called centro-affine geometry. While this geometry is classical, far more attention had been given to the study of affine geometry and equi-affine geometry. Theorems in centro-affine geometry were relatively scarce. Additionally, the existing theorems had to be adapted to their intended applications to the feedback equivalence problem. With that in mind, Robby established the basic theorems for curves and hyper-surfaces in centro-affine geometry [**GW97**].

Some applications of centro-affine geometry are found in [**GW93**] and [**GW96b**]. These papers revisit the geometric isomorphism between control systems and variational problems. They show that the feedback invariants for a co-dimension 1 control system are completely determined by the centro-affine geometry of the fibers. The evolution of the fiber geometry across the state space is uniquely determined. The relationship between these co-dimension 1 systems and Finsler geometry becomes clear. This connection shows how to transport the same analysis to Finsler structures. In this case the centro-affine fiber geometry determines Chern's connection for Finsler geometry. This geometric isomorphism with Finsler geometry also explains the appearance of pseudo-Riemannian metrics which were observed earlier.

The applications just cited use the centro-affine structure of hypersurfaces to study co-dimension 1 systems. The centro-affine structure of curves was being applied to the study of scalar control systems. That work is unfinished, but I intend to continue that project and report on this part of Robby's work.

7. Farewell to a Friend

My goal in this article has been to give the reader a sample of the substantial contributions Robby made to mathematics. I also want to take this opportunity to say something about Robby as a person. Many of us knew Robby through his passions: his love for his family, his loves of fine food and fine wine, his love of music, his love of travel, his ardent stamp collecting hobby, and his commitment to excellence in all things.

At the center of all these passions was a deeply caring person. This is what I think of first when I think of my friend. Robby's office door was always open, and this simple symbol reflects so much of Robby himself. I never saw him turn anyone away, regardless of how busy he was. Robby would guide the inexperienced, nurture the talented, and clear obstructions wherever he could. He was a marvelous teacher, mentor, colleague and friend. He touched the lives of so many people and in ways too numerous to list. Robby leaves a legacy of humanity as well as one of mathematics, and with very fond memories I bid him *adieu*.

References

[BCG+91] R. L. Bryant, S. S. Chern, R. B. Gardner, H. L. Goldschmidt, and P. A. Griffiths, *Exterior differential systems*, Mathematical Sciences Research Institute Publications, vol. 18, Springer-Verlag, New York, 1991.

[BG95] Robert L. Bryant and Robert B. Gardner, *Control structures*, Geometry in nonlinear control and differential inclusions (Warsaw, 1993), Banach Center Publ., vol. 32, Polish Acad. Sci., Warsaw, 1995, pp. 111–121.

[EGM+94] D. Eberly, R. Gardner, B. Morse, S. Pizer, and C. Scharlach, *Ridges for image analysis*, J. Math. Imaging and Vision **4** (1994), 353–373.

[Gar65] Robert B. Gardner, *Differential geometric methods in partial differential equations*, Ph.D. thesis, University of California at Berkeley, 1965.

[Gar67] Robert B. Gardner, *Invariants of Pfaffian systems*, Trans. Amer. Math. Soc. **126** (1967), 514–533.

[Gar68] Robert B. Gardner, *Some applications of the retraction theorem in exterior algebra*, J. Differential Geometry **2** (1968), 25–31.

[Gar69a] Robert B. Gardner, *The Cauchy problem for Pfaffian systems*, Comm. Pure Appl. Math. **22** (1969), 587–596.

[Gar69b] Robert B. Gardner, *A differential geometric generalization of characteristics*, Comm. Pure Appl. Math. **22** (1969), 597–626.

[Gar69c] Robert B. Gardner, *An integral formula for immersions in euclidean space*, J. Differential Geometry **3** (1969), 245–252.

[Gar70a] Robert B. Gardner, *A differential geometric approach to characteristics*, Global Analysis (Proc. Sympos. Pure Math., Vol. XVI, Berkeley, Calif., 1968), Amer. Math. Soc., Providence, R.I., 1970, pp. 179–186.

[Gar70b] Robert B. Gardner, *The Dirichlet integral in differential geometry*, Global Analysis (Proc. Sympos. Pure Math., Vol. XV, Berkeley, Calif., 1968), Amer. Math. Soc., Providence, R.I., 1970, pp. 231–237.

[Gar70c] Robert B. Gardner, *The geometry of subscalar pairs of metrics*, Proc. Carolina Conf. on Holomorphic Mappings and Minimal Surfaces (Chapel Hill, N.C., 1970), Dept. of Math., Univ. of North Carolina, Chapel Hill, N.C., 1970, pp. 29–42.

[Gar71] Robert B. Gardner, *The technique of integral formulas in the geometry of immersions*, Differentialgeometrie im Grossen (Tagung, Math. Forschungsinst., Oberwolfach, 1969), Bibliographisches Inst., Mannheim, 1971, pp. 127–144. Ber. Math. Forschungsinst. Oberwolfach, No. 4.

[Gar72a] Robert B. Gardner, *Lectures on exterior algebras over commutative rings*, Department of Mathematics, University of North Carolina, Chapel Hill, N. C., 1972.

[Gar72b] Robert B. Gardner, *Subscalar pairs of metrics and hypersurfaces with a nondegenerate second fundamental form*, J. Differential Geometry **6** (1972), 437–458, Collection of articles dedicated to S. S. Chern and D. C. Spencer on their sixtieth birthdays.

[Gar74] R. B. Gardner, *The influence of Carathéodory on differential systems*, Proceedings of the C. Carathéodory International Symposium (Athens, 1973), Greek Math. Soc., Athens, 1974, pp. 146–153.

[Gar75a] Robert B. Gardner, *The fundamental theorem of vector relative invariants*, J. Algebra **36** (1975), no. 2, 314–318.

[Gar75b] Robert B. Gardner, *Geometry of submanifolds of euclidean spaces*, Differential geometry (Proc. Sympos. Pure Math., Vol. XXVII, Stanford Univ., Stanford, Calif., 1973), Part 1, Amer. Math. Soc., Providence, R.I., 1975, pp. 125–134.

[Gar77a] Robert B. Gardner, *Differential geometric viewpoints on the development of shock waves*, The 1976 Ames Research Center (NASA) Conference on the Geometric Theory of Non-linear Waves (Moffett Field, Calif., 1976), Math. Sci. Press, Brookline, Mass., 1977, pp. 89–104. Lie Groups: History, Frontiers and Appl., Vol. VI.

[Gar77b] Robert B. Gardner, *New viewpoints in the geometry of submanifolds of* $\mathbf{R}^N$, Bull. Amer. Math. Soc. **83** (1977), no. 1, 1–35.

[Gar78] Robert B. Gardner, *Differential geometry and foliations: the Godbillon-Vey invariant and the Bott-Pasternack vanishing-theorems*, Differential topology, foliations and Gelfand-Fuks cohomology (Proc. Sympos., Pontifícia Univ. Católica, Rio de Janeiro, 1976), Springer, Berlin, 1978, pp. 75–94. Lecture Notes in Math., Vol. 652.

[Gar79] Robert B. Gardner, *Constructing Bäcklund transformations*, Partial differential equations and geometry (Proc. Conf., Park City, Utah, 1977), Lecture Notes in Pure and Appl. Math., vol. 48, Dekker, New York, 1979, pp. 77–89.

[Gar80] Robert B. Gardner, *Invariant theory*, Bull. Amer. Math. Soc. **12** (1980), 246–256.

[Gar83] Robert B. Gardner, *Differential geometric methods interfacing control theory*, Differential geometric control theory (Houghton, Mich., 1982), Progr. Math., vol. 27, Birkhäuser, Boston, Mass., 1983, pp. 117–180.

[Gar85] Robert B. Gardner, *Differential equations and the method of equivalence*, Proc. IEEE Conference on Decision and Control, December 1985, Fort Lauderdale, FL, pp. 1777–1778.

[Gar88] Robert B. Gardner, *The principal components construction applied to affine unimodular repère mobile*, Resultate Math. **13** (1988), no. 3-4, 235–239, Reprinted in *Affine Differentialgeometrie* [235–239, Tech. Univ. Berlin, Berlin, 1988].

[Gar89] Robert B. Gardner, *The method of equivalence and its applications*, CBMS-NSF Regional Conference Series in Applied Mathematics, vol. 58, Society for Industrial and Applied Mathematics (SIAM), Philadelphia, PA, 1989.

[Gar91] Robert B. Gardner, *The geometry of nonlinear control systems*, Differential geometry, Pitman Monographs Surveys Pure Appl. Math., vol. 52, Longman Sci. Tech., Harlow, 1991, pp. 179–198.

[Gar76] Robert B. Gardner, *Modules admitting determinants*, Linear and Multilinear Algebra **3** (1975/76), no. 3, 209–214.

[GK93] R. B. Gardner and N. Kamran, *Characteristics and the geometry of hyperbolic equations in the plane*, J. Differential Equations **104** (1993), no. 1, 60–116.

[GK95] R. B. Gardner and N. Kamran, *Normal forms and focal systems for determined systems of two first-order partial differential equations in the plane*, Indiana Univ. Math. J. **44** (1995), no. 4, 1127–1162.

[GK97] R. B. Gardner and N. Kamran, *Exterior differential systems and normal forms of parabolic systems*, Advances in mathematical sciences: CRM's 25 years (Montreal, PQ, 1994), CRM Proc. Lecture Notes, vol. 11, Amer. Math. Soc., Providence, RI, 1997, pp. 215–230.

[GKS95] Robert B. Gardner, Marcus Kriele, and Udo Simon, *Generalized spherical functions on projectively flat manifolds*, Results Math. **27** (1995), no. 1-2, 41–50, Festschrift dedicated to Katsumi Nomizu on his 70th birthday (Leuven, 1994; Brussels, 1994).

[GS85] Robert B. Gardner and William F. Shadwick, *Equivalence of one-dimensional Lagrangian field theories in the plane. I*, Global differential geometry and global analysis 1984 (Berlin, 1984), Lecture Notes in Math., vol. 1156, Springer, Berlin-New York, 1985, pp. 154–179.

[GS87a] R. B. Gardner and W. F. Shadwick, *Feedback equivalence of control systems*, Systems Control Lett. **8** (1987), no. 5, 463–465.

[GS87b] R. B. Gardner and W. F. Shadwick, *Overdetermined equivalence problems with an application to feedback equivalence*, Differential geometry: the interface between pure and applied mathematics (San Antonio, Tex., 1986), Contemp. Math., vol. 68, Amer. Math. Soc., Providence, RI, 1987, pp. 111–119.

[GS87c] R. B. Gardner and W. F. Shadwick, *A simple characterization of the contact system on $J^k(E)$*, Rocky Mountain J. Math. **17** (1987), no. 1, 19–21.

[GS90a] R. B. Gardner and W. F. Shadwick, *Feedback equivalence for general control systems*, Systems Control Lett. **15** (1990), no. 1, 15–23.

[GS90b] R. B. Gardner and W. F. Shadwick, *Symmetry and the implementation of feedback linearization*, Systems Control Lett. **15** (1990), no. 1, 25–33.

[GS91a] Robert B. Gardner and William F. Shadwick, *An algorithm for feedback linearization*, Differential Geom. Appl. **1** (1991), no. 2, 153–158.

[GS91b] Robert B. Gardner and William F. Shadwick, *An equivalence problem for a two-form and a vector field on $\mathbf{R}^3$*, Differential geometry, global analysis, and topology (Halifax, NS, 1990), CMS Conf. Proc., vol. 12, Amer. Math. Soc., Providence, RI, 1991, pp. 41–50.

[GS92] Robert B. Gardner and William F. Shadwick, *The GS algorithm for exact linearization to Brunovsky normal form*, IEEE Trans. Automat. Control **37** (1992), no. 2, 224–230.

[GSW89a] R. B. Gardner, W. F. Shadwick, and G. R. Wilkens, *Feedback equivalence and symmetries of Brunowski normal forms*, Dynamics and control of multibody systems (Brunswick, ME, 1988), Contemp. Math., vol. 97, Amer. Math. Soc., Providence, RI, 1989, pp. 115–130.

[GSW89b] Robert B. Gardner, William F. Shadwick, and George R. Wilkens, *A geometric isomorphism with applications to closed loop controls*, SIAM J. Control Optim. **27** (1989), no. 6, 1361–1368.

[GW93] Robert B. Gardner and George R. Wilkens, *Classical geometries arising in feedback equivalence*, Proc. IEEE Conference on Decision and Control, December 1993, San Antonio, TX, pp. 3437–3440.

[GW96a] Robert B. Gardner and George R. Wilkens, *Preface for "Applications of Finsler geometry to control theory"*, Finsler geometry (Seattle, WA, 1995), Contemp. Math., vol. 196, Amer. Math. Soc., Providence, RI, 1996, pp. 227–229.

[GW96b] Robert B. Gardner and George R. Wilkens, *A pseudo-group isomorphism between control systems and certain generalized Finsler structures*, Finsler geometry (Seattle, WA, 1995), Contemp. Math., vol. 196, Amer. Math. Soc., Providence, RI, 1996, pp. 231–243.

[GW97] Robert B. Gardner and George R. Wilkens, *The fundamental theorems of curves and hypersurfaces in centro-affine geometry*, Bull. Belg. Math. Soc. Simon Stevin **4** (1997), no. 3, 379–401.

[OPGP91] W. R. Oliver, S. M. Pizer, R. B. Gardner, and K. B. Pryzwansky, *Evaluation of cytologic deformation by boundary-driven energy minimization*, Prog. Clin. Biol. Res. **363** (1991), 355–368.

DEPARTMENT OF MATHEMATICS, UNIVERSITY OF HAWAII, 2565 THE MALL, HONOLULU, HI 96822–2273

E-mail address: `grw@math.hawaii.edu`

Proceedings of Symposia in Pure Mathematics
Volume **64**, 1999

A Hamiltonian approach to strong minima in optimal control

Andrei Agrachev, Gianna Stefani, and Pierluigi Zezza

ABSTRACT. We give a sufficient condition of geometric type for an optimal control problem on a manifold assuming that the maximized Hamiltonian is sufficiently smooth. The condition extends the classical condition of non-existence of conjugate (focal) points in the calculus of variations.

1. Introduction

We will state a sufficient condition for strong local optimality for an optimal control problem where the state belongs to a finite dimensional Riemannian manifold M and the time interval $I = [0, T]$ is fixed. The state end-point constraints are submanifolds of M, the control values belong to a subset of a finite dimensional manifold U. The problem can be described as

$$\textit{Minimize} \quad J(\xi, v) = c_0(\xi(0)) + c_T(\xi(T)) + \int_0^T f_t^0(\xi(t), v(t))\, dt$$

over all ξ, v satisfying the following control equation with end-point and control constraints

$$\dot{\xi}(t) = f_t(\xi(t), v(t)), \quad a.e.\ t \in I \tag{1.1}$$

$$\xi(0) \in N_0\ ,\ \xi(T) \in N_T \tag{1.2}$$

$$v(t) \in \Omega_t \subseteq U, \quad a.e.\ t \in I \tag{1.3}$$

where N_0, N_T are submanifolds of M. Further assumptions on the control problem are given in Section 2.

Let us first introduce the following definition; a couple (ξ, v) is called *admissible* if ξ is an absolutely continuous solution of (1.1), the constraints (1.2) and (1.3) are satisfied and the function $t \mapsto f_t^0(\xi(t), v(t)) \in L^1(I, \mathbb{R})$.

Our goal is to state conditions assuring that an admissible reference couple $(\widehat{\xi}, \widehat{v})$ is a strong local minimizer, that is, for some $\epsilon > 0$, $(\widehat{\xi}, \widehat{v})$ is a minimizer with respect to all the admissible couples (ξ, v) such that

$$\max_{t \in I}\ d(\xi(t), \widehat{\xi}(t)) < \epsilon,$$

where d is the distance on M induced by the Riemannian structure.

1991 *Mathematics Subject Classification.* Primary 49K15, 58F05.

Work partly supported by bilateral CNR research grant # 96.00184.01.

The optimality conditions are given in a geometric form that, in the classical cases, corresponds to the existence of a non self-intersecting field of extremals covering a neighbourhood of the reference state trajectory.

To this aim we assume that the maximized Hamiltonian is defined and sufficiently smooth around the reference adjoint covector; this enables us to use the symplectic properties of the Hamiltonian flow to construct a field of extremals starting from the Lagrangian submanifold of the initial transversality conditions.

The other assumptions (Assumptions *i* and *ii* of Theorem 2.3) extend the meaning of conjugate (focal) point and imply that the field has the required properties. They describe how the two flows corresponding to the maximized and reference Hamiltonians intersect along the reference covector.

We notice that these conditions make sense even if the second variation does not exist (see Example 2.4) and, when the second variation exists, they reduce to its coerciveness. The proof of the results is based on an auxiliary linear quadratic control problem obtained from the two Hamiltonians. Let us notice that this linear-quadratic problem is not the accessory problem even if the second variation exists, although, in this case, the maximized Hamiltonians of the two linear-quadratic problems coincided.

We include also the so called abnormal case. We obtain that an abnormal trajectory which satisfies the sufficient conditions is isolated among the neighbouring admissible ones.

Let us remark that using the *local* maximized Hamiltonian we can give a geometric proof of the second order sufficient conditions for weak local minima.

A similar approach has been used to state sufficient conditions for control problems in $\mathbb{R}^n$, see [**ASZ98**], where we state also conditions implying the smoothness of the maximized Hamiltonian.

Let us notice that the usual expression of the second variation written through second derivatives of the pre-Hamiltonian cannot be used for a problem on a manifold M; in fact the derivative with respect to the state variable (the *horizontal* derivative) does not have a meaning in T^*M, moreover the second derivative is not invariant for coordinate change because the first derivative is not, in general, equal to zero.

In a forthcoming paper we obtain the second variation as a linear quadratic problem on the tangent space to M, (see [**ASZb**]).

A deeper analysis of the role of Hamiltonian methods in deriving sufficient conditions for optimality will be carried out in [**ASZa**].

In Section 2 we state the main result whose proof is in Section 3.

2. Notations and statement of the results

For sake of simplicity, when the statement is clear, we do not write that a relation holds almost everywhere for $t \in I$.

Assume that M is a C^3 connected paracompact manifold, its Riemannian structure induces a Riemannian structure also on the associated tensor bundles so that a distance is well defined. For a Riemannian manifold X the space $L^\infty(I, X)$ is well defined and for $\xi \in C(I, X)$ and $\epsilon > 0$ the tube around ξ of radius ϵ is

$$\mathcal{T}(\xi, \epsilon) = \{(t, x) \in I \times X : d(x, \xi(t)) < \epsilon\} .$$

To state the regularity assumptions on the data let us introduce the following

DEFINITION 2.1. Assume that X, Y are finite dimensional C^k manifolds and let $\mathcal{T}$ be a tube around $\xi \in C(I, X)$. We will say that the map $G : \mathcal{T} \subseteq I \times X \to Y$ is quasi-C^p, $p \leq k-1$ if it satisfies the following

i. For each $t \in I$ the map G_t is C^p
ii. The tangent map $T_2^p G : (t, \nu) \mapsto T^p G_t(\nu)$ belongs to $L^\infty(I \times T^p X, T^p Y)$.

Regularity of the data. We assume that N_0, N_T are C^2 embedded submanifolds of M and that the functions

$$c_0 : N_0 \to \mathbb{R}, \; c_T : N_T \to \mathbb{R}$$

are C^2. We assume, moreover, that the reference vector field

$$\widehat{f}_t : (t, x) \mapsto f_t(x, \widehat{v}(t))$$

and the reference Lagrangian

$$\widehat{f}_t^0 : (t, x) \mapsto f_t^0(x, \widehat{v}(t))$$

are quasi-C^2 in a tube around $\widehat{\xi}$.

In what follows we will make use of the following fact, let X be a manifold and let 0_x be the zero of $T_x^* X$, then $T_{0_x} T^* X$ is canonically isomorphic to $T_x X \times T_x^* X$. Moreover for any C^2-function σ such that $d\sigma(x) = 0$ the second derivative $D^2\sigma(x)$ is well defined as a bilinear symmetric form on $T_x X$ and $d\sigma_* v = (v, D^2\sigma(x)(v, \cdot))$ for all $v \in T_x X$. Finally we identify any bilinear form Q on a vector space V with a linear form $Q : V \to V^*$ and we write

$$Q(v, w) = \langle Qv, w \rangle.$$

In this section we are going to use some basic symplectic methods on the symplectic manifold T^*M. For a general introduction to these methods and its application to variational problems we refer to [**Arn80**] while for more specific applications to optimal control we refer to [**AG97**].

Denote by $\pi : T^*M \to M$ the canonical projection. The space $T^*_{\pi \ell} M$ is canonically embedded in $T_\ell T^*M$ as the space of tangent vectors to vertical curves. Moreover any time dependent quasi-C^2 Hamiltonian $H_t : T^*M \to \mathbb{R}$ defines a Hamiltonian vector field which is denoted by $\vec{H}_t$.

The first natural assumption is that the reference couple $(\widehat{\xi}, \widehat{v})$ satisfies the Pontryagin maximum principle. To state it we introduce two Hamiltonians, the (control dependent) pre-Hamiltonian $h : I \times T^*M \times U \to \mathbb{R}$ which is defined as

$$h : (t, \ell, u) \mapsto \langle \ell, f_t(\pi(\ell), u) \rangle - p_0 f_t^0(\pi(\ell), u)$$

and the reference Hamiltonian $\widehat{H} : I \times T^*M \to \mathbb{R}$ which is defined as

$$\widehat{H} : (t, \ell) \mapsto h_t(\ell, \widehat{v}(t)).$$

From the assumptions on the reference vector field and Lagrangian it follows that also $\widehat{H}$ is quasi-C^2. Both the above Hamiltonians depend on the parameter p_0 which can take the values $\{0, 1\}$ characterizing, respectively, the abnormal and normal case.

The Pontryagin maximum principle. $(\widehat{\xi},\widehat{v})$ satisfies the Pontryagin maximum principle if there is a non-trivial couple $(p_0,\widehat{\lambda})$, where $\widehat{\lambda}: I \to T^*M$ is a lifting of $\widehat{\xi}$ to T^*M, satisfying the Hamiltonian system

$$\dot{\lambda}(t) = \vec{\widehat{H}}_t(\lambda(t)) \tag{2.1}$$

the transversality conditions

$$\widehat{\lambda}(0)_{|TN_0} = p_0\, dc_0(\widehat{\xi}(0))_{|TN_0} \tag{2.2}$$

$$-\widehat{\lambda}(T)_{|TN_T} = p_0\, dc_T(\widehat{\xi}(T))_{|TN_T} \tag{2.3}$$

and the maximization property

$$\widehat{H}_t(\widehat{\lambda}(t)) = \max_{u\in\Omega_t} h_t(\widehat{\lambda}(t),u).$$

$(\widehat{\xi},\widehat{v})$ is called *normal* if $p_0 = 1$, *abnormal* if $p_0 = 0$.

For simpler notations we set

$$x_0 = \widehat{\xi}(0),\ x_T = \widehat{\xi}(T),\ \ell_0 = \widehat{\lambda}(0),\ \ell_T = \widehat{\lambda}(T).$$

The two functions p_0c_0 and p_0c_T are defined on N_0 and N_T respectively but they can be extended to the whole manifold M in such a way that the transversality conditions (2.2) and (2.3) hold for the extended functions without restrictions to TN_0 and TN_T. Hence we denote by $\alpha,\beta : M \to \mathbb{R}$ two functions such that

$$\alpha_{|N_0} = p_0\, c_{0|N_0},\ \beta_{|N_T} = p_0\, c_{T|N_T} \tag{2.4}$$

$$\ell_0 = d\alpha(x_0),\ \ell_T = d\beta(x_T). \tag{2.5}$$

In the normal case α,β are cost functions equivalent, on the constraints, to the original ones while in the abnormal case they are extensions of the zero function on the constraints.

When $p_0 = 0$ all the costs disappear from our conditions and indeed we will study a problem with a zero cost. Proving that $\widehat{\xi}$ is a *strict* minimizer will imply that it is isolated among the admissible trajectories.

Let us now introduce our main assumption

ASSUMPTION 2.2. *There is $\epsilon > 0$ such that the Hamiltonian $H : \mathcal{T}(\widehat{\lambda},\epsilon) \subset I \times T^*M \to \mathbb{R}$ defined as*

$$H : (t,\ell) \mapsto \sup_{u\in\Omega_t} h_t(\ell,u),$$

is finite and quasi-C^2.

Under this assumption we can consider the Hamiltonian system

$$\dot{\lambda}(t) = \vec{H}_t(\lambda(t)), \tag{2.6}$$

and we denote by $\mathcal{H}_t(\ell)$ the solution of (2.6) with $\lambda(0) = \ell$, we will use analogous notations for the flows of other Hamiltonian systems.

As a consequence of the Pontryagin maximum principle, $\widehat{\lambda}$ is a solution of systems (2.6) and (2.1) hence, from the compactness of the time interval, it follows that there is an open neighbourhood $\mathcal{O}$ of ℓ_0 such that $\mathcal{H}$ and $\widehat{\mathcal{H}}$ are defined on $I\times\mathcal{O}$.

Moreover the end-points of $\widehat{\lambda}$ belong to the initial and final Lagrange manifolds in T^*M described by the transversality conditions, that is

$$\ell_0 \in L_0 = \{d\alpha(q) + \omega : q \in N_0, \omega \in T_q^{\perp} N_0\}$$
$$\ell_T \in L_T = \{-d\beta(q) + \omega : q \in N_T, \omega \in T_q^{\perp} N_T\}$$

where " $^{\perp}$" means orthogonal with respect to the dual coupling.

In the calculus of variations, when the final point is fixed, the idea underlying the proof of sufficient conditions for strong local optimality lies in constructing a field of extremals by projecting on M the Hamiltonian flow starting from L_0. If the initial point is not free then this projection is singular at ℓ_0 for $t = 0$ but it is not singular at $\widehat{\lambda}(t)$ for small $t > 0$. The non singularity of the projection on the whole interval corresponds to the possibility of extending the field of extremal until $t = T$ with no self-intersection and it is a sufficient condition for optimality. The previously described condition (which can be also phrased as : *there are no focal points in* $[0,T]$) can be written as

$$T_{\widehat{\lambda}(t)} \mathcal{H}_t(L_0) \cap T^*_{\widehat{\xi}(t)} M = \{0\}, \; 0 < t \leq T.$$

When the final point is fixed, we have that L_T is the vertical fiber over x_T, i.e.,

$$L_T = T^*_{x_T} M$$

and it remains vertical under the action of the backward reference flow, i.e.,

$$\widehat{\mathcal{H}}_t \widehat{\mathcal{H}}_T^{-1}(L_T) = T^*_{\widehat{\xi}(t)} M.$$

We end the analysis of this case with the following remark. Since $AC(I, \mathbb{R}^n) \cong \mathbb{R}^n \times L^1(I, \mathbb{R}^n)$ then, in the calculus of variation when the final point is fixed, we optimize in the space $L^1(I, \mathbb{R}^n)$ since any *control* $v \in L^1(I, \mathbb{R}^n)$ defines a unique state trajectory.

Here we consider a general optimal control problem and we do not assume that the final point is fixed, hence we have to take into account two phenomena: the projection might be singular on a non-trivial interval $[0, \tau]$, moreover, we have to optimize with respect to both the control function and the initial point.

The first phenomenon can be captured by considering the vertical component of $T_{\ell_0} L_0$ which is given by

$$\Pi_0 = T_{\ell_0} L_0 \cap T^*_{x_0} M = T^{\perp}_{x_0} N_0,$$

and the part of this subspace which remains vertical under the action of the tangent of the Hamiltonian flow $\mathcal{H}_t$ untill time τ

$$\Pi_\tau = \{\mathcal{H}_{\tau *}\ell : \ell \in \Pi_0 \;, \mathcal{H}_{t*}\ell \in T^*_{\widehat{\xi}(t)} M \;, \; t \in [0, \tau]\}.$$

Notice that this subspace has a dimension which is non-increasing in time.

REMARK. There are two important cases when this subspace is trivial. When there are no constraints on the initial point then $\Pi_\tau = \{0\}$ for all $\tau \in I$ because $\Pi_0 = \{0\}$. When the control problem admits a non-singular second variation and the linearized system is controllable then it is possible to show that $\Pi_\tau = \{0\}$ for $\tau > 0$ [**SZ97, ASZ98**].

As we mentioned, in problems where both end-points can vary, the initial condition plays the role of a control parameter. To investigate its effects we fix the control to be equal to the reference one and we vary only the initial condition, in this way, the control equation reduces to a two point boundary value problem depending on x, namely

$$\dot{\xi}(t) = \widehat{f}_t(\xi(t)) \tag{2.7}$$

$$\xi(0) = x \in N_0 \ , \ \xi(T) \in N_T, \tag{2.8}$$

which has x_0 has a solution. Denoting by

$$(t, x) \mapsto S_t(x)$$

the flow of the above differential equation, the functional $p_0 J$ reduces to a finite dimensional functional

$$J_0(x) = \alpha(x) + \gamma(x)$$

where

$$\gamma(x) = \beta(S_T(x)) + p_0 \int_0^T f_t^0(S_t(x), \widehat{v}(t))\, dt. \tag{2.9}$$

To describe the constraint set for this problem we pull back the final manifold N_T to the initial time by the reference flow S_t and we define

$$\widehat{N}_0 = S_T^{-1}(N_T),$$

hence the control equations (2.7), (2.8) can be equivalently written as

$$x \in N_0 \cap \widehat{N}_0.$$

Thanks to the particular choice of α and β (see equation (2.5)) and from Lemma 3.1 it follows that the first order extremality conditions hold on the whole space, that is

$$dJ_0(x_0) = 0.$$

Analogously to what we did for the final manifold N_T we pull back the final Lagrangian manifold L_T to the initial time by the reference Hamiltonian and we set

$$\widehat{L}_0 = \widehat{\mathcal{H}}_T^{-1}(L_T). \tag{2.10}$$

We can now state our main result which involves two assumptions: the first is a sufficient condition for J_0 to have a constrained strict local minimum, while the second one concerns the two manifolds $\mathcal{H}_t(L_0), \widehat{\mathcal{H}}_t(\widehat{L}_0)$. The condition is a geometric assumption on the way they intersect at $\widehat{\lambda}(t)$ and it is a natural extension of the conjugate (focal) point condition.

THEOREM 2.3. *Assume that $(\widehat{\xi}, \widehat{v})$ satisfies the Pontryagin maximum principle, that Assumption 2.2 holds and that*

i.) $D^2 J_0(x_0) > 0$ *on* $T_{x_0}N_0 \cap T_{x_0}\widehat{N}_0$

ii.) $T_{\widehat{\lambda}(t)}\mathcal{H}_t(L_0) \cap T_{\widehat{\lambda}(t)}\widehat{\mathcal{H}}_t(\widehat{L}_0) \subseteq \Pi_t,\ t \in I.$

If $p_0 = 1$ *then* $(\widehat{\xi}, \widehat{v})$ *is a strong local minimizer.*
If $p_0 = 0$ *then* $\widehat{\xi}$ *is a* C^0 *isolated trajectory among the admissible ones; in this case i) is equivalent to*

$$T_{x_0}N_0 \cap T_{x_0}\widehat{N}_0 = \{0\}.$$

REMARK. In [**ASZ98**] we prove a sufficient condition assuming that the second variation is coercive instead of i) and ii). Using the results is [**SZ97**], one can verify that the two conditions are equivalent when the second variation exists.

We end this section by an example where the second variation does not exists but the maximized Hamiltonian is smooth, so that we can check our conditions.

EXAMPLE 2.4. Let us consider the following problem where the state belongs to $\mathbb{R}^n$ and the controls to $\mathbb{R}^m$

$$\textit{Minimize} \qquad J(\xi, v) = \int_0^T \left\{ \|v(t)\|^{1+\epsilon} + \varphi(\xi(t)) \right\} dt$$

subject to

$$\dot{\xi}(t) = f(\xi(t)) + \sum_{i=1}^m v_i(t)\, f_i(\xi(t)).$$

Some easy computations yield that the maximized Hamiltonian is

$$H(q,p) = \langle p, f(q)\rangle + \frac{\epsilon}{(1+\epsilon)^{\frac{1+\epsilon}{\epsilon}}} \left(\sum_{i=1}^m \langle p, f_i(q)\rangle^2 \right)^{\frac{1}{2}+\frac{1}{2\epsilon}} - \varphi(q).$$

For $0 < \epsilon < 1$ although the Lagrangian is not C^2 the Hamiltonian H is C^2. For $\epsilon = 1$ we obtain a problem which is linear-quadratic in the control.

3. Proof of the result

Let us start by proving a technical Lemma which describes an important property of the reference Hamiltonian flow.

LEMMA 3.1. *For all $\ell \in \mathcal{O} \subseteq T^*M$ the following equality holds*

$$\widehat{\mathcal{H}}_t(\ell) = \left(\ell + p_0 \int_0^t d\widehat{f}_s^0(S_s(\pi\ell)) S_{s*}\, ds \right) S_{t*}^{-1}$$

PROOF. Thanks to the special structure of the reference Hamiltonian vector field we have that the projection on M of the flow of the reference Hamiltonian is the flow of the reference control system, i.e., $\pi\widehat{\mathcal{H}}_t(\ell) = S_t(\pi\ell)$ hence

$$\dot{\lambda}(t) = -\lambda(t)\widehat{f}_{t*} + p_0 d\widehat{f}_t^0(S_t(\pi\ell)).$$

The statement can be obtained by the variation of constants formula.

By setting $\lambda(t) = \omega(t) S_{t*}^{-1}$ one obtains

$$\omega(t) = \ell + p_0 \int_0^t d\widehat{f}_s^0(S_s(\pi\ell)) S_{s*}\, ds$$

and hence the statement follows. □

To prove Theorem 2.3 we need a third Hamiltonian which allows us to compare the cost along $(\widehat{\xi}, \widehat{v})$ with the neighbouring admissible trajectories (see Lemma 3.2). Consider the *relative* Hamiltonian G_t defined by

$$G_t = (H_t - \widehat{H}_t) \circ \widehat{\mathcal{H}}_t.$$

By the results in [**AG97**] the flow $\mathcal{G}_t$ of G_t is the pull-back of the Hamiltonian flow $\mathcal{H}_t$ by means of the reference one, i.e.

$$\mathcal{G}_t = \widehat{\mathcal{H}}_t^{-1} \circ \mathcal{H}_t$$

and hence $\mathcal{G}_t(\ell_0) = \ell_0$. Moreover, it follows from the relations

$$\widehat{H}_t(\ell) \le H_t(\ell)\ , \ \widehat{H}_t(\widehat{\lambda}(t)) = H_t(\widehat{\lambda}(t))$$

that G_t is a non-negative Hamiltonian defined on $\mathcal{O}$ and that it is identically zero at ℓ_0. As a consequence

$$dG_t(\ell_0) = 0 \ , \ D^2G_t(\ell_0) \geq 0.$$

It is very important to notice that $\mathcal{G}_{t*}$ is the Hamiltonian flow associated with the quadratic form $D^2G_t(\ell_0)$ as it can be easily verified in coordinates (see [**MR94**]).

We can now restate our theorem with the Hamiltonian G_t. Denote by

$$\Lambda_0 = T_{\ell_0}L_0 \text{ and } \widehat{\Lambda}_0 = T_{\ell_0}\widehat{L}_0,$$

the statement *ii)* of Theorem 2.3 becomes

$$ii)' \quad \mathcal{G}_{t*}\,\Lambda_0 \cap \widehat{\Lambda}_0 \subseteq \widehat{\mathcal{H}}_{t*}^{-1}(\Pi_t) = \{\mathcal{G}_{t*}\ell : \ell \in \Pi_0 \ , \mathcal{G}_{s*}\ell \in T^*_{x_0}M, s \in [0,t]\}.$$

Let $\boldsymbol{s}$ be the canonical form in $T^*M$and $\boldsymbol{\omega} = \boldsymbol{s} - G_t dt$ be the Cartan form on $I \times T^*M$ associated to G_t.

The link between the Cartan form and our problem is described by the following

LEMMA 3.2. *Let (ξ, v) be an admissible couple, define*

$$\eta(t) = S_t^{-1}(\xi(t)).$$

If λ is a lifting of η then

$$p_0 J(\xi, v) \geq \int_\lambda \boldsymbol{\omega} + \alpha(\eta(0)) + \gamma(\eta(T)),$$

where γ is defined in (2.9).

PROOF. To evaluate $\boldsymbol{\omega}$ along the curve λ we have to integrate on I the following expression which can be simplified by using Lemma 3.1

$$\begin{aligned}
&\langle \lambda(t), \dot\eta(t)\rangle - H_t(\widehat{\mathcal{H}}_t(\lambda(t))) + \widehat{H}_t(\widehat{\mathcal{H}}_t(\lambda(t))) = \\
&= \langle \widehat{\mathcal{H}}_t(\lambda(t)), S_{t*}\dot\eta(t)\rangle - p_0\int_0^t \langle d\widehat{f}^0_s(S_s(\eta(t))), S_{s*}\dot\eta(t)\rangle ds + \\
&\quad - H_t(\widehat{\mathcal{H}}_t(\lambda(t))) + \widehat{H}_t(\widehat{\mathcal{H}}_t(\lambda(t))) = \\
&= \langle \widehat{\mathcal{H}}_t(\lambda(t)), \dot\xi(t)\rangle - H_t(\widehat{\mathcal{H}}_t(\lambda(t))) - p_0\widehat{f}^0_t(\xi(t)) - p_0\int_0^t \langle d\widehat{f}^0_s(S_s(\eta(t))), S_{s*}\dot\eta(t)\rangle ds \leq \\
&\leq p_0 f^0_t(\xi(t), v(t)) - \frac{d}{dt} p_0 \int_0^t \widehat{f}^0_s(S_s(\eta(t)))\, ds.
\end{aligned}$$

By integration we have

$$\begin{aligned}
p_0 J(\xi, v) &\geq \int_\lambda \boldsymbol{\omega} + p_0\int_0^T \widehat{f}^0_s(S_s(\eta(T)))\, ds + p_0 c_0(\xi(0)) + p_0 c_T(\xi(T)) \\
&= \int_\lambda \boldsymbol{\omega} + \alpha(\eta(0)) + \gamma(\eta(T)),
\end{aligned}$$

which ends the proof. □

The flow $\mathcal{G}_t$ starting from L_0 generates the manifold

$$\mathcal{L} = \{(t, \mathcal{G}_t(\ell)) \ : \ t \in I \ , \ \ell \in \mathcal{O} \cap L_0\}$$

which is a Legendre submanifold of $I \times T^*M$.

Let us recall some results which are proved in [**ASZ98**] when $M = \mathbb{R}^n$ but which can be easily extended to our case.

LEMMA 3.3. *The differential form* $\boldsymbol{\omega}$ *is exact on* $\mathcal{L}$ *so that*

$$\boldsymbol{\omega} = d\varphi.$$

As a consequence the t-sections of $\mathcal{L}$ have some important properties

COROLLARY 3.4. *The Lagrangian submanifolds* $\mathcal{L}_t = \mathcal{G}_t(L_0)$ *are such that*

$$\mathcal{L}_0 = L_0 \ , \ d\,\varphi_0 = d(\alpha \circ \pi)$$
$$\boldsymbol{s}_{|\mathcal{L}_t} = d\,\varphi_t.$$

Moreover

$$d(\varphi_t \circ \pi_t^{-1}) = \pi_t^{-1},$$

at every point $t \in I$ *where* $\pi_t = \pi_{|\mathcal{L}_t}$ *is a diffeomorphism.*

To carry out the proof of Theorem 2.3 let us first consider the case when the initial point is free. In this case the reference extremal is always normal as it is easy to verify. We will show afterwards that we can reduce ourselves to the free initial-point case.

COROLLARY 3.5. *If* π_{t*} *is invertible for all* $t \in I$ *and* x_0 *is a strict local minimizer for* $\varphi_T \circ \pi_T^{-1} + \gamma$ *on* $\widehat{N}_0$ *then the reference couple* $(\widehat{\xi}, \widehat{v})$ *is a strict strong local minimizer.*

PROOF. The proof is the same as the one of Theorem 3.5 in [**ASZ98**], taking into account that in this case $J(\widehat{\xi}, \widehat{v}) = \alpha(x_0) + \gamma(x_0)$. □

When the initial point is not fixed we modify our problem by introducing exact penalties on the end-points and we consider a new problem with free end-points. Let $\chi_0, \widehat{\chi}_0$ be the square distance from N_0 and $\widehat{N}_0$ respectively, locally defined around x_0 by the Riemannian distance. For a sufficiently large ρ we consider the functional

$$J_\rho(\xi, v) = (\alpha + \rho\chi_0)(\xi(0)) + (\beta + \rho\widehat{\chi}_0 \circ S_T^{-1})(\xi(T)) + p_0 \int_0^T f_t^0(\xi(t), v(t))\, dt.$$

The idea of the proof is to obtain $D^2 G_t(\ell_0)$ as the Hamiltonian function associated to an auxiliary non-singular linear-quadratic problem in $T_{x_0}M$. To reach this goal we identify the space $T_{\ell_0}T^*M$, where $D^2 G_t(\ell_0)$ is defined, with $T_{x_0}M \times T^*_{x_0}M$, that is the space where is defined the Hamiltonian of a non-singular linear-quadratic problem having $T_{x_0}M$ as state space. To obtain this identification use an *horizontal section along* $\widehat{\lambda}(t)$ defined through the data of the problem.

The cost α defines the symplectic isomorphism

$$\imath : T_{x_0}M \times T^*_{x_0}M \to T_{\ell_0}T^*M$$
$$\imath : (v, \omega) \mapsto (d\alpha)_* v + \omega$$

and one can easily check

$$\imath^{-1}\,\delta = (\pi_*\,\delta, \delta - d(\alpha \circ \pi)_*\,\delta).$$

By means of this isomorphism we can consider the following identification of Lagrangian subspaces

$$\Lambda_0 = T_{x_0}N_0 \times T^{\perp}_{x_0}N_0$$
$$\widehat{\Lambda}_0 = \{(v, \omega + D^2\,J_0(x_0)(v, \cdot)) : (v, \omega) \in T_{x_0}\widehat{N}_0 \times T^{\perp}_{x_0}\widehat{N}_0\}.$$

The quadratic Hamiltonian $D^2 G_t(\ell_0)$ is a bilinear symmetric form on $T_{\ell_0}T^*M$, but by means of the previous identification we can think to it as a form on $T_{x_0}M \times T^*_{x_0}M$ or as a linear function

$$G''_t : T_{x_0}M \times T^*_{x_0}M \to T^*_{x_0}M \times T_{x_0}M.$$

Since it is non-negative then for any given symmetric positive definite

$$R : T_{x_0}M \times T^*_{x_0}M \to T^*_{x_0}M \times T_{x_0}M$$

it is uniquely defined a non-negative symmetric

$$\Sigma_t : T_{x_0}M \times T^*_{x_0}M \to T^*_{x_0}M \times T_{x_0}M$$

such that

$$G''_t = \frac{1}{2}\Sigma_t R^{-1}\Sigma_t.$$

Σ_t can be decomposed as

$$Q_t : T_{x_0}M \times T^*_{x_0}M \to T^*_{x_0}M$$
$$B_t : T_{x_0}M \times T^*_{x_0}M \to T_{x_0}M.$$

We can define a linear quadratic control problem on the vector space $T_{x_0}M$ with control functions $v \in L^2(I, T_{x_0}M \times T^*_{x_0}M)$ by means of

$$J''(\xi, v) = D^2 J_0(x_0)(\xi(T), \xi(T)) + \int_0^T \{\langle Rv(t), v(t)\rangle - 2\langle Q_t v(t), \xi(t)\rangle\}\, dt$$

$$\dot{\xi}(t) = 2B_t\, v(t) \tag{3.1}$$

$$\xi(0) \in T_{x_0}N_0 \ , \ \xi(T) \in T_{x_0}\widehat{N}_0. \tag{3.2}$$

The pre-Hamiltonian for this problem is

$$h''_t = \langle p, 2B_t u\rangle - \langle Ru, u\rangle + 2\langle Q_t u, q\rangle =$$
$$2\langle (q,p), \Sigma_t u\rangle - \langle Ru, u\rangle.$$

Therefore the maximizing control satisfies

$$\Sigma_t(q,p) = Ru$$

and easy computations show that the associated quadratic maximized Hamiltonian is G''_t. As usual we denote by $\mathcal{G}''_t$ the associate flow.

We are interested in the coerciveness of the above quadratic form on the subspace Z of $T_{x_0}M \times L^\infty(I, T_{x_0}M \times T^*_{x_0}M)$ described by the linear constraints (3.1) and (3.2). Since $\imath$ is a symplectic isomorphism we have that

$$\mathcal{G}''_t = \imath^{-1} \circ \mathcal{G}_{t*} \circ \imath,$$

hence one can check that the assumptions of Theorem 2.6 in [**SZ97**] are verified. In particular conditions $i)$ means that J''_0 is positive on $T_{x_0}N_0 \cap T_{x_0}\widehat{N}_0$ while $ii)'$ yields that there are no semi-conjugate points in I so that the linear quadratic form J'' is coercive on Z.

By Theorem 13.2 in [**Hes51**] there is a $\rho > 0$ such that the quadratic form

$$(x, v) \mapsto J''(x, v) + \rho\left(D^2\chi_0(x_0)(x, x) + D^2\widehat{\chi}_0(x_0)(\xi(T), \xi(T))\right)$$

is coercive on the whole space and hence also on the subspace defined by the constraint $\xi(T) = 0$. By applying the necessary conditions of Theorem 2.6 in [**SZ97**] and by using again the isomorphism $\imath$ we obtain

(a) $D^2 (J_0 + \rho\chi_0 + \rho\widehat{\chi}_0)(x_0) > 0$
(b) $\mathcal{G}_{t*}(\Gamma_0) \cap T^*_{x_0}M = \{0\}$, $t \in I$
(c) $\mathcal{G}_{t*}(\Gamma_0) \cap \Gamma = \{0\}$, $t \in I$

where

$$\Gamma_0 = \{d(\alpha + \rho\chi_0)_* v, v \in T_x M\}$$
$$\Gamma = \{-d(\gamma + \rho\widehat{\chi}_0)_* v, v \in T_x M\}.$$

Condition (b) shows that taking $L_0 = d(\alpha+\rho\chi_0)$ we have that π_{t*} is an isomorphism for all $t \in I$. To prove the theorem we have to show that x_0 is a strict local minimizer for $\varphi_T \circ \pi_T^{-1} + \gamma + \rho\widehat{\chi}_0$. Let us consider the t-dependent function

$$\varphi_t \circ \pi_t^{-1} + \gamma + \rho\widehat{\chi}_0.$$

For this function the first order extremality condition is verified at x_0 since

$$d(\varphi_t \circ \pi_t^{-1} + \gamma + \rho\widehat{\chi}_0)(x_0) = \ell_0 - \ell_0 = 0.$$

Consider now the quadratic form

$$q_t = D^2(\varphi_t \circ \pi_t^{-1} + \gamma + \rho\widehat{\chi}_0)(x_0)$$

which is continuously dependent on t. (a) implies that q_0 is positive definite and (c) implies that q_t is not degenerate for $t \in I$. Elementary arguments yield that q_T is positive definite hence our function has a strict local minimizer at x_0.

In the normal case the proof is complete by noting that on the admissible curves the penalties disappear and the new costs α, β coincide with the original ones.

In the abnormal case the functional J_ρ is zero on the admissible couples and hence $\widehat{\xi}$ being a strict local minimizer means that it is isolated.

References

[AG97] Andrei A. Agrachev and Revaz Gamkrelidze, *Symplectic methods for optimization and control*, Geometry of Feedback and Optimal Control (New York, New York) (B. Jacubczyk and W. Respondek, eds.), Marcel Dekker, 1997, 1–58.

[Arn80] Vladimir I. Arnold, *Mathematical methods in classical mechanics*, Springer, New York, 1980.

[ASZa] Andrei Agrachev, Gianna Stefani, and PierLuigi Zezza, *Coordinate free sufficient conditions for strong minima*, in progress.

[ASZb] Andrei Agrachev, Gianna Stefani, and PierLuigi Zezza, *An invariant second variation in optimal control*, submitted.

[ASZ98] Andrei Agrachev, Gianna Stefani, and PierLuigi Zezza, *Strong minima in optimal control*, Proceeding of the Steklov Mathematical Institute **220** (1998), to appear.

[Hes51] Magnus R. Hestenes, *Applications of the Theory of Quadratic Forms in Hilbert Space to Calculus of Variations*, Pacific J. Math. **1** (1951), 525–581.

[MR94] Jerrold E. Marsden and Tudor S. Ratiu, *Introduction to mechanics and symmetry*, Springer-Verlag, New York, 1994.

[SZ97] Gianna Stefani and PierLuigi Zezza, *Constrained regular LQ-control problems*, SIAM J. Control Optim. **35** (1997), 876–900.

Steklov Institute for Mathematics, 42, ul.Vavilova, MOSCOW, 117966, RUSSIA
E-mail address: agrachev@mian.su

Dipartimento di Matematica Applicata, "G. Sansone" , Università di Firenze, Via S. Marta 3, 50139 FIRENZE, ITALIA
E-mail address: stefani@dma.unifi.it

Dipartimento di Matematica DEFAS, Università di Firenze, Via C. Lombroso 6/17, 50134 FIRENZE, ITALIA
E-mail address: pzezza@facec.cce.unifi.it

Proceedings of Symposia in Pure Mathematics
Volume **64**, 1999

THE HEISENBERG GROUP H_3 IN TERMS OF THE ORTHONORMAL FRAME BUNDLE $O(R^2)$ AND THE CONTROL OF VARIATION OF THE HEISENBERG DIFFUSION

Helene Airault

ABSTRACT. One of our purposes is to show how the Heisenberg group H_3 covers the orthonormal frame bundle $O(\mathbb{R}^2)$ of the riemannian manifold $\mathbb{R}^2$ endowed with an appropriate connection. With this point of view, we do calculus of variations on H_3.

Introduction

On the plane $\mathbb{R}^2$,we consider the following connection: the parallel transport of a vector v along a parametrized curve $(x(t), y(t))$ is obtained after a rotation of the vector v by an angle $\theta(t)$ and $\theta(t)$ is given by the differential equation

$$d\theta(t) = C(x(t)dy(t) - y(t)dx(t)) \tag{0.1}$$

C is a constant. Let $(x_\epsilon(t), y_\epsilon(t))$ be a small perturbation of the curve $(x(t), y(t))$, we study the variation of $\theta_\epsilon(t)$ and we give a geometrical interpretation of this variation.

Let M be a n-dimensional riemannian manifold endowed with a riemannian connection Γ , i.e. a connection which is compatible with the riemannian metric on M (the parallel transport preserves the inner product). Along a curve $v(t)$ on M, we define a moving orthonormal frame $e(t) = (e_1(t), e_2(t), ..., e_n(t))$ such that

$$de^i_\alpha(t) = -\Gamma^i_{mk}(v(t))e^k_\alpha(t)dv^m(t) \tag{0.2}$$

Then $e(t)$ is a frame over $v(t)$; it defines an orthogonal map $e(t)$ from $\mathbb{R}^n$ to the tangent space $T_{v(t)}(M)$ at $v(t) \in M$. Equation (0.1) is deduced from (0.2) when $e(t) = e(\theta(t))$. The set of all orthonormal frames is the orthonormal frame bundle $O(M)$, it is a parallelized manifold: there are $n + (1/2)n(n-1)$ vector fields, $(L_i)_{i=1,..,n}$ and $(Y^i_j), (1 \leq i < j \leq n)$ which form a basis of the tangent space $T_r(O(M))$ at each $r \in O(M)$. The vector fields (Y^i_j) are vertical vector fields. These $n + (1/2)n(n-1)$ vector fields define a connection $\widetilde{\Gamma}$ on $O(M)$; the

1991 *Mathematics Subject Classification.* AMS subject classification 49J40, 49Q20, 60H07, 93A25.

connection $\widetilde{\Gamma}$ has a curvature equal to zero. The knowledge of the torsion of $\widetilde{\Gamma}$ is equivalent to the knowledge of the curvature and torsion of Γ. The Heisenberg group H_3 viewed in terms of $O(\mathbb{R}^2)$ is a perfect example of this situation. We show how to do calculus of variation on $O(\mathbb{R}^2)$ with the tools developed in [1] and deduce the calculus of variation on H_3.

The first paragraph is devoted to a general exposition concerning the geometry of the orthonormal frame bundle $O(M)$ of a riemannian manifold M.

We obtain conditions on the curvature of M which guarantee the hypoellipticity of the horizontal Laplacian on $O(M)$. In the second and the third paragraphs, we study the particular case of the Heisenberg group H_3 and its relationship to $O(\mathbb{R}^2)$. With the connection defined by (0.1) on $\mathbb{R}^2$, then $\mathbb{R}^2$ is a riemannian manifold of constant curvature equal to $2C$ and torsion tensor given by $T^1_{12} = -Cy$ and $T^2_{21} = Cx$. If we denote by $(L_i)_{i=1,2}$ the horizontal lifts of $(\partial/\partial x^i)_{i=1,2}$ with the connection (0.1), then the control system

$$d\psi(t) = L_1(\psi(t))dx(t) + L_2(\psi(t))dy(t) \tag{0.3}$$

where $\psi(t) = (v^{(1)}(t), v^{(2)}(t), \theta(t))$ can be written as (0.2) with $e(t) = e(\theta(t))$. The objective of the fourth paragraph is to do calculus of variation on H_3 and thus on $O(\mathbb{R}^2)$. Here, we shall consider the bundle of positevely oriented orthonormal frames on $\mathbb{R}^2$ and we shall verify in this particular case, some of the results of [1]. This shows how the theory of [1] can be adapted to the Heisenberg diffusion and thus to the hypoelliptic case.

Contents

1. The geometry of the orthonormal frame bundle $O(M)$ of a riemannian manifold M

Let M be a n-dimensional riemannian manifold. We denote by g_{ij} the riemannian metric on M. Let $x \in M$ and e be a linear map from $\mathbb{R}^n$ to the tangent space $T_x(M)$ at x. We say that e is an orthonormal frame at x if e is an orthogonal map; i.e. in local coordinates, we have

$$e^i_\alpha e^j_\alpha = g^{ij} \tag{1.1}$$

(g^{ij}) is the inverse matrix of (g_{ij}). We denote by $O(M)$ the set of all orthonormal frames (x, e). Let Γ be a riemannian connection on M; it means that Γ is compatible with the riemannian structure on M. We have

$$\partial_k g_{ij} = g_{lj}\Gamma^l_{ki} + g_{il}\Gamma^l_{kj} \tag{1.2}$$

On $O(M)$, the canonical form is given by

$$dx^m = e^m_\lambda d\omega^\lambda \tag{1.3}$$

The connection form associated to Γ is

$$dp^i_j = (e^{-1})^i_k(de^k_j + \Gamma^k_{ml} e^l_j dx^m) \tag{1.4}$$

See [7].

Since Γ is riemannian, and because of (1.1), we have

$$dp^i_j = -dp^j_i \tag{1.5}$$

Let $S = (S^l_j)$ be a $n \times n$ matrix. We denote by S^* the transposed matrix ($S^{*i}_{\ j} = S^j_i$). We assume that (e^i_α) satisfies (1.1), then the matrix $e^{-1}Se$ is antisymmetric if and only if

$$gS^* + Sg = 0 \tag{1.6}$$

where $g = (g^{ij})$. See for example [1] p.460.

1.a. Vertical and horizontal subspaces. The manifold $O(M)$ is parallelized.

Let $u = (u^j_i)$ be a $n \times n$ matrix with real coefficients; we put

$$\exp u = \sum_{n=0}^{\infty} \frac{u^n}{n!}$$

for the usual exponential of the matrix $u = (u^j_i)$. If u is an antisymmetric matrix, then $v = \exp u$ is in $O(n)$, ($v^* v = \mathbb{I}$). The endomorphism of $\mathbb{R}^n$ which has v as a matrix with respect to the canonical basis of $\mathbb{R}^n$ is also called v. If e is an orthonormal frame at $x \in M$, and if $v \in O(n)$, then $e \circ v$ is an orthonormal frame at $x \in M$. Let $u = (u^j_i)$ be an antisymmetric matrix, we define the vector field Y_u on $O(M)$ by

$$(Y_u f)(x, e) = \frac{d}{dt}_{t=0} f(x, e\exp(tu)) \tag{1.a.1}$$

LEMMA. *The expression of Y_u in local coordinates is*

$$(Y_u f)(x,e) = \sum_{i<j} u_j^i \Big(\sum_k (e_i^k \frac{\partial}{\partial e_j^k} - e_j^k \frac{\partial}{\partial e_i^k}) f \Big)(x,e) \tag{1.a.2}$$

PROOF.
We put $v(t) = \exp(tu)$, then $u = d/dt_{t=0} v(t)$. From (1.a.1), we deduce

$$(Y_u f)(x,e) = \sum_{s,k} \frac{\partial f}{\partial e_s^k} d/dt_{t=0} (e.v(t))_s^k$$
$$= \sum_{s,k,j} \frac{\partial f}{\partial e_s^k} (e_j^k . u_s^j)$$

Since $u_s^j = -u_j^s$, this gives (1.a.2).

DEFINITION. We put, for $i < j$,

$$Y_j^i = \sum_k (e_i^k \frac{\partial}{\partial e_j^k} - e_j^k \frac{\partial}{\partial e_i^k}) \tag{1.a.3}$$

REMARK. For $1 \leq i < j \leq n$, the $n(n-1)/2$ vector fields Y_j^i are linearly independent on $O(M)$. The subspace of $T_{(x,e)}(O(M))$ generated by the $n(n-1)/2$ vector fields $(Y_j^i(x,e))_{(1 \leq i<j \leq n)}$ is the vertical space at (x,e). See [4],[10].

The horizontal space at (x,e) is constructed by horizontal lift with the connection Γ, of vector fields on M. Let $X(x) = X^i(x) \frac{\partial}{\partial x^i}$ be a vector field on M. It defines the flow $(\Phi(t))_{t \geq 0}$ on M; we have

$$\frac{d\Phi(t)}{dt} = X(\Phi(t)) \tag{1.a.4}$$
$$\Phi(0) = x$$

Let $e(0) = e$ be an orthonormal frame over $x \in M$. We obtain the frames $e(t)$ over $\Phi(t)$ by parallel transport solving the equation

$$\Big((de(t)) e(t)^{-1} \Big)_j^k = -\Gamma_{mj}^k d\Phi(t)^m(t) \tag{1.a.5}$$
$$e(0) = e$$

The horizontal lift $\widetilde{X}$ on $O(M)$ of the vector field X is given by

$$(\widetilde{X} f)(x,e) = \frac{d}{dt}_{|t=0} f(\Phi(t), e(t)) \tag{1.a.6}$$

where $\Phi(t)$ is solution of (1.a.4) and $e(t)$ is solution of (1.a.5). In local coordinates,

$$\widetilde{X}(x,e) = X^i(x) \Big(\frac{\partial}{\partial x^i} - \Gamma_{i\alpha}^q e_p^\alpha \frac{\partial}{\partial e_p^q} \Big) \tag{1.a.7}$$

DEFINITION. We put $\widetilde{V}_j$, $j = 1,..,n$ for the horizontal vector fields on $O(M)$ given in local coordinates by

$$\widetilde{V}_j(x,e) = e^i_j\Big(\frac{\partial}{\partial x^i} - \Gamma^q_{il}e^l_p\frac{\partial}{\partial e^q_p}\Big) \tag{1.a.8}$$

PROPOSITION. *Let Γ be a connection compatible with the riemannian metric, then the vector fields $\widetilde{V}_j(x,e)$, $j = 1...,n$ and $Y^i_j(x,e)$,$(1 \le i < j \le n)$,see (1.a.8) and (1.a.3), form a basis of the tangent space $T_{(x,e)}(O(M))$ for any $(x,e) \in O(M)$*

PROOF.

The assertion results from the construction of $\widetilde{V}_j$ and Y^i_j. We shall give a direct proof. Let $(x(t), e(t))_{t\ge0}$ be a smooth curve on $O(M)$ such that $(x(0), e(0)) = (x, e)$. For a smooth function f defined on $O(M)$, we have

$$\frac{d}{dt}_{|t=0} f(x(t), e(t)) \in T_{(x,e)}(O(M))$$

and

$$\frac{d}{dt}_{|t=0} f(x(t), e(t)) = \sum_i \frac{\partial f}{\partial x^i}\frac{dx^i}{dt}_{|t=0} + \sum_{k,s}\frac{\partial f}{\partial e^k_s}\frac{de^k_s}{dt}_{|t=0}$$

Because of (1.3) and (1.4), we obtain

$$\begin{aligned}\frac{d}{dt}_{|t=0} f(x(t), e(t)) = \sum_m \Big(\frac{\partial f}{\partial x^m} - \Gamma^k_{ml}e^l_s\frac{\partial f}{\partial e^k_s}\Big)e^m_\lambda\frac{d\omega^\lambda(t)}{dt}_{|t=0} \\ + \sum_{k,s,r}(e^k_r\frac{\partial f}{\partial e^k_s})\frac{dp^r_s}{dt}_{|t=0}\end{aligned} \tag{1.a.9}$$

From (1.5), we deduce that the last term of (1.a.9) is

$$\sum_k\Big(\sum_{r<s}(e^k_r\frac{\partial}{\partial e^k_s} - e^k_s\frac{\partial}{\partial e^k_r})\Big)\frac{dp^r_s}{dt}_{|t=0} \tag{1.a.10}$$

This yields

$$\begin{aligned}\frac{d}{dt}_{|t=0} f(x(t), e(t)) = \sum_\lambda(\widetilde{V_\lambda}f)(x,e)\frac{d\omega^\lambda(t)}{dt}_{|t=0} \\ + \sum_{r<s}(Y^s_r f)(x,e)\frac{dp^r_s}{dt}_{|t=0}\end{aligned} \tag{1.a.11}$$

1.b. Hypoellipticity criteria for the horizontal laplacian on $O(M)$. Let R^i_{mjp} be the curvature tensor of the connection Γ; we have

$$R^i_{mjp} = \frac{\partial}{\partial x^j}\Gamma^i_{pm} - \frac{\partial}{\partial x^p}\Gamma^i_{jm} + (\Gamma^\alpha_{pm}\Gamma^i_{j\alpha} - \Gamma^\alpha_{jm}\Gamma^i_{p\alpha}) \tag{1.b.1}$$

Since Γ is compatible with the riemannian metric g_{ij}, we have

$$g_{\alpha k}R^k_{mjp} + g_{mk}R^k_{\alpha jp} = 0 \tag{1.b.2}$$

We denote T^k_{ij}, the torsion of Γ,

$$T^k_{ij} = \Gamma^k_{ij} - \Gamma^k_{ji} \tag{1.b.3}$$

PROPOSITION. *The bracket vector fields $[\widetilde{V_j}, \widetilde{V_s}]$ are determined by the curvature and the torsion of the connection Γ. We have*

$$[\widetilde{V_j}, \widetilde{V_s}] = \sum_{k<r} a^k_r Y^k_r + \sum_k b^k \widetilde{V_k} \tag{1.b.4}$$

with

$$\begin{aligned} a^k_r(x,e) &= e^i_j e^\alpha_s (e^{-1})^k_\mu R^\mu_{\lambda\alpha i}(x) e^\lambda_r \\ b^k(x,e) &= (1/2)(e^\alpha_s e^l_j - e^\alpha_j e^l_s) T^\mu_{\alpha l}(x)(e^{-1})^k_\mu \end{aligned} \tag{1.b.5}$$

PROOF.
With (1.a.8), we calculate that

$$\begin{aligned} [\widetilde{V_j}, \widetilde{V_s}](x,e) &= e^i_j e^\alpha_s R^q_{\lambda\alpha i}(x) e^\lambda_r \frac{\partial}{\partial e^q_r} \\ +(e^\alpha_s e^l_j - e^\alpha_j e^l_s)\Gamma^i_{\alpha l}(x) &\Big(\frac{\partial}{\partial x^i} - \Gamma^q_{il}(x) e^l_p \frac{\partial}{\partial e^q_p}\Big) \end{aligned} \tag{1.b.6}$$

We transform the first term

$$e^i_j e^\alpha_s R^q_{\lambda\alpha i}(x) e^\lambda_r \frac{\partial}{\partial e^q_r} \tag{1.b.7}$$

Denote $\delta_{\mu q}$ the Dirac tensor

$$\delta_{\mu q} = 1$$

if $\mu = q$ and

$$\delta_{\mu q} = 0 \tag{1.b.8}$$

if μ is different from q.

We find that (1.b.7) is equal to

$$e^i_j e^\alpha_s \delta_{\mu q} R^\mu_{\lambda\alpha i}(x) e^\lambda_r \frac{\partial}{\partial e^q_r} \tag{1.b.9}$$

In (1.b.9), we replace $\delta_{\mu q} = e^q_k (e^{-1})^k_\mu$. With the antisymetry property (1.b.2), we get for (1.b.9),

$$\sum_{k<r} a^k_r Y^k_r$$

On the other hand,

$$\begin{aligned} \Gamma^i_{\alpha l}\Big(\frac{\partial}{\partial x^i} - \Gamma^q_{il} e^l_p \frac{\partial}{\partial e^q_p}\Big) &= \Gamma^\mu_{\alpha l}\delta_{\mu i}\Big(\frac{\partial}{\partial x^i} - \Gamma^q_{il} e^l_p \frac{\partial}{\partial e^q_p}\Big) \\ &= \Gamma^\mu_{\alpha l} e^i_k (e^{-1})^k_\mu \Big(\frac{\partial}{\partial x^i} - \Gamma^q_{il} e^l_p \frac{\partial}{\partial e^q_p}\Big) \\ &= \Gamma^\mu_{\alpha l}(e^{-1})^k_\mu \widetilde{V_k} \end{aligned} \tag{1.b.10}$$

We replace (1.b.10) in (1.b.6). This proves (1.b.4)-(1.b.5).

PROPOSITION. *We assume that the curvature of M is non degenerate, i.e. we can calculate Y_r^k in terms of $[\widetilde{V_j}, \widetilde{V_s}]$ and $\widetilde{V_k}$ with (1.b.4), then the second order differential operator $\widetilde{L} = \sum_{j=1,..n}(\widetilde{V_j})^2$ is hypoelliptic on $O(M)$.*

PROOF.
It results from (1.b.4).

1.c. Zero curvature connection $\widetilde{\Gamma}$ on $O(M)$. The orthonormal frame bundle is a $n + \frac{n(n-1)}{2}$ dimensional parallelized manifold. At each point $(x, e) \in O(M)$, the vectors $(\widetilde{V_j}(x, e); j = 1..., n)$ and $Y_j^i(x, e)$, $(1 \leq i < j \leq n)$ form a basis of the tangent space $T_{(x,e)}(O(M))$. See (1.a.11). With this basis, we define a zero curvature connection $\widetilde{\Gamma}$ on $O(M)$ as follows: we take a basis $(\epsilon_i, i = 1, ..., n; E_j^i, 1 \leq i < j \leq n)$ in $\mathbb{R}^n \times \mathbb{R}^{n(n-1)/2}$ and we consider the linear isomorphism $b(x, e)$ from $\mathbb{R}^n \times \mathbb{R}^{n(n-1)/2}$ to $T_{(x,e)}(O(M))$ defined by

$$\begin{aligned} b(x,e)(\epsilon_i) &= \widetilde{V_i}(x,e) \qquad i = 1, ..., n \\ b(x,e)(E_j^i) &= Y_j^i \qquad 1 \leq i < j \leq n \end{aligned} \tag{1.c.1}$$

The covariant derivative associated to the connection $\widetilde{\Gamma}_{O(M)}$ is given by

$$(\widetilde{\nabla}_{\widetilde{X}}\widetilde{Y})(x,e) = \frac{d}{dt}_{t=0} b(x,e)b(x(t),e(t))^{-1}\widetilde{Y}(x(t),e(t)) \tag{1.c.2}$$

where $\widetilde{X}$,$\widetilde{Y}$ are two vector fields on $O(M)$ and $(x(t), e(t))$ is a curve on $O(M)$ such that

$$\frac{d}{dt}_{t=0}(x(t),e(t)) = \widetilde{X}(x,e) \tag{1.c.3}$$

and $(x(0), e(0)) = (x, e)$

We denote by $(L_i)_{i=1,..,n,..n+\frac{n(n-1)}{2}}$ the sequence of vector fields $\widetilde{V_i}(x, e)$, $i = 1, ..., n$,Y_j^i,$1 \leq i < j \leq n$.

From $\widetilde{\nabla}_{L_k} L_p = 0$ for any k, p, we deduce that the connection $\widetilde{\Gamma}_{O(M)}$ has zero curvature and that its torsion $T_{\widetilde{\Gamma}_{O(M)}}$ is given by

$$T_{\widetilde{\Gamma}_{O(M)}}(L_p, L_k) = [L_k, L_p] \tag{1.c.4}$$

PROPOSITION (1.C.5). *Consider the connection Γ compatible with the riemannian metric on M and let the zero curvature connection $\widetilde{\Gamma}_{O(M)}$ on $O(M)$ associated to the frames*

$$(\widetilde{V_j}(x,e)), \qquad (j = 1, ..., n); \qquad Y_j^i(x,e), \qquad (1 \leq i < j \leq n)$$

then it is equivalent to know the curvature and torsion of Γ or to know the torsion of $\widetilde{\Gamma}_{O(M)}$.

PROOF.
The assertion results from (1.c.4)-(1.b.5) and from the calculation of the Lie brackets $[Y_j^i, Y_s^r]$,$[Y_j^i, \widetilde{V_k}]$

2. The orthonormal frame bundle $O(\mathbb{R}^2)$. The connections Γ on $\mathbb{R}^2$ which are compatible with the metric δ_{ij}

Let $M = \mathbb{R}^2$ with the metric $g_{ij} = \delta_{ij}$, where δ_{ij} is the Dirac tensor (1.b.8). The orthonormal frame bundle $O(\mathbb{R}^2)$ can be identified with $\mathbb{R}^2 \times \mathbb{R}/2\pi\mathbb{Z}$ where $S^1 = \mathbb{R}/2\pi\mathbb{Z}$ is the circle group. We consider the map from $\mathbb{R}^2 \times \mathbb{R}$ to $O(\mathbb{R}^2)$ which associates to (x, θ) in $\mathbb{R}^2 \times \mathbb{R}$ the frame $(x, e(\theta))$ in $O(\mathbb{R}^2)$ where

$$e(\theta) = \begin{pmatrix} \cos(\theta) & -\sin(\theta) \\ sin(\theta) & \cos(\theta) \end{pmatrix} \tag{2.1}$$

is an orthonormal matrix; $e(\theta)$ is determined by the angle θ. This induces an isomorphism from the space $\mathbb{R}^2 \times \mathbb{R}/2\pi\mathbb{Z}$ to $O(\mathbb{R}^2)$. With this point of view, for a function f defined on $O(\mathbb{R}^2)$, we shall write $f(x, \theta)$ and we shall identify $f(x, \theta)$ and $f(x, e(\theta))$ where $e(\theta)$ is given by (2.1).

On $O(\mathbb{R}^2)$, consider the vector field

$$(Yf)(x, \theta) = \frac{d}{dt}_{t=0} f(x, e(\theta + t)) \tag{2.2}$$

See (1.a.1)-(1.a.2). Then

$$Yf = \frac{\partial f}{\partial \theta} \tag{2.3}$$

LEMMA. *On $O(\mathbb{R}^2)$, let Y_2^1 be the vector field defined by (1.a.3), then we have*

$$Y_2^1 = -\frac{\partial}{\partial \theta}$$

and the vertical subspace in $T_{(x,e(\theta))}(O(\mathbb{R}^2)$ is generated by Y_2^1.

PROOF.
Let

$$e(\theta) = \begin{pmatrix} e_1^1 & e_2^1 \\ e_1^2 & e_2^2 \end{pmatrix} \tag{2.4}$$

be the matrix (2.1). By (1.a.3), we have

$$Y_2^1 = e_1^1 \frac{\partial}{\partial e_2^1} + e_1^2 \frac{\partial}{\partial e_2^2} - e_2^1 \frac{\partial}{\partial e_1^1} - e_2^2 \frac{\partial}{\partial e_1^2}$$

We see that

$$Y_2^1 f = -\frac{\partial f}{\partial e_2^1}\frac{de_2^1}{d\theta} - \frac{\partial f}{\partial e_2^2}\frac{de_2^2}{d\theta} - \frac{\partial f}{\partial e_1^1}\frac{de_1^1}{d\theta} - \frac{\partial f}{\partial e_1^2}\frac{de_1^2}{d\theta}$$

We deduce that

$$Y_2^1 = -\frac{\partial}{\partial \theta}$$

REMARK. On $\mathbb{R}^2$, we consider connections Γ compatible with the euclidean inner product (δ_{ij}). The connection form associated to Γ is given by (1.4). For the Levi-Civita connection γ on $\mathbb{R}^2$, $(\gamma^l_{mk} = 0)$, its connection form is (see(1.4))

$$e(\theta)^{-1}de(\theta) = \begin{pmatrix} 0 & -1 \\ 1 & 0 \end{pmatrix} d\theta \tag{2.5}$$

2.a. Riemannian connections Γ on R^2. Let Γ be a connection compatible with the riemannian metric δ_{ij} on R^2. From (1.2), we see that

$$\Gamma^j_{ki} + \Gamma^i_{kj} = 0 \tag{2.a.1}$$

for any i, j, k. Since $\Gamma^1_{k1} = \Gamma^2_{k2} = 0$ and $\Gamma^1_{k2} = -\Gamma^2_{k1}$, all Christoffel symbols are determined by the two of them Γ^1_{12} and Γ^1_{22}.

LEMMA. *The curvature of Γ is given by*

$$R^1_{221} = \frac{\partial}{\partial x^2}\Gamma^1_{12} - \frac{\partial}{\partial x^1}\Gamma^1_{22}$$

$$R^1_{221} = R^2_{112} = -R^1_{212} = -R^2_{121} \tag{2.a.2}$$

and the others are zero. The torsion of Γ is given by

$$T^1_{12} = \Gamma^1_{12} \qquad \text{and} \qquad T^2_{12} = -\Gamma^2_{21} = \Gamma^1_{22} \tag{2.a.3}$$

PROOF.
See (1.b.1)

REMARK.
Let $x = (x^1, x^2)$. If $\Gamma^1_{12}(x) = -cx^2$, $\Gamma^1_{22}(x) = cx^1$, see (0.1), then $R^1_{221} = -2c$ and $T^1_{12} = -cx^2$, $T^2_{12} = cx^1$.

PROPOSITION.
Let Γ be a riemannian connection on $\mathbb{R}^2$, see (2.a.1), and let $(\phi(t))_{t\geq 0}$ be a differentiable curve on R^2. The parallel transport of a vector v along the curve $\phi(t)$ is given by

$$v(t) = e(\theta(t))v \tag{2.a.4}$$

where $e(\theta)$ is defined by (2.1) and $\theta(t)$ is solution of

$$d\theta(t) = \Gamma^1_{12}(\phi(t))d\phi^1(t) + \Gamma^1_{22}(\phi(t))d\phi^2(t) \tag{2.a.5}$$

$\theta(0) = 0$

PROOF.
We denote $e(t) = e(\theta(t))$. Then we have

$$(de(t))e(t)^{-1} = \begin{pmatrix} 0 & -1 \\ 1 & 0 \end{pmatrix} d\theta(t) \tag{2.a.6}$$

From (1.a.5) and (2.a.1), we deduce that

$$d\theta(t) = \Gamma^1_{m2}d\phi^m(t) \tag{2.a.7}$$

LEMMA.

For the connection Γ, the horizontal lifts $\widetilde{X^i},(i = 1,..,n)$ are given in local coordinates by

$$\widetilde{X^1} = \frac{\partial}{\partial x^1} + \Gamma^1_{12}\frac{\partial}{\partial\theta}$$
$$\widetilde{X^2} = \frac{\partial}{\partial x^2} + \Gamma^1_{22}\frac{\partial}{\partial\theta} \tag{2.a.8}$$

PROOF.

We use (1.a.6)-(1.a.7). On R^2, consider the curve $\phi(t) = (x_1 + t, x_2)$, then the equation (2.a.5)-(2.a.6) yield

$$d\theta(t) = \Gamma^1_{12}dt$$

This proves (2.a.8) for $\widetilde{X^1}$, since we have (1.a.6). Similarly, we obtain $\widetilde{X^2}$ with the curve $\phi(t) = (x_1, x_2 + t)$.

REMARK. Let $\psi(t) = (\phi^1(t), \phi^2(t), \theta(t))$, then the equation (2.a.5) can be written

$$d\psi(t) = \widetilde{X^1}(\psi(t))d\phi^1(t) + \widetilde{X^2}((\psi(t))d\phi^2(t) \tag{2.a.9}$$

DEFINITION.

We call the equation (2.a.9), control equation associated to the connection Γ on $O(R^2)$.

REMARK. We may consider (2.a.9) as an Hamiltonian system. See [6] where the geometry of the Heisenberg group is studied.

In the following, we shall see in the particular case of $O(R^2)$, hypoellipticity conditions for horizontal laplacians on $O(M)$, see (1.b).

PROPOSITION.

We have

$$[\widetilde{X^1}, \widetilde{X^2}] = (\frac{\partial}{\partial x^1}\Gamma^1_{22}(x) - \frac{\partial}{\partial x^2}\Gamma^1_{12}(x))\frac{\partial}{\partial\theta} \tag{2.a.10}$$

The connection Γ has non zero curvature if and only if $[\widetilde{X^1}, \widetilde{X^2}]$ is not zero.

PROOF. This results from (2.a.8)-(2.a.2). Compare with (1.b.4)-(1.b.5).

COROLLARY.

The operator

$$\mathcal{L} = (\widetilde{X^1})^2 + (\widetilde{X^2})^2 \tag{2.a.11}$$

is hypoelliptic on $R^3 = R^2 \times R$ if and only if the connection Γ on R^2 has non zero curvature.

DEFINITION.

On $O(R^2)$, we consider the vector fields , see(1.a.8)

$$\widetilde{V_j}(x,\theta) = e^i_j(\theta)(\frac{\partial}{\partial x^i} + \Gamma^1_{i2}(x)\frac{\partial}{\partial\theta}) \tag{2.a.12}$$

where the $e^i_j(\theta)$ are given by the matrix (2.1)-(2.4).

LEMMA.

$$[\widetilde{V_1}, \widetilde{V_2}] = R^2_{121}\frac{\partial}{\partial\theta} + T^{\mu}_{12}(e^{-1})^{\alpha}_{\mu}\widetilde{V_\alpha} \tag{2.a.13}$$

PROOF.
By direct computation, we get

$$[\widetilde{V_1}, \widetilde{V_2}] = (\frac{\partial}{\partial x^1}\Gamma^1_{22} - \frac{\partial}{\partial x^2}\Gamma^1_{12})\frac{\partial}{\partial\theta} + \Gamma^1_{i2}(\frac{\partial}{\partial x^i} + \Gamma^1_{i2}\frac{\partial}{\partial\theta}) \tag{2.a.14}$$

Compare with (1.b.4)-(1.b.5).

COROLLARY.
The operator

$$\widetilde{\mathcal{L}} = (\widetilde{V_1})^2 + (\widetilde{V_2})^2 \tag{2.a.15}$$

is hypoelliptic on R^3 if and only if the connection Γ has non zero curvature.

REMARK. It is easy to define connections on R^n, which are compatible with the euclidean metric $g_{ij} = \delta_{ij}$. For example, we can do as follows.

PROPOSITION. *Let $f^j_{mi}(x)$,$(1 \leq i < j \leq n)$,$m = 1,..,n$ be a family of $n^2(n-1)/2$ differentiable functions on R^n. There exists a connection Γ, compatible with the metric δ_{ij} on R^n such that*

$$\Gamma^s_{ij}(x) = f^j_{is}(x) \qquad if \qquad s < j$$

and

$$\Gamma^s_{ij}(x) = -f^s_{ij}(x) \qquad if \qquad s > j \tag{2.a.16}$$

PROOF.
For a differentiable curve $(\phi(t))_{t\geq 0}$ on R^n, we define the functions $p_{ij}(t)$; $1 \leq i < j \leq n$ through the differential equations

$$\frac{dp_{ij}(t)}{dt} = \sum_{m=1}^{n} f^j_{mi}(\phi(t))\frac{d\phi^m}{dt}(t) \tag{2.a.17}$$

$p_{ij}(0) = 0$ where $1 \leq i < j \leq n$. Let $u = (u^j_i)$ be the antisymmetric matrix defined by

$$u^j_i(t) = p_{ij}(t) \qquad if \qquad i < j$$

and

$$u^j_i(t) = -p_{ji}(t) \qquad if \qquad j < i$$

We put

$$e_\phi(t) = e\exp(u^j_i(t))$$

The matrix $e_\phi(t)$ defines the parallel transport along the curve $(\phi(t))$ for the connection Γ.

2.b. Connections on $O(R^2)$ associated to a frame. These connections associated to a frame have a curvature equal to zero. We shall follow (1.c).

The connection $\Gamma^\sharp$ induced by the frame $(\widetilde{X_1}, \widetilde{X_2}, \frac{\partial}{\partial\theta})$. The connection $\Gamma^\sharp$ will be constructed on $\mathbb{R}^3 = \mathbb{R}^2 \times \mathbb{R}$. For (x, θ) in $\mathbb{R}^2 \times \mathbb{R}$, we consider the matrix $b^\sharp(x)$ defined by

$$b^\sharp(x) = \begin{pmatrix} 1 & 0 & 0 \\ 0 & 1 & 0 \\ \Gamma^1_{12}(x) & \Gamma^1_{22}(x) & 1 \end{pmatrix} \tag{2.b.1}$$

The vector fields $\widetilde{X_1}$, $\widetilde{X_2}$ given by (2.a.8) are the two first columns of the matrix $b^\sharp(x)$.

Let $\Gamma^\sharp$ be the connection on R^3 whose covariant derivative is

$$\nabla^\sharp{}_{\widetilde{X}} \widetilde{Y}(x,\theta) = \frac{d}{dt}_{|t=0} (b^\sharp(\phi(t)))^{-1} \widetilde{Y}(\phi(t)) \tag{2.b.2}$$

where $(\phi(t))$ is a curve on $\mathbb{R}^2 \times \mathbb{R}$ such that $\phi(0) = (x, \theta)$ and $\frac{d}{dt}_{|t=0}\phi(t) = \widetilde{X}(x, \theta)$.

LEMMA.
The Christoffel symbols for the connection $\Gamma^\sharp$ are

$$\Gamma^{\sharp\cdot}{}_{11} = (0, 0, \frac{\partial}{\partial x^1}\Gamma^1_{12}) \qquad \Gamma^{\sharp\cdot}{}_{12} = (0, 0, \frac{\partial}{\partial x^1}\Gamma^1_{22})$$
$$\Gamma^{\sharp\cdot}{}_{21} = (0, 0, \frac{\partial}{\partial x^2}\Gamma^1_{12}) \qquad \Gamma^{\sharp\cdot}{}_{22} = (0, 0, \frac{\partial}{\partial x^2}\Gamma^1_{22})$$

PROOF.
We have

$$\Gamma^{\sharp j}_{pq} = -\sum_\alpha (\partial_p b^{\sharp j}{}_\alpha)({b^\sharp}^{-1})^\alpha_q \tag{2.b.3}$$

LEMMA.
The torsion of $\Gamma^\sharp$ is given by

$$T_{\Gamma^\sharp}(\widetilde{X_1}, \widetilde{X_2}) = [\widetilde{X_2}, \widetilde{X_1}] \tag{2.b.4}$$

$$T_{\Gamma^\sharp}(\frac{\partial}{\partial\theta}, \widetilde{X_i}) = 0 \tag{2.b.5}$$

It is equivalent to know the curvature of Γ or the torsion of $\Gamma^\sharp$.

PROOF.
From (2.b.4) and (2.a.2)-(2.a.10).

The connection $\widetilde{\Gamma}$ induced by the frame $(\widetilde{V_1}, \widetilde{V_2}, \frac{\partial}{\partial\theta})$.

The vector fields $(\widetilde{V_1}, \widetilde{V_2}, \frac{\partial}{\partial\theta})$ are defined by (2.a.12). We define the matrix $\widetilde{b}(x, \theta)$ whose columns are the components of the vector fields $(\widetilde{V_1}, \widetilde{V_2}, \frac{\partial}{\partial\theta})$. We have

$$\widetilde{b}(x,\theta) = b^\sharp(x) R_\theta \tag{2.b.6}$$

where

$$R_\theta = \begin{pmatrix} \cos(\theta) & -\sin(\theta) & 0 \\ \sin(\theta) & \cos(\theta) & 0 \\ 0 & 0 & 1 \end{pmatrix} \tag{2.b.7}$$

We obtain the Christoffel symbols associated to $\widetilde{\Gamma}$ with

$$\widetilde{\Gamma}^j_{pq} = -\sum_\alpha (\partial_p \widetilde{b}^j_\alpha)(\widetilde{b}^{-1})^\alpha_q \tag{2.b.8}$$

This yields $\widetilde{\Gamma}^j_{iq} = \Gamma^{\sharp j}_{iq}$ for $i = 1, 2$. See (2.b.3). Since $(\partial_\theta R_\theta) R_\theta{}^{-1}$ is equal to the matrix

$$\begin{pmatrix} 0 & -1 & 0 \\ 1 & 0 & 0 \\ 0 & 0 & 0 \end{pmatrix} \tag{2.b.9}$$

we obtain

$$(\widetilde{\Gamma}^j_{3q}) = \begin{pmatrix} 0 & -1 & 0 \\ 1 & 0 & 0 \\ \Gamma^1_{22} & -\Gamma^1_{12} & 0 \end{pmatrix} \tag{2.b.10}$$

The torsion of $\widetilde{\Gamma}$ is $\widetilde{T}_{12} = T^\sharp{}_{12}$ and $\widetilde{T}_{13} = -(0, 1, \Gamma^1_{22})$, $\widetilde{T}_{23} = (1, 0, \Gamma^1_{12})$

REMARK. The connections $\Gamma^\sharp$ and $\widetilde{\Gamma}$ defined on $\mathbb{R}^2 \times \mathbb{R}$ respectively with (2.b.2) and (2.b.8) induce naturally connections on $O(\mathbb{R}^2) = \mathbb{R}^2 \times \mathbb{R}/2\pi\mathbb{Z}$, since the matrix $b^\sharp(x)$ is independent of θ and the matrix $\widetilde{b}(x, \theta)$ is given for θ (mod 2π). Thus, we can verify the proposition (1.c.5) on the example of the connection $\widetilde{\Gamma}$ induced on $O(\mathbb{R}^2)$ by the frame $(\widetilde{V_1}, \widetilde{V_2}, \frac{\partial}{\partial\theta})$. See (2.b.8).

3. The Heisenberg group H_3 and the orthonormal frame bundle $O(R^2)$

We consider the group H_3 of 3×3 matrices of the form

$$g = \begin{pmatrix} 1 & x & z \\ 0 & 1 & y \\ 0 & 0 & 1 \end{pmatrix} \tag{3.1}$$

and the map from R^3 to H_3 defined by

$$(x, y, z) \to g \tag{3.2}$$

The multiplication of matrices induces the following law on R^3

$$(x, y, z) * (x', y', z') = (x + x', y + y', z + z' + xy') \tag{3.3}$$

The Lie algebra $\mathcal{H}_3$ is generated by the matrices

$$P = \partial_1 = \begin{pmatrix} 0 & 1 & 0 \\ 0 & 0 & 0 \\ 0 & 0 & 0 \end{pmatrix}, \qquad Q = \partial_2 = \begin{pmatrix} 0 & 0 & 0 \\ 0 & 0 & 1 \\ 0 & 0 & 0 \end{pmatrix}, \qquad R = \partial_3 = \begin{pmatrix} 0 & 0 & 1 \\ 0 & 0 & 0 \\ 0 & 0 & 0 \end{pmatrix} \tag{3.4}$$

With the multiplication of matrices, we find that the vector fields

$$Y_g^{(1)} = g\partial_1 = \partial_1, \qquad Y_g^{(2)} = g\partial_2 = \partial_2 + x\partial_3, \qquad Y_g^{(3)} = g\partial_3 = \partial_3 \tag{3.5}$$

are left invariant, and the vector fields

$$X_g^{(1)} = \partial_1 g = \partial_1 + y\partial_3, \qquad X_g^{(2)} = \partial_2 g = \partial_2, \qquad X_g^{(3)} = \partial_3 g = \partial_3 \tag{3.6}$$

are right invariant. Thus, with the law (3.3), the vector fields

$$\frac{\partial}{\partial x}, \qquad \frac{\partial}{\partial y} + x\frac{\partial}{\partial z}, \qquad \frac{\partial}{\partial z} \tag{3.7}$$

are left invariant on R^3; the vector fields

$$\frac{\partial}{\partial x} + y\frac{\partial}{\partial z}, \qquad \frac{\partial}{\partial y}, \qquad \frac{\partial}{\partial z} \tag{3.8}$$

are right invariant.

3.a. Connections on R^2 and the Heisenberg group H_3.
On R^3, we consider the vector fields

$$L_1 = \frac{\partial}{\partial x} + \alpha y\frac{\partial}{\partial z}, \qquad L_2 = \frac{\partial}{\partial y} + \beta x\frac{\partial}{\partial z} \tag{3.a.1}$$

where α and β are constants. For $i = 1, 2$, the vector field L_i is a linear combination of the left invariant and the right invariant vector fields generated by ∂_i.

The connection Γ on R^2 is constructed such that L_i, $i = 1, 2$ are the horizontal lifts on $O(R^2)$ of the vector fields ∂_i, $i = 1, 2$ on R^2. See(2.a.5)-(2.a.8).

To construct Γ, we define the parallel transport of a vector v along a differentiable curve $\phi(t) = (\phi^1(t), \phi^2(t))$ by $v(t) = e(\theta(t))v$ where

$$d\theta(t) = \alpha\phi^2(t)d\phi^1(t) + \beta\phi^1(t)d\phi^2(t) \tag{3.a.2}$$

$\theta(0) = 0$.

We deduce from (3.a.2) the following lemma.

LEMMA. *For the connection Γ defined on R^2 by (3.a.2), we have*

$$\Gamma^2_{11} = -\Gamma^1_{12} = -\alpha y, \qquad \Gamma^1_{22} = -\Gamma^2_{21} = \beta x \tag{3.a.3}$$

and all the others Γ^j_{is} are zero. The curvature of Γ is given by

$$R^1_{221} = R^2_{112} = -R^1_{212} = -R^2_{121} = \alpha - \beta \tag{3.a.4}$$

and all the others R^r_{iks} are zero. The torsion of Γ is given by

$$T^2_{12} = \beta x, \qquad T^1_{12} = \alpha y \tag{3.a.5}$$

COROLLARY. *The operator $\mathcal{L} = (L_1)^2 + (L_2)^2$ on R^3 is hypoelliptic if and only if the connection Γ on R^2 has a curvature different from zero.*

PROOF.
From (3.a.1), we calculate $[L_1, L_2] = \beta - \alpha$ and then we use (3.a.4)

3.b. Connections on $O(\mathbb{R}^2)$. The connection $\Gamma^\sharp$ associated to the frame $(L_1, L_2, \frac{\partial}{\partial\theta})$ on R^3. This connection has a curvature equal to zero. We see from (2.b.8) that the Christoffel symbols of $\Gamma^\sharp$ are

$$\Gamma^{\sharp\cdot}_{\ 12} = -(0,0,\beta), \qquad \Gamma^{\sharp\cdot}_{\ 21} = (0,0,-\alpha) \tag{3.b.1}$$

and all the others are zero.

REMARK1.
As we have seen in (2.b.4), the operator $\mathcal{L} = (L_1)^2 + (L_2)^2$ is hypoelliptic on R^3 if and only if the connection $\Gamma^\sharp$ associated to the frame (L_1, L_2, ∂_3) has a torsion different from zero.

REMARK2.
With (3.3), we identify the group H_3 with R^3. We consider the parallel transport along a curve $g(t)$ on H_3 defined by

$$V(t) = g(t)^n V_0 g(t)^p \tag{3.b.2}$$

where

$$V_0 = \begin{pmatrix} 0 & \alpha & \gamma \\ 0 & 0 & \beta \\ 0 & 0 & 0 \end{pmatrix}$$

n, p are integers, $n \in Z, p \in Z$, $g(t)$ is a 3×3 matrix in H_3. The connection obtained with this parallel transport is associated to the frame (L_1, L_2, ∂_3), with $L_1 = \partial_1 + py\partial_3$ and $L_2 = \partial_2 + nx\partial_3$.

4. Variation of a diffusion on the Heisenberg group H_3. Perturbation of the control equation

We shall interpret in the case of the Heisenberg group H_3, the calculus of variation developed in [1]. For that purpose, we recall some notations and results in [1].

Let M be a n-dimensional riemannian manifold endowed with a riemannian connection Γ. Let $\omega(\tau)$ be a brownian curve on R^n. By rolling M on R^n with Γ, and keeping the contact point on the curve $\omega(\tau)$, we obtain a curve $r(\tau) = (v(\tau), e(\tau))$ on $O(M)$. See [5]. We consider the curve $\omega^\epsilon(\tau)$ to be a perturbation of $\omega(\tau)$. By "rolling" M on R^n with contact points $(\omega^\epsilon(\tau))$, we get the perturbed curve

$$r^\epsilon(\tau) = (v^\epsilon(\tau), e^\epsilon(\tau)) \tag{4.1}$$

on $O(M)$. We have

$$dv^\epsilon(\tau)^i = (e^\epsilon)^i_\alpha(\tau) o (d\omega^\epsilon)^\alpha \tag{4.2}$$
$$(de^\epsilon)^i_\alpha(\tau) = -\Gamma^i_{mk}(v^\epsilon(\tau))(e^\epsilon)^k_\alpha(\tau) o (dv^\epsilon(\tau))^m$$

where o denotes the Stratanovich differential. To study the two vectors

$$\frac{d}{d\epsilon}_{|\epsilon=0} v^\epsilon(\tau) \in T_{v(\tau)}(M) \tag{4.3}$$

$$\frac{d}{d\epsilon}_{|\epsilon=0} r^\epsilon(\tau) \in T_{r(\tau)}(O(M)) \tag{4.4}$$

we consider the differential forms

the canonical form

$$\Theta : T(O(M)) \to R^n$$

the connection form

$$\varpi : T(O(M)) \to So(n)$$

See (1.3)-(1.4). We put

$$\beta(\tau) = e(\tau)^{-1} \frac{d}{d\epsilon}_{|\epsilon=0} v^\epsilon(\tau) = \Theta_{r(\tau)}(\frac{d}{d\epsilon}_{|\epsilon=0} r^\epsilon(\tau)) \tag{4.5}$$

and

$$\begin{aligned}\rho(\tau) &= \varpi_{r(\tau)}(\frac{d}{d\epsilon}_{|\epsilon=0} r^\epsilon(\tau)) \\ &= e(\tau)^{-1} \frac{d}{d\epsilon}_{|\epsilon=0} e^\epsilon(\tau) + e(\tau)^{-1} B e(\tau)\end{aligned} \tag{4.6}$$

where

$$B_r^i = \Gamma_{kr}^i \frac{d}{d\epsilon}_{|\epsilon=0} (v^\epsilon(\tau))^{(k)}$$

We have $\rho(\tau) \in So(n)$. Let $\Gamma^\dagger$ be the connection on M associated to Γ with $(\Gamma_{ij}^\dagger = \Gamma_{ji})$; the connection $\Gamma^\dagger$ is not necessarily riemannian. We denote by $e^\dagger(\tau)$ the parallel transport with $\Gamma^\dagger$. We put

$$\beta^\dagger(\tau) = e(\tau)^{\dagger^{-1}} \frac{d}{d\epsilon}_{|\epsilon=0} v^\epsilon(\tau) \tag{4.7}$$

and

$$\rho^\dagger(\tau) = \varpi^\dagger{}_{r(\tau)}(\frac{d}{d\epsilon}_{|\epsilon=0} r^\epsilon(\tau)) \tag{4.8}$$

where $\varpi^\dagger$ is the connection form associated to $\Gamma^\dagger$. The expressions of $d\beta(\tau)$, $d\beta^\dagger(\tau)$, $d\rho(\tau)$, $d\rho^\dagger(\tau)$ are given in terms of the curvature and torsion of Γ. See p.470 [1], and [2],[3].

4.a. Perturbation of the Heisenberg diffusion. In the following, we shall establish for $O(\mathbb{R}^2)$ and thus for the Heisenberg group, the formulas giving the differentials of $\beta(\tau)$, $\beta^\dagger(\tau)$, $\rho(\tau)$, $\rho^\dagger(\tau)$ in terms of the curvature and torsion of the connection Γ. We shall compare the results with those obtained in [1]. As a perturbation of $\omega(\tau)$, we take a simple case: If $\omega(\tau) = (\omega^1(\tau), \omega^2(\tau))$ is a brownian curve on R^2, we consider the perturbation $\omega_\epsilon(\tau)$ given by

$$\begin{aligned}\omega_\epsilon^1(\tau) &= \cos(\epsilon\mu)\omega^1(\tau) - \sin(\epsilon\mu)\omega^2(\tau) + \epsilon h^1(\tau) \\ \omega_\epsilon^2(\tau) &= \sin(\epsilon\mu)\omega^1(\tau) + \cos(\epsilon\mu)\omega^2(\tau) + \epsilon h^2(\tau)\end{aligned} \tag{4.a.1}$$

where μ is a constant. On $H_3 = R^3$, we take the vector fields

$$L_1(x, y, \theta) = (1, 0, \alpha y) \qquad \textit{and} \qquad L_2(x, y, \theta) = (0, 1, \beta x) \tag{4.a.2}$$

Let $u(t) = (v^1(t), v^2(t), \theta(t))$ be the diffusion defined by the Stratanovich equations

$$du(t) = L_1(u(t))o(e(\theta(t))d\omega(t))^{(1)} + L_2(u(t))o(e(\theta(t))d\omega(t))^{(2)} \tag{4.a.3}$$

$u(0) = 0.$

The diffusion $u(t) = (v^1(t), v^2(t), \theta(t))$ defined by (4.a.3) is a diffusion on the Heisenberg group H_3. We put $v(t) = (v^1(t), v^2(t))$ and we denote by $e(\theta)$ the rotation of angle θ on R^2. See (2.1). Let $u_\epsilon(t) = (v^1_\epsilon(t), v^2_\epsilon(t), \theta_\epsilon(t))$ be the perturbation of $u(t)$ defined by

$$du_\epsilon(t) = L_1(u_\epsilon(t))(e(\theta_\epsilon(t))d\omega_\epsilon(t))^{(1)} + L_2(u_\epsilon(t))(e(\theta_\epsilon(t))d\omega_\epsilon(t))^{(2)} \qquad (4.a.4)$$

and $u_\epsilon(0) = 0$. We put $v_\epsilon(t) = (v^1_\epsilon(t), v^2_\epsilon(t))$. We wish to study the vectors

$$\gamma(t) = \frac{d}{d\epsilon}_{|\epsilon=0} v_\epsilon(t) \qquad (4.a.5)$$

$$\kappa(t) = \frac{d}{d\epsilon}_{|\epsilon=0} \theta_\epsilon(t) \qquad (4.a.6)$$

4.b. The first method consists in "going back" to $T_{v(0)}(R^2)$ with the parallel transport $e(\theta(t))$, (see(3.a.2)) along the curve $v(t) = (v^1(t), v^2(t))$ where

$$d\theta(t) = \alpha v^2(t)dv^1(t) + \beta v^1(t)dv^2(t) \qquad (4.b.1)$$

$\theta(0) = 0$.

The connection Γ defined by (4.b.1)-(2.a.4) satisfies (3.a.3)- (3.a.4)-(3.a.5).

THEOREM. *Let*

$$\beta(t) = e(\theta(t))^{-1}\frac{d}{d\epsilon}_{|\epsilon=0} v_\epsilon(t) \qquad (4.b.2)$$

then $\beta(t)$ satisfies the equations

$$d\beta(t) = e(\theta(t))^{-1}[(\rho + q)odv(t) + T(odv, \gamma)] + dh(t) \qquad (4.b.3)$$

$$d\rho(t) = \Omega(odv(t), e(\theta(t))\beta(t)) \qquad (4.b.4)$$

where q is the matrix

$$q = \begin{pmatrix} 0 & -\mu \\ \mu & 0 \end{pmatrix}$$

Ω is the curvature tensor and T is the torsion of the connection Γ.

Compare with [1] p.470. To prove (4.b.3)-(4.b.4), we use several lemmas.

LEMMA. *Let $\gamma(t)$ and $\kappa(t)$ the vectors given by (4.a.5)-(4.a.6), then*

$$d\gamma^{(1)} = -(\kappa + \mu)dv^{(2)} + (e(\theta(t))dh(t))^{(1)}$$

$$d\gamma^{(2)} = (\kappa + \mu)dv^{(1)} + (e(\theta(t))dh(t)^{(2)} \qquad (4.b.5)$$

$$d\kappa = \alpha\gamma^{(2)}(t)dv^{(1)}(t) + \alpha v^{(2)}(t)d\gamma^{(1)}(t) + \beta\gamma^{(1)}(t)dv^{(2)}(t) + \beta v^{(1)}(t)d\gamma^{(2)}(t) \quad (4.b.6)$$

PROOF.

We differentiate (4.a.4) with respect to ϵ, and make $\epsilon = 0$.

LEMMA. *Let $\beta(t)$ be given by (4.b.2). We have*

$$\begin{aligned} d\beta(t) &= -\begin{pmatrix} 0 & -1 \\ 1 & 0 \end{pmatrix} \beta(t) o d\theta(t) \\ +(\kappa+\mu)\begin{pmatrix} 0 & -1 \\ 1 & 0 \end{pmatrix} e(\theta(t))^{-1} o &\begin{pmatrix} dv^{(1)}(t) \\ dv^{(2)}(t) \end{pmatrix} + dh(t) \end{aligned} \tag{4.b.7}$$

PROOF.
We calculate $d\beta(t)$ using (4.b.5)-(4.b.6)-(4.b.2).

LEMMA. *Let $r_\epsilon(\tau) = (v_\epsilon(\tau), e(\theta_\epsilon(\tau)))$ be the frame over $v_\epsilon(\tau)$; denote by ϖ_r the connection form on $O(M)$. If we put*

$$\rho(\tau) = \varpi_{r(\tau)}\Big(\frac{d}{d\epsilon}_{\epsilon=0} r^\epsilon(\tau)\Big) \tag{4.b.8}$$

then, we have

$$\rho(\tau) = \begin{pmatrix} 0 & -1 \\ 1 & 0 \end{pmatrix} \left[\kappa(\tau) - (\alpha v^{(2)}(\tau)\gamma^{(1)}(\tau) + \beta v^{(1)}(\tau)\gamma^{(2)}(\tau))\right] \tag{4.b.9}$$

and

$$d\rho(\tau) = (\alpha-\beta)\begin{pmatrix} 0 & -1 \\ 1 & 0 \end{pmatrix} \left[\gamma^{(2)}(\tau) dv^{(1)}(\tau) - \gamma^{(1)}(\tau) dv^{(2)}(\tau)\right] \tag{4.b.10}$$

The constant $(\alpha-\beta)$ is the curvature of the connection Γ.

PROOF. From [1], see also (4.6), we have

$$\rho(\tau) = e(\theta(\tau))^{-1} \frac{d}{d\epsilon}_{|\epsilon=0} e(\theta_\epsilon(\tau)) + e(\theta(\tau))^{-1} B e(\theta(\tau)) \tag{4.b.11}$$

where

$$B^i_r = \Gamma^i_{kr} \frac{d}{d\epsilon}_{|\epsilon=0} (v^\epsilon(\tau))^{(k)} \tag{4.b.12}$$

We calculate that

$$e(\theta(\tau))^{-1} \frac{d}{d\epsilon}_{|\epsilon=0} e(\theta_\epsilon(\tau)) = \begin{pmatrix} 0 & -1 \\ 1 & 0 \end{pmatrix} \kappa \tag{4.b.13}$$

Moreover, we have

$$\begin{aligned} B^1_1 = 0, \qquad B^2_1 &= -\alpha v^{(2)}\gamma^{(1)} - \beta v^{(1)}\gamma^{(2)} \\ B^1_2 = -B^2_1, \qquad B^2_2 &= 0 \end{aligned} \tag{4.b.14}$$

We replace (4.b.13)-(4.b.14) into (4.b.11). This gives (4.b.9). When we differentiate (4.b.9), we obtain (4.b.10).

COROLLARY. *We have*

$$d\rho(t) = \Omega(odv(t), e(\theta(t)\beta(t)) \tag{4.b.15}$$

where Ω is the curvature of the connection Γ.

PROOF.
It comes from (4.b.10)-(3.a.4) since $\gamma(t) = e(\theta(t))\beta(t)$.

Proof of the theorem (4.b.3)-(4.b.4). The differentials $d\beta(t)$, $d\rho(t)$
From (4.b.9), we obtain

$$q + \rho = (\kappa + \mu)\begin{pmatrix} 0 & -1 \\ 1 & 0 \end{pmatrix} - (\alpha v^{(2)}\gamma^{(1)} + \beta v^{(1)}\gamma^{(2)})\begin{pmatrix} 0 & -1 \\ 1 & 0 \end{pmatrix} \tag{4.b.16}$$

From (4.b.7), we deduce that

$$\begin{aligned} d\beta(t) = &-\begin{pmatrix} 0 & -1 \\ 1 & 0 \end{pmatrix}\beta(t)d\theta(t) + (q+\rho)e(\theta(t))^{-1}o\begin{pmatrix} dv^{(1)}(t) \\ dv^{(2)}(t) \end{pmatrix} \\ &+\begin{pmatrix} 0 & -1 \\ 1 & 0 \end{pmatrix}(\alpha v^{(2)}\gamma^{(1)} + \beta v^{(1)}\gamma^{(2)})e(\theta(t))^{-1}o\begin{pmatrix} dv^{(1)}(t) \\ dv^{(2)}(t) \end{pmatrix} + dh(t) \end{aligned} \tag{4.b.17}$$

Since

$$\begin{pmatrix} 0 & -1 \\ 1 & 0 \end{pmatrix}e(\theta(t))^{-1} = e(\theta(t))^{-1}\begin{pmatrix} 0 & -1 \\ 1 & 0 \end{pmatrix}$$

we deduce that

$$(q+\rho)e(\theta(t))^{-1} = e(\theta(t))^{-1}(q+\rho)$$

we obtain from (4.b.17) that

$$d\beta(t) = e(\theta(t))^{-1}[(q+\rho)odv(t) + H(t)] + dh(t) \tag{4.b.18}$$

with

$$\begin{aligned} H(t) = &\begin{pmatrix} 0 & -1 \\ 1 & 0 \end{pmatrix}o(\alpha v^{(2)}\gamma^{(1)} + \beta v^{(1)}\gamma^{(2)})\begin{pmatrix} dv^{(1)}(t) \\ dv^{(2)}(t) \end{pmatrix} \\ &-\begin{pmatrix} 0 & -1 \\ 1 & 0 \end{pmatrix}\gamma(t)d\theta(t) \end{aligned} \tag{4.b.19}$$

To prove (4.b.3), it is enough to identify $H(t)$ with the torsion term $T(odv, \gamma)$. Because of (4.b.1), we get

$$H = \begin{pmatrix} 0 & -1 \\ 1 & 0 \end{pmatrix}\begin{pmatrix} \beta v^{(1)} \\ -\alpha v^{(2)} \end{pmatrix}(\gamma^{(2)}dv^{(1)} - \gamma^{(1)}dv^{(2)}) \tag{4.b.20}$$

The result follows from (3.a.5).

4.c. The second method to study

$$\gamma(t) = \frac{d}{d\epsilon}_{|\epsilon=0} v_\epsilon(t)$$

consists in "going back" to $T_{v(0)}(R^2)$ using the parallel transport defined with the connection $\Gamma^\dagger$, $({\Gamma^\dagger}_{ij} = \Gamma_{ji})$. In [9], Malliavin used the connection on the right to do the calculus of variation on the left. See [1] p.452-470-474. "Going back" with the connection $\Gamma^\dagger$ extends to the Heisenberg group the Malliavin's point of view.

LEMMA. *Let $\Gamma^\dagger$ be the connection on R^2 defined by $\Gamma^\dagger_{ij} = \Gamma_{ji}$ where Γ is the connection (4.b.1). The parallel transport $e^\dagger(t)$ of a vector $X = (x^{(1)}, x^{(2)})$ along a differentiable curve $\phi(t) = (\phi^{(1)}(t), \phi^{(2)}(t))$ is given by*

$$\begin{pmatrix} x^{(1)}(t) \\ x^{(2)}(t) \end{pmatrix} = e^\dagger(t) \begin{pmatrix} x^{(1)} \\ x^{(2)} \end{pmatrix} \tag{4.c.1}$$

with

$$\begin{aligned} \frac{dx^{(1)}}{dt} &= (-\alpha\phi^{(2)}x^{(1)} - \beta\phi^{(1)}x^{(2)})\frac{d\phi^{(2)}}{dt} \\ \frac{dx^{(2)}}{dt} &= (\alpha\phi^{(2)}x^{(1)} + \beta\phi^{(1)}x^{(2)})\frac{d\phi^{(1)}}{dt} \end{aligned} \tag{4.c.2}$$

PROOF.
We deduce (4.c.2) from

$$dx^{(i)}(\tau) = -{\Gamma^\dagger}^i_{mk}(\phi(\tau))x^{(k)}(\tau)d\phi^{(m)}(\tau) \tag{4.c.3}$$

and from (3.a.3).

REMARK. The parallel transport $e^\dagger(t)$ does not preserve the metric.

LEMMA. *We have*

$$[de^\dagger(t)]e^\dagger(t)^{-1} = \begin{pmatrix} -\alpha\phi^{(2)}d\phi^{(2)} & -\beta\phi^{(1)}d\phi^{(2)} \\ \alpha\phi^{(2)}d\phi^{(1)} & \beta\phi^{(1)}d\phi^{(1)} \end{pmatrix} \tag{4.c.4}$$

PROOF.
We deduce (4.c.4) from (4.c.2).

DEFINITION. We put

$$\beta^\dagger(t) = e^\dagger(t)^{-1}\frac{d}{d\epsilon}_{|\epsilon=0} v_\epsilon(t) \tag{4.c.5}$$

THEOREM. *The vector $\beta^{\dagger}(t)$ satisfies*

$$d\beta^{\dagger}(t) = {e^{\dagger}(t)}^{-1}[q + \rho(t)]odv(t) + dh(t) \tag{4.c.6}$$

where $\rho(t)$ is the image of

$$\frac{d}{d\epsilon}_{|\epsilon=0} r^{\epsilon}(t)$$

through the connection form ϖ. See (4.b.8). We denote by q the matrix

$$q = \begin{pmatrix} 0 & -\mu \\ \mu & 0 \end{pmatrix}$$

REMARK. There is no torsion term in (4.c.6). Compare with [1] and with (4.b.3).

We first prove a lemma.

LEMMA.

$$d\beta^{\dagger}(t) = {e^{\dagger}(t)}^{-1} dz(t) \tag{4.c.7}$$

with

$$dz^1 = (\alpha v^{(2)}\gamma^{(1)} + \beta v^{(1)}\gamma^{(2)})dv^{(2)} + d\gamma^{(1)}$$

$$dz^2 = -(\alpha v^{(2)}\gamma^{(1)} + \beta v^{(1)}\gamma^{(2)})dv^{(1)} + d\gamma^{(2)}$$

PROOF.

$$\beta^{\dagger}(t) = {e^{\dagger}(t)}^{-1}\gamma(t) \tag{4.c.8}$$

thus

$$d\beta^{\dagger}(t) = {e^{\dagger}(t)}^{-1}[-de^{\dagger}(t){e^{\dagger}(t)}^{-1}\gamma(t) + d\gamma(t)]$$

We replace

$$[de^{\dagger}(t)]{e^{\dagger}(t)}^{-1}$$

by (4.c.4). This proves (4.c.7).

Proof of (4.c.6). We have, by (4.b.16)

$$[q + \rho(\tau)]odv(\tau) = [(\kappa + \mu) - (\alpha v^{(2)}\gamma^{(1)} + \beta v^{(1)}\gamma^{(2)})] \begin{pmatrix} 0 & -1 \\ 1 & 0 \end{pmatrix} odv(\tau) \tag{4.c.9}$$

Using (4.b.5), we obtain (4.c.6).

4.d. The third method consists in estimating a perturbation of the Heisenberg diffusion by means of a connection associated to a frame on $O(\mathbb{R}^2)$.

We consider the vector fields

$$L_1 = \frac{\partial}{\partial x} + \alpha y \frac{\partial}{\partial \theta}$$

$$L_2 = \frac{\partial}{\partial y} + \beta x \frac{\partial}{\partial \theta} \tag{4.d.1}$$

See (4.a.2). We define the frame

$$r = (L_1, L_2, \frac{\partial}{\partial\theta}) \tag{4.d.2}$$

This defines a frame on R^3 and also on $O(\mathbb{R}^2)$ since (L_1, L_2) are the horizontal lifts on $O(\mathbb{R}^2)$ of the vector fields ∂_i on R^2, $i = 1,2$ and since $\frac{\partial}{\partial\theta}$ generates the vertical space in $T_{(x,e(\theta))}(O(R^2))$; See (2.3), (3.a.1) and (3.b).

Let $\omega(\tau) = (\omega^{(1)}(\tau), \omega^{(2)}(\tau))$ be a brownian motion on R^2; we perturb the control equation

$$dv(\tau) = L_1(v(\tau))od\omega^{(1)}(\tau) + L_2(v(\tau))od\omega^{(2)}(\tau) \tag{4.d.3}$$

where we have $v(\tau) = (v^{(1)}(\tau), v^{(2)}(\tau), \theta(\tau))$ and where o denotes the Stratanovich differential.

REMARK. The stochastic differential equation (4.d.3) defines a diffusion on $H_3 = R^3$; it defines as well a diffussion on $O(R^2)$, i.e. the horizontal diffusion associated to the connection Γ given by (4.b.1).

We consider the perturbation $\omega_\epsilon(\tau)$ of the brownian motion on R^2 given by (4.a.1). Let

$$v_\epsilon(\tau) = (v_\epsilon^{(1)}(\tau), v_\epsilon^{(2)}(\tau), \theta_\epsilon(\tau))$$

be the solution of

$$dv_\epsilon(\tau) = L_1(v_\epsilon(\tau))od\omega_\epsilon^{(1)}(\tau) + L_2(v_\epsilon(\tau))od\omega_\epsilon^{(2)}(\tau) \tag{4.d.4}$$

with $v_\epsilon(0) = 0$.

REMARK. (4.d.4) is slightly different from (4.a.4). Both perturbations can be studied in the same way.

DEFINITION. We consider the frame (4.d.2); we put

$$r(x,y,\theta) = \begin{pmatrix} 1 & 0 & 0 \\ 0 & 1 & 0 \\ \alpha y & \beta x & 1 \end{pmatrix} \tag{4.d.5}$$

and

$$\delta(t) = \frac{d}{d\epsilon}_{|\epsilon=0} v_\epsilon(t) \tag{4.d.6}$$

$$z(t) = r(v(t))^{-1}\delta(t) \tag{4.d.7}$$

THEOREM. *We have*

$$dz(t) = A_2(z(t))od\omega^{(1)}(t) + A_1(z(t))od\omega^{(2)}(t) + h(t) \tag{4.d.8}$$

with $h(t) = (h^{(1)}(t), h^{(2)}(t), 0)$ *and*

$$A_1 = -\mu\frac{\partial}{\partial x} + (\beta-\alpha)x\frac{\partial}{\partial\theta}$$

$$A_2 = \mu\frac{\partial}{\partial y} - (\beta-\alpha)y\frac{\partial}{\partial\theta} \tag{4.d.9}$$

REMARK. The equations (4.d.8) can be written

$$z^{(1)}(t) = -\mu\omega^{(2)}(t) + h^{(1)}(t)$$

$$z^{(2)}(t) = \mu\omega^{(1)}(t) + h^{(2)}(t)$$

$$dz^{(3)}(t) = (\beta - \alpha)[z^{(1)}(t) o d\omega^{(2)}(t) - z^{(2)}(t) o d\omega^{(1)}(t)] \qquad (4.d.10)$$

The last equation (4.d.10) is the equation corresponding to (4.b.4) and $(\beta - \alpha)$ is the curvature of Γ. With (1.c.5), we can interpret (4.d.8)-(4.d.10) in terms of curvature.

COROLLARY. *When* $h = 0$, $\mu = 1$,*then*

$$z(\tau) = r(v(\tau))^{-1} \frac{d}{d\epsilon}_{|\epsilon=0} v^{\epsilon}(\tau)$$

is a diffusion on R^3 *with infinitesimal generator*

$$\mathcal{L} = (1/2)[(A_1)^2 + (A_2)^2]$$

$$= (1/2)[\frac{\partial^2}{\partial x^2} + \frac{\partial^2}{\partial y^2} - 2(\beta - \alpha)(x\frac{\partial}{\partial x} + y\frac{\partial}{\partial y})\frac{\partial}{\partial \theta} + (\beta - \alpha)^2(x^2 + y^2)\frac{\partial^2}{\partial \theta^2} - 2(\beta - \alpha)\frac{\partial}{\partial \theta}]$$

We have

$$z^{(1)}(t) = -\omega^{(2)}(t)$$

$$z^{(2)}(t) = \omega^{(1)}(t)$$

$$z^{(3)}(t) = (1/2)(\alpha - \beta)[z^{(1)}(t)^2 + z^{(2)}(t)^2]$$

Proof of theorem (4.d.8).

From (4.d.3), we deduce that

$$v^{(1)}(\tau) = \omega^{(1)}(\tau), \qquad v^{(2)}(\tau) = \omega^{(2)}(\tau)$$

$$dv^{(3)}(\tau) = \alpha\omega^{(2)}(\tau) o d\omega^{(1)}(\tau) + \beta\omega^{(1)}(\tau) o d\omega^{(2)}(\tau)$$

We have for $v_\epsilon(\tau)$

$$v_\epsilon^{(1)}(\tau) = \cos(\epsilon\mu)\omega^{(1)}(\tau) - \sin(\epsilon\mu)\omega^{(2)}(\tau) + \epsilon h^{(1)}(\tau)$$

$$v_\epsilon^{(2)}(\tau) = \sin(\epsilon\mu)\omega^{(1)}(\tau) + \cos(\epsilon\mu)\omega^{(2)}(\tau) + \epsilon h^{(2)}(\tau)$$

$$dv_\epsilon^{(3)}(\tau) = \alpha v_\epsilon^{(2)}(\tau) o dv_\epsilon^{(1)}(\tau) + \beta v_\epsilon^{(1)}(\tau) o dv_\epsilon^{(2)}(\tau) \qquad (4.d.11)$$

We take the derivative with respect to ϵ at $\epsilon = 0$, and we consider $\delta(t)$ given by (4.d.6)

From(4.d.11), we obtain

$$\delta^{(1)} = -\mu\omega^{(2)} + h^1, \qquad \delta^{(2)} = \mu\omega^{(1)} + h^2$$

$$d\delta^{(3)} = \alpha\delta^{(2)}dv^{(1)} + \alpha v^{(2)}d\delta^{(1)} + \beta\delta^{(1)}dv^{(2)} + \beta v^{(1)}d\delta^{(2)}$$

From (4.d.5), we deduce that

$$r(v(t))^{-1} = \begin{pmatrix} 1 & 0 & 0 \\ 0 & 1 & 0 \\ -\alpha v^{(2)}(t) & -\beta v^{(1)}(t) & 1 \end{pmatrix}$$

Since $z(t) = r(v(t))^{-1}\delta(t)$, we obtain

$$z^{(1)} = -\mu\omega^{(2)} + h^1 = \delta^{(1)}$$

$$z^{(2)} = \mu\omega^{(1)} + h^2 = \delta^{(2)}$$

$$z^{(3)} = -\alpha v^{(2)}\delta^{(1)} - \beta v^{(1)}\delta^{(2)} + \delta^{(3)}$$

$$dz^{(3)} = (\alpha - \beta)(\delta^{(2)}dv^{(1)} - \delta^{(1)}dv^{(2)})$$

This yields (4.d.8).

References

1. H. Airault , and P. Malliavin., *Integration by parts formulas and diffusion vector fields on elliptic probability spaces*, Probab. Theory Relat. Fields 106 (447-494) (1996).
2. B.K. Driver., *A Cameron-Martin quasi-invariance theorem for brownian motion on a compact manifold*, J. Funct. Anal. 110 (603- 608) (1992).
3. S. Fang, and P. Malliavin., *Stochastic analysis on the path space of a riemannian manifold. Markovian stochastic calculus.*, J. Funct. Anal. 118 (249-274)(1993).
4. S. Fang., *Rotations et quasi-invariance sur l'espace des chemins*, Potential Anal. 4 (1995), 67-77.
5. N. Ikeda and S. Watanabe., *Stochastic differential equations and diffusion processes*, North Holland, 2nd edition, Amsterdam (1989).
6. V. Jurdjevic, *Geometric control theory,*, Studies in Advances Mathematics 51 (1997).
7. S. Kobayashi and K. Nomizu, *Foundations of differential geometry*, New York. Intersciences publishers (1969).
8. P. Malliavin, *Formule de la moyenne, calcul de perturbations et théorème d'annulation pour les formes harmoniques*, J. Funct. Anal. 17 (274-291) (1974).
9. P. Malliavin, *Naturality of quasi-invariance of some measures*, Stochastic analysis and applications. In: Cruzeiro,A.B. and Zambrini, J.C; Proceedings of the 1989 Lisbon Conference (144-154).
10. S. Watanabe, *Levy's stochastic area formula and Brownian motion on compact Lie groups, Ito's stochastic calculus and probability theory.*, . Eds: N. Ikeda, S. Watanabe, M. Fukushima, H. Kunita. Conference dedicated to K. Ito (1996) Springer-Verlag (p.401-411).

INSSET , Université de Picardie Jules Verne, 48 rue Raspail, F-02100 Saint-Quentin (Aisne) France

Proceedings of Symposia in Pure Mathematics
Volume **64**, 1999

Linear Control Systems on Lie Groups and Controllability

Victor Ayala and Juan Tirao

ABSTRACT. In this paper, we deal with a linear control system Σ defined on a Lie group G with Lie algebra $\mathfrak{g}$. The dynamic of Σ is determined by a drift vector field, which is an element in the normalizer of $\mathfrak{g}$ in the Lie algebra of all smooth vector fields on G, and by the control vectors, which are elements in $\mathfrak{g}$ considered as left-invariant vector fields. We characterize the normalizer of $\mathfrak{g}$ as the semidirect product of $\mathfrak{g}$ with the Lie algebra of $\mathrm{Aut}(G)$, and we give an explicit expression for the integrable curves of each element in the normalizer. If the drift vector field has a singularity, then the Lie algebra rank condition is necessary for the controllability property. But in general this condition is not sufficient. We prove, however, that the (ad-)rank condition is sufficient for the locally controllability of Σ and also for the global controllability in the compact case. In particular, we extend the well-known Kalman rank condition when G is an abelian connected Lie group. Our work is related to earlier work of L. Markus.

1. Introduction

The twofold purpose of this paper is to generalize the notion of linear control systems from $\mathbb{R}^n$ to a connected Lie group G and to analyze some algebraic conditions which give information about controllability for this particular class of control systems. By definition in the context of Lie groups a *linear control system* is determined by the pair

$$\Sigma = (G, \mathcal{D}),$$

where the state space is a real finite dimensional Lie group G with Lie algebra $\mathfrak{g}$ and the dynamic $\mathcal{D}$ is given by the family of differential equations on G

$$\dot{x} = X(x) + \sum_{j=1}^{\kappa} \mu_j Y^j(x).$$

Here the drift vector field X is an element of the normalizer of $\mathfrak{g}$ in the Lie algebra $X(G)$ of all C^∞ vector fields on G, and the control vectors Y^j, $j = 1, 2, \ldots, \kappa$, belong to $\mathfrak{g}$, where $\mathfrak{g}$ denotes the Lie algebra of left-invariant vector fields. The input functions $u = (u_1, u_2, \ldots, u_\kappa)$ belong to $\mathcal{U}$, the class of unrestricted admissible controls. The elements of $\mathcal{U}$ are piecewise constant functions of the form

$$u : [0, \infty) \to \mathbb{R}^\kappa.$$

1991 *Mathematics Subject Classification.* 93B05, 22E15.

Support: Work supported by Proyecto Fondecyt No 1960641 and International Centre for Theoretical Physics.

The collection $\mathcal{D}$ is the family of vector fields associated with Σ, i.e.,

$$\mathcal{D} = \{X + \sum_{j=1}^{\kappa} u_j Y^j | u \in \mathbb{R}^{\kappa}\}.$$

First we study in detail *linear vector fields* on Lie groups, i.e., the elements in the normalizer

$$\eta := \mathrm{norm}_{X(G)}(\mathfrak{g}).$$

We show that for a connected Lie group G the centralizer $\mathfrak{z}(\mathfrak{g})$ in $X(G)$ consists of all right invariant vector fields on G. Let $\mathfrak{h}$ denote the subalgebra of all $X \in \mathrm{norm}_{X(G)}(\mathfrak{g})$ such that $X_e = 0$. We show that $\mathfrak{z}(\mathfrak{g}) \otimes \mathfrak{h} \cong \mathrm{norm}_{X(G)}(\mathfrak{g})$ represents $\mathrm{norm}_{X(G)}(\mathfrak{g})$ as a semidirect product of $\mathfrak{h}$ and $\mathfrak{z}(\mathfrak{g})$ and that $X \mapsto \mathrm{ad}(X)_{\mathfrak{g}}$ is a Lie algebra isomorphism from $\mathfrak{h}$ onto the image of $\mathrm{norm}_{X(G)}(\mathfrak{g})$ in $\partial\mathfrak{g}$, the Lie algebra of derivations of $\mathfrak{g}$. Let $\Phi : \mathrm{norm}_{X(G)}(\mathfrak{g}) \mapsto \mathfrak{g}\times \partial\mathfrak{g}$ be defined by $\Phi(Z + X) = (Z', \mathrm{ad}(X)_{\mathfrak{g}})$ for all $Z \in \mathfrak{z}(\mathfrak{g})$ and all $X \in \mathfrak{h}$, where Z' is the unique left invariant vector field on G such that $Z'_e = -Z_e$. In this paper we will prove the following theorems.

Theorem 2.2. (i) If G is a connected Lie group, then $\Phi : \mathrm{norm}_{X(G)}(\mathfrak{g}) \mapsto \mathfrak{g}\times \partial\mathfrak{g}$ is an injective homomorphism.
(ii) If G is connected and simply connected then $\Phi : \mathrm{norm}_{X(G)}(\mathfrak{g}) \mapsto \mathfrak{g}\times \partial\mathfrak{g}$ is a surjective isomorphism. □

Theorem 2.3. Let $\pi : G_1 \to G$ be a universal covering homomorphism of a connected Lie group G. Then Φ maps $\mathrm{norm}_{X(G)}(\mathfrak{g})$ isomorphically onto $\mathfrak{g} \otimes \mathrm{aut}(G)$. Moreover, if $i : \mathfrak{g} \otimes \mathrm{aut}(G) \to \mathfrak{g} \otimes \partial(\mathfrak{g})$ denotes the inclusion map, then we have the following

$$\begin{array}{ccc} \mathrm{norm}_{X(G_1)}(\mathfrak{g}) & \longrightarrow & \mathfrak{g} \otimes \partial(\mathfrak{g}) \\ \iota \uparrow & & \uparrow i \\ \mathrm{norm}_{X(G)}(\mathfrak{g}) & \longrightarrow & \mathfrak{g} \otimes \mathrm{aut}(G) \end{array}$$

□

We find it convenient to identify vector fields on G with C^∞-functions defined on G into the Lie algebra $\mathfrak{g}$ via the identification of $\tilde{F} \in X(G)$ with $F \in C^\infty(G, \mathfrak{g})$ defined by

$$\tilde{F}_x = dL_x(F(x)), \quad x \in G,$$

where as usual L_x denotes the left translation by x.

Theorem 2.5. If $F \in C^\infty(G, \mathfrak{g})$, then $\tilde{F} \in \mathrm{norm}_{X(G)}(\mathfrak{g})$ if and only if for each $Y \in \mathfrak{g}$

$$F(x \exp Y) = \left(e^{-\,\mathrm{ad}\, Y}\right) F(x) + \left(\frac{1 - e^{-\,\mathrm{ad}\, Y}}{\mathrm{ad}\, Y}\right) \left(Y(F) + [Y, F(e)]\right).$$

□

Corollary 2.6. Let G be a connected Lie group. An element $\tilde{F} \in \mathrm{norm}_{X(G)}(\mathfrak{g})$ is characterized by the pair $(F(e), dF_e)$. □

In Theorem 2.8 we characterize the one parameter group τ_t of diffeomorphisms of G associated to $X \in \mathrm{norm}_{X(G)}(\mathfrak{g})$. It follows in particular that any member of $\mathrm{norm}_{X(G)}(\mathfrak{g})$ has a global analytic flow on G associated with it. We further obtain in Theorem 2.9 an expression for the integrable curves of $\tilde{F} \in \eta$.

Let $\Sigma = (G, \mathcal{D})$ be a linear control system. This class of systems generalizes (i) linear control systems on $\mathbb{R}^n$ [4], linear control systems on a matrix Lie group [5], and the class of invariant control systems. We study the controllability property of Σ when the drift vector field X in the normalizer has a singularity and we show that it is possible to reduce the study to the case $X_e = 0$, i.e., the one-parameter group T of G-diffeomorphisms induced by X is a subgroup of $\mathrm{Aut}(G)$. We consider the Lie algebra generated by the control vectors

$$\mathcal{H} = Span_{\mathcal{L}.A.}\{Y^1, \ldots, Y^\kappa\}$$

and

$$\langle X|\mathcal{H}\rangle := \text{the smallest } \mathrm{ad}(X)\text{-invariant subalgebra of } \mathfrak{g} \text{ containing } \mathcal{H}.$$

Denote by H and $\langle X|H\rangle$ the connected Lie subgroups of G with Lie algebras $\mathcal{H}$ and $\langle X|\mathcal{H}\rangle$ respectively. The next theorem may be deduced from standard facts.

Theorem 3.3. (i) The orbit of Σ through the identity element e is $\langle X|H\rangle$.
(ii) Σ is transitive if and only if

$$dim.Span_{\mathcal{L}.A.}\{Y^j, \mathrm{ad}^i(X)(Y^j)|0 \le i \le dim(G), 1 \le j \le \kappa\} = dim(G).$$

□

Unfortunately, Theorem 3.3 does not characterize controllability. To show this, we give an example on the simply connected nilpotent Heisenberg Lie group. On the other hand, we can deduce local controllability.

Theorem 3.5. Let G be a connected Lie group and $\Sigma = (G, D)$ be a transitive linear control system. If

$$dim.Span\{Y^j, ad^i(X)(Y^j)|0 \le i \le dim(G), 1 \le j \le \kappa\} = dim(G),$$

then Σ is locally controllable. □

Corollary 3.6. Let G be a connected Lie group and $\Sigma = (G, \mathcal{D})$ be a transitive linear control system such that X is an infinitesimal automorphism.

1. If G is abelian,

$$\Sigma \text{ is controllable} \Leftrightarrow \Sigma \text{ satisfies the rank condition.}$$

2. If G is compact,

$$\Sigma \text{ satisfies the rank condition} \Rightarrow \Sigma \text{ is controllable.}$$

□

In particular, Corollary 3.6 generalizes to abelian connected Lie groups the well-known Kalman rank condition [4]. Global controllability of Σ follows from Theorem 3.5 if G is compact. On the other hand, if the positive orbit $S_\Sigma(e)$ of e is a semigroup, then Theorem 3.5 gives a sufficient condition for global controllability.

We conjecture that $S_\Sigma(e)$ is not a semigroup. Our work is also related to a paper of L. Markus [5].

This paper is organized as follows: Section 2 contains a characterization of the normalizer of $\mathfrak{g}$, Section 3 contains the controllability results, and in Section 4 we give some examples.

2. Linear Vector Fields

Let $\Sigma = (G, \mathcal{D})$ be a linear control system on the Lie group G, as defined in the introduction. By definition the drift vector field X of Σ is an element in the normalizer of $\mathfrak{g}$ in the Lie algebra of all C^∞ vector fields on G. A vector field X on $\mathbb{R}^n$ is called *linear* if $X_p = A(p) + u$ for all $p \in \mathbb{R}^n$, where $A : \mathbb{R}^n \to \mathbb{R}^n$ is a linear map and u is a fixed vector of $\mathbb{R}^n$. Such linear vector fields on $\mathbb{R}^n$ can be characterized in the following way: a C^∞ vector field X on $\mathbb{R}^n$ is linear if and only if $[X, U]$ is a constant vector field for every constant vector field U on $\mathbb{R}^n$ (see Corollary 2.7). Therefore the set of all linear vector fields on $\mathbb{R}^n$ is the normalizer of the Lie algebra of $\mathbb{R}^n$ (which consists of all constant vector fields) in the Lie algebra $X(\mathbb{R}^n)$ of all C^∞ vector fields on $\mathbb{R}^n$. Thus the notion we have introduced of a linear control system on a Lie group is a generalization from linear systems on $\mathbb{R}^n$ to arbitrary Lie groups.

If G is a Lie group, the Lie algebra $\mathfrak{g}$ of G is henceforth identified with the Lie algebra of all left-invariant vector fields on G. Also $X(G)$ will denote the Lie algebra of all C^∞ vector fields on G. Let

$$\mathrm{norm}_{X(G)}(\mathfrak{g}) := \{X \in X(G) \mid [X, Y] \in \mathfrak{g}, \text{ for all } Y \in \mathfrak{g}\}.$$

DEFINITION 2.1. The elements in $\mathrm{norm}_{X(G)}(\mathfrak{g})$ are called *(group)linear vector fields* on G. □

The Jacobi identiy implies that $\mathrm{norm}_{X(G)}(\mathfrak{g})$ is a Lie subalgebra of $X(G)$, and that the map $X \mapsto \mathrm{ad}(X)_\mathfrak{g}$ is a Lie algebra homomorphism of $\mathrm{norm}_{X(G)}(\mathfrak{g})$ into the derivation algebra $\partial\mathfrak{g}$ of $\mathfrak{g}$. The kernel of this homomorphism is clearly the centralizer $\mathfrak{z}(\mathfrak{g})$ of $\mathfrak{g}$ in $X(G)$. Using the standard fact that two vector fields commute if and only if the mappings in their corresponding local flows commute and that the flow is given by $p \mapsto p\exp(tX)$ for $X \in \mathfrak{g}$ we obtain:

$$Y \in \mathfrak{z}(\mathfrak{g}) \Leftrightarrow \tau_Y(p\exp(tX), s) = \tau_Y(p, s)\exp(tX), \ \forall X \in \mathfrak{g}, \ p \in G, \ s, t \in \mathbb{R}.$$

It follows for a connected Lie group that $\tau_Y(pg, s) = \tau_Y(p, s)g$ for any $g \in G$, since any $g \in G$ is a finite product of exponentials. But this is precisely the property that the local flow of the vector field Y is right invariant. In fact, if $\tau_Y(pg, s) = \tau_Y(p, s)g$ for any $g \in G$, then

$$Y_{pg} = \left.\frac{d}{dt}\right|_{t=0} \tau(pg, t) = \left.\frac{d}{dt}\right|_{t=0} \tau(p, t)g = (dR_g)_p(Y_p),$$

i.e., Y is a right invariant vector field. Conversely if Y is a right invariant vector field, then

$$\frac{d}{dt}\bigg|(\tau_Y(p, t)g) = (dR_g)(\tau_Y(p, t))(Y_{\tau_Y(p,t)}) = Y_{\tau_Y(p,t)g}$$

and $\tau_Y(p, 0)g = pg = \tau_Y(pg, 0)$. Thus $\tau_Y(p, t)g$ and $\tau_Y(pg, t)$ are both solutions of the same first order ordinary differential equation and both coincide at $t = 0$, and hence are equal as we wanted to prove.

Therefore we have seen that for a connected Lie group G the centralizer $\mathfrak{z}(\mathfrak{g})$ consists of all right invariant vector fields on G. Now let us consider the set $\mathfrak{h}$ of all $X \in \mathrm{norm}_{X(G)}(\mathfrak{g})$ such that $X_e = 0$. It follows readily that $\mathfrak{h}$ is a vector subspace and from computation of the Lie product from the corresponding flows that $\mathfrak{h}$ is a Lie subalgebra. Also since $\mathfrak{z}(\mathfrak{g})$ is a Lie algebra ideal in $\mathrm{norm}_{X(G)}(\mathfrak{g})$, it follows that $[\mathfrak{z}(\mathfrak{g}), \mathfrak{h}] \subseteq \mathfrak{z}(\mathfrak{g})$. Since a right invariant vector field which vanishes at a point is the zero vector field, it follows that $\mathfrak{h} \cap \mathfrak{z}(\mathfrak{g}) = \{0\}$, and hence that we may form the semidirect product $\mathfrak{z}(\mathfrak{g}) \otimes \mathfrak{h}$.

Given $X \in \mathrm{norm}_{X(G)}(\mathfrak{g})$ let Y be the right invariant vector field on G such that $Y_e = X_e$. Then $X - Y \in \mathfrak{h}$, and $\mathrm{ad}(X)_{\mathfrak{g}} = \big(\mathrm{ad}(X-Y)\big)_{\mathfrak{g}}$, since $Y \in \mathfrak{z}(\mathfrak{g})$ the kernel of $X \mapsto \mathrm{ad}(X)_{\mathfrak{g}}$. It follows that $\mathrm{norm}_{X(G)}(\mathfrak{g}) = \mathfrak{z}(\mathfrak{g}) \oplus \mathfrak{h}$ and $X \mapsto \mathrm{ad}(X)_{\mathfrak{g}} : \mathfrak{h} \to \partial\mathfrak{g}$ is a Lie algebra isomorphism onto the image of $\mathrm{norm}_{X(G)}(\mathfrak{g})$ in $\partial\mathfrak{g}$. It now follows readily that $(Z, Y) \mapsto Z + Y : \mathfrak{z}(\mathfrak{g}) \otimes \mathfrak{h} \to \mathrm{norm}_{X(G)}(\mathfrak{g})$ is a Lie algebra isomorphism.

If Z is a right invariant vector field on G let Z' be the left invariant vector field on G such that $Z'_e = -Z_e$. Then the linear map $Z \mapsto Z'$ is a Lie algebra isomorphism from $\mathfrak{z}(\mathfrak{g})$ onto $\mathfrak{g}$. Let $\Phi : \mathrm{norm}_{X(G)}(\mathfrak{g}) \to \mathfrak{g} \otimes \partial\mathfrak{g}$ (where the latter is a semidirect product of Lie algebras determined by the action of $\partial\mathfrak{g}$ on $\mathfrak{g}$) be defined by $\Phi(Z + X) = (Z', \mathrm{ad}(X)_{\mathfrak{g}})$ for all $Z \in \mathfrak{z}(\mathfrak{g})$ and all $X \in \mathfrak{h}$. We have thus shown the first part of the following theorem.

THEOREM 2.2. *(i) If G is a connected Lie group, then $(Z, Y) \mapsto Z + Y$: $\mathfrak{z}(\mathfrak{g}) \otimes \mathfrak{h} \to norm_{X(G)}(\mathfrak{g})$ is an isomorphism and $\Phi : \mathrm{norm}_{X(G)}(\mathfrak{g}) \to \mathfrak{g} \otimes \partial\mathfrak{g}$ is an injective homomorphism of Lie algebras.*
(ii) If G is connected and simply connected then $\Phi : \mathrm{norm}_{X(G)}(\mathfrak{g}) \to \mathfrak{g} \otimes \partial\mathfrak{g}$ is surjective, hence an isomorphism.

Proof. (i) This part was previously established.

(ii) The only thing that remains to be proved is the surjectivity assertion. Given $D \in \partial\mathfrak{g}$ then $\exp(tD)$ is an automorphism of $\mathfrak{g}$ for any $t \in \mathbb{R}$. Since G is assumed to be simply connected, there exists a unique automorphism φ_t of G such that $(d\varphi_t)_e = \exp(tD)$. It is easy to see that the map $(g, t) \mapsto \varphi_t(g)$ is a C^∞ flow on G. Let X be the associated vector field.

Let us prove that $X \in \mathrm{norm}_{X(G)}(\mathfrak{g})$. In fact if $Y \in \mathfrak{g}$ then

$$\begin{aligned} [X, Y]_g &= \left.\frac{d}{dt}\right|_{t=0} d\varphi_{-t} dL_{\varphi_t(g)} Y_e = \left.\frac{d}{dt}\right|_{t=0} dL_g d\varphi_{-t} Y_e \\ &= dL_g \left.\frac{d}{dt}\right|_{t=0} \exp(-tD) Y_e = -(DY)_g. \end{aligned}$$

Hence $X \in \mathrm{norm}_{X(G)}(\mathfrak{g})$ and $\mathrm{ad}(X)_{\mathfrak{g}} = -D$. Moreover, $X_e = \left.\frac{d}{dt}\right|_{t=0} \varphi_t(e) = 0$. This completes the proof of the theorem. □

Let G be a connected and simply connected Lie group. Then the map δ : $\mathrm{Aut}(G) \to \mathrm{Aut}(\mathfrak{g})$ given by $\delta(\varphi) = (d\varphi)_e$ is a surjective isomorphism. Since $\mathrm{Aut}(\mathfrak{g})$ is a closed subgroup of $GL(\mathfrak{g})$ it has a unique structure of a topological Lie subgroup of $GL(\mathfrak{g})$. Moreover its Lie algebra is $\partial\mathfrak{g}$. Hence $\mathrm{Aut}(G)$ inherits a unique Lie group structure making δ an isomorphism of Lie groups. We shall identify the Lie algebra of $\mathrm{Aut}(G)$ with $\partial\mathfrak{g}$ via $(d\delta)_e$, and we shall denote by $\mathrm{Exp} : \partial\mathfrak{g} \to \mathrm{Aut}(G)$ the corresponding exponential map. Now let $\pi : G_1 \to G$

be a universal covering homomorphism of a connected Lie group G and let $K = \ker \pi$. Then the group $\mathrm{Aut}(G)$ of all Lie automorphisms of G can be identified with$\{\varphi \in \mathrm{Aut}(G_1) : \varphi(K) \subset K\}$. In this way $\mathrm{Aut}(G)$ inherits a Lie structure which makes it isomorphic to $\{\varphi \in \mathrm{Aut}(G_1) : \varphi(K) \subset K\}$ with its unique structure of a closed Lie subgroup of $\mathrm{Aut}(G_1)$. We identify the Lie algebra $\mathrm{aut}(G)$ of $\mathrm{Aut}(G)$ with a Lie subalgebra of $\partial\mathfrak{g}$ and we also denote by $\mathrm{Exp} : \mathrm{aut}(G) \to \mathrm{Aut}(G)$ the corresponding exponential map. Then for any $D \in \mathrm{aut}(G)$ and $X \in \mathfrak{g}$ we have

$$(d\,\mathrm{Exp}\,D)_e = e^D,$$

and

$$(\mathrm{Exp}\,D)(\exp X) = \exp\big((d\,\mathrm{Exp}\,D)_e X\big) = \exp(e^D(X)).$$

Let $\zeta : X(G) \to X(G_1)$ be defined by $\zeta(X) = X_1$ where X_1 is the unique vector field on G_1 which is π-related to X. Then from the invariance of the Lie bracket it follows that

$$\iota(\mathrm{norm}_{X(G)}(\mathfrak{g})) \subset \mathrm{norm}_{X(G_1)}(\mathfrak{g}).$$

THEOREM 2.3. *Let $\pi : G_1 \to G$ be a universal covering homomorphism of a connected Lie group G. Then Φ maps* $\mathrm{norm}_{X(G)}(\mathfrak{g})$ *isomorphically onto* $\mathfrak{g}\otimes\mathrm{aut}(G)$. *Moreover if $i : \mathfrak{g} \otimes \mathrm{aut}(G) \to \mathfrak{g} \otimes \partial(\mathfrak{g})$ denotes the inclusion map then we have the following commutative diagram:*

$$\begin{array}{ccc} norm_{X(G_1)}(\mathfrak{g}) & \longrightarrow & \mathfrak{g}\otimes\partial(\mathfrak{g}) \\ \iota\uparrow & & \uparrow i \\ norm_{X(G)}(\mathfrak{g}) & \longrightarrow & \mathfrak{g}\otimes aut(G). \end{array}$$

Proof. From the invariance of the Lie bracket we see that the following diagram commutes:

$$\begin{array}{ccc} \mathrm{norm}_{X(G_1)}(\mathfrak{g}) & \longrightarrow & \mathfrak{g}\otimes\partial(\mathfrak{g}) \\ \iota\uparrow & & \uparrow id \\ \mathrm{norm}_{X(G)}(\mathfrak{g}) & \longrightarrow & \mathfrak{g}\otimes\partial(\mathfrak{g}). \end{array}$$

Thus the only thing that remains to be proved is that $\Phi(\mathfrak{h}) = \mathrm{aut}(\mathfrak{g})$. If $X \in X(G)$ then it is clear that $\pi(\tau_{\iota(X)}(g,t)) = \tau_X(\pi(g),t)$ for all $g \in G_1$ and all $t \in \mathbb{R}$. Now if $X \in \mathfrak{h}$ and $D = \Phi(\iota(X))$ then $\pi\big(\mathrm{Exp}(-tD)(g)\big) = \pi(\tau_{\iota(X)}(g,t)) = \tau_X(\pi(g),t)$. In particular if we put $g = k \in K$, we get $\pi\big(\mathrm{Exp}(-tD)(k)\big) = \tau_X(e,t) = e$ because $X_e = 0$. Hence $\mathrm{Exp}(tD)(K) \subset K$, which in turns implies that $D \in \mathrm{aut}(G)$. In exactly the same way one proves that if $D \in \mathrm{aut}(G)$, then the vector field $X_1 \in \mathrm{norm}_{X(G_1)}(\mathfrak{g})$ such that $\Phi(X_1) = D$ (see Theorem 2.2) is π-related to a unique vector field $X \in \mathfrak{h}$. This completes the proof of the theorem. □

Let us denote by G_e the tangent space of G at e. If $F \in G_e$ we denote by $\tilde{F}$ the left invariant vector field on G such that $\tilde{F}_e = F$. Also, as usual, we identify G_e with the Lie algebra $\mathfrak{g}$ through the isomorphism $F \to \tilde{F}$. Moreover we find it convenient to identify vector fields on G with functions of G into G_e. If $F : G \to \mathfrak{g}$ let

$$\tilde{F}_x = dL_x(F(x)) = \widetilde{F(x)}_x, \quad x \in G.$$

Clearly the map $F \mapsto \tilde{F}$ is a linear isomorphism of $C^\infty(G, G_e)$ onto $X(G)$. Observe that the constant function $F(x) = Y$ corresponds to the left invariant vector field $\tilde{Y}$.

LEMMA 2.4. *Given* $Y \in G_e$ *and* $F \in C^\infty(G, G_e)$*, let* $H \in C^\infty(G, G_e)$ *be defined by*

$$H(x) = \tilde{Y}_x(F) + [Y, F(x)], \quad x \in G. \tag{2.1}$$

Then $\tilde{H} = [\tilde{Y}, \tilde{F}]$*. Moreover, for* $\tilde{F} \in norm_{X(G)}(\mathfrak{g})$*,* $\Phi(\tilde{F}) = (-F(e), \mathrm{ad}(F(e)) - dF_e)$*.*

Proof. We have

$$\left([\tilde{Y}, \tilde{F}]f\right)(x) = \left(\tilde{Y}\left(\tilde{F}f\right)\right)(x) - \left(\tilde{F}\left(\tilde{Y}f\right)\right)(x)\ ,\ f \in C^\infty(G);$$

$$\left(\tilde{Y}\left(\tilde{F}f\right)\right)(x) = (\frac{d}{dt})_{t=0}\left(\tilde{F}f\right)(x\exp tY) = (\frac{d}{dt})_{t=0}\left(\widetilde{F(x\exp tY)}f\right)(x\exp tY)$$

$$= \left(\widetilde{(\tilde{Y}F)(x)}f\right)(x) + \left(\tilde{Y}\left(\widetilde{F(x)}f\right)\right)(x);$$

$$\left(\tilde{F}\left(\tilde{Y}f\right)\right)(x) = \tilde{F}_x\left(\tilde{Y}f\right) = \left(\widetilde{F(x)}\left(\tilde{Y}f\right)\right)(x);$$

$$\left([\tilde{Y}, \tilde{F}]f\right)(x) = \left(\widetilde{(\tilde{Y}F)(x)}f\right)(x) + \left([\tilde{Y}, \widetilde{F(x)}]f\right)(x)$$
$$= \left(\widetilde{(\tilde{Y}F)(x)}f\right)(x) + \left(\widetilde{[Y, F(x)]}f\right)(x).$$

Therefore

$$[\tilde{Y}, \tilde{F}]_x = \widetilde{\tilde{Y}_x(F)}_x + \widetilde{[Y, F(x)]}_x = \left(\widetilde{\tilde{Y}_x(F) + [Y, F(x)]}\right)_x.$$

Now we compute $\Phi(\tilde{F})$. Let $\tilde{F} = Z + X$ with $Z \in \mathfrak{z}(\mathfrak{g})$ and $X \in \mathfrak{h}$. Then $\Phi(\tilde{F}) = (Z_e, \mathrm{ad}(X)_\mathfrak{g}) = (-F(e), \mathrm{ad}(X)_\mathfrak{g})$. Now given $Y \in G_e$ we have

$$[X, \tilde{Y}]_e = [\tilde{F}, \tilde{Y}]_e = -\tilde{Y}_e(F) - [Y, F(e)] = -dF_e(Y) + [F(e), Y].$$

Thus $\mathrm{ad}(X)_\mathfrak{g} = \mathrm{ad}(F(e)) - dF_e$. □

THEOREM 2.5. *If* $F \in C^\infty(G, \mathfrak{g})$*, then* $\tilde{F} \in \mathrm{norm}_{X(G)}(\mathfrak{g})$ *if and only if for each* $Y \in \mathfrak{g}$*,*

$$F(x\exp Y) = \left(e^{-\mathrm{ad}\, Y}\right)F(x) + \left(\frac{1 - e^{-\mathrm{ad}\, Y}}{\mathrm{ad}\, Y}\right)\left(Y(F) + [Y, F(e)]\right), \tag{2.2}$$

where $Y(F) = dF_e(Y)$*.*

Proof. If $F \in C^\infty(G, \mathfrak{g})$, let $B(Y) = Y(F) + [Y, F(e)]$. Then from equation (2.1) we get $\tilde{F} \in \mathrm{norm}_{X(G)}(\mathfrak{g})$ if and only if

$$\tilde{Y}_x(F) + [Y, F(x)] = B(Y) \quad \text{for all } x \in G,$$

if and only if

$$(\frac{d}{dt})F(x\exp tY) = -[Y, F(x\exp tY)] + B(Y) \quad \text{for all } x, t.$$

Let $f(t) = F(x\exp tY)$ and set $g(t) = \left(e^{-t\,\mathrm{ad}\, Y}\right)F(x) + (\frac{1-e^{-t\,\mathrm{ad}\, Y}}{\mathrm{ad}\, Y})B(Y)$. Then one verifies directly that $g(t)$ also satisfies the preceding differential equation and that $f(0) = g(0)$. By uniqueness of solutions, we conclude that $f(t) = g(t)$ for all t. Taking $t = 1$ yields the theorem. □

COROLLARY 2.6. *For G a connected Lie group, an element $\tilde{F} \in \mathrm{norm}_{X(G)}(\mathfrak{g})$ is characterized by the pair $(F(e), dF_e)$.*

Proof. Let $U = \exp(G)$. We prove by induction on n that the function F is determined by the pair $(F(e), dF_e)$ on U^n. Since $G = \bigcup_n U^n$ the corollary will follow. We have from equation (2.2)

$$F(\exp Y) = \left(e^{-\mathrm{ad}Y}\right) F(e) + \left(\frac{1-e^{-\mathrm{ad}Y}}{\mathrm{ad}Y}\right)(dF_e(Y) + [Y, F(e)]),$$

which shows that F is determined on U. Now assume that F is determined on U^n $n \geq 1$ and take $x \in U^n$. Then

$$F(x \exp Y) = \left(e^{-\mathrm{ad}Y}\right) F(x) + \left(\frac{1-e^{-\mathrm{ad}\,Y}}{\mathrm{ad}\,Y}\right)(dF_e(Y) + [Y, F(e)]),$$

which shows that F is determined on U^{n+1}. □

COROLLARY 2.7. *Let $G = \mathbb{R}^n$. Then $X \in \mathrm{norm}_{X(G)}(\mathfrak{g})$ if and only if $X_p = A(p) + u$ for all $p \in \mathbb{R}^n$, where $A : \mathbb{R}^n \to \mathbb{R}^n$ is a linear map and $u \in \mathbb{R}^n$.*

Proof. Under the usual identification of the tangent space $\mathbb{R}^n_p$ with $\mathbb{R}^n$ the vector field $\tilde{F}$ becomes the same as the function $F \in C^\infty(\mathbb{R}^n, \mathbb{R}^n)$. Thus if we replace x by 0 and Y by p in equation (2.2) we get $F(p) = dF_0(p) + F(0)$. □

Let $Diff(G)$ denote the group of all diffeomorphisms of G. Let us consider the map $\alpha : G \otimes \mathrm{Aut}(G) \to Diff(G)$ defined by $\alpha(x, \Gamma) = R_{x^{-1}} \cdot \Gamma$, and let $\mathrm{Exp} : \mathrm{norm}_{X(G)}(\mathfrak{g}) \to Diff(G)$ be defined by requiring the commutativity of the following diagram:

$$\begin{array}{ccc} \mathrm{norm}_{X(G)}(\mathfrak{g}) & \longrightarrow & \mathfrak{g} \otimes \mathrm{aut}(\mathfrak{g}) \\ \mathrm{Exp} \downarrow & & \downarrow \mathrm{Exp} \\ Diff(G) & \longleftarrow & G \otimes \mathrm{Aut}(G). \end{array}$$

THEOREM 2.8. *(i) The map $\alpha : G \otimes \mathrm{Aut}(G) \to Diff(G)$ is an injective group homomorphism.*
(ii) For $X \in \mathrm{norm}_{X(G)}(\mathfrak{g})$, the one parameter group τ_t of diffeomorphisms of G associated to X is given by

$$\tau_t = \mathrm{Exp}(-tX).$$

In particular any member of $\mathrm{norm}_{X(G)}(\mathfrak{g})$ has a global analytic flow on G associated with it.

Proof. Let us prove (i):

$$\alpha((x, \Gamma)(y, \Psi)) = \alpha(x\Gamma(y), \Gamma\Psi) = R_{(x\Gamma(y))^{-1}}\Gamma\Psi,$$

$$R_{(x\Gamma(y))^{-1}} = R_{x^{-1}} R_{\Gamma(y)^{-1}} = R_{x^{-1}}(\Gamma R_{y^{-1}} \Gamma^{-1}).$$

Hence,

$$\alpha((x, \Gamma)(y, \Psi)) = R_{x^{-1}}(\Gamma R_{y^{-1}} \Gamma^{-1})\Gamma\Psi = \alpha(x, \Gamma)\alpha(y, \Psi).$$

Moreover α is injective since $\alpha(x, \Gamma) = I$ implies $e = R_{x^{-1}}\Gamma(e) = x^{-1}$, and therefore $I = \alpha(e, \Gamma) = \Gamma$.

(ii) Since $\mathrm{Exp}(-tX)$ is a one parameter group of diffeomorphisms of G, it is enough to compute the associated vector field $\tilde{H} \in X(G)$. Take $g = \exp Y$, $Y \in \mathfrak{g}$

and consider $g(t) = \mathrm{Exp}(-tX)(g)$. Let $\Phi(X) = (Z, D)$ and let $\mathrm{Exp}(-t(Z, D)) = (x(t), \Gamma(t))$. Then

$$g(t) = \mathrm{Exp}(-t(Z,D))(g) = \left(R_{x(t)^{-1}}\Gamma(t)\right)(g) = (\Gamma(t)(g))x(t)^{-1}. \tag{2.3}$$

Now let $\mu : G \times \mathrm{Aut}(G) \to G$ be defined by $\mu(x, \Gamma) = \Gamma(x)$. Then,

$$\left(\frac{d}{dt}\right)_{t=0} \Gamma(t)(g) = d\mu_{(g,I)}(0, \dot{\Gamma}(0)) = d\mu_{(g,I)}(0, -D)$$

$$= \left(\frac{d}{dt}\right)_{t=0} \mu(g, \mathrm{Exp}(-tD)) = \left(\frac{d}{dt}\right)_{t=0} (\mathrm{Exp}(-tD))(\exp Y)$$

$$= \left(\frac{d}{dt}\right)_{t=0} \exp\left(e^{-tD}(Y)\right) = d\exp_Y(-D(Y)).$$

Going back to equation (2.3) and applying Theorem 1.7 of [2]:

$$\dot{g}(0) = \left(\frac{d}{dt}\right)_{t=0} \Gamma(t)(g) + dL_g(Z) = (dL_g)_e \left(\frac{1 - e^{-\operatorname{ad} Y}}{\operatorname{ad} Y}\right)(-D(Y) + (dL_g)_e(Z)).$$

Then the analytic function $H \in C^\infty(G, g)$ which corresponds to the vector field $\tilde{H}$ satisfies

$$H(\exp Y) = \left(\frac{1 - e^{-\operatorname{ad} Y}}{\operatorname{ad} Y}\right)(D(Y) + Z), \quad Y \in \mathfrak{g}.$$

From Theorem 2.5 and Corollary 2.6 the analytic function F corresponding to $\tilde{F}$ also satisfies

$$F(\exp Y) = \left(\frac{1 - e^{-\operatorname{ad} Y}}{\operatorname{ad} Y}\right)(D(Y) + Z), \quad Y \in \mathfrak{g}.$$

Therefore $F(x) = H(x)$ for all $x \in G$. This completes the proof of the theorem. □

For $\tilde{F} \in \mathrm{norm}_{X(G)}(\mathfrak{g})$ let $\Phi(\tilde{F}) = (E, \phi)$. Using the Baker-Campbell-Hausdorff formula [7], we obtain an explicit expression for the integrable curves of $\tilde{F}$.

To this end let $g(z) = (1 - e^{-z})/z$. Then g is an entire function on $\mathbb{C}$. Put $f(z) = g(z)^{-1} - z/2$. Then f is analytic in a neighborhood of $z = 0$, $f(0) = 1$, and f is even. We write

$$f(z) = 1 + \sum_{p=1}^{\infty} K_{2p} z^{2p}.$$

Then it is easy to verify that the K_{2p}'s are all rational numbers. Now define $d_1(E, \phi) = E$ and $d_n(E, \phi)$, $n \geq 1$ by the recursion formula

$$\begin{aligned}(n+1)d_{n+1}(E,\phi) &= \frac{1}{2}[E, d_n(E,\phi)] + \phi(d_n(E,\phi)) \\ &+ \sum_{2\leq 2p\leq n} K_{2p} \sum_{\substack{k_1,\dots,k_{2p}>0 \\ k_1+\cdots+k_{2p}=n}} [d_{k_1}(E,\phi), [\cdots [d_{k_{2p}}(E,\phi), E]\cdots].\end{aligned} \tag{2.4}$$

Then for each $n \geq 1$, d_n is a polynomial map of $\mathfrak{g} \otimes \partial\mathfrak{g}$ into $\mathfrak{g}$ homogeneous of degree n. In particular we have

$$d_2(E,\phi) = \tfrac{1}{2}\phi(E), \quad d_3(E,\phi) = \tfrac{1}{12}[E, \phi(E)] + \tfrac{1}{6}\phi^2(E),$$
$$d_4(E,\phi) = \tfrac{1}{24}[E, \phi^2(E)] + \tfrac{1}{24}\phi^3(E).$$

Moreover the series

$$\sum_{n=1}^{\infty} d_n(E, \phi)$$

converges absolutely in a neighborhood of $(0,0) \in \mathfrak{g} \otimes \partial\mathfrak{g}$.

THEOREM 2.9. *The integrable curve $x(t)$ of $\tilde{F} \in norm_{X(G)}(\mathfrak{g})$ through the point $x \in G$ is given in a neighborhood of $t = 0$ by*

$$x(t) = (\operatorname{Exp}(-t\phi)(x)) \exp\left(\sum_{n\geq 1}(-1)^{n+1}t^n d_n(E,\phi)\right).$$

In particular if $x = \exp X$, then

$$x(t) = \exp\left(e^{-t\phi}(X)\right) \exp\left(\sum_{n\geq 1}(-1)^{n+1}t^n d_n(E,\phi)\right).$$

Proof. Let $\operatorname{Exp}(-t(E,\phi)) = (g(t), \Gamma(t))$. Then from equation (2.3) we have $x(t) = (\Gamma(t)(x))g(t)^{-1}$. Now we use the Baker-Campbell-Hausdorff formula to compute

$$\operatorname{Exp}(E,\phi)\operatorname{Exp}(0,-\phi) = \operatorname{Exp} C((E,\phi),(0,-\phi)),$$

where $C((E,\phi),(0,-\phi)) = \sum_{n\geq 1} c_n((E,\phi),(0,-\phi))$. From Lemma 2.15.3 of [7] we know that

$$c_1((E,\phi),(0,-\phi)) = (E,0)$$

and that for $n \geq 1$

$$(n+1)c_{n+1}((E,\phi),(0,-\phi)) = \frac{1}{2}[(E,2\phi), c_n((E,\phi),(0-\phi))]$$
$$+ \sum_{2\leq 2p\leq n} K_{2p} \sum_{\substack{k_1,\ldots,k_{2p}>0\\ k_1+\cdots+k_{2p}=n}} [c_{k_1}((E,\phi),(0,-\phi)),[\cdots[c_{k_{2p}}((E,\phi),(0,-\phi)),(E,0)]\cdots].$$

By induction on n it follows easily that $c_n((E,\phi),(0-\phi)) = (d_n(E,\phi),0)$ with $d_n(E,\phi) \in \mathfrak{g}$. Hence $d_1(E,\phi) = E$. Moreover, using the semidirect product in $\mathfrak{g} \otimes \partial\mathfrak{g}$, we obtain

$$\begin{aligned}(n+1)d_{n+1}(E,\phi) &= \frac{1}{2}[E, d_n(E,\phi)] + \phi(d_n(E,\phi)) \\ &+ \sum_{2\leq 2p\leq n} K_{2p} \sum_{\substack{k_1,\ldots,k_{2p}>0,\\ k_1+\cdots+k_{2p}=n}} [d_{k_1}(E,\phi),[\cdots[d_{k_{2p}}(E,\phi),E]\cdots],\end{aligned}$$

which coincides with equation (2.4). Now

$$\begin{aligned}\operatorname{Exp}(-t(E,\phi)) &= \operatorname{Exp} C\left(-t(E,\phi),(0,t\phi)\right)\operatorname{Exp}(0,-t\phi) \\ &= \left(\exp(\left(\textstyle\sum_{n\geq 1}(-t)^n d_n(E,\phi)\right), I\right)(e, \operatorname{Exp}(-t\phi)).\end{aligned}$$

Therefore $g(t) = \exp\left(\sum_{n\geq 1}(-t)^n d_n(E,\phi)\right)$ and $\Gamma(t) = \operatorname{Exp}(-t\phi)$. This completes the proof of the theorem. □

3. Controllability

This section is devoted to the statement and proof of some controllability results for a linear control system $\Sigma = (G, \mathcal{D})$ such that the drift vector field $X \in \mathrm{norm}_{X(G)}(\mathfrak{g})$ has a singularity. First of all, we restrict the study to linear control systems $\Sigma = (G, \mathcal{D})$ determined by

$$\dot{x} = X(x) + \sum_{j=1}^{k} u_j Y^j(x),$$

where the control vectors $Y^j \in \mathfrak{g}$, for $j = 1, \ldots, \kappa$ are considered as left-invariant vector fields and the flow generated by $X \in \mathrm{aut}(G)$ and denoted by

$$T = \{X_t | t \in \mathbb{R}\}$$

is a one-parameter subgroup of $\mathrm{Aut}(G)$. Later we will extend these results. Of course, X has a singularity at the identity element e of G. We call X an *infinitesimal automorphism* of G.

The system Σ induces the following objects: the group

$$G_\Sigma := \{\theta^1_{t_1} \circ \theta^2_{t_2} \circ \cdots \circ \theta^\nu_{t_\nu} | \theta^j \in \mathcal{D},\ t_j \in \mathbb{R},\ \nu \in \mathbb{N}\}$$

and the semigroup

$$S_\Sigma := \{\theta^1_{t_1} \circ \theta^2_{t_2} \circ \cdots \circ \theta^\nu_{t_\nu} | \theta^j \in \mathcal{D},\ t_j \geq 0,\ \nu \in \mathbb{N}\}$$

of global diffeomorphisms on G.

DEFINITION 3.1. A linear control system $\Sigma = (G, \mathcal{D})$ is said to be
a) *controllable* if $S_\Sigma(e) = G$, and
b) *transitive* if $G_\Sigma(e) = G$. □

The Lie algebra induced by Σ is defined by $L(\Sigma) = \mathrm{Span}_{\mathcal{L}.A.}(\mathcal{D})$. It is well known that for analytical control systems the Σ-orbits $G_\Sigma(x)$, $x \in G$ are the integral manifolds of the distribution generated by $L(\Sigma)$ [6].

We recall that there exists a correspondence between connected Lie subgroups of G and Lie subalgebras of $\mathfrak{g}$. Under this correspondence normal subgroups are associated with ideals [7]. And if H is a connected Lie subgroup of G, then each $h \in H$ is a finite product of elements of the form $\exp(tY)$ where Y belongs to the Lie algebra $\mathcal{H}$ of H and $t \in \mathbb{R}$ [3].

We denote by $\mathcal{H}$ the Lie subalgebra of $\mathfrak{g}$ generated by the control vectors of Σ, i.e.,

$$\mathcal{H} = \mathrm{span}_{\mathcal{L}.A.}\{Y^1, \ldots, Y^\kappa\}\ .$$

Since X is an infinitesimal automorphism it follows from Theorem 2.3 that $X \in \mathrm{norm}_{X(G)}(\mathfrak{g})$. So, by definition $\mathfrak{g}$ is an $\mathrm{ad}(X)$-invariant subalgebra of $X(G)$. A more direct proof follows.

LEMMA 3.2. *If X is an infinitesimal automorphism, then $\mathfrak{g}$ is $\mathrm{ad}^i(X)$-invariant for each $i \geq 0$.*

Proof. It is enough to prove the lemma for $i = 1$. Let us take $Y \in \mathfrak{g}$; then

$$[X, Y](e) = -Y_e X = -\frac{d}{ds}\bigg|_{s=0} X_{exp(sY)}.$$

On the other hand,

$$[X,Y](x) = \frac{d}{dt}\bigg|_{t=0} (X_{-t})_*(Y_{X_t(x)}) = \frac{d}{dt}\bigg|_{t=0} \frac{d}{ds}\bigg|_{s=0} X_{-t} \circ Y_s \circ X_t(x)$$

$$= \frac{d}{dt}\bigg|_{t=0} \frac{d}{ds}\bigg|_{s=0} L_x \circ X_{-t}(\exp sY) = -\frac{d}{ds}\bigg|_{s=0} (L_x)_* X_{\exp(sY)} = (L_x)_*[X,Y](e).$$

Thus $[X,Y]$ is a left-invariant vector field in $X(G)$; consequently $[X,Y] \in \mathfrak{g}$ for all $Y \in \mathfrak{g}$. □

We denote by $\langle X \mid \mathcal{H}\rangle$ the smallest $\mathrm{ad}(X)$-invariant subalgebra of $\mathfrak{g}$ containing $\mathcal{H}$ and by H and $\langle X|H\rangle$ the connected Lie subgroups of G with Lie algebras $\mathcal{H}$ and $\langle X|\mathcal{H}\rangle$ respectively.

Remark. Since the vector field X normalizes the subalgebra $\langle X \mid \mathcal{H}\rangle$, then it is standard that the members of its flow normalize the members of the flow arising from vector fields in this subalgebra.

Additionally X has a singularity at e. This allows us to conclude that the orbit of the identity element of G with respect to the group generated by all $\exp(tX)$ and all $\exp(sY)$ is the same as the orbit with respect to the group generated by all $\exp(sY)$. So, in particular, we have:

THEOREM 3.3. *a) The orbit of Σ through the identity element e is $\langle X|H\rangle$.*
b) Σ is transitive if and only if

$$dim.Span_{\mathcal{L}.A.}\{Y^j, ad^i(X)(Y^j) | 0 \le i \le dim(G), 1 \le j \le \kappa\} = dim(G). \qquad \square$$

Furthermore, we can explicitly write the Lie algebra $L(\Sigma)$ of our system as a semidirect product of Lie algebras.

THEOREM 3.4. *Let $\Sigma = (G, D)$ be a linear control system such that X is an infinitesimal automorphism. Then*

$$L(\Sigma) \cong \langle X|\mathcal{H}\rangle \otimes L(T)\ .$$

Proof. We have the following facts:

i) Since $X_t \in \mathrm{Aut}(\langle X|H\rangle)$, we can consider the representation

$$T \xrightarrow{\rho} \mathrm{Aut}(\langle X|H\rangle)$$

to construct the semidirect product of $\langle X|H\rangle$ with T relative to ρ [7]. We denote this Lie group by

$$S := \langle X|H\rangle \otimes T\ .$$

ii) $d\rho$ is a representation of the Lie algebra $L(T)$ of T on $\langle X|\mathcal{H}\rangle$ such that

$$d\rho(L(T)) \subset \mathrm{Der}(\langle X|\mathcal{H}\rangle)\ .$$

iii) On the product $\langle X|\mathcal{H}\rangle \times L(T)$ the bracket

$$[(Y_1,W_1),(Y_2,W_2)] = ([Y_1,Y_2] + d\rho(W_1)(Y_2) - d\rho(W_2)(Y_1), [W_1,W_2])$$

is well defined and transforms this product into a Lie algebra called the semidirect product of $\langle X|\mathcal{H}\rangle$ with $L(T)$ relative to ρ [7]. Let us denote this Lie algebra by

$$\mathfrak{s} =: \langle X|\mathcal{H}\rangle \otimes L(T)\ .$$

Since $L(T)$ is an abelian algebra we have $[\mathfrak{s},\mathfrak{s}] \subset \langle X|\mathcal{H}\rangle$. In order to verify that the canonical map

$$L(\Sigma) \to \langle X|\mathcal{H}\rangle \otimes L(T)$$

is an isomorphism of Lie algebras we take elements on $\mathcal{D}$ and then we compute the bracket. For this it is enough to consider the generators of $L(\Sigma)$; since

$$L(\Sigma) = \text{Span}_{\mathcal{L}.A.}\{X + \sum_{j=1}^{\kappa} u_j Y^j | j = 1, \dots, \kappa\},$$

we may choose two elements of the form $X + Y^i$, $X + Y^j$. The bilinear property of the bracket in $X(G)$ shows that

$$[X + Y^i, X + Y^j] = [Y^i, Y^j] + [X, Y^j] + [Y^i, X] .$$

From this and the definition of bracket in $\mathfrak{s}$ the proposition follows. □

Remarks. 1. Since $\mathfrak{s}$ is the Lie algebra of S, Theorem 3.3 follows also from Theorem 3.4.

2. Since $S_\Sigma(e) \subset G_\Sigma(e)$, the Lie algebra rank condition given in Theorem 3.4 is necessary for the controllability of Σ. Unfortunately, this condition does not characterize the controllability property for linear control systems on an arbitrary connected Lie group G as in the linear case on $\mathbb{R}^n$. In the next section we give an example in the nilpotent simply connected Heisenberg Lie group of dimension 3 (see Example 2).

Define inductively the $\text{ad}(X)(\mathcal{H})$-sequence by:

$$\mathcal{H}_0 = \mathcal{H}$$

$$\mathcal{H}_i = \mathcal{H}_{i-1} + \text{ad}^i(X)(\mathcal{H}), \quad i \in \mathbb{N},$$

where

$$\text{ad}^i(X)(\mathcal{H}) = \{[X, \text{ad}^{i-1}(X)(Y)] | Y \in \mathcal{H}\}.$$

Since we consider finite dimensional Lie groups and $\mathcal{H} \neq 0$, from Lemma 3.2 there exists a smallest integer p, $0 \leq p < \dim(G)$ such that the $\text{ad}(X)(\mathcal{H})$-sequence stabilizes at p, i.e.

$$\mathcal{H}_p = \mathcal{H}_{p+q}, \ \forall q \in \mathbb{N}.$$

In particular we have,

$$\mathcal{H}_p = Span\{Y^j, \text{ad}^i(X)(Y^j) | 1 \leq i \leq p, \ 1 \leq j \leq \kappa\} \subset \langle X|\mathcal{H}\rangle \ \subset \mathfrak{g}.$$

We say that Σ *satisfies the rank* (ad-*rank*) *condition*, if

$$dim(\mathcal{H}_p) = dim(G).$$

We have the following result.

THEOREM 3.5. *Let G be a connected Lie group and let $\Sigma = (G, \mathcal{D})$ be a transitive linear control system such that the drift vector field X is an infinitesimal automorphism. If Σ satisfies the rank condition, then Σ is locally controllable.*

Proof. We denote by $x(t,u)$ the solution of the differential equation

$$\dot{x} = X(x) + \sum_{j=1}^{k} u_j Y^j(x)$$

determined by the control $u \in \mathcal{U} \subset L_\infty([0,t], \mathbb{R}^k)$ and the initial condition e. Let us consider for each $t > 0$ the end point map

$$E_t : \mathcal{U} \to G$$

defined by $E_t(u) = x(t,u)$. This map is smooth [1] and its derivative $(F_t)_0$ at 0 is defined in a neighborhood B of the constant control $u \equiv 0$ and is given by

$$(F_t)_0(u(\cdot)) = \int_0^t e^{(t-\tau)\,\mathrm{ad}(X)}(\sum_{j=1}^{k} u_j(\tau) Y_e^j) d\tau.$$

Next we will suppose that there exists a vector w in the dual space G_e^* of the tangent space G_e such that

$$\langle w, (F_t)_0(u(\cdot))\rangle = 0, \ \forall u(\cdot) \in B.$$

In particular,

$$\int_0^t \sum_{j=1}^{k} \langle w, e^{(t-\tau)\,\mathrm{ad}(X)}(Y_e^j)\rangle u_j(\tau) d\tau = 0 \ .$$

Since the last expression is true for every piecewise constant function

$$u : [0,t] \to \mathbb{R}^k,$$

we can conclude

$$\langle w, e^{(t-\tau)\,\mathrm{ad}(X)} Y_e^j\rangle = 0, \quad \forall \tau \in [0,t] \ .$$

By derivation we obtain for each $i \geq 0$ and $j = 1, \ldots, k$

$$\langle w, \mathrm{ad}^i(X) Y_e^j\rangle = 0 \ .$$

This contradicts our rank condition assumption. So, the linear map $(F_t)_0$ is onto.

From the implicit function theorem we can conclude that there is a neighborhood U of the identity element e in G such that $U \subset S_\Sigma(e)$, and consequently Σ is locally controllable at e. Since Σ is transitive, $(L_x)_* \mathcal{H}_p$ generates the tangent space of G in every point $x \in G$. By analogous arguments it follows that Σ is locally controllable on G. □

COROLLARY 3.6. *Let G be a connected Lie group and $\Sigma = (G, \mathcal{D})$ be a transitive linear control system such that X is an infinitesimal automorphism.*

1. If G is abelian,

$$\Sigma \text{ is controllable} \Leftrightarrow \Sigma \text{ satisfies the rank condition.}$$

2. If G is compact,

$$\Sigma \text{ satisfies the rank condition} \Rightarrow \Sigma \text{ is controllable}$$

Proof. 1. By assumption the Lie algebra $\mathfrak{g}$ of G is abelian and $X_t \in \mathrm{Aut}(G)$, $\forall t \in \mathbb{R}$. In particular, the positive orbit $S_\Sigma(e)$ is a semigroup. Since $\mathcal{H}_p = \mathfrak{g}$, it follows from Theorems 3.3 and 3.5 that Σ is locally controllable if and only if Σ satisfies the rank condition. Denote by U the neighborhood of e contained in $S_\Sigma(e)$. Since G is connected

$$\cup_m U^m = G,$$

and we obtain the desired conclusion.

2. From Theorem 3.5, there are neighborhoods $U_1, U_2, ..., U_m$ of G such that $G = U_1 \cup U_2 \cup ... \cup U_m$, and Σ is locally controllable on each U_j, $j = 1, 2, ..., m$. In particular, Σ is controllable □

Remark. By the Corollary 3.6, we can decide the controllability property with the rank condition for any Lie group

$$G = T^n \times \mathbb{R}^m, \quad n, m \in \mathbb{N},$$

where $T^n = S^1 \times \cdots \times S^1$, ($n$-times), is a torus.

Extension. Let G be a connected Lie group and $X \in \mathrm{norm}_{X(G)}(\mathfrak{g})$. We have studied the controllability property for linear control systems such that the drift vector field X has a singularity at the identity element e of G. It is possible to extend these results to the case that X has a singularity.

Let $\tilde{F} \in \mathrm{norm}_{X(G)}(\mathfrak{g})$ and $x_0 \in G$ with $\tilde{F}_{x_0} = 0$. We define the vector field $\tilde{F}^{L(x_0)}$ by

$$(\tilde{F}^{L(x_0)})y = dL_{x_0^{-1}}(\tilde{F}_{x_0 y}), \quad y \in G \ .$$

The left-invariance of the Lie bracket shows that

$$\tilde{F}^{L(x_0)} \in \mathrm{norm}_{X(G)}(\mathfrak{g}) \ .$$

Of course $\tilde{F}^{L(x_0)}$ has a singularity at e and by Theorem 2.3 we obtain that there exists $\phi \in \mathrm{aut}(G)$ such that $\Phi(\tilde{F}^{L(x_0)}) = (0, -\phi)$.

Thus if $X = \tilde{F}$ has a singularity at $x_0 \subset G$, by Theorem 2.3 it follows that $X^{L(x_0)}$ is an infinitesimal automorphism of G, i.e.,

$$X_t^{L(x_0)} = \mathrm{Exp}(t\phi) \in \mathrm{Aut}(G), \ \forall t \in \mathbb{R}.$$

Let $\Sigma = (G, \mathcal{D})$ be a linear control system. This class of system generalizes:

1. Linear control systems on $\mathbb{R}^n$.

In fact, if L is an unrestricted time-invariant linear control system on $\mathbb{R}^n$ then L is determined by the dynamic

$$\dot{x} = Ax + Bu,$$

where $x \in \mathbb{R}^n$, $u \in \mathcal{U}$ and A and B are constant matrices of appropriate dimensions. Denote by $b_1, b_2, \ldots b_\kappa$ the columns of B; then we can describe L by

$$\dot{x} = Ax + \sum_{j=1}^{\kappa} u_j b_j \ .$$

It is well known that the constant vector b_j defines a left invariant vector field Y^j on $\mathbb{R}^n$, given by

$$Y^j(x) = b_j, \quad x \in \mathbb{R}^n.$$

Moreover, the flow of the linear vector field A given by $A_t = e^{tA}, \quad t \in \mathbb{R}$ belong to $GL_n(\mathbb{R})$, the Lie group of all $\mathbb{R}^n$-automorphisms. In this case $\eta = \mathbb{R}^n \otimes M_n(\mathbb{R})$. So for $X = (0, A) \in \eta$, then $L = (G, \mathcal{D})$, where G is the commutative connected Lie group $\mathbb{R}^n$ and

$$\mathcal{D} = \{X + \sum_{j=1}^{\kappa} u_j Y^j | u \in R^\kappa\} .$$

Corollary 3.6 then shows

THEOREM 3.7. *(Kalman R., Ho Y., Narendra K., [4]) The system L is controllable if and only if*

$$dim.Span\{b_1, \ldots, b_p, Ab_1, \ldots, Ab_p, \ldots, A^{n-1}b_1, \ldots, A^{n-1}b_\kappa\} = n. \qquad \square$$

Remark. We know that for $G = \mathbb{R}^n$ an element of the normalizer has the form :

$$b + A \in \mathbb{R}^n \otimes M_n(\mathbb{R}) .$$

In particular,

$$b + A \text{ has a singularity} \Leftrightarrow b \in Im(A).$$

2. Linear control system on a matrix Lie group.

Markus introduces in [5] the notion of a linear control system Σ on a q^2-dimensional matrix Lie subgroup G of $GL_n(\mathbb{R})$ with dynamic defined by a family of differential equations on G of the form

$$\dot{P} = X^*(P) + \sum_{j=1}^{\kappa} u_j Y^j P .$$

Here the drift vector field X^* defined by $X^*(P) = AP - PA$ is induced by an element A in the Lie algebra $M_n(\mathbb{R})$ of all $n \times n$ real matrices and for each $j = 1, 2, \ldots, \kappa$, Y^j is defined by left multiplication. In this case, for each $t \in \mathbb{R}$, the flow

$$X_t^*(P) = e^{tA} P e^{-tA}, \quad \forall t \in \mathbb{R}$$

is an automorphism of G. Again, $X = (0, X^*) \in \eta$. For this class of systems the author obtains the following results:

Theorem. (Markus L. [5]) If Σ is controllable on G, then Σ satisfies the Lie algebra rank condition. $\square$

This result follows from Theorem 3.3.

4. Examples

In this section we compute some examples. We use the Lie algebra rank condition, Theorem 3.3 and Theorem 3.5 to study controllability.

1. Let $G = SL_2(\mathbb{R})$ be the Lie subgroup of $GL_2(\mathbb{R})$ whose elements are matrices of determinant 1. The Lie algebra $sl_2(\mathbb{R})$ of G is given by

$$L(G) = \text{Span}_{\mathcal{L}.A.} \left\{ Y^1 = \begin{pmatrix} 0 & 1 \\ -1 & 0 \end{pmatrix}, Y^2 = \begin{pmatrix} 0 & 1 \\ 0 & 0 \end{pmatrix} \right\}$$

where $[Y^1, Y^2] = Y^3 = \begin{pmatrix} 1 & 0 \\ 0 & -1 \end{pmatrix}$.

We consider the matrix $A = \frac{1}{2}\begin{pmatrix} 0 & 1 \\ 1 & 0 \end{pmatrix} \in M_2(\mathbb{R})$ and the one-parameter group of automorphism (X_t) of G induced by A, i.e.,

$$X_t(x) = e^{tA}xe^{-tA}, \ t \in \mathbb{R}, \ x \in G .$$

For the linear control system $\Sigma = (G, \mathcal{D})$ with

$$\mathcal{D} = \{X + uY^2 | u \in \mathbb{R}\}$$

we have:

$$\mathrm{ad}(X)(Y^2) = Y^3 \quad \text{and} \quad \mathrm{ad}^2(X)(Y^2) = Y^1 .$$

Since

$$\mathrm{ad}(X)(\mathcal{H}) = sl_2(\mathbb{R}),$$

then by Theorem 3.5 Σ is locally controllable.

2. Let us take the simply connected nilpotent Heisenberg Lie group G of dimension 3 with the Lie algebra

$$\mathfrak{g} = \mathbb{R}Y^1 + \mathbb{R}Y^2 + \mathbb{R}Y^3,$$

with generators defined by
$Y^1 = \frac{\partial}{\partial x_1}$
$Y^2 = \frac{\partial}{\partial x_2} + x_1 \frac{\partial}{\partial x_3}$
$Y^3 = \frac{\partial}{\partial x_3}$.

We have $[Y^1, Y^2] = Y^3$ and the other Lie brackets are null. Of course, G is a nilpotent nonabelian Lie group homeomorphic to $\mathbb{R}^3$. We consider the linear control system Σ on G defined by

$$\dot{x}(t) = X(x(t)) + u(t)Y^2(x), \ u(t) \in \mathbb{R},$$

where X is the infinitesimal automorphism associated with the derivation

$$D = \begin{pmatrix} 0 & 1 & 0 \\ 1 & 0 & 0 \\ 0 & 0 & 0 \end{pmatrix} \in \partial\mathfrak{g}.$$

Here $\mathcal{H} = Span\{Y^2\}$. A computation shows that Σ is given by
$\dot{x_1}(t) = x_2(t)$
$\dot{x_2}(t) = x_1(t) + u(t)$
$\dot{x_3}(t) = 0,$

Since $[X, Y^2] = Y^1$ and $[X, Y^1] = Y^2$, we have

$$\mathrm{ad}(X)(\mathcal{H}) = Span\{Y^1, Y^2\} \underset{\neq}{\subset} \mathfrak{g}$$

while

$$< X \mid \mathcal{H} >= Span_{\mathcal{L}.A.}\{Y^1, Y^2\} = Span\{Y^1, Y^2, Y^3\} = \mathfrak{g}.$$

The system Σ given above is obviously not locally controllable although it is transitive. In fact, none of the elements in $S_\Sigma(e)$ can leave the $(x_1, x_2)-$ plane.

References

[1] A. A. Agrachev, R. V. Gamkrelidze and A. V. Sarychev, "*Local Invariants of Smooth Control Systems*", Acta Applicandae Mathematicae 14 (1989), 191-237.

[2] S. Helgason, "*Differential Geometry and Symmetric Spaces*", Academic Press, 1962.

[3] V. Jurdjevic and H. Sussmann, "*Control systems on Lie groups*", Differential Equations, **12** (1972), pp. 313-329.

[4] R. Kalman, Y. Ho, K. Narendra, "*Controllability of linear dynamical systems*", Contrib. to Diff. Equations **1** (2), 1962, pp. 189-213.

[5] L. Markus, "*Controllability of multi-trajectories on Lie groups*", Proceedings of Dynamical Systems and Turbulence, Warwick 1980, Lecture Notes in Mathematics 898, pp. 250-265.

[6] H. Sussmann, "*Orbits of families of vector fields and integrability of distributions*, Trans. Amer. math. soc. **180** (1973) 171-188.

[7] V. Varadarajan, "*Lie groups, Lie algebras and their Representations*, Prentice Hall, Inc. 1974.

Departamento de Matematica, Universidad Catolica del Norte, Casilla 1280, Antofagasta, Chile

FAMAF, Universidad Nacional de Cordoba, 5000 Cordoba, Argentina

Proceedings of Symposia in Pure Mathematics
Volume **64**, 1999

Real Analytic Geometry and Local Observability

Zbigniew Bartosiewicz

ABSTRACT. The language of real analytic geometry is used to give necessary and sufficient conditions of stable local observability of analytic systems. A full proof is provided in the most general case. Globalization of this result yields a new characterization of local observability at large which uses the concept of sheaf.

1. Introduction

The modern real analytic geometry started 50 years ago with the fundamental paper of Henri Cartan [**Ca**]. It was the first time that essential differences between real and complex analytic geometries were discovered (e.g lack of coherence in the real case). The language of germs and sheaves, algebras and ideals, commonly used in modern geometry, nicely fits in the study of observability of real analytic (control) systems. We use it to express and prove a characterization of stable local observability at a distinguished point, which leads to a new necessary and sufficient condition of local observability at large. This is a continuation of work started in papers of D. Mozyrska and mine [**Mo, MB1, MB2**], where preliminary results were presented under some regularity assumptions.

In Section 2 we recall basic definitions and facts from real analytic geometry. In particular, the interplay between geometry and algebra is presented in detail as we are going to use it extensively throughout the paper. Real radical, one of the main tools to be exploited in the paper, is of special interest as it exhibits the peculiarities of real geometry.

Section 3 contains the main result, a characterization of stable local observability, which is preceded by necessary definitions and a short review of known facts. Stable local observability was introduced in [**Ba2**]. It holds at a point x_0 if the system is locally observable at every point of some neighborhood of x_0.

We globalize the main result in Section 4 using real analytic sets and corresponding sheaves of ideals of germs of real analytic functions. This leads to a new characterization of local observability understood globally, i.e. at every point. We compare this with our former criterion [**Ba1**].

The proof of the main result is contained in Section 5. It is complete and without additional assumptions of regularity. In particular, coherence of the constructed ideals, long thought by the author to be essential in the proof, need not be

1991 *Mathematics Subject Classification.* Primary 93B07, 14P15; Secondary 93B29, 93B27.
This work was supported by KBN grant No. 8 T11A 011 12.

assumed. Though this strips the theory off a potential deepness, but the theorem will definitely be easier to use in practice.

2. Basic real geometry

We assume that the reader is familiar with the concepts of germs of functions and sets, and with fundamentals of the sheaf theory and theory of analytic sets. Necessary definitions can be found e.g. in [**GR, GMT**]. If B is a "global" object (a set, a function, or a family of functions), B_x will always denote its germ at the point x (but the precise meaning of the germ will depend on the meaning of the object). If C is a germ, $\tilde{C}$ will denote one of its representatives. By $\mathcal{O}_x$ we denote the algebra of germs of real analytic functions at x, where $x \in \mathbb{R}^n$ (n fixed throughout the paper), and by m_x the (only) maximal ideal in $\mathcal{O}_x$, consisting of all germs in $\mathcal{O}_x$ that vanish at x. By $\mathcal{O}$ we denote the sheaf of germs of real analytic functions on $\mathbb{R}^n$.

If U is an open subset in $\mathbb{R}^n$ then $\mathcal{O}_U$ will mean the algebra of real analytic functions on U. Thus $U \mapsto \mathcal{O}_U$ is the presheaf of partially defined real analytic functions on $\mathbb{R}^n$. If A is a subalgebra of $\mathcal{O}_U$ and $x_0 \in U$ then A_{x_0} means the set of germs at x_0 of functions from A. Of course A_{x_0} is again an algebra over $\mathbb{R}$. If I is an ideal in $\mathcal{O}_U$ then I_{x_0} means the ideal in $\mathcal{O}_{x_0}$ generated by the germs at x_0 of function from I.

Consider a set-germ A in $\mathbb{R}^n$ (at some point x). Then $J(A)$ denotes the ideal in $\mathcal{O}_x$ of germs (at x) of real analytic functions that vanish on A. If I is an ideal in $\mathcal{O}_x$ then $Z(I)$ will denote the zero set-germ of I (at x). Let us recall that $Z(I)$ is defined as the intersection of the set-germs $Z(\varphi_i)$, $i = 1, \ldots, k$, where $\varphi_1, \ldots, \varphi_k$ are generators of the ideal I. Since only finite intersections of set-germs are defined, we must use here the property that $\mathcal{O}_x$ is Noetherian.

We have a natural duality between ideals and set-germs. If $I_1 \subset I_2$ then $Z(I_2) \subset Z(I_1)$

Let P be any commutative ring with a unit and I be an ideal in P. Then the *real radical of* I, denoted by $\sqrt[\mathbb{R}]{I}$, is the set of all $a \in P$ for which there is $m \in \mathbb{N}$, $k \in \mathbb{N} \cup \{0\}$ and $b_1, \ldots, b_k \in P$ such that

$$a^{2m} + b_1^2 + \ldots + b_k^2 \in I.$$

The real radical is an ideal in P and it contains I. If I is a proper ideal in P, then also $\sqrt[\mathbb{R}]{I}$ is proper. An ideal I is called *real* if $\sqrt[\mathbb{R}]{I} = I$.

THEOREM 2.1 ([**Ri**]). *Let $x \in \mathbb{R}^n$. If I is an ideal in $\mathcal{O}_x$ then* $J(Z(I)) = \sqrt[\mathbb{R}]{I}$. □

Theorem 2.1 implies that there is a 1:1 correspondence between germs of analytic sets at x and real ideals in $\mathcal{O}_x$.

Let $\mathcal{G}$ be a family of analytic functions on some open set $U \subset \mathbb{R}^n$. Denote by $S_x(\mathcal{G})$ the germ at x of the level set of $\mathcal{G}$ that passes through x. In other words

$$S_x(\mathcal{G}) = \{y \in U : \varphi(x) = \varphi(y) \text{ for } \varphi \in \mathcal{G}\}_x.$$

The set-germ $S_x(\mathcal{G})$ is a germ of analytic set. One of the representatives of $S_x(\mathcal{G})$ is the analytic set in U: $\{y \in U : \varphi(x) = \varphi(y) \text{ for } \varphi \in \mathcal{G}\}$.

3. Observability

Let us consider a control system with output

$$(3.1) \qquad (\Sigma) \quad \begin{array}{rcl} \dot{x} & = & f(x,u) \\ y & = & h(x), \end{array}$$

where $x(t) \in \mathbb{R}^n$, $y(t) \in \mathbb{R}^r$ and $u(t) \in \Omega$ - arbitrary set. If $\omega \in \Omega$ then f_ω, defined by $f_\omega(x) = f(x,\omega)$, is a vector field on $\mathbb{R}^n$. We assume that the map h and the vector field f_ω for every $\omega \in \Omega$ are analytic, and that controls u are piecewise constant functions of time. By $\gamma(t, x_0, u)$ we denote the solution of the equation $\dot{x} = f(x,u)$ corresponding to control u and the initial condition $x(0) = x_0$, and evaluated at time t.

Let $x_1, x_2 \in \mathbb{R}^n$. Then they are *indistinguishable* if

$$h(\gamma(t, x_1, u)) = h(\gamma(t, x_2, u))$$

for every control u and every $t \geq 0$ for which both sides of the equation are defined. Otherwise x_1 and x_2 are *distinguishable.*

We say that Σ is *locally observable at* x_0 (LO(x_0)) if there is a neighborhood U of x_0 such that for every $x \in U$, x and x_0 are distinguishable. We say that Σ is *stably locally observable at* x_0 (SLO(x_0)) if there is a neighborhood U of x_0 such that Σ is locally observable at x for every $x \in U$. We call Σ *locally observable* (LO) or *stably locally observable* (SLO) when it is locally observable or stably locally observable at every $x \in \mathbb{R}^n$.

By the observation algebra of system Σ, denoted by $\mathcal{H}(\Sigma)$, we mean the smallest algebra over $\mathbb{R}$ of real analytic functions on $\mathbb{R}^n$ containing the components of the map h and closed under the Lie derivatives with respect to the vector fields f_ω, $\omega \in \Omega$.

Let us recall the main results concerning local observability. From here, whenever Σ is fixed, we replace $\mathcal{H}(\Sigma)$ by $\mathcal{H}$.

THEOREM 3.1 ([**HK**]).
a) Points x_1 and x_2 are indistinguishable iff $\varphi(x_1) = \varphi(x_2)$ for every $\varphi \in \mathcal{H}$.
b) If $\dim d\mathcal{H}(x_0) = n$ *then Σ is LO(x_0).*
c) $(\forall x_0 \in \mathbb{R}^n\colon \dim d\mathcal{H}(x_0) = n) \Rightarrow (\forall x_0 \in \mathbb{R}^n\colon \Sigma \text{ is } LO(x_0)) \Rightarrow$
$(\exists X$ *- a real analytic set in* $\mathbb{R}^n\colon \forall x_0 \notin X\colon \dim d\mathcal{H}(x_0) = n)$. □

Let us denote the condition $\dim d\mathcal{H}(x_0) = n$ by HK(x_0) (Hermann-Krener condition at x_0). The last part of Theorem 3.1 says that if we are interested in local observability at large, i.e. at each point, then condition HK(x_0) is satisfied almost everywhere, so the gap between sufficient condition and necessary condition for local observability at large is quite narrow. However, when one is interested in local observability at a particular point of the state space, the Hermann-Krener condition may be far from being necessary (see [**Ba1**]).

We say that $\mathcal{H}$ is *injective at* x_0 if there is a neighborhood U of x_0 such that for every $x_1, x_2 \in U$: the condition $\varphi(x_1) = \varphi(x_2)$ for every $\varphi \in \mathcal{H}$ implies $x_1 = x_2$. Then we have a slightly stronger statement.

PROPOSITION 3.2. *HK(x_0) $\Rightarrow$ $\mathcal{H}$ is injective at x_0 $\Rightarrow$ Σ is SLO(x_0) $\Rightarrow$ Σ is LO(x_0).* □

None of the implications in Proposition 3.2 may be, in general, reversed. But we have important, though obvious, global equivalence.

PROPOSITION 3.3. *Σ is SLO(x_0) for every $x_0 \in \mathbb{R}^n$ iff Σ is LO(x_0) for every $x_0 \in \mathbb{R}^n$. In other words, Σ is LO iff Σ is SLO.* □

Let J_x be the ideal in $\mathcal{O}_x$ generated by germs at x_0 of functions from $\mathcal{H}$ that vanish at x.

THEOREM 3.4 ([**Ba1**]).
a) Σ is LO(x_0) iff $\sqrt[\mathbb{R}]{J_{x_0}} = m_{x_0}$.
b) HK(x_0) iff $J_{x_0} = m_{x_0}$. □

If G is a subset of $\mathcal{O}_x$ then $D(G)$ will denote the ideal in $\mathcal{O}_x$ generated by Jacobians $\frac{\partial(\varphi_1,\dots,\varphi_n)}{\partial(x_1,\dots,x_n)}$, where $\varphi_i \in G$. Observe that these Jacobians are well defined on germs of functions. If $\mathcal{G}$ is a family of real analytic functions on an open set U in $\mathbb{R}^n$ then similarly we define the Jacobian ideal $D(\mathcal{G})$ in $\mathcal{O}_U$. We shall need also extensions of these concepts to the case where x belongs to an analytic manifold M and U is an open set in M with a global coordinate system (i.e. diffeomorphic to an open subset of $\mathbb{R}^n$). Though Jacobians will then depend on a chosen coordinate system, Jacobian ideals $D(G)$ or $D(\mathcal{G})$ will not. Furthermore, there is a simple relation between Jacobian ideals for functions and germs of functions. If $\mathcal{G}$ is a family of analytic functions on U and $x_0 \in U$ then we have

$$D(\mathcal{G})_{x_0} = D(\mathcal{G}_{x_0}). \tag{3.2}$$

Now, for a point $x \in \mathbb{R}^n$, we define a sequence of ideals in $\mathcal{O}_x$. Let $I_x^{(0)} = (0)$ and $I_x^{(k+1)} = \sqrt[\mathbb{R}]{D(\mathcal{H}_x \cup I_x^{(k)})}$. It is clear that instead of $\mathcal{H}_x$ in this definition one can take previously defined ideal J_x, i.e. $I_x^{(k+1)} = \sqrt[\mathbb{R}]{D(J_x \cup I_x^{(k)})}$. This leads to the following generalization of statement b) of Theorem 3.4.

COROLLARY 3.5. *HK(x_0) $\Leftrightarrow J_{x_0} = m_{x_0} \Leftrightarrow I_{x_0}^{(1)} = \mathcal{O}_{x_0}$.* □

The ideals $I_x^{(k)}$ will be the main tools in studying stable local observability. First we prove the basic fact.

PROPOSITION 3.6. *For any $k \geq 0$, $I_x^{(k)} \subset I_x^{(k+1)}$, and there is $s \geq 0$ such that $I_x^{(s)} = I_x^{(s+1)}$.*

PROOF. To prove the first part we proceed by induction. It is clear that $I_x^{(0)} \subset I_x^{(1)}$, so assume that $I_x^{(k)} \subset I_x^{(k+1)}$ for some $k \geq 0$. Then also $\mathcal{H}_x \cup I_x^{(k)} \subset \mathcal{H}_x \cup I_x^{(k+1)}$ and $D(\mathcal{H}_x \cup I_x^{(k)}) \subset D(\mathcal{H}_x \cup I_x^{(k+1)})$, so finally $\sqrt[\mathbb{R}]{D(\mathcal{H}_x \cup I_x^{(k)})} \subset \sqrt[\mathbb{R}]{D(\mathcal{H}_x \cup I_x^{(k+1)})}$. This means that $I_x^{(k+1)} \subset I_x^{(k+2)}$. Since the ring $\mathcal{O}_x$ is Noetherian the sequence of ideals $I_x^{(k)}$ must stabilize at some s. □

Now we can state the main result of the paper.

THEOREM 3.7. *System Σ is SLO(x_0) iff $I_{x_0}^{(s)} = \mathcal{O}_{x_0}$ for some $s > 0$.*

The proof of Theorem 3.7 is contained in Section 5.

The statements of Corollary 3.5, Proposition 3.6 and Theorem 3.7 can be translated into the language of germs of analytic sets. Let $X_x^{(k)} = Z(I_x^{(k)})$. Because the ideals $I_x^{(k)}$ are real we also get $J(X_x^{(k)}) = I_x^{(k)}$, so there is one-to-one correspondence between the ideals and the set-germs. We have then the following:

COROLLARY 3.8. *Let* $x_0 \in \mathbb{R}^n$.
a) $HK(x_0)$ *iff* $X^{(1)}_{x_0} = \emptyset$.
b) For every $k \geq 0$, $X^{(k)}_{x_0} \supset X^{(k+1)}_{x_0}$, *and for some* $s \geq 0$, $X^{(s)}_{x_0} = X^{(s+1)}_{x_0}$.
c) Σ *is* $SLO(x_0)$ *iff* $X^{(s)}_{x_0} = \emptyset$. □

EXAMPLE 3.9 ([**MB2**]). Consider the following system

$$\begin{aligned} \dot{x}_1 &= x_1 \\ \dot{x}_2 &= x_2 \\ \dot{x}_3 &= 0 \\ y &= x_1^2 + x_2^2 + x_3^2 \end{aligned}$$

and choose $x_0 = 0$. Then $\mathcal{H} = \mathbb{R}[x_1^2, x_2^2, x_3^2]$ and $I^{(1)}_{x_0} = J_{x_0} = (x_1x_2x_3)$. So $X^{(1)}_{x_0}$ is the germ of a union of three planes intersecting at 0. In the next step we obtain $I^{(2)}_{x_0} = \sqrt[\mathbb{R}]{(x_1^2x_2^2, x_2^2x_3^2, x_1^2x_3^2)} = (x_1x_2, x_2x_3, x_1x_3)$. A quick calculation shows that X_{x_0} is the germ of a union of three lines intersecting at 0. Finally, $I^{(3)}_{x_0} = m_{x_0}$ and then, inevitably, $I^{(4)}_{x_0} = \mathcal{O}_{x_0}$, so Σ is stably locally observable at x_0.

4. Globalization

Using the language of sheaves we can nicely sew up local characterizations of stable local observability into a global criterion. Let $I^{(k)}$ be a sheaf of ideals in $\mathcal{O}$ whose stalk at $x \in \mathbb{R}^n$ is equal $I^{(k)}_x$. In other words, we put $I^{(0)} = (0)$ - the sheaf of zero-ideals, and define $I^{(k)}$ inductively by

$$I^{(k+1)} = \sqrt[\mathbb{R}]{D(\mathcal{H} \cup I^{(k)})}$$

where $\mathcal{H}$ is identified with the set of sections of the sheaf $x \mapsto \mathcal{H}_x$ and operations D and $\sqrt[\mathbb{R}]{\ }$ are performed germwise.

Each sheaf of ideals corresponds to an analytic set. So we can define the set $X^{(k)}$ as follows:

$$x \in X^{(k)} \Leftrightarrow I^{(k)}_x \neq \mathcal{O}_x.$$

It's clear that the germ of $X^{(k)}$ at x is precisely $X^{(k)}_{x_0}$.

Since local observability and stable local observability are equivalent, we have a new criterion of local observability. It's a consequence of Theorem 3.7.

THEOREM 4.1. Σ *is LO iff* $I^{(s)} = \mathcal{O}$ *for some* $s > 0$. □

REMARK 4.2. Part a) of Theorem 3.4 gives also necessary and sufficient condition of local observability at large. Namely: Σ is LO iff $\forall x \in \mathbb{R}^n$, $\sqrt[\mathbb{R}]{J_x} = m_x$. However neither the ideals J_x nor m_x form a sheaf, so we have to check this condition at every point separately.

Translating ideals into sets we obtain the following:

COROLLARY 4.3. Σ *is LO iff* $X^{(k)} = \emptyset$ *for some* $k > 0$. □

EXAMPLE 4.4. Let the observation algebra of a system Σ on $\mathbb{R}^3$ be generated by 3 functions: $\varphi_1 = x_1x_3 + x_2^2/2$, $\varphi_2 = x_1^2/2 + x_2^2/2$ and $\varphi_3 = x_2(x_1 - x_3)$. Then $I^{(1)} = \sqrt[\mathbb{R}]{D(\{\varphi_1, \varphi_2, \varphi_3\})} = \sqrt[\mathbb{R}]{(x_3(x_1^2 + x_2^2) - x_1^3)}$. The set $X^{(2)}$ is the famous Cartan's umbrella which serves as a simple example of a lack of coherence (see [**Ca, GMT**]). In the next step we get $I^{(2)} = (x_2, x_1(x_1 - x_3))$. The corresponding set is a union of two lines. The ideals stabilize at $k = 3$: $I^{(3)} = (x_1, x_2) = I^{(4)}$ and $X^{(3)}$ is a line - the stick of the umbrella - consisting of singular points of $X^{(1)}$. Thus Σ is not locally observable.

5. Proof of Theorem 3.7

Sufficiency.

LEMMA 5.1. *Let $U \subset \mathbb{R}^n$ be an open set and X be an analytic set in U. Consider a family $\mathcal{G}$ of analytic functions on U. If $\forall x \notin X$: $S_x(\mathcal{G}) = \{x\}$ then $\forall x \in X$: $S_x(\mathcal{G}) \subset X_x$.*

PROOF. Suppose that there is $x \in X$ such that $S_x(\mathcal{G}) \not\subset X_x$. This means that for every representatives $\tilde{S}_x(\mathcal{G})$ and $\tilde{X}_x$ we have $\tilde{S}_x(\mathcal{G}) \not\subset \tilde{X}_x$. Take $\tilde{X}_x := X$ and arbitrarily small neighborhood V of x in U. Let $\tilde{S}_x(\mathcal{G})$ be a representative of $S_x(\mathcal{G})$ in V. Then there is $y \in \tilde{S}_x(\mathcal{G})$ such that $y \notin X$. Take a sequence (y_n) of such points converging to x. We may assume that all these points belong to some (large enough) representative Y of $S_x(\mathcal{G})$ and that Y is an analytic set. Only finite number of points y_n may be isolated points of Y. This means that arbitrarily close to x there is a point y_n for which $S_{y_n}(\mathcal{G}) \neq \{y_n\}$. Thus we get a contradiction. □

LEMMA 5.2. *Let U be an open set in $\mathbb{R}^n$ and $\mathcal{G}$ be a family of analytic functions on U. For every $x \in U$: if $x \notin Z(D(\mathcal{G}))$ then $S_x(\mathcal{G}) = \{x\}$.*

PROOF. If $x \notin Z(D(\mathcal{G}))$ then there are functions $\varphi_1, \dots, \varphi_n \in G$ such that

$$\frac{\partial(\varphi_1, \dots, \varphi_n)}{\partial(x_1, \dots, x_n)}(x) \neq 0.$$

Thus the map $\bar{x} \mapsto (\varphi_1(\bar{x}), \dots, \varphi_n(\bar{x}))$ is injective in a neighborhood of x. This implies that $S_x(\mathcal{G}) = \{x\}$. □

LEMMA 5.3. *Assume that $X_{x_0}^{(k)} \neq X_{x_0}^{(k+1)}$ for some $k \geq 0$. Then there is a neighborhood U of x_0 in $\mathbb{R}^n$ and a representative $\tilde{X}_{x_0}^{(k+1)}$ of $X_{x_0}^{(k+1)}$ in U such that for every $x \in U$:*

$$x \notin \tilde{X}_{x_0}^{(k+1)} \Rightarrow S_x(\mathcal{H}) = \{x\}. \tag{5.1}$$

PROOF. First observe that $Z(D(\mathcal{H}_{|U}))$ is a representative of $Z(D(\mathcal{H}_{x_0}))$. We procede by induction. Let $k = 0$ and assume that $X_{x_0}^{(1)} \neq X_{x_0}^{(0)} = \mathbb{R}^n_{x_0}$. Take any neighborhood U of x_0. Since $X_{x_0}^{(1)} = Z(D(\mathcal{H}_{x_0}))$ then $Z(D(\mathcal{H}_{|U})) =: \tilde{X}_{x_0}^{(1)}$ is a representative of $X_{x_0}^{(1)}$ in U. If $x \notin \tilde{X}_{x_0}^{(1)}$ then, by Lemma 5.2, $S_x(\mathcal{H}) = S_x(\mathcal{H}_{|U}) = \{x\}$.

Now assume that the statement of the theorem holds for $k - 1 \geq 0$. Hence, there is a neighborhood U of x_0 and a representative $\tilde{X}_{x_0}^{(k)} = Z(\varphi_1, \dots, \varphi_s)$ of $X_{x_0}^{(k)}$ such that if $x \in U$ and $x \notin \tilde{X}_{x_0}^{(k)}$ then $S_x(\mathcal{H}) = \{x\}$. The functions $\varphi_1, \dots, \varphi_s$ are representatives on U of generators of $I_{x_0}^{(k)}$. The set $\tilde{X}_{x_0}^{(k+1)} := Z(D(\mathcal{H}_{|U} \cup \{\varphi_1, \dots, \varphi_s\}))$ is a representative of $X_{x_0}^{(k+1)}$. Clearly $\tilde{X}_{x_0}^{(k+1)} \subset \tilde{X}_{x_0}^{(k)}$. Take $x \in U$ such that $x \notin \tilde{X}_{x_0}^{(k+1)}$. If $x \notin \tilde{X}_{x_0}^{(k)}$ then $S_x(\mathcal{H}) = \{x\}$. Assume then that $x \in \tilde{X}_{x_0}^{(k)}$. From Lemma 5.1 we get $S_x(\mathcal{H}) \subset (\tilde{X}_{x_0}^{(k)})_x$. Then $S_x(\mathcal{H}) = S_x(\mathcal{H}) \cap (\tilde{X}_{x_0}^{(k)})_x = S_x(\mathcal{H}_{|U} \cup \{\varphi_1, \dots, \varphi_s\})$. Since $x \notin Z(D(\mathcal{H}_{|U} \cup \{\varphi_1, \dots, \varphi_s\}))$ then, by Lemma 5.2, $S_x(\mathcal{H}) = S_x(\mathcal{H}_{|U} \cup \{\varphi_1, \dots, \varphi_s\}) = \{x\}$. This finishes the inductive step of the proof. □

To prove sufficiency assume that $I^{(s)}_{x_0} = \mathcal{O}_{x_0}$ for some $s > 0$. This means that $X^{(s)}_{x_0} = \emptyset$. From Lemma 5.3 it follows that nontrivial set-germs $S_x(\mathcal{H})$ (i.e. different from $\{x\}$) may be found only in $\tilde{X}^{(s)}_{x_0}$ — some representative of $X^{(s)}_{x_0}$. But $X^{(s)}_{x_0} = \emptyset$ so in some neighborhood of x_0 the level set-germs of $\mathcal{H}$ must be trivial. This means that Σ is stably locally observable at x_0.

Necessity.

LEMMA 5.4. *Let M be a real analytic manifold, $x_0 \in M$, U be a neighborhood of x_0 diffeomorphic to an open set in $\mathbb{R}^n$, and $\mathcal{G}$ be a family of analytic functions on U. If $D(\mathcal{G}) = 0$ (zero ideal) then arbitrarily close to x_0 there is $x \in U$ such that $S_x(\mathcal{G}) \neq \{x\}$.*

PROOF. We may assume that $M = \mathbb{R}^n$ and $U \subset \mathbb{R}^n$. Let

$$s = \max_{\substack{x \in U \\ \varphi_i \in \mathcal{G}}} \operatorname{rank} \frac{\partial(\varphi_1, \dots, \varphi_n)}{\partial(x_1, \dots, x_n)}(x).$$

Then $s < n$ and arbitrarily close to x_0 there is $x \in U$ and $\varphi_1, \dots, \varphi_n \in \mathcal{G}$ such that rank $\frac{\partial(\varphi_1,\dots,\varphi_n)}{\partial(x_1,\dots,x_n)}(\bar{x}) = s$. This rank is preserved in some neighborhood V of $\bar{x}$. Thus we may assume that gradients of $\varphi_1, \dots, \varphi_s$ are linearly independent at every point of V and span $d\mathcal{G}(x)$ for $x \in V$. Then by Frobenius Theorem V is a union of integral manifolds of codistribution $d\mathcal{G}$. The integral manifolds are the level sets of $\mathcal{G}$ and have dimension greater than or equal 1. This means that $S_{\bar{x}}(\mathcal{G}) \neq \{\bar{x}\}$. □

LEMMA 5.5. *Let U be an open subset of $\mathbb{R}^n$ and $\varphi_1, \dots, \varphi_k$ be analytic functions on U. Assume that $Y = Z(\varphi_1, \dots, \varphi_k)$ is an analytic manifold in U. Let $x_0 \in Y$ and $\mathcal{G}$ be a family of analytic functions on U.*

If $Y \subset Z(D(\mathcal{G} \cup \{\varphi_1, \dots, \varphi_k\}))$ then arbitrarily close to x_0 there is $x \in Y$ such that $S_x(\mathcal{G}) \neq \{x\}$.

PROOF. We may assume that $\nabla\varphi_1, \dots, \nabla\varphi_k$ are linearly independent at every point of U. Changing the coordinates we can obtain $\varphi_i = x_i$, $i = 1, \dots, k$. Let $\psi_1, \dots, \psi_{n-k} \in \mathcal{G}$. Then for $x \in Y$ we have

$$0 = \det \begin{bmatrix} \nabla\varphi_1(x) \\ \vdots \\ \nabla\varphi_k(x) \\ \nabla\psi_1(x) \\ \vdots \\ \nabla\psi_{n-k}(x) \end{bmatrix} = \det \begin{bmatrix} I & 0 \\ \multicolumn{2}{c}{\left(\frac{\partial\psi_i}{\partial x_j}(x)\right)_{\substack{i=1,\dots,n-k \\ j=1,\dots,n}}} \end{bmatrix}$$

$$= \det \left(\frac{\partial\psi_i(x)}{\partial x_j}\right)_{\substack{i=1,\dots,n-k \\ j=k+1,\dots,n}}.$$

Hence, after restricting to manifold Y, we get $D(\mathcal{G}_{|Y}) = 0$. From Lemma 5.4, arbitrarily close to x_0 there is $x \in Y$ such that $S_x(\mathcal{G}_{|Y}) \neq \{x\}$. But $S_x(\mathcal{G}_{|Y}) \subset S_x(\mathcal{G})$ so also $S_x(\mathcal{G}) \neq \{x\}$. □

LEMMA 5.6. *Assume that $X^{(s)}_{x_0} = X^{(s+1)}_{x_0}$ for some $s \geq 0$ and $X^{(s)}_{x_0} \neq \emptyset$. Then in every neighborhood of x_0 there is x such that $S_x(\mathcal{H}) \neq \{x\}$.*

PROOF. Let $\varphi_1, \dots, \varphi_k$ be representatives of generators of ideal $I_{x_0}^{(s)}$, defined on some common neighborhood U of x_0. Then $\tilde{X}_{x_0}^{(s)} := Z(\varphi_1, \dots, \varphi_k)$ is a representative of $X_{x_0}^{(s)}$ in U. In every neighborhood of x_0 one can find a regular point of the analytic set $\tilde{X}_{x_0}^{(s)}$ (see [**Na, Ca**]). Let $\bar{x}$ be such a point and let V be a neighborhood of $\bar{x}$ in $\mathbb{R}^n$ such that $Y := V \cap \tilde{X}_{x_0}^{(s)}$ is an analytic manifold. Then $Y = Z(\varphi_{1|V}, \dots, \varphi_{k|V})$. Let $x \in Y$. Then $x \in \tilde{X}_{x_0}^{(s+1)} := Z(D(\mathcal{H}_{|U} \cup \{\varphi_1, \dots, \varphi_k\}))$ so $Y \subset Z(D(\mathcal{H}_{|V} \cup \{\varphi_{1|V}, \dots, \varphi_{k|V}\}))$. The statement of the lemma follows now from Lemma 5.5. □

Lemma 5.6 shows that if $\tilde{X}_{x_0}^{(s)} = \tilde{X}_{x_0}^{(s+1)}$ for some $s \geq 0$ then Σ is not stably locally observable.

References

[Ba1] Z. Bartosiewicz, *Local observability of nonlinear systems*, Systems & Control Letters 25 (1995), 295-298.

[Ba2] ———, *Remarks on local observability of nonlinear systems*, in: Proceedings of the First International Symposium on Mathematical Models in Automation and Robotics, September 1-3, 1994, Miedzyzdroje, Poland, Technical University of Szczecin Press, 41-44.

[Ca] H. Cartan, *Variété analytiques réelles et variété analytiques complexes*, Bull. Soc. Math. France 85 (1957), 77-99.

[GMT] F. Guaraldo, P. Macri and A. Tancredi, *Topics on Real Analytic Spaces*, Vieweg, Braunschweig, 1986.

[GR] R.C.Gunning and H.Rossi, *Analytic Functions of Several Complex Variables*, Prentice-Hall, Englewood Cliffs, 1965.

[HK] R. Hermann and A. Krener, *Nonlinear controllability and observability*, IEEE Transactions AC-22 (1977), 728-740.

[Mo] D. Mozyrska, *Rank condition of stable local observability of analytic systems on* $\mathbb{R}^2$, Zeszyty Naukowe Politechniki Białostockiej. Matematyka-Fizyka-Chemia (Białystok) 16 (1996), 143-150.

[MB1] D. Mozyrska and Z. Bartosiewicz, *Rank condition of stable local observability of analytic systems on* $\mathbb{R}^3$, Proceedings of International Conference on Control '96, Exeter, September 1996,

[MB2] ———, *Algebraic criteria for stable local observability of analytic systems on* $\mathbb{R}^n$, Proceedings of European Control Conference ECC-97, Brussels, July 1997,

[Na] R. Narasimhan, *Introduction to the Theory of Analytic Spaces*, LNM 25, Springer-Verlag, Berlin, 1966.

[Ri] J.-J. Risler, *Le thèoréme des zèros en gèomètries algèbrique et analytique rèelles*, Bull. Soc. Math. France 104 (1976), 113-127.

POLITECHNIKA BIAŁOSTOCKA, WIEJSKA 45, BIAŁYSTOK, POLAND

E-mail address: bartos@cksr.ac.bialystok.pl

Proceedings of Symposia in Pure Mathematics
Volume **64**, 1999

External Dynamic Feedback Equivalence of Observable Discrete-Time Control Systems

Ewa Pawłuszewicz and Zbigniew Bartosiewicz

This paper is dedicated to Joseph Johnson

ABSTRACT. The problem of external dynamic feedback equivalence of nonlinear discrete-time systems on smooth and analytic manifolds is studied. The main result says that two systems are equivalent in such sense iff their difference output universes are isomorphic. This criterion is then used to give a characterization of external dynamic feedback linearization of nonlinear systems.

1. Introduction

In early nineties B. Jakubczyk proposed a clear definition of dynamic (state) feedback equivalence of nonlinear continuous-time systems and gave a characterization of that equivalence [**Ja1, Ja2**]. The main concept used in that characterization was the differential algebra of a control system. Jakubczyk's results were then transferred to the discrete-time case by Bartosiewicz, Jakubczyk and Pawłuszewicz [**BJP**]. The notion of difference algebra of a discrete-time system replaced the differential original allowing to state the following result: two systems are dynamically feedback equivalent iff their difference algebras are isomorphic.

The language of difference algebra (as a mathematical theory) was first used in control literature by M. Fliess [**F1, F2**], who was also the initiator and a strong advocate of differential-algebraic methods for continuous-time systems. But in this paper we steal more from Jakubczyk than from Fliess.

In [**BJP**] only systems without outputs and only state feedbacks were studied. We relax that restriction here adding the output part and using it in our dynamic transformations. But to be able to transfer external trajectories of one system into external trajectories of the other, we need to pass through the states. To achieve this we assume observability. To have smooth or analytic dependence of states on outputs and controls we add a rank condition for the state-ouput maps. At this point the notion of difference algebra of a system appears to be insufficient for our purposes. We need to compose smooth or analytic maps and to amalgamate (glue up)

1991 *Mathematics Subject Classification.* Primary 93B17, 93B18; Secondary 93B25, 93C10, 93C35.

The work of the first author was supported by KBN under the grant No. 8 T11A 005 13.

The work of the second author was supported by KBN under the Technical University of Białystok grant No. W/IMF/1/96 (1996-98).

families of "matching" partially defined functions. Function universes, introduced by J. Johnson [**Jh**], are exactly what we need. They are far reaching generalizations of algebras, especially useful in differential-geometric problems. In fact, universe spaces are good alternatives for differential manifolds in cases where singularities appear. Function universes and universe spaces were earlier used in control theory to study observability of continuous-time systems [**BJ**]. Pawłuszewicz examined relations between difference algebras and difference universes in [**P1**]. Preliminary results on external dynamic feedback equivalence were presented in [**PB**].

One of the main reasons to study dynamic equivalence is the problem of dynamic linearization (see [**AKM, G, Ja, JC, LM, SNT**] for static linearization). We solve this problem in the case when feedback transformations that linearize a nonlinear system depend on past (or future) outputs and controls. A necessary and sufficient condition for such linearizability is that the difference output universe be free. On the differential level this property is related to so called flatness of the system (see e.g. [**MMR**]). It is known that many interesting and important systems are not statically linearizable. Thus dynamic feedback extends the class of linarizable systems.

Another motivation to study equivalence of systems, in particular equivalence to linear systems, is stabilization and regulation of systems [**BL, LB**]. When output of the system, instead of state, is used to achieve the required properties, the output equivalence is a natural problem to consider. Dynamic equivalence is again more powerful than the static one.

2. Preliminaries

2.1. Sequences. Let A be any set and r be any integer. Let us consider the set of infinite sequences $Z_r = \{z(r), z(r+1), \dots\}$ where $z(i) \in A$. We will denote this set by $J_r(A)$. Similarly $J_r(A_1, \dots, A_k)$ will denote the set of sequences $(Z_r^1, \dots, Z_r^k)$, where $Z_r^i \in J_r(A_i)$ and A_i is a set, $i = 1, \dots, k$. Let

$$J(A_1, \dots, A_k) = \dot{\bigcup_{r\in\mathbb{Z}}} J_r(A_1, \dots, A_k)$$

where $\dot{\bigcup}$ means a disjoint sum. We write $J_r(s_1, \dots, s_k)$ for $J_r(\mathbb{R}^{s_1}, \dots, \mathbb{R}^{s_k})$ and similarly for $J(\mathbb{R}^{s_1}, \dots, \mathbb{R}^{s_k})$. The map $c_i : J(A) \to J(A)$, $i \geq 0$, defined by $c_i(Z_r) = Z_{r+i}$, where $Z_{r+i} = \{z(r+i), z(r+i+1), \dots\}$ for $Z_r \in J(A)$, will be called a *restriction operator*. A *shift operator* is a map $s_k : J(A) \to J(A)$: $s_k(Z_r) = Y_{r+k}$, $k \in \mathbb{Z}$, where $y(i) = z(i-k)$ for $i \geq r+k$. It is clear that $c_i(J_r(A)) = J_{r+i}(A)$ and $s_k(J_r(A)) = J_{r+k}(A)$. Moreover, for any $i \geq 0$ and $k \in \mathbb{Z}$ the restriction and shift operators commute, i.e. $s_k c_i = c_i s_k$. The maps c_i and s_k may be defined similarly on spaces $J(A_1, \dots, A_s)$. The same properties hold.

If $\gamma : J(A_1, \dots, A_s) \to B$ then the *extension of* γ is the map
$\Gamma : J(A_1, \dots, A_s) \to J(B)$ defined by

$$\Gamma(Z_r^1, \dots, Z_r^s) = \tilde{Z}_r,$$

where $\tilde{Z}_r(i) = \gamma(c_{i-r}(Z_r^1, \dots, Z_r^s))$ for $i \geq r$. We easily get the following property of extensions.

PROPOSITION 2.1. *If Γ is the extension of a map $\gamma : J(A_1, \dots, A_s) \to B$, $i \geq 0$ and $k \in \mathbb{Z}$, then:*

(i) $s_k\Gamma = \Gamma s_k$
(ii) $c_k\Gamma = \Gamma c_k$. □

The definition and the properties of extension naturally generalize to the case where $\gamma : J(A_1, \dots, A_s) \to B_1 \times \cdots \times B_p$. Then the extension of γ is the map $\Gamma : J(A_1, \dots, A_s) \to J(B_1, \dots, B_p)$.

Let A be a set and let A^∞ denote the set of sequences $(z(0), z(1), \dots)$, $z(i) \in A$. Observe that A^∞ may be identified with $J_0(A)$. We have natural projections: $\Pi_s : A^\infty \to A^s$, $s \geq 1$, $\Pi_s(z(0), z(1), \dots) = (z(0), \dots, z(s-1))$.

Now let A be a topological space. We define the topology on A^∞ as the weakest topology for which all maps Π_s are continuous. The basis $\mathcal{B}$ of this topology consists of the sets defined in the following way: a set V is an element of $\mathcal{B}$ if and only if it is the inverse image of an open set in A^s under map Π_s for some $s \in \mathbb{N}$. For products of finitely many spaces of type A^∞ we choose the product topology.

LEMMA 2.2. *If A is Hausdorff then also A^∞ is Hausdorff.*

PROOF. Let X, Y be different points in A^∞ and assume that they differ on $s-1$ position, i.e. $X = (x(0), \dots, x(s-2), x(s-1), x(s), \dots)$, $Y = (x(0), \dots, x(s-2), y(s-1), y(s), \dots)$ and $x(s-1) \neq y(s-1)$. Let $\Pi_s(X) = x$, $\Pi_s(Y) = y$. Then $x \neq y$ and $x, y \in A^s$. But A^s is a Hausdorff space, so one can choose disjoint neighborhoods U_x and U_y of points x and y, respecively. Relation $x \in U_x$ implies $X \in \Pi_s^{-1}(U_x)$, $y \in U_y$ implies $Y \in \Pi_s^{-1}(U_y)$. Since $U_x \cap U_y = \emptyset$, then $\Pi_s^{-1}(U_x) \cap \Pi_s^{-1}(U_y) = \emptyset$. Hence A^∞ is Hausdorff. □

Let A and B be sets and $V \subset A^\infty$. We say that a function $\varphi : V \to B$ *depends on finite number of variables* if there exist $k \geq 1$ and a function $\tilde{\varphi} : A^k \to B$ such that $\varphi = \tilde{\varphi} \circ \Pi_{k|V}$.

Now let A be a topological space. A function $\varphi : A^\infty \to B$ *locally depends on finite number of variables* if for any $X \in A^\infty$ there exists an open neighborhood V of X such that $\varphi_{|V}$ depends on finite number of variables.

Let A and B be manifolds of class C^s, $s = 0, 1, \dots, \infty, \omega$ ($s = 0$ means topological, $s = \infty$ - smooth, $s = \omega$ - analytic). A function $\varphi : A^\infty \to B$ *is of class* C^s if φ locally depends on finite number of variables and locally there exist $k \in \mathbb{N}$ and a function $\tilde{\varphi} : A^k \to B$ of class C^s such that $\varphi = \tilde{\varphi} \circ \Pi_k$.

These definitions can be easily generalized to functions defined on finite products of spaces of type A^∞.

PROPOSITION 2.3. *An analytic function $\varphi : \mathbb{R}^\infty \to \mathbb{R}$ depends (globally) on finite number of variables.*

PROOF. (sketch) First we show that if we have two local representations $\tilde{\varphi}_1$ and $\tilde{\varphi}_2$ of φ, corresponding to open sets $V_1, V_2 \subset \mathbb{R}^\infty$ whose intersection is not empty, then they must depend on the same number of variables. This follows from uniqueness of power series representation for analytic functions. Then we use the fact that the segment joining two points in $\mathbb{R}^\infty$ is compact. This allows us to show that for any two points in $\mathbb{R}^\infty$ we have local representations depending on the same number of variables. □

Let A be a topological space. As A^∞ may be identified with $J_0(A)$, the topology on A^∞ determines the topology on $J_0(A)$. Similarly $J_0(A_1, \dots, A_s)$ is identified with $A_1^\infty \times \dots \times A_s^\infty$. The topology on $J(A_1, \dots, A_s)$ is defined as follows. A set

$V \in J(A_1, \dots, A_s)$ is *open* if for any $k \in \mathbb{Z}$, $s_k(V) = V$ and $V \cap J_0(A_1, \dots, A_s)$ is open. Observe that $J(A)$ is not Hausdorff.

We shall assume from now on that all the maps defined on $J(A_1, \dots, A_s)$ with values in a set B (but not $J(B)$) are shift invariant, i.e. $\varphi(s_k(Z_r^1, \dots, Z_r^p)) = \varphi(Z_r^1, \dots, Z_r^p)$ for any $r, k \in \mathbb{Z}$. This allows us to define them only for sequences starting at 0. Moreover, whenever A_i and B are topological spaces, we assume that such maps depend locally on finite number of variables. If $\varphi : J(A) \to B$ depends on $z(0), z(1), \dots, z(q)$ in a neighborhood of Z_0, we say that it *is of order less or equal* q in this neighborhood. We also say then that φ *is locally of finite order.* Similarly in the more general case where the domain of φ is $J(A_1, \dots, A_s)$.

2.2. Trajectories and behaviors. Let $\mathcal{M}$ be an n-dimensional (Hausdorff) manifold of class C^s, $s = \infty$ or $s = \omega$. Let us consider a nonlinear discrete-time control system defined on this manifold:

$$\Sigma : \begin{aligned} x(k+1) &= f(x(k), u(k)) \\ y(k) &= h(x(k), u(k)) \end{aligned} \tag{2.1}$$

where $k \in \mathbb{Z}$, $x(k) \in \mathcal{M}$, $u(k) \in \mathbb{R}^m$ and $y(k) \in \mathbb{R}^p$. We assume that functions f and $h = (h_1, \dots, h_p)$ are of the same class as manifold $\mathcal{M}$. The function f will be called the *dynamics* of system Σ, the function h - the *output function.*

Let $r \in \mathbb{Z}$. Consider the following sequences:

$$\begin{aligned} X_r &= \{x(r), x(r+1), \dots\} \in J_r(\mathcal{M}), \\ U_r &= \{u(r), u(r+1), \dots\} \in J_r(m), \\ Y_r &= \{y(r), y(r+1), \dots\} \in J_r(p). \end{aligned}$$

A *trajectory* of system Σ is any triple (X_r, U_r, Y_r) that satisfies (2.1) for $k \geq r$. An *inner trajectory* of system Σ is any pair (X_r, U_r) that satisfies the first equation of the system. An *external trajectory* is any pair (Y_r, U_r) for which there is X_r such that (X_r, U_r, Y_r) is a trajectory of Σ. It is clear that if (X_r, U_r, Y_r) is a trajectory of system Σ then also $(X_{r+k}, U_{r+k}, Y_{r+k}) := c_k(X_r, U_r, Y_r)$ is a trajectory of Σ for any $r, k \in \mathbb{Z}, k \geq 0$. A similar fact holds for inner and external trajectories. The shifts s_k preserve trajectories as well. The set of all trajectories of system Σ forms the *behavior* of the system. We denote it by $B(\Sigma)$. Analogously, the set of all inner trajectories of Σ forms the *inner behavior*, denoted by $B^i(\Sigma)$. It is a subset of $J(\mathcal{M}, m)$. The set of all external trajectories of system Σ will be called the *external behavior* of system Σ and denoted by $B^e(\Sigma)$. It is a subset of space $J(p, m)$. Then

$$B^e(\Sigma) = \bigcup_{r \in \mathbb{Z}} B^e_r(\Sigma)$$

where $B^e_r(\Sigma)$ is the set of external trajectories of system Σ starting at moment r. Similar identities hold for behaviors and inner behaviors.

We also have for any $k \in \mathbb{Z}$ and $i \geq 0$

$$s_k(B_r(\Sigma)) = B_{r+k}(\Sigma), \quad s_k(B^e_r(\Sigma)) = B^e_{r+k}(\Sigma), \quad s_k(B^i_r(\Sigma)) = B^i_{r+k}(\Sigma)$$

and

$$c_i(B_r(\Sigma)) \subset B_{r+i}(\Sigma), \quad c_i(B^e_r(\Sigma)) \subset B^e_{r+i}(\Sigma), \quad c_i(B^i_r(\Sigma)) \subset B^i_{r+i}(\Sigma).$$

2.3. External dynamic feedback equivalence. Let $\mathcal{M}$ and $\tilde{\mathcal{M}}$ denote n and $\tilde{n}$-dimensional manifolds of class C^s. Let us consider two nonlinear discrete-time systems

$$\Sigma : \begin{array}{c} x(k+1) = f(x(k), u(k)) \\ y(k) = h(x(k), u(k)) \end{array} \quad \text{and} \quad \tilde{\Sigma} : \begin{array}{c} \tilde{x}(k+1) = \tilde{f}(\tilde{x}(k), \tilde{u}(k)) \\ \tilde{y}(k) = \tilde{h}(\tilde{x}(k), \tilde{u}(k)) \end{array}$$

where $x(k) \in \mathcal{M}$, $\tilde{x}(k) \in \tilde{\mathcal{M}}$, $u(k), \tilde{u}(k) \in \mathbb{R}^m$, $y(k) \in \mathbb{R}^p$, $\tilde{y}(k) \in \mathbb{R}^{\tilde{p}}$. Let us consider the following maps of class C^s:

$$\phi^e : J(\tilde{p}, m) \to \mathbb{R}^p, \ \psi^e : J(\tilde{p}, m) \to \mathbb{R}^m \tag{2.2}$$

$$\tilde{\phi}^e : J(p, m) \to \mathbb{R}^{\tilde{p}}, \ \tilde{\psi}^e : J(p, m) \to \mathbb{R}^m. \tag{2.3}$$

The extensions Φ^e, Ψ^e, $\tilde{\Phi}^e$, $\tilde{\Psi}^e$ of $\phi^e, \psi^e, \tilde{\phi}^e, \tilde{\psi}^e$ give:

$$\chi^e = (\Phi^e, \Psi^e) : J(\tilde{p}, m) \to J(p, m) : (\tilde{Y}_r, \tilde{U}_r) \mapsto (\Phi^e(\tilde{Y}_r, \tilde{U}_r), \Psi^e(\tilde{Y}_r, \tilde{U}_r))$$

$$\tilde{\chi}^e = (\tilde{\Phi}^e, \tilde{\Psi}^e) : J(p, m) \to J(\tilde{p}, m) : (Y_r, U_r) \mapsto (\tilde{\Phi}^e(Y_r, U_r), \tilde{\Psi}^e(Y_r, U_r)).$$

From Proposition 2.1 it follows that $\chi^e s_i = s_i \chi^e$.

Let A and B be topological spaces and $q^- \in \mathbb{N} \cup \{0\}$. If $\phi : J(A) \to B$ is a map of locally finite order then *the dynamic transformation of type q^- associated with* ϕ is the map $D_\phi^{q^-} : J(A) \to J(B)$ defined by

$$D_\phi^{q^-}(Z_r) = (s_{q^-} \Phi)(Z_r).$$

Then $D_\phi^{q^-}(J_r(A)) \subset J_{r+q^-}(B)$. The definition can be easily extended to the case where $\varphi : J(A_1, \dots, A_s) \to B_1 \times \cdots \times B_p$.

We shall be interested in transformations of the form

$$D_{(\phi^e, \psi^e)}^{q^-} : J(\tilde{p}, m) \to J(p, m)$$

associated with (ϕ^e, ψ^e) given by (2.2).

If

$$D_{(\phi^e, \psi^e)}^{q^-}(B_r^e(\tilde{\Sigma})) \subset B_{r+q^-}^e(\Sigma) \tag{2.4}$$

for $r \in \mathbb{Z}$, then $D_{(\phi^e, \psi^e)}^{q^-}$ will be called an *external dynamic transformation of type q^- of system $\tilde{\Sigma}$ into system Σ*. Then the external trajectories locally in the neighborhoods of the points $(Y, U) \in J(p, m)$ and $(\tilde{Y}, \tilde{U}) \in J(\tilde{p}, \tilde{m})$ are related in the following way

$$y(k) = \phi^e(\tilde{y}(k - q^-), \dots, \tilde{y}(k + q^+), \tilde{u}(k - q^-), \dots, \tilde{u}(k + q^+)) \tag{2.5a}$$

$$u(k) = \psi^e(\tilde{y}(k - q^-), \dots, \tilde{y}(k + q^+), \tilde{u}(k - q^-), \dots, \tilde{u}(k + q^+)). \tag{2.5b}$$

where $q^- + q^+ = q$ and (ϕ^e, ψ^e) is of order less or equal q in the neighborhood of $(\tilde{X}, \tilde{U})$.

PROPOSITION 2.4. *If there exist an external dynamic feedback transformation of type q^- associated with (ϕ^e, ψ^e) that transforms system $\tilde{\Sigma}$ into system Σ, then for every $p^- \in \mathbb{N} \cup \{0\}$ there also exists an external dynamic feedback transformation of type p^-, that transforms $\tilde{\Sigma}$ into Σ. Moreover, this transformation is associated with (ϕ^e, ψ^e).*

PROOF. Let us assume that system $\tilde{\Sigma}$ is transformed into system Σ by an external dynamic transformation of type q^-. This implies that there exist maps ϕ^e, ψ^e given by (2.2) such that the condition (2.4) is satisfied. This implies that

$$D^{p^-}_{(\phi^e,\psi^e)}(B^e_r(\tilde{\Sigma})) = (s_{p^-}\chi^e)(B^e_r(\tilde{\Sigma})) = (s_{q^-}\chi^e s_{p^- - q^-})(B^e_r(\tilde{\Sigma}))$$

$$= (s_{q^-}\chi^e)(B^e_{r+p^- - q^-}(\tilde{\Sigma})) \subset B^e_{r+p^- - q^- + q^-}(\Sigma) = B^e_{r+p^-}(\Sigma).$$

So maps ϕ^e, ψ^e define a dynamic transformation of type p^-. □

For $q^- > 0$ this dynamic transformation will be called *external dynamic delay feedback transformation.* Then

$$y(k) = \phi^e(s_{q^-}(\tilde{Y}_k, \tilde{U}_k)), \qquad u(k) = \psi^e(s_{q^-}(\tilde{Y}_k, \tilde{U}_k)).$$

When $q^- > 0$, q is globally bounded, and $q^+ \le 0$ globally, we call $D^{q^-}_{(\phi^e,\psi^e)}$ *strictly delay* transformation. Then the current values of controls and outputs in Σ depend on past and current values of controls and outputs in $\tilde{\Sigma}$.

In the case when $q^- = 0$ and $q^+ > 0$ we have the *external dynamic anticipating feedback transformations* which lead to equations:

$$y(k) = \phi^e(\tilde{Y}_k, \tilde{U}_k), \qquad u(k) = \psi^e(\tilde{Y}_k, \tilde{U}_k),$$

$$\tilde{y}(k) = \tilde{\phi}^e(Y_k, U_k), \qquad \tilde{u}(k) = \tilde{\psi}^e(Y_k, U_k).$$

Now current outputs and controls of one system depend on current and future values of outputs and controls of the second system.

DEFINITION 2.5. The systems Σ and $\tilde{\Sigma}$ are *externally dynamically feedback equivalent* if there exist $q^-, \tilde{q}^- \ge 0$ and maps (ϕ^e, ψ^e) and $(\tilde{\phi}^e, \tilde{\psi}^e)$ of locally finite order, defined on external behaviors of the systems, such that they have local extensions on open neighborhoods of each trajectory and the associated maps satisfy the equations:

$$D^{q^-}_{(\phi^e,\psi^e)}(B^e_r(\tilde{\Sigma})) = B^e_{r+q^-}(\Sigma), \quad D^{\tilde{q}^-}_{(\tilde{\phi}^e,\tilde{\psi}^e)}(B^e_r(\Sigma)) = B^e_{r+\tilde{q}^-}(\tilde{\Sigma})$$

and

$$D^{q^-}_{(\phi^e,\psi^e)} D^{\tilde{q}^-}_{(\tilde{\phi}^e,\tilde{\psi}^e)|B^e(\Sigma)} = s_{q^- + \tilde{q}^-|B^e(\Sigma)}, \quad D^{\tilde{q}^-}_{(\tilde{\phi}^e,\tilde{\psi}^e)} D^{q^-}_{(\phi^e,\psi^e)|B^e(\tilde{\Sigma})} = s_{q^- + \tilde{q}^-|B^e(\tilde{\Sigma})}.$$

COROLLARY 2.6. *The systems Σ and $\tilde{\Sigma}$ are externally dynamically delay feedback equivalent iff they are externally dynamically anticipating feedback equivalent.* □

REMARK 2.7. When the orders of dynamic transformations are bounded, we may replace in Corollary 2.6 delay feedback by strictly delay feedback. This is the case of analytic systems defined on $\mathbb{R}^n$. In general, the anticipating part in the feedback transformations may be unbounded and then cannot be shifted to the past as this would require infinite units of time. In future calculations we'll be using, for simplicity, the anticipating feedback.

3. Function and difference universes

3.1. Function universes. In order to work with input-output nonlinear control sytems we need to substitute to analytic (smooth) functions and to amalgamate functions. These operations are naturally defined in function universe. Here we present only basic ideas of theory of function universes. More details can be found in [**Jh, BJ**].

Let X, Y be sets. A *partially defined function* on X (with values in Y) is any map $\varphi : A \to Y$, where $A \subset X$ is called domain of φ and denoted by $\operatorname{dom}\varphi$. Let Y_X be the set of all partially defined functions on X. One can extend any function $\varphi \in Y_X$ to the one defined on X by assigning $\varphi(x) = \emptyset_0$ for $x \notin \operatorname{dom}\varphi$. We call $\emptyset_0$ the *phantom* (in Y) [**BJ**]. Now $\operatorname{dom}\varphi = \{x \in X : \varphi(x) \neq \emptyset_0\}$. If $a \in Y$, $x \in X$ then we set $a_X(x) := a$. If $A \subset X$ then 0_A denotes the zero function with domain A. If $0\varphi = 0_X$ then φ is a *global function.* Let A_n denote the set of functions of class C^s ($s = \infty$ or ω), partially defined on open subsets in $\mathbb{R}^n$. In particular A_0 can be identified with $\mathbb{R} \cup \emptyset_0$. If X is a manifold of class C^s, then

$$A_X := \{\varphi \in \mathbb{R}_X : \operatorname{dom}\varphi \text{ is open and } \varphi_{|\operatorname{dom}\varphi} \text{ is of class } C^s\}.$$

So one has $A_{\mathbb{R}^n} = A_n$. The topology in A_0 is defined as follows: a subset $B \subset A_0$ is open if $B = A_0$ or B is an open subset in $\mathbb{R}$.

The functions $\varphi, \psi \in Y_X$ are *matching*, if they take on the same values on $\operatorname{dom}\varphi \cap \operatorname{dom}\psi$. Let us consider a set $M \subset Y_X$ of functions that are matching (*matching set*) and define the function $\underline{M} \in A_0$, in the following way: $\underline{M}(x) = \emptyset_0$ if no function in M is defined at the point x and $\underline{M}(x) = \varphi(x)$ for any function $\varphi \in M$ defined at the point x. The process of constructing $\underline{M}$ is called the *amalgamation* of the functions of M.

Let $\varphi_1, \dots, \varphi_k \in \mathbb{R}_X$ and $F \in A_k$. Then $F \circ (\varphi_1, \dots, \varphi_k)$ is a partially defined function on X given by

$$(F \circ (\varphi_1, \dots, \varphi_k))(x) = F(\varphi_1(x), \dots, \varphi_k(x))$$

for $x \in X$. If $\varphi_i(x) = \emptyset_0$ or $(\varphi_1(x), \dots, f_k(x)) \notin \operatorname{dom} F$, then $F(\varphi_0(x), \dots, \varphi_n(x)) = \emptyset_0$. The map $(\varphi_1, \dots, \varphi_k) \mapsto F \circ (\varphi_1, \dots, \varphi_k)$ is called a *substitution.*

A set $\mathcal{U} \subset Y_X$ containing 0_X and closed under analytic (smooth) substitutions and amalgamation is called an *analytic (smooth) function universe on the set* X [**Jh**]. A *function subuniverse* of the universe $\mathcal{U}$ is a subset $\mathcal{C} \subset \mathcal{U}$ that is a function universe on X. Let $\mathcal{H} \subset \mathbb{R}_X$. The *function subuniverse generated by* $\mathcal{H}$ is the smallest function subuniverse containig $\mathcal{H}$ [**BJ**].

In a natural way the function universe $\mathcal{U}$ induces the topology on X. The open sets have the form $\operatorname{dom}\varphi : \varphi \in \mathcal{U}$. If $\mathcal{U}$ is a function universe on X then the pair $(X, \mathcal{U})$ is called a *universal space* [**BJ**].

Let $\mathcal{U}_1, \mathcal{U}_2$ denote the function universes defined on X_1 and X_2, respectively. A map $\tau : \mathcal{U}_1 \to \mathcal{U}_2$ is a *homomorphism* of function universes $\mathcal{U}_1$ and $\mathcal{U}_2$ if

1. $\tau(F \circ (\varphi_1, \dots, \varphi_k)) = F(\tau\varphi_1, \dots, \tau\varphi_k)$ for $\varphi_1, \dots, \varphi_k \in \mathcal{U}_1$, $F \in A_k$,
2. $\tau(\underline{M}) = \underline{\tau(M)}$ for any matching set $M \subset \mathcal{U}_1$,
3. $\tau(0_{X_1}) = 0_{X_2}$.

If $\mathcal{U}$ is a function universe on X then the set of all homomorphisms $\mathcal{U} \to A_0$ will be denoted by $S\mathcal{U}$. Let us consider the map $t_{\mathcal{U}} : \mathcal{U} \to \mathbb{R}_{S\mathcal{U}}$, $t_{\mathcal{U}}(\varphi) = \hat{\varphi}$, where $\hat{\varphi} : S\mathcal{U} \to A_0$, $\hat{\varphi}(P) := P(\varphi)$ for $P \in S\mathcal{U}$. Let $\hat{\mathcal{U}} = \operatorname{im} t_{\mathcal{U}}$. $\hat{\mathcal{U}}$ is a function universe on $\mathcal{SU}$. It defines the topology on $\mathcal{SU}$: the open sets have the form $\operatorname{dom}\hat{\varphi} = \{P \in \mathcal{SU} : P(\varphi) \neq \emptyset_0\}$, where $\hat{\varphi} \in \hat{\mathcal{U}}$.

Let us consider the following map of universal spaces: $t_X : X \to \mathcal{SU}$, $x \mapsto \hat{x}$, where $\hat{x}(\varphi) := \varphi(x)$.

LEMMA 3.1 (**[Jh, BJ]**). *If X is a Hausdorff space then the map t_X is a homeomorphism.* □

LEMMA 3.2. *Let $(X_1,\mathcal{U}_1)$, $(X_2,\mathcal{U}_2)$ be universal spaces. Suppose that topologies on X_1 and X_2 are Hausdorff. If $\tau : \mathcal{U}_1 \to \mathcal{U}_2$ is a homomorphism of function universes then there exists a map $\sigma : X_2 \to X_1$ such that $\sigma^* = \tau$, where $\sigma^*(\varphi) = \varphi \circ \sigma$ for $\varphi \in \mathcal{U}_1$. Moreover, σ is a continuous map.*

PROOF. Let us consider maps $t_{X_1} : X_1 \to \mathcal{SU}_1$ and $t_{X_2} : X_2 \to \mathcal{SU}_2$. Define $\mu : \mathcal{SU}_2 \to \mathcal{SU}_1$ by $\mu(P_2)(\varphi_1) := P_2(\tau\varphi_1)$ where $P_2 \in \mathcal{SU}_2$. By Lemma 3.1, $P_2 = \hat{x}_2$ for some $x_2 \in X_2$.

Let $V = \operatorname{dom}\hat{\varphi}_1 \subset \mathcal{SU}_1$, $\varphi_1 \in \mathcal{U}_1$, and $\varphi_2 = \tau\varphi_1$. Then

$$\mu^{-1}(V) = \{P_2 \in \mathcal{SU}_2 : \hat{\varphi}_1(\mu(P_2)) \neq \emptyset_0\} = \{P_2 \in \mathcal{SU}_2 : \mu(P_2)(\varphi_1) \neq \emptyset_0\}$$

$$= \{P_2 \in \mathcal{SU}_2 : P_2(\tau\varphi_1) \neq \emptyset_0\} = \{P_2 \in \mathcal{SU}_2 : \hat{\varphi}_2(P_2) \neq \emptyset_0\} \subset \mathcal{SU}_2.$$

This means that the inverse image of an open set under the map μ is an open set, hence μ is a continuous map.

Since t_{X_i}, $i = 1,2$, is a bijective map, then $t_{X_i}^{-1}$ is well defined. Let us define a map $\sigma : X_2 \to X_1$ in the following way: $\sigma := t_{X_1}^{-1} \circ \mu \circ t_{X_2}$. Then σ is a continuous map of spaces X_1 and X_2. Moreover, for $\varphi_1 \in \mathcal{U}_1$, $x_2 \in X_2$ one obtains

$$(\sigma^*(\varphi_1))(x_2) = (\varphi_1 \circ \sigma)(x_2) = (\varphi_1 \circ (t_{X_1}^{-1} \circ \mu \circ t_{X_2}))(x_2)$$

$$= \varphi_1(t_{X_1}^{-1}(\mu(\hat{x}_2))) = \mu(\hat{x}_2)(\varphi_1) = \hat{x}_2(\tau\varphi_1) = (\tau\varphi_1)(x_2).$$

This ends the proof. □

REMARK 3.3. For any sets X, Y and $\sigma : X \to Y$ the map $\sigma^* : \mathbb{R}_Y \to \mathbb{R}_X$ is a homomorphism of function universes. If X, Y are manifolds and σ is of class C^s, then the restriction of σ^* to A_Y is a homomorphism of universes A_Y and A_X.

If a homomorphism $\tau : \mathcal{U}_1 \to \mathcal{U}_2$ is a bijective map then it is an *isomophism of function universes.*

REMARK 3.4. If $\tau : \mathcal{U}_1 \to \mathcal{U}_2$ is an isomorphism then there exists homeomorphism $\sigma : X_2 \to X_1$ such that $\sigma^* = \tau$.

3.2. Difference universes. Let $k, l \in \mathbb{Z}$. Consider sequences $Y_k \in J_k(m)$, $Y_l \in J_l(m)$. Let

$$Y_k \simeq Y_l \iff s_{l-k}(Y_k) = Y_l.$$

Then $\simeq$ is an equivalence relation in $J(m)$.

It's easy to see that the quotient space $J(m)/\simeq$ is homeomorphic with $J_0(m)$. Then the spaces $\mathcal{M} \times J_0(m)$ and $\mathcal{M} \times J(m)/\simeq$ are homeomorphic as well. This implies that $\mathcal{M} \times J(m)/\simeq$ is Hausdorff.

Let $\mathcal{U}(\mathcal{M}, J(m))$ denote the set of shift invariant functions of class C^s that are partially defined on open subsets of $\mathcal{M} \times J(m)$ and locally depend on finite number of variables. Then $\mathcal{U}(\mathcal{M}, J(m))$ is a function universe on $\mathcal{M} \times J(m)$. It is isomorphic with $\mathcal{U}(\mathcal{M}, J_0(m))$ — the function universe of partially defined C^s functions on $\mathcal{M} \times J_0(m)$.

PROPOSITION 3.5. *If $\mathcal{M}$ and $\tilde{\mathcal{M}}$ are n and $\tilde{n}$-dimensional manifolds of class C^s, then each homomorphism of function universes $\tau : \mathcal{U}(\mathcal{M}, J(m)) \to \mathcal{U}(\tilde{\mathcal{M}}, J(\tilde{m}))$ is of the form $\tau = \sigma^*$ for $\sigma : \tilde{\mathcal{M}} \times J(\tilde{m}) \to \mathcal{M} \times J(m)$ and $\sigma^*(\varphi) = \varphi \circ \sigma$ for $\varphi \in \mathcal{U}(\mathcal{M}, J(m))$. Moreover σ is a continuous map. If τ is an isomorphisn then σ is a homeomorphism.*

PROOF. Let $\xi : \mathcal{U}(\mathcal{M}, J(m)) \to \mathcal{U}(\mathcal{M}, J_0(m))$ and $\tilde{\xi} : \mathcal{U}(\tilde{\mathcal{M}}, J(\tilde{m})) \to \mathcal{U}(\tilde{\mathcal{M}}, J_0(\tilde{m}))$ be isomorphisms of function universes. Since $\tau : \mathcal{U}(\mathcal{M}, J(m)) \to \mathcal{U}(\tilde{\mathcal{M}}, J(\tilde{m}))$ is a homomorphism then also $\overline{\tau} = \tilde{\xi} \circ \tau \circ \xi^{-1}$, $\overline{\tau} : \mathcal{U}(\mathcal{M}, J_0(m)) \to \mathcal{U}(\tilde{\mathcal{M}}, J_0(\tilde{m}))$, is a homomorphism. Lemma 3.2 implies existence of a map $\overline{\sigma} : \mathcal{M} \times J_0(m) \to \tilde{\mathcal{M}} \times J_0(\tilde{m})$ such that $\overline{\sigma}^* = \overline{\tau}$.

Let us consider a map $\sigma : \mathcal{M} \times J(m) \to \tilde{\mathcal{M}} \times J(\tilde{m})$ given by

$$\sigma(x, s_k U_0) := (\overline{\sigma}_1(x, U_0), (s_k \circ \overline{\sigma}_2)(x, U_0))$$

where $\overline{\sigma} = (\overline{\sigma}_1, \overline{\sigma}_2)$ and $x \in \mathcal{M}$, $U_0 \in J_0(m)$. A simple calculation shows that σ is the required map. □

A *difference universe* is a function universe $\mathcal{U}$ together with a homomorphism d of $\mathcal{U}$ into itself. If $(\mathcal{U}_1, d_1), (\mathcal{U}_2, d_2)$ are two difference universes then a map $\tau : \mathcal{U}_1 \to \mathcal{U}_2$ is a *homomorphism of difference universes* if it is a homomorphism of function universes and satisfies the condition:

$$d_2 \circ \tau = \tau \circ d_1.$$

If additionaly τ is bijective, then τ is an *isomorphism of difference universes.*

The *difference operator* associated with system Σ is the map $d_\Sigma : \mathcal{U}(\mathcal{M}, J(m)) \to \mathcal{U}(\mathcal{M}, J(m))$ defined by

$$(d_\Sigma \varphi)(x, U_r) = \varphi(f(x, u(r)), c_1 U_r) \tag{3.1}$$

where f is the dynamics of system Σ and $\varphi \in \mathcal{U}(\mathcal{M}, J(m))$. If $\varphi \in \mathcal{U}(\mathcal{M}, J(m))$ and $l \geq 2$, then $d^l_\Sigma \varphi := d_\Sigma(d^{l-1}_\Sigma \varphi)$. When $\varphi : \mathcal{M} \times J(m) \to \mathbb{R}^s$ we put

$$(d_\Sigma \varphi(x, U_r)) = ((d_\Sigma \varphi_1)(x, U_r), \dots, (d_\Sigma \varphi_s)(x, U_r)).$$

PROPOSITION 3.6. *The function universe $\mathcal{U}(\mathcal{M}, J(m))$ together with the operator d_Σ forms a difference universe.*

PROOF. We have to check that the map $d_\Sigma : \mathcal{U}(\mathcal{M}, J(m)) \to \mathcal{U}(\mathcal{M}, J(m))$ is a homomorphism. Let $\varphi_1, \dots, \varphi_k \in \mathcal{U}(\mathcal{M}, J(m))$ and $(x, U_r) \in \mathcal{M} \times J(m)$. Then

$$d_\Sigma(F \circ (\varphi_1, \dots, \varphi_k))(x, U_r) = (F \circ (\varphi_1, \dots, \varphi_k))(f(x, u(r)), c_1 U_r)$$

$$= F(\varphi_1(f(x, u(r)), c_1 U_r), \dots, \varphi_k(f(x, u(r)), c_1 U_r)$$

$$= F(d_\Sigma \varphi_1)(x, U_r), \dots, (d_\Sigma \varphi_k)(x, U_r)) = (F \circ (d_\Sigma \varphi_1, \dots, d_\Sigma \varphi_k))(x, U_r).$$

Hence d_Σ commutes with substitutions.

If $M \subset \mathcal{U}(\mathcal{M}, J(m))$ is a matching set and $\varphi, \psi \in M$, then $(d_\Sigma \varphi)(x, U_r)$ and $(d_\Sigma \psi)(x, U_r)$ are matching, so $d_\Sigma(M)$ is also a matching set. Moreover, if $(x, U_r) \in \mathcal{M} \times J(m)$, then for some $\varphi \in M$ one has

$$\underline{(d_\Sigma M)}(x, U_r) = \underline{\{d_\Sigma \varphi : \varphi \in M\}}(x, U_r) = (d_\Sigma \varphi)(x, U_r)$$

$$= \varphi(f(x,u(r)),c_1U_r) = (\underline{M})(f(x,u(r)),c_1U_r) = (d_\Sigma(\underline{M})(x,U_r).$$

Hence d_Σ commutes with amalgamation. The third condition of the definition is obvious. □

The function universe $(\mathcal{U}, d_\Sigma)$ will be called *the difference universe of the system* Σ and denoted by $\mathcal{U}_\Sigma$.

4. The main results

4.1. Observability. Let us consider a nonlinear control system Σ defined by (2.1)). Let $g : \mathcal{M}\times J(m)\to\mathcal{M}\times J(m)$ be defined by $g(x(0),U)= (f(x(0),u(0)),c_1U)$. If (X,U) is an inner trajectory of the system Σ, then for $k \geq 0$, $(x(k),U_k) = g^k(x(0),U)$, where $g^k = \underbrace{g\circ g\circ\ldots\circ g}_{k \text{ times}}$.

We say that points $x, z \in \mathcal{M}$ are *indistinguishable*, if for any nonnegative integer k and for any control U

$$\forall\, i = 1,\ldots,p \;\; (h_i\circ g^k)(x,U) = (h_i\circ g^k)(z,U).$$

The definitions of g and d_Σ imply that if points $x, z \in \mathcal{M}$ are indistinguishable, then for any control U one has

$$d^l_\Sigma h(x,U) = d^l_\Sigma h(z,U), \;\; l = 0,1,\ldots$$

If two points $x, z \in \mathcal{M}$ aren't indistinguishable, then they are *distinguishable.*

The system Σ is *observable* if any two distinct points $x, z \in \mathcal{M}$ are indistinguishable. The system Σ is *locally observable at* $x \in M$ if there exists a neighborhood W of x such that any point $z \in W$, $z \neq x$, is distinguishable from x. The system Σ is *locally observable*, if it is locally observable at each point of the manifold $\mathcal{M}$. Hence observability of system Σ implies local observability of this system.

Let $T : \mathcal{M}\times J(m) \to J(p,m)$ be a map defined by

$$T(x,U) = (h, d_\Sigma h, d^2_\Sigma h,\ldots, u, d_\Sigma u, d^2_\Sigma u,\ldots)^T(x,U)$$

where $(\ldots)^T$ means transposition of vector $(\ldots)$.

PROPOSITION 4.1. $d_\Sigma T = c_1 T$

PROOF. Definitions of map T and operator d_Σ imply that

$$(d_\Sigma T)(x(r),U_r) = T(f(x(r),u(r)),c_1U_r) = T(x(r+1),c_1U_r)$$

$$= (h(x(r+1),c_1U_r),(d_\Sigma h)(x(r+1),c_1U_r),\ldots,$$

$$u(x(r+1),c_1U_r),(d_\Sigma u)(x(r+1),c_1U_r),\ldots)$$

$$= (Y_{r+1},U_{r+1}) = c_1(Y_r,U_r) = (c_1T)(x(r),U_r)$$

□

Let $T_{|k} : \mathcal{M}\times J(m) \to (\mathbb{R}^p)^k \times (\mathbb{R}^m)^k$ be defined by

$$T_{|k}(x,U) = (h, d_\Sigma h,\ldots, d^{k-1}_\Sigma h, u, d_\Sigma u,\ldots, d^{k-1}_\Sigma u)^T(x,U).$$

For a fixed control U we also define the map $T_U : \mathcal{M} \to J(p)$:

$$T_U(x) := (h, d_\Sigma h, d^2_\Sigma h,\ldots)^T(x,U).$$

Let us take the first k components of map T_U and define

$$T_{U|k}(x) := (h, d_\Sigma h, \dots, d_\Sigma^{k-1} h)(x).$$

We say that the map T_U is an *immersion at point* $x \in \mathcal{M}$ if there exists $k \in \mathbb{N}$ such that the map $T_{U|k}$ is an immersion at this point. T_U is an *imbedding* if it is injective and it is an immersion at every $x \in \mathcal{M}$.

THEOREM 4.2. *If for some $U \in J(m)$ the map T_U is an immersion at $x \in \mathcal{M}$ then system Σ is locally observable at x.*

PROOF. Since T_U is an immersion at x, then for some k, in local coordinates around x, the rank of the matrix $\frac{\partial T_{U|k}}{\partial x}(x, U)$ is full, i.e. equal n. The Inverse Function Theorem implies that $T_{U|k}$ is locally injective, hence T_U is also injective.

Suppose that the system Σ isn't observable at the point $x \in \mathcal{M}$. Then in any neighborhood W of x there is point z, $z \neq x$, that is indistinguishable from x. So for any $U \in J(m)$ and $k = 0, 1, \dots$ one has

$$(h \circ g^k)(x, U) = (h \circ g^k)(z, U)$$

This implies that $d_\Sigma^k h(x, U) = d_\Sigma^k h(z, U)$ for $k = 0, 1, \dots$ Hence $T_U(x) = T_U(z)$, which contradicts local injectivity shown above. □

From now on we shall assume that the following conditions are satisfied:

A. For any $x \in \mathcal{M}$ and $U \in J(m)$ there exists $k \in \mathbb{N}$ such that the map $T_{|k}$ is proper on some neighborhood of point (x, U).

B. For any $U \in J(m)$, $T_U : \mathcal{M} \to J(p)$ is an imbedding.

Observe that condition **B** implies observability (and local observability) of Σ. In fact, it implies that any two states can be distinguish by every control U. Though this is much stronger than ordinary observability, this property is essential in proving the following theorem.

THEOREM 4.3. *If conditions* **A** *and* **B** *are fulfilled, then T is a left invertible map and its left inverse $S : \operatorname{im} T \to \mathcal{M} \times J(m)$ is of the form $S = (S_1, \mathrm{id}_{J(m)})$, where $S_1 : \operatorname{im} T \to \mathcal{M}$ is of class C^s (so locally it depends on finite number of variables).*

PROOF. T is left invertible from condition **B**.

Let $(\overline{x}_0, \overline{U}) \in \mathcal{M} \times J(m)$ be any fixed point. Let us consider the map $S = T^{-1}$ defined on $\operatorname{im} T$, $S = (S_1, \mathrm{id}_{J(m)})$, where $\mathrm{id}_{J(m)} : J(p, m) \to J(m)$ is the projection onto the space $J(m)$. Let us consider also the map $T_1 : \mathcal{M} \times J(m) \to J(p)$ and take k such that condition **B** is fulfilled for $\overline{x}_0$ and $\overline{U}$. Let $T_{1|k}(x_0, U) =: Y_{|k}$. Since $Y_{|k}$ depends only on x_0 and $U_{|k} = (u(0), \dots, u(k-1))$, then we have $T_{1|k} : \mathcal{M} \times (\mathbb{R}^m)^k \to (\mathbb{R}^p)^k$. Hence $T_{|k} : \mathcal{M} \times (\mathbb{R}^m)^k \to (\mathbb{R}^p)^k \times (\mathbb{R}^m)^k$, $T_{|k}(x_0, U_{|k}) = (Y_{|k}, U_{|k})$ and rank of this map is full (it equals $n + km$) in some neighborhood V of the point $(x_0, U_{|k})$. The Inverse Function Theorem and condition **A** imply that $x_0 = S_{1|k}(Y_{|k}, U_{|k})$, where $S_{1|k} : (\mathbb{R}^p)^k \times (\mathbb{R}^m)^k \supset \operatorname{im} T_{|k} \to \mathcal{M}$ and $S_{1|k}$ is of class C^s (i.e. has a C^s extension on a neighborhood of $\operatorname{im} T_{|k}$; see the proof of Theorem 4.6). Since the maps $S_{|k}$ and $T_{|k}$ are mutually inverse, so $S_{|k} \circ T_{|k} = \mathrm{id}_V$ and for any $(x_0, U_{|k}) \in V$ one has $(S_{1|k} \circ T_{|k})(x_0, U_{|k}) = x_0$. For any U there also holds $(S_1 \circ T)(x_0, U) = x_0$. So for any (x_0, U) such that $(x_0, U_{|k}) \in V$ one obtains

$$(S_{1|k} \circ T_{|k})(x_0, U) = (S_1 \circ T)(x_0, U)$$

This implies that $S_{1|k}(Y_{|k}, U_{|k}) = S_1(Y, U)$. □

Let $\mathcal{U}_\Sigma$ be a difference universe of system Σ. By $\mathcal{H}_\Sigma$ we will denote the difference subuniverse of $\mathcal{U}_\Sigma$ generated (as a function universe) by functions $h_i, d_\Sigma^k h_i, u_j^{(k)} \in \mathcal{U}(\mathcal{M}, J(m))$, where $i = 1, \dots, p$, $j = 1, \dots, m$, $k = 0, 1, \dots$ and $u_j^{(k)}$ is a coordinate function, $u_j^{(k)} : \mathcal{M} \times J(m) \to \mathbb{R}$, $u_j^{(k)}(x, U) = u_j(k)$. This subuniverse will be called the *difference output universe of system* Σ (or shortly: *output universe*). It is clear that $d_\Sigma \mathcal{H}_\Sigma \subset \mathcal{H}_\Sigma$.

THEOREM 4.4. *If Σ satisfies conditions* **A** *and* **B**, *then* $\mathcal{H}_\Sigma = \mathcal{U}_\Sigma$.

PROOF. The inclusion $\mathcal{H}_\Sigma \subset \mathcal{U}_\Sigma$ is obvious. The inclusion $\mathcal{U}_\Sigma \subset \mathcal{H}_\Sigma$ follows from Theorem 4.3 , because for any function $\varphi \in \mathcal{U}_\Sigma$ one has

$$\varphi = \varphi \circ \mathrm{id}_{\mathcal{M}\times J(m)} = \varphi \circ S \circ (h, d_\Sigma h, \dots, u, d_\Sigma u, \dots) = \Psi(h, d_\Sigma h, \dots, u, d_\Sigma u, \dots)$$

where $\Psi = \varphi \circ S$ is of class C^s and locally depends on finite number of variables. □

4.2. Dynamic equivalence. Let us consider two nonlinear control systems defined on n and $\tilde{n}$-dimensional manifolds of class C^s:

$$(4.1) \qquad \Sigma : \begin{array}{r} x(k+1) = f(x(k), u(k)) \\ y(k) = h(x(k), u(k)) \end{array} \quad \text{and} \quad \tilde{\Sigma} : \begin{array}{r} \tilde{x}(k+1) = \tilde{f}(\tilde{x}(k), \tilde{u}(k)) \\ \tilde{y}(k) = \tilde{h}(\tilde{x}(k), \tilde{u}(k)) \end{array}$$

where $u(k), \tilde{u}(k) \in \mathbb{R}^m$, $y(k) \in \mathbb{R}^p$, $\tilde{y}(k) \in \mathbb{R}^{\tilde{p}}$.

PROPOSITION 4.5. *Assume that systems Σ and $\tilde{\Sigma}$ satisfy conditions* **A** *and* **B**. *If Σ and $\tilde{\Sigma}$ are externally dynamically feedback equivalent and $\sigma : \tilde{M} \times J(m) \to \mathcal{M} \times J(m)$ is defined by $\sigma := S \circ \chi^e \circ \tilde{T}$, then the map $\sigma^* : \mathcal{U}_\Sigma \to \mathcal{U}_{\tilde{\Sigma}}$, $\sigma^*(\varphi) = \varphi \circ \sigma$ for $\varphi \in \mathcal{U}_\Sigma$, is an isomorphism of difference universes.*

PROOF. By Remark 3.3 σ^* is a homomorphism of function universes. Let $(\tilde{X}_r, \tilde{U}_r, \tilde{Y}_r)$ be a trajectory of system $\tilde{\Sigma}$. Let $(Y_r, U_r) = \chi^e(\tilde{Y}_r, \tilde{U}_r)$ and for $k \geq r$, $x(k) = S(Y_k, U_k)$, where $Y_k = c_{k-r} Y_r$, $U_k = c_{k-r} U_r$. If $\varphi \in \mathcal{U}_\Sigma$ then for $k \geq r$ one has

$$(d_{\tilde{\Sigma}} \circ \sigma^*)(\varphi)(\tilde{x}(k), \tilde{U}_k) = \sigma^*(\varphi)(\tilde{f}(\tilde{x}(k), \tilde{u}(k)), c_1\tilde{U}_k) = (\varphi \circ \sigma)(\tilde{x}(k+1), c_1\tilde{U}_k)$$

$$= \varphi(S \circ \chi^e \circ \tilde{T})(\tilde{x}(k+1), c_1\tilde{U}_k) = \varphi(x(k+1), c_1 U_k) = (d_\Sigma \varphi)(x(k), U_k)$$

$$= (d_\Sigma \varphi)(\sigma(\tilde{x}(k), \tilde{U}_k)) = \sigma^*(d_\Sigma \varphi)(\tilde{x}(k), \tilde{U}_k),$$

so σ^* commutes with difference operator d_Σ.

Dynamic (anticipating) equivalence of systems Σ and $\tilde{\Sigma}$ implies

$$\chi^e \tilde{\chi}^e_{|B^e(\tilde{\Sigma})} = \mathrm{id}_{B^e(\tilde{\Sigma})} \quad \text{and} \quad \tilde{\chi}^e \chi^e_{|B^e(\Sigma)} = \mathrm{id}_{B^e(\Sigma)},$$

so

$$\sigma \circ \tilde{\sigma} = S \circ \chi^e \circ \tilde{T} \circ \tilde{S} \circ \tilde{\chi}^e \circ T = \mathrm{id}_{\mathcal{M}\times J(m)}$$

$$\tilde{\sigma} \circ \sigma = \tilde{S} \circ \tilde{\chi}^e \circ T \circ S \circ \chi^e \circ \tilde{T} = \mathrm{id}_{\mathcal{M}\times J(m)}$$

and for $\varphi \in \mathcal{U}_{\tilde{\Sigma}}$,

$$(\sigma^* \circ \tilde{\sigma}^*)(\tilde{\varphi}) = (\tilde{\varphi} \circ \tilde{\sigma}) \circ \sigma = \tilde{\varphi} \circ (\tilde{\sigma} \circ \sigma) = \tilde{\varphi}.$$

Similarly $(\tilde{\sigma}^* \circ \sigma^*)(\varphi) = \varphi$. Hence the map σ^* is bijective. □

THEOREM 4.6. *Systems Σ and $\tilde{\Sigma}$ satisfying conditions* **A** *and* **B** *are externally dynamically feedback equivalent if and only if the difference output universes $\mathcal{H}_\Sigma$ and $\mathcal{H}_{\tilde{\Sigma}}$ are isomorphic.*

PROOF. "$\Rightarrow$" If systems Σ and $\tilde{\Sigma}$ are externally dynamically feedback equivalent then the map σ^* defined in Proposition 4.5 is an isomorphism of difference universes $\mathcal{U}_\Sigma$ and $\mathcal{U}_{\tilde{\Sigma}}$. Theorem 4.4 implies $\mathcal{U}_\Sigma = \mathcal{H}_\Sigma$ and $\mathcal{U}_{\tilde{\Sigma}} = \mathcal{H}_{\tilde{\Sigma}}$, hence $\tilde{\sigma}^*$ is a required isomorphism of difference output universes $\mathcal{H}_\Sigma$ i $\mathcal{H}_{\tilde{\Sigma}}$.

"$\Leftarrow$" Suppose that difference output universes $\mathcal{H}_\Sigma = \mathcal{U}_\Sigma$ and $\mathcal{H}_{\tilde{\Sigma}} = \mathcal{U}_{\tilde{\Sigma}}$ are isomorphic, and $\tau : \mathcal{H}_\Sigma \to \mathcal{H}_{\tilde{\Sigma}}$ is an isomorphism. So there exists a map $\sigma : \tilde{\mathcal{M}} \times J(m) \to \mathcal{M} \times J(m)$ such that $\tau = \sigma^*$. Let $(\tilde{x}, \tilde{U}) \in \tilde{\mathcal{M}} \times J(m)$. Let us consider the external trajectory $(\tilde{Y}, \tilde{U}) = \tilde{T}(\tilde{x}, \tilde{U})$ of system $\tilde{\Sigma}$. One has $(\tilde{Y}_k, \tilde{U}_k) := c_k(\tilde{Y}, \tilde{U})$ and $(\tilde{x}(k), \tilde{U}_k) = \tilde{S}(\tilde{Y}_k, \tilde{U}_k)$ for $k \geq 0$. The maps $\phi^e = (\phi^e_1, \dots, \phi^e_n)$, $\psi^e = (\psi^e_1, \dots, \psi^e_m)$ where

$$\begin{aligned} \phi^e_l &:= \tau h_l \circ \tilde{S}, \quad l = 1, \dots, p \\ \psi^e_j &:= \tau u_j \circ \tilde{S}, \quad j = 1, \dots, m \end{aligned} \tag{4.2}$$

are well defined on external trajectories of $\tilde{\Sigma}$. Since functions τh_l, τu_j and $\tilde{S}_1$ locally depend on finite number of variables, so do the functions ϕ^e_l and ψ^e_j. We are going to show that the pair $(Y, U) := (\Phi^e(\tilde{Y}, \tilde{U}), \Psi^e(\tilde{Y}, \tilde{U}))$, where Φ^e, Ψ^e are extensions of ϕ^e, ψ^e, is an external trajectory of system Σ. Let $(x(k), U_k) := \sigma(\tilde{x}(k), \tilde{U}_k)$ and $y(k) := h(x(k))$ for $k \geq 0$. To show that both definitions of U coincide we prove that

$$\sigma_2(\tilde{x}(0), \tilde{U}) = \Psi^e(\tilde{Y}, \tilde{U}), \tag{4.3}$$

where $\sigma = (\sigma_1, \sigma_2)$. Let $u^{(k)} : (x, U) \mapsto u(k)$, $k = 0, 1, \dots$, denote the function that chooses k-th component of the sequence U. Then equivalently the equation (4.3) can be written as

$$u^{(k)}(\sigma_2(\tilde{x}(0), \tilde{U})) = u^{(k)}(\Psi^e(\tilde{Y}, \tilde{U})), \ k \geq 0.$$

But one has

$$\begin{aligned} u^{(k)}(\sigma_2(\tilde{x}(0), \tilde{U})) &= \tau(u^{(k)})(\tilde{x}(0), \tilde{U}) = \tau(d^k_\Sigma u^{(0)}(\tilde{x}(0), \tilde{U}) \\ &= d^k_{\tilde{\Sigma}}(\tau u^{(0)})(\tilde{x}(0), \tilde{U}) = (\tau u^{(0)}(\tilde{x}(k), \tilde{U}_k) \end{aligned}$$

and

$$u^{(k)}(\Psi^e(\tilde{Y}, \tilde{U})) = (\tau u^{(0)} \circ \tilde{S} \circ c_k)(\tilde{Y}, \tilde{U}) = (\tau u^{(0)} \circ \tilde{S})(\tilde{Y}_k, \tilde{U}_k) = (\tau u^{(0)})(\tilde{x}(k), \tilde{U}_k).$$

Hence the equation (4.3) is true.

If $\varphi \in \mathcal{U}_\Sigma$ and $k \geq 0$ then

$$\begin{aligned} \varphi(x(k+1), U_{k+1}) &= \varphi(\sigma(\tilde{x}(k+1), \tilde{U}_{k+1})) = \sigma^*(\varphi)(\tilde{x}(k+1), \tilde{U}_{k+1}) \\ &= (\tau\varphi)(\tilde{x}(k+1), c_1\tilde{U}_k) = d_{\tilde{\Sigma}}(\tau\varphi)(\tilde{x}(k), \tilde{U}_k) = \tau(d_\Sigma\varphi)(\tilde{x}(k), \tilde{U}_k) \\ &= \sigma^*(d_\Sigma\varphi)(\tilde{x}(k), \tilde{U}_k) = (d_\Sigma\varphi)(\sigma(\tilde{x}(k), \tilde{U}_k)) = (d_\Sigma\varphi)(x(k), U_k) \\ &= \varphi(f(x(k), u(k)), c_1 U_k). \end{aligned}$$

Since functions from $\mathcal{U}_\Sigma$ distinguish points of $\mathcal{M} \times J(m)$, (X, U) is an inner trajectory of system Σ. Similarly

$$y_l(k) = \phi_l^e(\tilde{Y}_k, \tilde{U}_k) = \tau h_l(\tilde{S}(\tilde{Y}_k, \tilde{U}_k)) = (\tau h_l)(\tilde{x}(k), \tilde{U}_k)$$

$$= (h_l \circ \sigma)(\tilde{x}(k), \tilde{U}_k) = h_l(\sigma(\tilde{x}(k), \tilde{U}_k) = h_l(x(k), U_k) = h_l(x(k), u(k))$$

so (Y, U) is an external trajectory of system Σ. In the analogous way one can construct maps $\tilde{\phi}^e$, $\tilde{\psi}^e$ and show that if (Y, U) is an external trajectory of Σ then $(\tilde{Y}, \tilde{U})$ is an external trajectory of $\tilde{\Sigma}$.

Because $\chi^e = S \circ \sigma \circ \tilde{T}$, $\tilde{\chi}^e = \tilde{S} \circ \tilde{\sigma} \circ T$ and $\tau : \mathcal{H}_\Sigma \to \mathcal{H}_{\tilde{\Sigma}}$ is an isomorphism, so $\tilde{\sigma} \circ \sigma = \mathrm{id}_{\tilde{\mathcal{M}} \times J(m)}$ and $\sigma \circ \tilde{\sigma} = \mathrm{id}_{\mathcal{M} \times J(m)}$. Hence $\chi^e \circ \tilde{\chi}^e = \mathrm{id}$ and $\tilde{\chi}^e \circ \chi^e = \mathrm{id}$ on external trajectories, so maps $\chi^e = (\Phi^e, \Psi^e)$ and $\tilde{\chi}^e = (\tilde{\Phi}^e, \tilde{\Psi}^e)$ are mutually inverse on external behaviors of given systems.

Now let us fix an external trajectory $(\overline{Y}, \overline{U}) = T(\overline{x}_0, \overline{U})$ of system Σ. Conditions **A** and **B** imply existence of a positive integer k and neighborhoods $W \subset \mathcal{M} \times (\mathbb{R}^m)^k$ of $(\overline{x}_0, \overline{U}|_k)$ and $V \subset (\mathbb{R}^p)^k \times (\mathbb{R}^m)^k$ of $(\overline{Y}_{|k}, \overline{U}_{|k})$ such that the map $T_{|k} : W \to V$ is proper and is an immersion at each point of the set W. We can assume that $T_{|k}$ is injective. Hence $\mathcal{N} = T_{|k}(W)$ is a regular closed manifold in V [**Pl**]. We can assume that on manifold $\mathcal{N}$ maps ϕ^e, ψ^e are of order less or equal k. Hence Grauert's theorem (for C^ω) or Whitney's theorem (for C^∞) implies that ϕ^e, ψ^e can be extended onto V. Moreover, this neighborhood induces an open neighborhood of trajectory $(\overline{Y}, \overline{U})$ in $J(p, m)$ onto which ϕ^e and ψ^e are extended. □

Theorem 4.4 and Proposition 2.6 imply:

COROLLARY 4.7. *Systems Σ and $\tilde{\Sigma}$ satisfying conditions* **A** *and* **B** *are externally dynamically feedback equivalent if and only if the difference universes $\mathcal{U}_\Sigma$ and $\mathcal{U}_{\tilde{\Sigma}}$ are isomorphic.* □

REMARK 4.8. Observability of the systems is not essential in the characterization of the external dynamic feedback equivalence. In a future paper we are going to get rid of this assumption. Setting the systems on manifolds is not necessary either. The best choice would be a universe space, but it's not easy to transfer the rank condition to this case. This condition seems to be essential as it allows to express states as smooth (analytic) functions of outputs and controls. Nevertheless, some "regular" universe spaces might work here.

4.3. External dynamic linearization. As an example of applications of dynamic feedback equivalence of control systems we examine the linearization problem. Before studying it we shall introduce the concept of free difference universe. It follows free differential algebras used by Jakubczyk [**Ja1**].

A difference universe $(\mathcal{U}, d)$ (defined on a set X) is *free* if there exist a finite number of global functions $\omega_1, \ldots, \omega_r \in \mathcal{U}$ such that

1. for any function $\nu : J(r) \to A_0$ of class C^s, (locally depending on finite number of variables)

$$\nu \circ W = 0(\nu \circ W) \Rightarrow \nu = 0\nu,$$

   where $W = \{d^j w_i\}_{\substack{i=1,\ldots,r \\ j=0,1,\ldots}}$

2. for any function $\varphi \in \mathcal{U}$ there exists a (partially defined) function $\nu : J(r) \to A_0$ of class C^s (locally depending on finite number of variables) such that $\varphi = \nu \circ W$.

Functions $\omega_1, \ldots, \omega_r$ are *free generators* of difference universe $(\mathcal{U}, d)$.

LEMMA 4.9. *If $\varphi_1, \ldots, \varphi_n \in \mathcal{U}$ and $\psi_1, \ldots, \psi_m \in \mathcal{U}$ are free generators of $(\mathcal{U}, d_\Sigma)$ then $n = m$.*

PROOF. The proof is elementary but tedious so we skip it. The basic idea is to express the first generators by the others and get some identities which cannot hold if the numbers of generators are different. □

Let us recall that two linear systems

$$\Lambda : \begin{array}{r} x(k+1) = Ax(k) + Bu(k) \\ y(k) = Cx(k) + Du(k) \end{array} \quad \text{and} \quad \tilde{\Lambda} : \begin{array}{r} \tilde{x}(k+1) = \tilde{A}\tilde{x}(k) + \tilde{B}\tilde{u}(k) \\ \tilde{y}(k) = \tilde{C}\tilde{x}(k) + \tilde{D}\tilde{u}(k) \end{array}$$

(or shortly: systems (A, B, C, D) and $(\tilde{A}, \tilde{B}, \tilde{C}, \tilde{D})$) are equivalent with respect to a linear static state feedback transformations if there are matrices T, F, G - T, G invertible - such that trajectories $(\tilde{X}, \tilde{U})$ of $\tilde{\Lambda}$ are transformed onto trajectories (X, U) of Λ by transformations

$$x(k) = T\tilde{x}(k) \text{ and } u(k) = F\tilde{x}(k) + G\tilde{u}(k) \text{ for } k \geq 0$$

LEMMA 4.10. *If two observable linear systems (A, B, C, D) and $(\tilde{A}, \tilde{B}, \tilde{C}, \tilde{D})$ are equivalent with respect to a linear static state feedback transformations then they are equivalent with respect to a linear output anticipating transformations.*

PROOF. Let

$$K = \begin{pmatrix} C \\ CA \\ CA^2 \\ \vdots \end{pmatrix}.$$

We can compute the output sequence of system Λ corresponding to control U and the initial condition $x(0)$ as $Y = Kx(0) + s_1(K * BU) + DU$, where "$*$" means the operator of discrete convolution and s_1 denotes the one-unit right shift. K is an injective matrix (i.e. the operator associated with this matrix is injective). Take a matrix Z that is a left inverse of K and has only finitely many nonzero elements. Then $x(0) = ZY - ZDU - Zs_1(K * BU) = \gamma(Y, U)$ and γ depends on finitely many variables. Then $X = \Gamma(Y, U)$. Similarly for system $\tilde{\Sigma}$.

Let us consider linear static state feedback transformations

$$x = T\tilde{x}, \; u = F\tilde{x} + G\tilde{u} \tag{4.4}$$

that give the equivalence of the systems. Then the corresponding output anticipating feedback transformations are

$$y(k) = CT\tilde{\Gamma}(\tilde{Y}_k, \tilde{U}_k) + DF\tilde{\Gamma}(\tilde{Y}_k, \tilde{U}_k) + DG\tilde{u}(k)$$

and

$$u(k) = F\tilde{\Gamma}(\tilde{Y}_k, \tilde{U}_k) + G\tilde{u}(k).$$

Similarly we get the transformations in the other direction. □

LEMMA 4.11. *The difference output universe of a minimal linear system Λ is free. The number of free generators of this universe is equal to the number of controls.*

PROOF. Choosing appropriate static feedback transformations (4.4), we obtain the Brunovsky canonical form of the dynamics [**Br**]:

$$\overline{\Lambda}: \begin{aligned} \tilde{x}_{i,j}(k+1) &= \tilde{x}_{i,j+1}(k) \\ \tilde{x}_{i,\kappa_i}(k+1) &= \tilde{u}_i(k) \end{aligned} \tag{4.5}$$

where $j = 1, \dots, \kappa_{i-1}$, $i = 1, \dots, m$.

Moreover for the system $\overline{\Lambda}$ one has $d_\Sigma x_{i1} = x_{i2}, \dots, d_\Sigma^{\kappa_i} x_{i1} = u_i^{(0)}, d_\Sigma^{\kappa_i+1} x_{i1} = u_i^{(1)}, \dots$, hence any functions x_{ij} and $u_i^{(j)}$ can be represented as $\nu(W)$, where $\nu : J(p) \to \mathbb{R}$ and $W = \{x_{11}, \dots, x_{m1}, d_\Sigma x_{11}, \dots, d_\Sigma x_{m1}, \dots\}$. Let $x_i = \nu_i(W)$ and $u_k^{(j)} = g_{kj}(W)$. This means that:

1. any function $\varphi \in \mathcal{U}_{\bar{\Lambda}}$ can be expressed as

$$\begin{aligned} \varphi &= \varphi \circ (x_1, \dots, x_n, u_1^{(0)}, \dots, u_m^{(0)}, u_1^{(1)}, \dots) \\ &= \varphi \circ (\nu_1(W), \dots, \nu_n(W), g_{10}(W), \dots) = \nu \circ W \end{aligned}$$

2. the condition $\nu \circ W = 0$ means

$$\begin{aligned} &\nu \circ (x_{11}, x_{21}, \dots, x_{m1}, x_{12}, \dots, \\ &x_{m2}, \dots, u_1^{(0)}, \dots, u_m^{(0)}, \dots) \equiv 0 \end{aligned}$$

hence $\nu \equiv 0$ on its domain.

So the difference output universe $\mathcal{H}_{\overline{\Lambda}} = \mathcal{U}_{\overline{\Lambda}}$ is free and its generators are functions x_{ij}. Hence the output universe $\mathcal{H}_\Lambda$ is also free and number of its generators is equal to the number of controls of system Λ. □

A nonlinear system

$$\Sigma: \begin{aligned} x(k+1) &= f(x(k), u(k)) \\ y(k) &= h(x(k)) \end{aligned}$$

where $k \in \mathbb{Z}$, $x(k) \in \mathcal{M}$, $u(k) \in \mathbb{R}^m$, $y(k) \in \mathbb{R}^p$, is *externally dynamically feedback linearizable* if it is externally dynamically feedback equivalent to a minimal linear control system

$$\tilde{\Lambda}: \begin{aligned} \tilde{x}(k+1) &= A\tilde{x}(k) + B\tilde{u}(k) \\ \tilde{y}(k) &= C\tilde{x}(k) \end{aligned}$$

where $\tilde{x}(k) \in \mathbb{R}^{\tilde{n}}$, $\tilde{u}(k) \in \mathbb{R}^m$, $\tilde{y}(k) \in \mathbb{R}^{\tilde{p}}$.

PROPOSITION 4.12. *Assume that system Σ satisfies conditions* **A** *and* **B**. *If $\mathcal{H}_\Sigma$ is free and the number of its free generators is equal to the number of controls then Σ is externally dynamically feedback linearizable.*

PROOF. Assume that $\mathcal{H}_\Sigma$ has one free generator φ. Then Σ is equivalent to the system $\tilde{\Lambda}$: $\tilde{x}(k+1) = \tilde{u}(k)$, $\tilde{y}(k) = \tilde{x}(k)$. The isomorphism τ between $\mathcal{H}_\Sigma$ and $\mathcal{H}_{\tilde{\Lambda}}$ is given by $\tau(\varphi) = \tilde{x}$ and $\tau(d_\Sigma^k \varphi) = u^{(k-1)}$ for $k > 0$. If there are more generators, we repeat this procedure for every generator φ_i, each time choosing a new variable x_i and a new control u_i. □

Collecting the results of this section we get the following characterization:

THEOREM 4.13. *If an observable system Σ satisfies conditions* **A** *and* **B** *then it is externally dynamically feedback linearizable if and only if its difference output universe $\mathcal{H}_\Sigma$ is free and the number of its free generators is equal to the number of controls.* □

References

[AKM] E. Aranda-Bricaire, U. Kotta, C.H. Moog, *Accessibility and feedback linearization of discrete-time systems*, Proceedings of the 33rd Conference on Decision and Control, Lake-Buena Vista, December 1994

[BJ] Z. Bartosiewicz, J. Johnson, *Systems on universe spaces*, Acta Applicandae Mathematicae 34 (1994), 233–266

[BJP] Z. Bartosiewicz, B. Jakubczyk, E. Pawłuszewicz, *Dynamic feedback equivalence of nonlinear discrete-time systems*, in: Proc. First Internat. Symp. on Mathematical Models in Automation and Robotics, Sept. 1–3, 1994, Międzyzdroje, Poland, Tech. Univ. of Szczecin Press, pp. 37–40

[Br] P. Brunovský, *A classification of linear controllable systems*, Kybernetika 3 (1970)

[BL] C. I. Byrnes, W. Lin, *Losslessness, feedback equivalence and global stabilization of discrete-time nonlinear systems*, IEEE Trans. Aut. Contr. 39 (1984), 83–98

[F1] M. Fliess, *Esquisses pour une theorie des systems non lineaires en temps discret*, in: Rediconti del Seminario Matematico, Universitàe Politecnico Torino, Fasciolo speciale 1987

[F2] ———, *Automatique en temps discret at algèbre aux différences*, Forum Mat. 2 (1990), 213–232

[G] J. W. Grizzle, *Feedback linearization of discrete-time systems*, in: A. Bensoussan, J. L. Lions, Eds., Analysis and Optimization of Systems, Springer-Verlag, 1986

[Ja] B. Jakubczyk, *Feedback linearization of discrete-time systems*, Syst. Contr. Letts. 9 (1987), 411–416

[Ja1] ———, *Remarks on equivalence and linearization of nonlinear systems*, in: Proc. Nonlinear Control Systems Design Symposium IFAC, Bordeaux, France, 1992

[Ja2] ———, *Dynamic feedback equivalence of nonlinear control systems*, preprint

[JC] G. Jayaraman, H. J. Chizeck, *Feedback linearization of discrete-time systems*, in: Proc. of 32nd IEEE CDC, San Antonio, 1993, 2972–2977

[Jh] J. Johnson, *A generalized global differential calculus I*, Cahiers Top. et Geom. Diff. XXVII (1986), 25–83

[LM] H. G. Lee, S. I. Marcus, *On input-output linearization of discrete-time nonlinear systems*, Systems & Control Letters 8 (1987)

[LB] W. Lin, C. I. Byrnes, *Design of discrete-time nonlinear control systems via smooth feedback*, IEEE Trans. Aut. Contr. 39 (1994), 2340–2346

[MMR] Ph. Martin, R. M. Murray, P. Rouchon, *Flat systems*, Plenary Lectures and Mini-Courses, European Control Conference ECC'97, G. Bastin and M. Gevers, Eds., Brussels, 1997

[Pl] R. S. Palais, *Real Algebraic Differential Topology, Part I*; Mathematics Lecture Series 10, Publish or Perish, Inc., Wilmington, 1980

[PB] E. Pawłuszewicz, Z. Bartosiewicz, *Dynamic feedback equivalence of nonlinear input-output discrete-time control systems*, Invited Lectures and Short Communications Delivered at the Sixth International Colloquim on Differential Equations, August 18–23, 1995, Plovdiv, Bulgaria, 1996

[P1] E. Pawłuszewicz, *On difference algebras and difference universes*, Zeszyty Naukowe Politechniki Białostockiej. Matematyka-Fizyka-Chemia (Białystok) 16 (1996)

[P2] ———, *Dynamic output linearization of nonlinear discrete-time control systems*, UKACC International Conference on CONTROL'96, Exeter, UK, 2–5 Sep.1996

[SNT] C. Simoes, H. Nijmeijer, J. Tsinias, *Nonsmooth stabilizability and feedback linearization of discrete-time nonlinear systems*, Internat. J. Robust Nonlinear Control (6) 1996, 171–188

Politechnika Białostocka, Wiejska 45, Białystok, Poland
E-mail address: `epaw@cksr.ac.bialystok.pl`

Politechnika Białostocka, Wiejska 45, Białystok, Poland
E-mail address: `bartos@cksr.ac.bialystok.pl`

Proceedings of Symposia in Pure Mathematics
Volume **64**, 1999

Good Needle-like Variations

R.M. Bianchini

ABSTRACT. We select some needle-like variations of a reference trajectory with the property that the variations at the same time can be continuously summed. This property implies that they can be used both for stating an high order Maximum Principle for the Mayer control problem with constraints on the end-points, than for proving the local controllability property of the trajectory.

1. Introduction

The variational approach to the optimization control problems consists in approximating the reachable set of a control system near the final point $x^*(t_1)$ of a trajectory x^* by means of the tangent vectors to the reachable set at $x^*(t_1)$. However the set of tangent vectors may be too large to provide sufficient information on the reachable set when one deals with one of the following problems: Mayer control problems with constraints on the final point, local controllability problems, time optimal problems [**1**, Ex. 3.1]. In these cases one has to look for subsets of the set of tangent vectors with suitable properties. In [**1**] it is proved that local regular tangent cones suit this purpose.

Local regular tangent cones to the reachable set have been constructed by means of the variations of the reference control, for example the set of Pontriagyn-McShane variations [**2**] is a local regular tangent cone. The trajectory variations are tangent vectors to a curve at $x^*(t_1)$ obtained by slightly modifying the reference control. Not any subset of trajectory variations is a regular tangent cone, see counterexample in [**3**], and one has to look for further conditions to select among the variations the "good" ones. Almost all the known local regular tangent cones to the reachable set are sets of needle-like variations, that is the variation of the reference control takes place on a small interval. However H. Sussmann has found a counterexample to the conjecture that any needle-like variation is a good one.

The aim of this paper is to select particular needle-like variation such that the set of tangent vectors generated by these variations is a local regular tangent cone to the reachable set at the final point of the trajectory. By means of these variations we define a cone of tangent vectors, the variational cone of the reference trajectory, which gives rise to a high order Maximum Principle which generalizes the known ones. The trajectory variations we define are simpler and more general than the

---

1991 *Mathematics Subject Classification.* Primary 49B10, 49B15.
This research was supported by a grant from Ministero dell'Universitá e della Ricerca Scientifica e Tecnologica.

ones introduced in [2], [4], [5], [7]. Notice that it is difficult to compare these variations and therefore it is interesting in itself to find their common properties.

The following notations will be used throughout the paper.
M denotes a differentiable finite dimensional paracompact C^2 manifold.
Given $x \in M$, we will denote by $o(\epsilon)$ and by $O(\epsilon)$ the maps defined on an Euclidian space with values in T_xM such that $\lim_{\epsilon \to 0} \frac{o(\epsilon)}{\|\epsilon\|} = 0$ and $\lim_{\epsilon \to 0} O(\epsilon) = 0$ respectively.
For $v \in T_xM$, we will write

$$y = x + \epsilon\, v + o(\epsilon)$$

if the equality holds in a chart at x and hence in any chart.
The symbols $int\, K$, $cl\, K$, ∂K denote respectively the interior, the closure and the boundary of a subset K of a topological space X. In the case where X is a vector space, $co\, K$ denotes the convex hull of K. A subset A of a vector space is a cone if $\alpha \geq 0$ and $y \in A$ implies $\alpha\, y \in A$.
Given a convex subset K of T_xM, we denote by $int_{rel}K$ the relative interior of K. B stands for the unit closed ball of T_xM, $D_ig(y_1, \ldots, y_r)$ for the partial derivative with respect to the i-th variable of a map g at the point $(y_1, \ldots, y_r)$.

We end this introduction by recalling the definition of local regular tangent cone to a set Y at a point $x \in Y$.

Definition 1.1. A cone K is a *regular tangent cone* [8] to $Y \subset M$ at x if for each $\epsilon \in [0, \bar{\epsilon}]$ there exists a continuous map $r_\epsilon : co\, K \cap B \to T_xM$ such that $lim_{\epsilon \to 0^+} \|r_\epsilon(y)\|/\epsilon = 0$ uniformly with respect to y and

$$x + \epsilon y + r_\epsilon(y) \in Y \quad , \quad \forall y \in co\, K \cap B.$$

Definition 1.2. A cone of tangent vectors K is said a *local regular tangent cone* to Y at x if for each vector $v \in int_{rel} co\, K$ there exists a convex cone, W, containing v in its interior, such that $W \cap K$ is a regular tangent cone to Y at x.

2. Variations

Let us consider a control process on M

$$\dot{x} = f(t,\, x,\, u) \tag{1}$$

where $f : J \times M \times U \to TM$ and $U \subset \mathbb{R}^m$. We assume that:

A): the control maps $u(\cdot)$ belong to a given set $\mathcal{U}$, the set of admissible controls; each control map $u(\cdot) \in \mathcal{U}$ is defined on a compact interval, I_u, which may depend on u;

B): J is an open interval of $\mathbb{R}$ and f is such that for all $u(\cdot) \in \mathcal{U}$ the time-dependent vector field $(t, x) \mapsto f(t, x, u(t))$ is a C^1 Caratheodory vector field, i.e.

1. $\forall x \in M$, $t \mapsto f(t, x, u(t))$ is measurable
2. *a.e.* $t \in J$, $x \mapsto f(t, x, u(t))$ is C^1
3. in any chart $\Big(\varphi,\, V\Big)$ for each compact set $K \subset \varphi(V)$ there exists an integrable map ψ such that

$$\|f(t, x, u(t))\| + \|D_2 f(t, x, u(t))\| \leq \psi(t)$$

a.e. $t \in [t_0, t_1]$, $\forall x \in K$.

C): if $u_1 : [a,b] \to U$ and $u_2 : [b,c] \to U$ are admissible controls, then the map $u : [a,c] \to U$ defined by

$$u(t) = \begin{cases} u_1(t) & a < t < b \\ u_2(t) & b < t < c \end{cases}$$

belongs to $\mathcal{U}$

D): any restriction to a compact subinterval of an admissible control belongs to $\mathcal{U}$.

Hypothesis B), see **[9]**, implies that for each $u \in \mathcal{U}$ and each $(t_0, x_0) \in I_u \times M$ there exists an unique maximal solution of (1), $S(\cdot, t_0, x_0, u)$, such that $S(t_0, t_0, x_0, u) = x_0$. Moreover the map $x \mapsto S(t, t_0, x, u)$ is C^1 and its derivative is the solution of the linear differential equation

$$\begin{cases} \dot{y} = D_2 f(t, S(t, t_0, x, u), u(t))\, y \\ y(t_0) = Id. \end{cases} \tag{2}$$

Let $X \subset M$; for all $(t_0, t) \in J \times J$ we will denote by $R(t, t_0, X)$ the set of points reachable at time t by means of the solutions of (1) which at time t_0 belong to X, that is

$$R(t, t_0, X) = \{x(t)\ ,\ x(\cdot) \quad \text{solution of (1) such that} \quad x(t_0) \in X\}.$$

Notice that $R(t, t_0, X)$ may be the empty set for some t.

Let $x^* : [t_0, t_1] \to M$ be a trajectory of (1) relative to the control u^*. Let us start by recalling the definition of needle-like variation of the pair (x^*, u^*). Roughly speaking a vector $v \in T_{x^*(\tau)}M$ is a needle-like variation at τ if for each ϵ sufficiently small it is possible to change the reference control u^* in a small interval near τ in such a way that if u_ϵ denotes the perturbed control then the two curves $\epsilon \mapsto S(t_1, t_0, x_0, u_\epsilon)$ and $\epsilon \mapsto S(t_1, \tau, x^*(\tau) + \epsilon v, u^*)$ have the same tangent vector at $\epsilon = 0$.

For sake of simplicity we introduce the following notations: let u be an admissible control defined on the interval $[a,b]$ and let $\tau \in [t_0, t_1]$; we set

$$S^*(x, \tau, u) = S(\tau, b, S(b, a, x, u), u^*)$$
$$S^*(\tau, u) = S^*(x^*(a), \tau, u)$$

when they are defined. Notice that if $a = \tau$, then $S^*(\tau, u)$ is the value at time of the solution relative to the control u of the pull back system associated to u^*, **[7]**.

DEFINITION 2.1. A vector $v \in T_{x^*(\tau)}M$ is an *needle-like variation* of (x^*, u^*) at $\tau \in [t_0, t_1]$ if there exist a positive number $\bar{\epsilon}$ and a family of control maps, $\{\nu(\epsilon)\, ,\, \epsilon \in [0, \bar{\epsilon}]\,\} \subset \mathcal{U}$ with the following properties:

1. $\nu(\epsilon)$ is defined on the interval $[a(\epsilon), b(\epsilon)] \subset [t_0, t_1]$ and $\lim_{\epsilon\to 0} a(\epsilon) = \lim_{\epsilon \to 0} b(\epsilon) = \tau$
2. there exists a neighborhood V of $x^*(\tau)$ such that $(x, \epsilon) \mapsto S^*(x, \tau, u(\epsilon))$ is continuous on $V \times [0, \bar{\epsilon}]$
3. $\epsilon \mapsto D_1 S^*\Big(x^*(a(\epsilon)), \tau, u(\epsilon)\Big)$ is continuous
4. $S^*(\tau, u(\epsilon)) = x^*(\tau) + \epsilon v + o(\epsilon)$.

We say that v has order k if both $\big(b(\epsilon) - \tau\big)$ and $\big(\tau - a(\epsilon)\big)$ goes to 0 as ϵ^k.

REMARK 2.2. Without loss of generality we can always suppose that the number $\bar{\epsilon}$ in the previous definition is equal to 1.

Let $G(t,s)$ denote the transport along the reference flow from time s up to time t, i.e. $G(t,s) = D_3S(t,s,x^*(s),u^*)$, and let v be a needle-like variation at τ. By the definition of needle-like variation it follows that $G(t_1,\tau)v$ is a tangent vector to $R(t_1,t_0,x_0)$ at $x^*(t_1)$. This type of tangent vectors will be named *tangent vectors generated at* τ. It is known that any collection of tangent vectors generated at times one different from the others is a derived cone **[14]** and hence a regular tangent cone, but the converse is not true. We want to select particular needle like variations in such a way that any collection of tangent vectors generated by them is a regular tangent cone. This goal will be achieved by suitable hypothesis on the interval on which the perturbation takes place. One of these is that the point τ has to be external to the interval $[a(\epsilon),b(\epsilon)]$; therefore we will consider two types of variations, the left and the right ones, depending whether the perturbation takes place before or after τ.

DEFINITION 2.3. A vector $v \in T_{x^*(\tau)}M$ is a *left variation (right variation) of order* k of (x^*,u^*) at $\tau \in [t_0,t_1]$ if there exist positive numbers $\bar{c}$, $\bar{\epsilon}$ and for each $\epsilon \in [0,\bar{\epsilon}]$ a family of control maps, $\{u_\epsilon(c)\,,\, c \in [0,\bar{c}]\ \} \subset \mathcal{U}$ with the following properties:

1. $u_\epsilon(c)$ is defined on the interval $[\tau - a\epsilon^k, \tau - \epsilon^k]$ ($[\tau + \epsilon^k, \tau + a\epsilon^k]$)
2. there exists a neighborhood V of $x^*(\tau)$ such that $(x,c) \mapsto S^*(x,\tau,u_\epsilon(c))$ is continuous on $V \times [0,\bar{c}]$
3. $D_1S^*(x^*(\tau - a\epsilon^k),\tau,u_\epsilon(c))$ is continuous in c and goes to the identity as ϵ goes to 0
4. $S^*(\tau,u_\epsilon(c)) = x^*(\tau) + \epsilon c\, v + o(\epsilon)$ uniformly w.r.t. c.

Both left than right variations will be named *g-variations*. By definition the g-variations are needle-like variations. The following Proposition shows that if a needle-like variation can be obtained on an interval, then it is a g-variation.

We recall that τ is a Lebesgue point of a map $g \in L^1([t_0,t_1])$ if

$$\int_{\tau-\epsilon}^{\tau} g(s)\,ds = \epsilon\, g(\tau) + o(\epsilon) \quad \text{and} \quad \int_{\tau}^{\tau+\epsilon} g(s)\,ds = \epsilon\, g(\tau) + o(\epsilon)$$

It is known that the set of Lebesgue points has full measure; for further properties of Lebesgue points we refer to **[10]**.

PROPOSITION 2.4. Let I be an interval contained in $[t_0,t_1]$ and let $g \in L^1(I)$ be such that $g(t)$ is a needle-like variation at t for each t in the set L of Lebesgue points of g. For each $t \in L$ let $[a_t(\epsilon),b_t(\epsilon)]$ be the interval as in Definition 2.1 relatively to the needle-like variation $g(t)$. If there exists positive numbers N and s such that for each $\tau \in L$

$$[a_\tau(\epsilon),b_\tau(\epsilon)] \subset [\tau - (N\,\epsilon)^s, \tau + (N\,\epsilon)^s] \tag{3}$$

then for each $t \in L$ $g(t)$ is both a right than a left variation at t of order s.

PROOF. We will only prove that $g(t)$ is a right variation; in an analogous way it can be proved that it is a left variation.

Choose $0 < \epsilon < 1$ so small that $[t,\, t + 4\epsilon^s] \subset I$. Since $t \in L$, it is possible to choose $t' \in L$ such that $t + 2(\epsilon)^s < t' < t + 3(\epsilon)^s$ and $\lim_{\epsilon\to 0} g(t') = g(t)$. Since $g(t')$ is a needle-like variation at t', for each $\sigma \in [0,\overline{\sigma}]$ there exists a control map u_σ defined on the interval $[a_{t'}(\sigma),\, b_{t'}(\sigma)]$ such that

$$S^*(t',u_\sigma) = x^*(t') + \sigma g(t') + o(\sigma). \tag{4}$$

By (3) it follows that if $\sigma < \epsilon/N$ then the control u_σ is defined on a subset of the interval $[t+\epsilon^s,\, t+4\,\epsilon^s]$. Let $0 \le c \le N^{-1}$ and let $\nu_\epsilon(c)$ be the control map defined on the interval $[\tau + (\epsilon)^s,\, \tau + 4(\epsilon)^s]$ which coincides with $u_{t'}(\epsilon c)$ on the interval in which it is defined and coincides with u^* otherwise. It is easy to verify that $\nu_\epsilon(c)$ verifies the hypothesis of definition 2.3. Moreover

$$S^*(t, \nu_\epsilon(c)) = S\Big(t, t', S^*(t', u_{t'}(\epsilon c)), u^*\Big) =$$
$$x^*(t) + \epsilon c G(t, t')\, g(t') + f(\epsilon c) = x^*(t) + \epsilon\, c\, g(t) + h(\epsilon,\, c)$$

Since t' does not depend on c and f is continuous, $h(\epsilon,\, c)$ is continuous w.r.t. c . Moreover $f(\sigma) = o(\sigma)$ and $G(t, t')\, g(t') - g(t)$ goes to 0 as ϵ goes to 0, therefore $h(\epsilon, c) = o(\epsilon)$ uniformly in c. The proposition is proved. □

We want to prove that the g-variations at the same time τ can be "summed" in a continuous way. Let us set the following definition:

DEFINITION 2.5. An *r-parametric control variation of u^* at τ* is a family of controls maps $\{\nu(\epsilon, \mathbf{c})\,,\, (\epsilon, \mathbf{c}) \in [0, \bar{\epsilon}] \times [0, \bar{c}]^r\} \subset \mathcal{U}$ with the following properties:

1. $\nu(\epsilon, \mathbf{c})$ is defined on the interval $[a(\epsilon), b(\epsilon)] \subset [t_0, t_1]$ and $\lim_{\epsilon\to 0} a(\epsilon) = \lim_{\epsilon\to 0} b(\epsilon) = \tau$ uniformly in $\mathbf{c}$
2. there exists a neighborhood V of $x^*(\tau)$ such that $(x, \mathbf{c}) \mapsto S^*(x, \tau, \nu(\epsilon, \mathbf{c}))$ is continuous on $V \times [0, \bar{c}]^r$
3. $D_1 S^*\Big(x^*(a(\epsilon)), \tau, \nu(\epsilon, \mathbf{c})\Big)$ is continuous in $\mathbf{c}$ and goes to the identity as ϵ goes to 0 uniformly in $\mathbf{c}$

Let $u_i(\epsilon, \overline{c_i})\,,\, i = 1, 2$ be r_i- parametric control variations concentrated at τ and let u_i be defined on the interval $[a_i(\epsilon), b(\epsilon_i)]$; if $b_1(\epsilon) \le a_2(\epsilon)$ for each ϵ, then it is possible to define a $(r_1 + r_2)$- parametric control variation at τ, $u_1\#u_2$ on the interval $[a_1(\epsilon), b_2(\epsilon)]$, in the following way:

$$u_1\#u_2\Big(\epsilon, (\mathbf{c_1}, \mathbf{c_2})\Big)(t) = \begin{cases} u_1(\epsilon, \mathbf{c_1})(t)\ ,\ t \in [a_1(\epsilon), b_1(\epsilon)) \\ u^*(t)\ ,\ t \in [b_1(\epsilon), a_2(\epsilon)] \\ u_2(\epsilon, \mathbf{c_2})(t)\ ,\ t \in (a_2(\epsilon), b_2(\epsilon)] \end{cases}$$

the control map $u_1\#u_2$ will be named concatenation of u_1 and u_2 . The concatenation of two control variations has the following property:

LEMMA 2.6. *Let $u_i(\epsilon, \mathbf{c_i})\,,\, i = 1, 2$ be two parametric control variations which can be concatenated. If*

$$S^*(\tau, u_i(\epsilon, \mathbf{c_i})) = x^*(\tau) + \epsilon\varphi_i(\mathbf{c_i}) + o(\epsilon)$$

uniformly in $\mathbf{c_i}$, then

$$S^*\Big(\tau, u_1\#u_2(\epsilon, (\mathbf{c_1}, \mathbf{c_2}))\Big) = x^*(\tau) + \epsilon\Big(\varphi_1(\mathbf{c_1}) + \varphi(\mathbf{c_2})\Big) + o(\epsilon) \tag{5}$$

uniformly in $(\mathbf{c})$.

PROOF. For sake of simplicity we will omit the parameters ϵ and $\mathbf{c_i}$ when no confusion may occurs.

$$S^*(\tau,\, u_1\# u_2) = S^*\Big(S\big(a_2,\tau,S^*(\tau,u_1),u^*\big),\tau,u_2\Big) =$$

$$S^*(\tau,u_2) + D_1S^*(x^*(a_2),\tau,u_2)\Big(S\big(a_2,\tau,S^*(\tau,u_1),u^*\big) - x^*(a_2)\Big) + o(\epsilon) =$$

$$S^*(\tau,u_2) + \big(Id + O(\epsilon)\big)\big(\epsilon\, G(a_2,\tau)\,\varphi(\mathbf{c_1}) + o(\epsilon)\big) + o(\epsilon) =$$

$$x^*(\tau) + \epsilon\big(\varphi(\mathbf{c_1}) + \varphi(\mathbf{c_2})\big) + o(\epsilon).$$

□

PROPOSITION 2.7. Let $v_1,\dots,v_p$ be left (right) variations of order k at τ. There exists $b>1$ and a p-parametric control variation $\nu(\epsilon,\mathbf{c})$ defined on the interval $[\tau - b\epsilon^{1/k}, \tau - \epsilon^{1/k}]$ ($[\tau + \epsilon^{1/k}, \tau + b\epsilon^{1/k}]$) such that

$$S^*(\tau,\nu(\epsilon,\mathbf{c})) = x^*(t_1) + \epsilon\sum_{i=1}^{p} c_i v_i + o(\epsilon)$$

uniformly in $\mathbf{c}$.

PROOF. We will provide the proof only for right variations; for left variations the proof is analogous. By variation definition, the Proposition is true for $p=1$. Let the property hold for $p=s$ and let $v_1,\dots,v_{s+1}$ be right variations of order k at τ. By induction hypothesis there exist $b'>1$, an s-parametric control variation $\nu_s(\epsilon,\mathbf{c})$ defined on the interval $[\tau+\epsilon^{1/k}, \tau+b'\epsilon^{1/k}]$ and such that

$$S^*(\tau,\nu_s(\epsilon,\mathbf{c})) = x^*(\tau) + \epsilon\sum_{i=1}^{s} c_i v_i + o(\epsilon).$$

Since v_{s+1} is an right variation of order k, there exists $a'>1$, a control variation $\mu(\epsilon,c)$ defined on the interval $[\tau+\epsilon^{1/k}, \tau+a'\epsilon^{1/k}]$ such that

$$S^*(\tau,\mu(\epsilon,c)) = x^*(t_1) + \epsilon\, c v_{s+1} + o(\epsilon).$$

The control variation $\mu\Big((b')^k\,\epsilon,\; c_{s+1}/(b')^k\Big)$ is defined on $[\tau+b'\epsilon^{1/k}, \tau+a'b'\epsilon^{1/k}]$, therefore ν_s and μ can be concatenated. Set $\nu_s\#\mu = \nu_{s+1}$. $\nu_{s+1}(\epsilon,\mathbf{c})$ is an (s+1)-parametric control variation defined on the interval $[\tau+\epsilon^{1/k}, \tau+a'b'\epsilon^{1/k}]$. By lemma 2.6

$$S^*\big(\tau,\nu_{s+1}(\epsilon,\mathbf{c})\big) = x^*(\tau) + \epsilon\sum_{i=1}^{s} c_i v_i + \epsilon\, c_{s+1}\, v_{s+1} + o(\epsilon).$$

By induction arguments the Proposition is proved. □

PROPOSITION 2.8. Let $\{v_1,\dots,v_r\}$ be g-variations at τ. There exists an r-parametric control variation $\mu(\epsilon,\mathbf{c})$ at τ such that

$$S(\tau,\mu(\epsilon,\mathbf{c})) = x^*(\tau) + \epsilon\sum_{i=1}^{r} c_i\, v_i + o(\epsilon) \qquad (6)$$

uniformly in $\mathbf{c}$.

PROOF. Let $\{I_j\}$ be a partition of the set $\{1,\dots,r\}$ such that:
if $i \in I_j$, $j = 1,\dots,s$, then v_i is a left variation of order k_j and $k_1 < k_2 < \dots < k_s$; if $i \in I_j$, $j = s+1,\dots,p$ then v_i is a right variation of order k_j and $k_{s+1} > k_{s+2} > \dots > k_p$.

Let $\nu_j(\epsilon,\mathbf{c})$ be the control variations of Proposition 2.7 relative to the g-variations of order k_j, $\{v_i, i \in I_j\}$. If $j \le s$, then $\nu_j(\epsilon,\mathbf{c})$ is defined on the interval $[\tau - b_j\epsilon^{k_j}, \tau - \epsilon^{k_j}]$ and if $j > s$, then $\nu_j(\epsilon,\mathbf{c})$ is defined on the interval $[\tau + \epsilon^{k_j}, \tau + b_j\epsilon^{k_j}]$. Therefore if ϵ is sufficiently small, then $\nu_1(\epsilon,\mathbf{c}), \nu_2(\epsilon,\mathbf{c}),\dots,\nu_r(\epsilon,\mathbf{c})$ can be concatenated and their concatenation is an r-parametric control variation at τ which verify (6). □

Let us consider the collection of the transport at time T of the g-variations at any $\tau \in [t_0, T]$.

DEFINITION 2.9. Let $T \in (t_0, t_1]$. The variational cone of the pair (x^*, u^*) at T, $\mathcal{K}(T)$, is the union of the left variations at T and of the transport along the reference flow from time τ up to the time T of the g-variations at time $\tau \in [t_0, T)$

$$\mathcal{K}(T) = \Big\{G(T,\tau)v \text{ , } v \text{ is a g-variation of } (x^*,u^*) \text{ at } \tau \in [t_0,T)\Big\} \cup \Big\{w \text{ , } w \text{ is a left variation of } (x^*,u^*) \text{ at } T\Big\}.$$

The cone $\mathcal{K}(T)$ is a cone contained in $T_{x^*(T)}M$ and its elements are tangent vectors to $R(T, x_0)$ at $x^*(T)$. Moreover the following property holds: for each couple (t,s), $t_0 < s < t \le t_1$

$$G(t,s)\mathcal{K}(s) \subseteq \mathcal{K}(t). \tag{7}$$

THEOREM 2.10. *The variational cone $\mathcal{K}(T)$ is a local regular tangent cone of $R(T,x_0)$ at $x^*(T)$.*

PROOF. Let $,v$, $\|v\| = 1$ be a tangent vector belonging to the relative interior of the convex hull of $\mathcal{K}(T)$. Let $\{w_1,\dots,w_p\} \subset \mathcal{K}(T)$ be such that $\|w_i\| = 1$ and $v \in int_{rel}co\{w_1,\dots,w_p\}$ and let $\mathcal{C}$ be the cone generated by the $w_i's$. We are going to prove that $\mathcal{C}$ is a regular tangent cone.

For each w_i there exists a time τ_i and a g-variation v_i, at τ_i such that $w_i = G(T,\tau_i)\,v_i$. Possibly by reordering the variations we can suppose that the only distinct times among the $\tau_i's$ are $\tau_1,\dots,\tau_r$. Let $\mu_i(\epsilon,\mathbf{c}_i)$, $\epsilon \in [0,\bar\epsilon_i]$, $c \in [0,(\bar c_i)^{j_i}]$ be the control variations at τ_i relative to the g-variations at τ_i as in Proposition 2.8. There exists $\bar\epsilon$ such that the control variations μ_i are defined on disjoint intervals $I_i(\epsilon)$ $\forall \epsilon \in [0,\bar\epsilon]$. Let $a = min\,\overline{c_i}$; for each $\epsilon \in [0,\bar\epsilon]$ and each $\mathbf{c} \in ([0,a])^p$, set

$$u(\epsilon,\mathbf{c})(t) = \begin{cases} \mu_i(\epsilon,\mathbf{c}_i)(t) \text{ , } t \in I_i(\epsilon) \\ u^*(t) \text{ , } \quad \text{otherwise } t \in [t_0,T] \end{cases}$$

It is not difficult to see that

$$S(T,t_0,x_0,u(\epsilon,\mathbf{c})) = x^*(T) + \epsilon\sum_{i=1}^{p} c_i\,w_i + o(\epsilon) \tag{8}$$

uniformly in $\mathbf{c}$ and $o(\epsilon)$ is continuous w.r.t. $\mathbf{c}$.

Each point in $\mathcal{C}$ is linear combination of the $w_i's$ and the coefficients can be chosen continuous, i.e. there exists p continuous maps, $\psi_i : \mathcal{C} \to [0,1]$ such that

$y = \sum_{i=1}^{p} \psi_i(y) w_i$. If $\epsilon \in [0, \bar{\epsilon} a]$, then $u\Big(\epsilon/a, a(\psi_1(y), \dots, \psi_p(y)\Big)$ is well defined and therefore by (8) it follows that

$$x^*(T) + \epsilon\, a\, y + g_\epsilon(y) \in R(T, t_0, x_0)$$

g_ϵ continuous map and $o(\epsilon)$ uniformly in y. The Theorem is proved. □

In some optimization problems the interest is focused on the reachable set $R(t, t_0, S_0)$ near the final point, $x^*(t)$, of a trajectory starting from S_0. It is obvious that the cone $\mathcal{K}(t)$ previously defined is a cone of tangent vectors of $R(t, t_0, S_0)$ at $x^*(t)$, but a larger cone of tangent vectors to this set can be defined taking into account that in this case also the initial point of the trajectory can be modified.

DEFINITION 2.11. Let S_0 be a subset of M and let the reference pair (x^*, u^*) be such that $x_0 = x^*(t_0) \in S_0$. For each local regular tangent cone N of S_0 at x_0 we denote by $\mathcal{K}(t, N)$ the set

$$\mathcal{K}(t, N) = \mathcal{K}(t) \cup \{\, G(t, t_0) w \;,\; w \in N \,\}. \tag{9}$$

Arguments similar to the ones used in the proof of Theorem 2.10 prove the following theorem:

THEOREM 2.12. *$\mathcal{K}(t, N)$ is a local regular tangent cone to $R(t, t_0, S_0)$ at $x^*(t)$.*

Let us prove that, if in addition to hypothesis A), B), C) we assume

D): for each $\omega \in U$ the constant map $u(t) \equiv \omega$ belongs to $\mathcal{U}$

then the Pontryagin-McShane variations are g-variations.

PROPOSITION 2.13. If $f(t, x, u)$ is a continuous map, then for each Lebesgue point $\tau \in [t_0, t_1]$ of u^* and for each $\omega \in U$

$$\big[f\big(\tau, x^*(\tau), \omega\big) - f\big(\tau, x^*(\tau), u^*(\tau)\big)\big] \tag{10}$$

is both a left and a right variation of (x^*, u^*) at τ.

PROOF. It is known [**2**] that (10) is a needle-like variation at each Lebesgue point of u^*; moreover the control variations which produce it verifies hypothesis (3) of Proposition 2.4. The set of Lebesgue points of u^* coincides with the set of Lebesgue points of the map $t \mapsto \big[f\big(\tau, x^*(\tau), \omega\big) - f\big(\tau, x^*(\tau), u^*(\tau)\big)\big]$, therefore the proof is complete. □

REMARK 2.14. More generally if we do not assume the continuity of the map f, then it can be proved that (10) is both a left and a right variation for each τ in a set of full measure which may depend on ω.

The results in [**1**] together with Theorem 2.12 imply that $\mathcal{K}(t, N)$ provides new conditions for the adjoint variable of the Maximum Principle in an optimal control problem with constraints on the end-points.

3. Examples of g-variations

In this section we are going to present some further examples of g-variations of a reference pair.

EXAMPLE 3.1. Knobloch and Krener variations

Let $\varphi(\cdot)$ be a continuous function defined on an interval $I \subset [t_0, t_1]$; let us suppose that for each $\tau \in I$ there exists a family of control maps $\{\nu_\tau(\epsilon)\}$ such that:

1. $\nu_\tau(\epsilon)(t) = u^*(t)$ outside an interval which goes to τ as ϵ goes to zero
2. the map
$$(\epsilon, x) \mapsto S(t_1, x, \nu_\tau(\epsilon))$$
is C^1 on the set $[0, \bar{\epsilon}] \times \{\|x - x_0\| \leq \alpha\}$
3. $S(t_1, x_0, \nu_\tau(\epsilon)) = x^*(t_1) + \epsilon^k \varphi(\tau) + o(\epsilon)$

These variations are essentially the variations defined by Knobloch in [**5**]. By definition 2.1 it follows that for each $\tau \in I$, $\varphi(\tau)$ is a needle-like variation, therefore if the hypothesis (3) of Proposition 2.4 is fulfilled, then $\varphi(\tau)$ is a g-variation.

The needle-like variations introduced by Krener in [**4**], are particular cases of Knobloch's variations which verify hypothesis (3), so that they are g-variations.

EXAMPLE 3.2. Bianchini-Stefani variations

Let us assume that the set of admissible controls $\mathcal{U}$ is endowed by a topology Π with respect to which the solution map $(t, \tau, x, u) \mapsto S(t, \tau, x, u)$ is continuous and C^1 in x. Examples of topology with this property can be found in [**11**]. In [**7**] the following definition of a reference trajectory variation is given:

DEFINITION 3.3. A vector $v \in T_{x^*(\tau)}M$ is a right BS-variation of order k of (x^*, u^*) at time $\tau \in [t_0, t_1)$ if there are:

1. positive numbers $\bar{\epsilon}$, $\bar{c}$, $\bar{\gamma}$
2. a map $\eta : [0, \bar{\epsilon}] \times [0, \bar{c}] \times [0, \gamma] \to \mathcal{U}$ continuous in the Π topology such that
$$S(t + \epsilon + \epsilon\gamma, t + \epsilon\gamma, x^*(t + \epsilon\gamma), \eta(\epsilon, c, \gamma)) = x^*(t + \epsilon(1 + \gamma)) + \epsilon^k cv + o(\epsilon^k)$$
uniformly with respect to and c.

Analogous is the definition of left variations.

By definition the BS-variations are g-variations. In [**6**], [**7**] several example of BS- variations can be found; they are variations of the trajectory relative to the drift term of an affine system

$$\dot{x} = f_0(x) + \sum_{i=1}^{m} u_i f_i(x) \qquad |u_i| \leq 1$$

and they are constructed by means of the relations in the Lie algebra generated by the $f_i's$ at a point of the reference trajectory. This proves that the g-variations need not to be the limit of sequences of needle-like variations each occurring at a different time, see the quoted paper for the details.

EXAMPLE 3.4. Kawski variations

In [**12**] Kawski constructs a variation of the stationary trajectory of an affine control system. We are going to prove that Kawski's variation is a g-variation. He considers the following system:

$$\begin{cases} \dot{x}_1 = & u \\ \dot{x}_2 = & x_1 \\ \dot{x}_3 = & x_1^3 \\ \dot{x}_4 = & x_3^2 - x_2^7 \end{cases}$$

$|u| \leq 1$. Let $x^*(t) = 0$ and $u^*(t) = 0$. Let T and $\overline{u}$ be such that $S(T, 0, 0, \overline{u}) = (0, a, 0, b)$ with $a > 0$. Let $\overline{u}^1$ denote the time-reversed control defined by $\overline{u}^1(t) = \overline{u}(T - t)$. Define inductively via concatenation $u_1 = \overline{u}^1 \# \overline{u}$ and $u_k = \overline{u} \# u_{k-1} \# \overline{u}^1$; $u_k : [0, 2kT] \to [1, 1]$. Let $k(\epsilon)$ be the smallest integer greater than or equal to ϵ^{-7} and set

$$\nu(\epsilon, c)(t) = \begin{cases} c^{1/7} u_{k(\epsilon)}(2kT(t - \epsilon)/\epsilon) & t \in [\epsilon, 2\,\epsilon] \\ u^*(t) \quad \text{otherwise} \end{cases}$$

For each fixed ϵ the map $\nu(\epsilon, \cdot) : [0, 1] \to L^1\Big([0, T], [-1, 1]\Big)$ is continuous and $\nu(\epsilon, c)$ converges to u^* as ϵ goes to 0 uniformly in c. Since the system is an affine one, the family $\{\nu(\epsilon, c)\}$ is an admissible control variation of u^*. Moreover direct calculation implies that

$$S(2\epsilon, \epsilon, 0, \nu(\epsilon, c)) = \Big(0, 0, 0, c\epsilon^{64} A + o(\epsilon^{64})\Big)$$

with $A < 0$. Therefore $(0, 0, 0, A)$ is a g-variation.

EXAMPLE 3.5. Let

$$\dot{x} = f_0(x) + \sum_{i=0}^{m} u_i\, f_i(x) \qquad \mathbf{u} \in \Omega \tag{11}$$

f_i C^∞ vector fields on M and Ω subset of $\mathbb{R}^m$ with no empty interior.
Let x^* be a solution of the system relative to the control u^*. Let us suppose that $u^*(t) \equiv \omega_0$ on an interval I and that $\omega_0 \in int\,\Omega$. Set $g_0 = f_0 + \sum_{i=0}^m \omega_i\, f_i$; for each $i \in (1, \ldots, m)$ and for each $t \in I$

$$\Big[f_i, \Big[f_i, g_0\Big]\Big]\; x^*(t)$$

is a g-variation at t.

We are going to provide the proof only for analytic systems; the general results can be obtained by standard arguments, see [**13**].

Let us rewrite the system (11) in the following way

$$\dot{x} = g_0(x) + \sum_{i=0}^{m} u_i\, f_i(x) \qquad \mathbf{u} \in \Omega'$$

with $g_0 = f_0 + \sum_{i=0}^m \omega_i\, f_i$ and $\Omega' = \Omega - \omega_0$. Since 0 belongs to the interior of Ω' there exists $a > 0$ such that if $|u_i| \leq a$, then $\mathbf{u} \in \Omega'$. Fix $\tau \in I$ and

$i \in (1,\ldots,m)$. Let $u(\epsilon,b)$: $[\tau+\epsilon, \tau+(\epsilon+8b\epsilon)] \to \Omega'$ be the piecewise control whose i-th component is defined by

$$(u(\epsilon,b))_i(t) = \begin{cases} a & t \in [\tau+\epsilon+n\epsilon\, b, \tau+\epsilon+(n+1)\epsilon\, b] \quad 0,3,5,6 \\ -a & \text{otherwise} \end{cases}$$

and the other components are equal to 0.

$$S^*(\tau, u(\epsilon,b)) = \exp -(8b+1)\epsilon\, g_0 \circ Z(\epsilon,b) \circ \exp \epsilon\, g_0\,(x^*(\tau))$$

where $Z(\epsilon,b)$ is

$$Z(\epsilon,b) = \exp \epsilon\, b(g_0 - a\,f_i) \circ \exp 2\epsilon\, b(g_0 + a\,f_i) \circ \exp \epsilon\, b(g_0 - a\,f_i)$$

$$\circ \exp \epsilon\, b(g_0 + a\,f_i) \circ \exp 2\epsilon\, b(g_0 - a\,f_i) \circ \exp \epsilon\, b(g_0 + a\,f_i)$$

The Campbell-Hausdorff formula implies that

$$Z(\epsilon,b) = \exp\Big(8b\epsilon g_0 + \frac{2}{3}c^3\epsilon^3 a^2\,[f_i,[f_i,g_0]] + o(\epsilon^3)\Big).$$

Therefore

$$S^*(\tau, u(\epsilon,b)) == x^*(\tau) + \frac{2}{3}c^3\epsilon^3 a^2 [f_i,[f_i,g_0]]\,(x^*(\tau)) + o($$

It is easy to verify that the control map $u(\sqrt[3]{\epsilon}, \sqrt[3]{c}/a^2)$ has the property required by Definition 2.3.

References

[1] Bianchini, R.M. , *Variational Approach to some Optimization Control Problems*, in *Geometry in Nonlinear Control and Differential Inclusions*, Banach Center Publication, Vol 32, pp. 83-94, Warszawa 1995.

[2] Lee, E.B. , and Markus, L. , *Foundations of Optimal Control Theory*, J. Wiley & Sons, New York 1967.

[3] Bressan, A., *A high order test for optimality of bang-bang controls*, SIAM Journal on Control and Optimization, Vol 23, pp. 38-48, 1985.

[4] Krener, A. , *The High Order Maximal Principle and its Applications to Singular Extremals*, SIAM Journal on Control and Optimization, Vol 15, pp. 256-292, 1977.

[5] Knobloch, H.W. , *Higher Order Necessary Conditions in Optimal Control Theory*, Lecture Notes in Control and Information Sciences, No 34, Springer Verlag, Berlin, 1981.

[6] Bianchini, R.M. , and Stefani, G. , *Controllability along a Reference Trajectory: a Variational Approach*, SIAM Journal on Control and Optimization, Vol 31, pp. 900-927, 1993.

[7] Bianchini, R.M. , and Stefani, G. , *Graded Approximations and Controllability Along a Trajectory*, SIAM Journal on Control and Optimization, Vol 28, pp. 903-924, 1990.

[8] Polovinkin, E.S. , and Smirnov, G.V. , *An Approach to the Differentiation of Many-valued Mappings and Necessary Conditions for Optimization of Solutions of Differential Inclusions*, [in russian], Differencial'nye Uravneija, Vol. 22, pp. 944-954, 1986.

[9] McShane, E.J. , *Unified Integration*, Academic Press, INC., Orlando, 1983.

[10] Hewitt, E. , and Stromberg, K. , *Real and Abstract Analysis*, Springer-Verlag, Berlin Heidelberg, 1975.

[11] Bianchini, R.M. , and Margheri, A. *First Order Differentiability of the Flow of a Control System with L^p Controls*, to appear in Journal of Optimization Theory and Application.

[12] Kawski, M. , *Control Variations with an Increasing Number of Switching*, Bulletin American Mathematical Society, Vol 18, pp. 149-152, 1988.

[13] Sussmann, H. , *A General Theorem on Local Controllability*, SIAM J. Control Optim., Vol. 25,pp. 158-194, 1987.

[14] Hestenes, M., *Calculus of Variations and Optimal Control Theory*, Wiley, New York, 1967.

DIP. DI MATEMATICA U. DINI, VIALE MORGAGNI 67/A, 50134 FIRENZE, ITALY.

E-mail address: rosabian@udini.math.unifi.it

Proceedings of Symposia in Pure Mathematics
Volume **64**, 1999

Representations of Dirac Structures on Vector Spaces and Nonlinear L-C Circuits

Anthony M. Bloch and Peter E. Crouch

ABSTRACT. In this paper we review Dirac structures on vector spaces and derive a parameterization of Dirac structures on a vector space as the graph of a certain linear map. We then derive the equations of motion of nonlinear LC circuits with input and output ports and show that they correspond to the equations of motion of a differential system defined by a Dirac structure. No assumptions are placed upon the topology of the network. Canonical coordinates are described in which the equations of motion demonstrate their underlying Hamiltonian nature together with constrained variables and constants of motion.

1. Introduction

In this paper we discuss Dirac structures on vector spaces and their application to the modeling of nonlinear L-C electrical networks with input-output ports. The principal references to Dirac structures are I. Dorfman [**6**] and T. Courant [**4**]. The latter reference is concerned with Dirac structures on finite dimensional differential systems, whereas the former reference treats a more general context. In finite dimensions Dirac structures may be defined generally on manifolds and have application to the formulation of many physical systems. This has been the focus of work by van der Schaft et al. [**9, 10, 11, 12**], although the relevance of Dirac structures to nonlinear electrical networks was pointed out initially by the authors who, in turn, were prompted to consider Dirac structures by Professor J. Marsden, with reference to the paper by Courant [**4**].

The main interest in Dirac structures in the context of this paper is their ability to deal naturally with Hamiltonian like systems with constraints. Indeed, one of the main results in [**4**], in the context of a differential system on a manifold defined by a Dirac structure which enjoys an associated integrability condition, is that locally the system can be written as a Hamiltonian system, involving a subset of the coordinates, while the remaining coordinates are either constant under the flow, or are fixed by algebraic constraints. Indeed, it is the ability to simultaneously include

1991 *Mathematics Subject Classification.* Primary 34A34; Secondary 70H05.

Research partially supported by National Science Foundation PYI grant DMS-91-57556 and grant DMS-9803181, AFOSR grant F49620-96-1-0100, a Guggenheim Fellowship and the Institute for Advanced Study.

Research partially supported by NATO grant CRG 910926.

both constraints and constants of motion, capturing qualities of systems defined by both (degenerate) pre-symplectic structures and Poisson structures, respectively, that is of direct importance to many applications, including those of nonlinear L-C electrical circuits.

In the case of nonlinear L-C electrical circuits the Dirac structure involved in effectively modeling the corresponding flow is defined on a vector space of branch currents and voltages, and is therefore constant, which greatly simplifies much of the necessary analysis. Indeed, the modeling is so elegant that we can include in this formalism circuits which do not obey the usual topological restrictions placed on them; that is, circuits which do not include capacitor loops, or inductor cut-sets. These observations have been made in some detail in works by Van der Schaft et al. [**9, 10, 11, 12**]. However, even on vector spaces, Dirac structures have many different representations. In this paper we concentrate on a representation which does not seem to have received any prior attention, and which seems to the authors to have particular significance in the context of nonlinear L-C electrical networks. Indeed we are able to exhibit the natural canonical coordinates for the underlying Hamiltonian flow quite naturally in terms of the representation we introduce here, and without reference to the general representation result referenced above.

The effectiveness of Dirac structures in this regard is easily understood by comparing our approach with the works by others [**3**], [**7**], on the Hamiltonian and Lagrangian formulation of L-C and R-L-C electrical circuits.

In section 3 we discuss representations of constant Dirac structures on vector spaces. In section 4 we briefly describe the equations of motion described by differential systems and defined by constant Dirac structures. In section 5 we discuss the application of Dirac structures in modeling nonlinear L-C electrical networks with input-output ports. In section 6 we give an example.

2. Poisson and Pre-Symplectic Systems

Many dynamical systems evolve according to the classical Hamiltonian equations on an even-dimensional vector space $\mathbf{R}^{2m}$: let $(q_1, q_2, \ldots, q_n, p_1, \ldots, p_n)$ be coordinates on $\mathbf{R}^{2n}$ and $H(\mathbf{q}, \mathbf{p})$ a smooth function on $\mathbf{R}^{2m}$. Then Hamilton's equations are given by

$$\begin{aligned} \dot{q}_i &= \frac{\partial H}{\partial p_i} \\ \dot{p}_i &= -\frac{\partial H}{\partial q_i}, \qquad i = 1, \ldots, n. \end{aligned} \tag{2.1}$$

The equations (2.1) may also be written as

$$\dot{\mathbf{x}} = J \nabla H \tag{2.2}$$

where $\mathbf{x} = (\mathbf{q}, \mathbf{p})$, ∇ is the gradient with respect to the standard metric on $\mathbf{R}^{2n}$, and J is the skew symmetric tensor

$$J = \begin{bmatrix} 0 & I \\ -I & 0 \end{bmatrix}, \tag{2.3}$$

I the $n \times n$ identity matrix.

This flow can also be represented in the following fashion:

Let

$$w = \sum_{i=1}^{n} dq_i \Lambda dp_i \tag{2.4}$$

be a (closed nondegenerate) 2-form on $\mathbf{R}^{2n}$ and ξ denote a vector in $\mathbf{R}^{2n}$. Let X_H be the vector field defined by

$$w(X_H, \xi) = dH.\xi \tag{2.5}$$

ξ an arbitrary $2n$ vector and $\cdot$ the natural pairing between vectors and forms. Then X_H is just the Hamiltonian vector field (2.1), as a sample computation shows.

This picture can be generalized in many ways (see e.g. Arnold [**2**], Abraham and Marsden [**1**] and Marsden and Ratiu [**8**]). As indicated above we are concerned here with the generalization to so-called Dirac structures which is useful for describing certain interconnected systems such as LC circuits. There are two special cases of Dirac structures which are instructive to recall briefly first (see e.g. Courant [**4**], Marsden and Ratiu [**8**]).

Firstly we note that the notion of Hamiltonian system is generalizable to a system on any symplectic manifold M^{2n}, a manifold endowed with a nondegenerate closed 2-form w: the Hamiltonian vector field is defined as before by

$$w(X_H, \xi) = dH.\xi,$$

ξ an arbitrary vector field on M. Darboux's theorem tells us that the flow is locally of the classical form (2.1).

The form w induces an invertible map $w^b : TM \to T^*M$. Denoting the inverse of this map locally by w^{ij}, the flow may be written locally as

$$\dot{x}_j = w^{ji} \frac{\partial H}{\partial x_i} \tag{2.6}$$

w^{ji} being the Poisson tensor.

In some examples however w^{ji} need not be invertible, such systems may still be viewed as Hamiltonian.

More precisely, let P be a Poisson manifold, a manifold endowed with a bilinear operation $\{\ ,\ \}$ on $C^\infty(P)$ such that (i) $(C^\infty(P), \{\ \})$ is a Lie algebra and (ii) $\{FG, H\} = \{F, H\}G + F\{G, H\}$ for all $F, G, H \in C^\infty(P)$.

Then for $H \in C^\infty(P)$ the Hamiltonian vector field X_H on P is defined uniquely by the requirement that

$$X_H[G] = \{G, H\} \qquad \forall G \in C^\infty(P)$$

The flow may be written locally as

$$\dot{x}_j = B^{ij} \frac{\partial H}{\partial x_j} \tag{2.7}$$

B^{ij} is not necessarily invertible and is called the Poisson structure. Indeed in the case the system is symplectic B^{ij} is the negative inverse of the matrix representing the symplectic form (see Marsden and Ratiu [**8**]).

Globally B is the two-tensor in $\Lambda^2(T^*P)$ defined by

$$B(z)(dF(z), dG(z)) = \{F, G\}(z) \tag{2.8}$$

and $\{x_i, x_j\} = B^{ij}(z)$.

An important class of such systems are the so-called Lie-Poisson systems defined on the dual of a Lie algebra, see again Marsden and Ratiu [**8**]. An example is the classical Euler equation for the rigid body.

Finally we note also that there exists a generalization of the symplectic case in the "opposite" direction, to the case where w is degenerate. In this case the relation $w(X_H, \xi) = dH.\xi$ does not define X_H uniquely, the induced map $w^b : TM \to T^*M$ has a nontrivial kernel and H defines X_H up to $X_H + Z$, Z in the kernel of the map w^b. w is called a presymplectic form.

Below we shall consider (on vector spaces) Dirac systems which contain the Poisson and presymplectic systems as special cases.

3. Representations of Dirac Structures on Vector Spaces

Let V^n be an n-dimensional vector space, V^* its dual, and let $\langle \cdot, \cdot \rangle$ be the natural pairing between them. We have the following fundamental definition.

DEFINITION 1. [**4, 6**] *A subspace $L \subset V \oplus V^*$ is a Dirac structure on V if*

(i) $\langle v, \bar{\alpha} \rangle + \langle \bar{v}, \alpha \rangle = 0$, $\forall (v, \alpha)$, $(\bar{v}, \bar{\alpha}) \in L$.

(ii) *L is a maximal subspace which satisfies (i).*

One can easily show that a Dirac structure L on V^n is in fact an n-dimensional subspace. One can also show that (i) is equivalent to the condition

$$\langle v, \alpha \rangle = 0; \qquad \forall (v, \alpha) \in L. \tag{3.1}$$

We will give three examples of Dirac structures on V through the following three propositions.

PROPOSITION 1. *Let $E^m \subset V^n$ be an m-dimensional subspace of V, $m \leq n$, and $J : E \to E^*$ a skew symmetric linear mapping. Then $L = \{(v, \alpha); v \in E, \alpha \in V^*, \alpha|_E = Jv\}$ is a Dirac structure on V.*

Proof. If $F \subset V$ is a subspace, we set

$$F^\perp = \{\alpha \in V^*; \langle \alpha, f \rangle = 0, \quad f \in F\} \subset V^*.$$

Clearly the subspace L satisfies (3.1), by skew symmetry of J. L also contains (E, JE) and $(0, E^\perp)$ so L has at least dimension n. Thus L has dimension n and is a Dirac structure. ∎

We introduce a little notation. If U, W, V_1, V_2 are vector spaces and $A : W \to V_1$, $B : W \to V_2$, $C : V_1 \to U$, $D : V_2 \to U$ are linear maps, we set

$$(A, B) : W \longrightarrow V_1 \oplus V_2; \qquad (A, B)w = (Aw, Bw)$$
$$(C + D) : V_1 \oplus V_2 \longrightarrow U; \qquad (C + D)(v_1, v_2) = (Cv_1 + Dv_2)$$

Notice that $(A, B)^* = A^* + B^*$ and $(C + D)^* = (C^*, D^*)$, where A^* denotes the dual operator of A.

PROPOSITION 2. *Let W^n be an n-dimensional vector space and $A : W \to V^*$, $B : W \to V$ linear maps satisfying:*

$$Rank\ (A, B) = Rank\ (A^* + B^*) = n \tag{3.2}$$

$$A^*B + B^*A = 0. \tag{3.3}$$

Then $L = Range\ (B, A) = \{(v, \alpha); v = Bw, \alpha = Aw, w \in W\}$ is a Dirac structure on V. Alternatively, we may write $L = \ker(A^* + B^*) = \{(v, \alpha); A^*v + B^*\alpha = 0\}$.

Proof. Clearly L has dimension n, by condition 2. L satisfies (3.1) by condition (3.3). ∎

PROPOSITION 3. *Let $W^n \subset V \oplus V^*$ be an n-dimensional vector space and $\mathbb{J} : W \to W^*$ a skew symmetric linear map. Then $L =$ graph $\mathbb{J}$ is a Dirac structure on V.*

Proof. If $W^n \subset V \oplus V^*$, then we may write $W = V_1 \oplus V_2^*$, where V_1 has dimension m, and V_2 has dimension $n-m$, so that $V \cong V_1 \oplus V_2$. If $w \in W$ we write $w = (v_1, \alpha_2)$, $v_1 \in V_1$, $\alpha_2 \in V_2^*$. Similarly, we may write $W^* = V_1^* \oplus V_2 \cong (V_1 \oplus V_2^*)^*$. If $w^* \in W^*$ we write $w^* = (\alpha_1, v_2), \alpha_1 \in V_1^*$, $v_2 \in V_2$. Relative to these decompositions, we may express $\mathbb{J}$ in terms of mappings $\mathbb{J}_{11} : V_1 \to V_1^*$, $\mathbb{J}_{21} : V_1 \to V_2$, $\mathbb{J}_{12} : V_2^* \to V_1^*$ and $\mathbb{J}_{22} : V_2^* \to V_2$ so that if $(\alpha_1, v_2) = \mathbb{J}(v_1, \alpha_2)$ we have

$$\alpha_1 = \mathbb{J}_{11} v_1 + \mathbb{J}_{12}\alpha_2; \qquad v_2 = \mathbb{J}_{21} v_1 + \mathbb{J}_{22}\alpha_2. \tag{3.4}$$

If $\mathbb{J}$ is skew symmetric, we have

$$(\mathbb{J}w_1)(w_2) = -(\mathbb{J}w_2)(w_1), \qquad w_1, w_2 \in W.$$

Thus, if $w_1 = (v_1, \alpha_2)$, $w_2 = (u_1, \beta_2)$, we have

$$\begin{aligned} \langle \mathbb{J}_{11} v_1, u_1 \rangle + \langle \mathbb{J}_{11} u_1, v_1 \rangle &= 0 \\ \langle \mathbb{J}_{12}\alpha_2, u_1 \rangle + \langle \alpha_2, \mathbb{J}_{21} u_1 \rangle &= 0 \\ \langle \beta_2, \mathbb{J}_{22}\alpha_2 \rangle + \langle \mathbb{J}_{22}\beta_2, \alpha_2 \rangle &= 0. \end{aligned} \tag{3.5}$$

Now

$$\begin{aligned} L = \text{graph } \mathbb{J} &= \{(w, \mathbb{J}w); w \in W\} \\ &= \{(v, \alpha); v = (v_1, v_2) \in V, \alpha = (\alpha_1, \alpha_2) \in V^*, \text{satisfying } 4\}. \end{aligned}$$

Thus if $(v, \alpha) \in L \subset V \oplus V^*$, $\langle \alpha, v \rangle = \langle \alpha_1, v_1 \rangle + \langle \alpha_2, v_2 \rangle = \langle \mathbb{J}_{11} v_1 + \mathbb{J}_{12}\alpha_2, v_1 \rangle + \langle \alpha_2, \mathbb{J}_{21} v_1 + \mathbb{J}_{22}\alpha_2 \rangle = 0$, by (3.5). Thus $L =$ graph $\mathbb{J}$ does indeed satisfy 1, and is clearly a subspace of $V \oplus V^*$ of dimension n. Hence L is a Dirac structure on V. ∎

Both propositions 1 and 3 can be cast in a dual setting through maps $J : E \to E^*$; $E \subset V^*$, and maps $\mathbb{J} : W^* \to W$; $W = V_1 \oplus V_2^*$, $V = V_1 \oplus V_2$. We do not write down the obvious details here.

We say that a Dirac structure $L \subset V \oplus V^*$ has a representation 1, 2, or 3 if it is presented in the form of propositions 1, 2, or 3, respectively, (and also the dual forms in propositions 1 and 3).

THEOREM 4. *If $L \subset V \oplus V^*$ is a Dirac structure, then it possesses a representation in each of the forms 1, 2 and 3.*

Proof. That every Dirac structure has a representation in the form 1 or 2 is demonstrated in [**6**] and [**12**]. We show that every Dirac structure has a representation 3. We start with L in representation 2, as in proposition 2; $L = \ker(A^* + B^*)$, where $A : W \to V^*$, $B : W \to V$ satisfy conditions (3.2) and (3.3). Assume that A has rank m, and find a subspace $V_1 \subset V$ such that $A^*|_{V_1}$ has rank m. Thus we can find a complementary subspace V_2 such that $V = V_1 \oplus V_2$. Write $A_1 = A^*|_{V_1}$, $A_2 = A^*|_{V_2}$, $B_1 = B^*|_{V_1^*}$, $B_2 = B^*|_{V_2^*}$. Thus we can write

$$\begin{aligned} A^* &= A_1 + A_2, \qquad & B^* &= B_1 + B_2 \\ A &= (A_1^*, A_2^*), \qquad & B &= (B_1^*, B_2^*) \end{aligned}$$

We may therefore write condition (3.3) in the form:

$$A_1 B_1^* + A_2 B_2^* + B_1 A_1^* + B_2 A_2^* = 0. \tag{3.6}$$

Now $A_1^*: W \to V_1$ has full rank and range $A_2 \subset$ range A_1, so range $B_1 \subset$ range A_1+ range B_2. By condition 2, Rank $(A^*+B^*) = n$, so Rank $((A_1+B_2)+(B_1+A_2)) = n$, and hence Rank $(A_1 + B_2) = n$, and $A_1 + B_2 : V_1 \oplus V_2^* \to W^*$ is invertible. We may write

$$\begin{aligned} L = \{(v_1, v_2, \alpha_1, \alpha_2); (v_1, v_2, \alpha_1, \alpha_2) &\in V_1 \oplus V_2 \oplus V_1^* \oplus V_2^* \\ A_1 v_1 + A_2 v_2 + B_1 \alpha_1 + B_2 \alpha_2 &= 0\}. \end{aligned}$$

Hence $(v_1, v_2, \alpha_1, \alpha_2) \in L$ if and only if

$$(A_1 + B_2)(v_1, \alpha_2) + (B_1 + A_2)(\alpha_1, v_2) = 0.$$

Hence

$$(v_1, \alpha_2) = -(A_1 + B_2)^{-1}(B_1 + A_2)(\alpha_1, v_2).$$

Let $\mathbb{J} = -(A_1 + B_2)^{-1}(B_1 + A_2)$, $\overline{W} = V_1^* \oplus V_2$, and write $w = (\alpha_1, v_2) \in \overline{W}$, $w^* = (v_1, \alpha_2) \in \overline{W}^*$. We deduce that $(v_1, v_2, \alpha_1, \alpha_2) \in L$ if and only if $w^* = \mathbb{J}w$. We need to show that $\mathbb{J}$ is skew symmetric as a map $\mathbb{J} : \overline{W} \to \overline{W}^* \cong V_1 \oplus V_2^*$. From (3.6) we have

$$(A_1 + B_2)(B_1^*, A_2^*) + (B_1 + A_2)(A_1^*, B_2^*) = 0$$

or

$$(A_1 + B_2)(B_1 + A_2)^* + (B_1 + A_2)(A_1 + B_2)^* = 0$$

or

$$(A_1 + B_2)^{-1}(B_1 + A_2) = -(B_1 + A_2)^*(A_1 + B_2)^{*-1}.$$

Hence $\mathbb{J} = -\mathbb{J}^*$ as required and L is indeed the graph of $\mathbb{J} : \overline{W} \to \overline{W}^*$. ∎

It is clear that the proof of theorem 4 provides a mechanism of moving from the representation 2 to the representation 3. Indeed we may move from representation 3 to representation 2 easily as we now show.

Let the Dirac structure L be represented as in proposition 3 by the graph of $\mathbb{J} : W \to W^*$ where $W = V_1 \oplus V_2^*$. $(w, w^*) \in$ graph $\mathbb{J}$ if $w = (v_1, \alpha_2)$, $w^* = (\alpha_1, v_2)$ satisfy the relationship (3.4). Let $a : W \to V^*$, $b : W \to V$ be defined by

$$\begin{aligned} a(v_1, \alpha_2) &= (\mathbb{J}_{11} v_1 + \mathbb{J}_{12}\alpha_2, \alpha_2) \in V_1^* \oplus V_2^* \cong V^* \\ b(v_1, \alpha_2) &= (v_1, \mathbb{J}_{21} v_1 + \mathbb{J}_{22}\alpha_2) \in V_1 \oplus V_2 \cong V \end{aligned} \tag{3.7}$$

Thus in a suitable matrix representation we can write

$$a = \begin{pmatrix} \mathbb{J}_{11} & \mathbb{J}_{12} \\ 0 & I \end{pmatrix}, \qquad b = \begin{pmatrix} I & 0 \\ \mathbb{J}_{21} & \mathbb{J}_{22} \end{pmatrix}$$

Now

$$\begin{aligned} w^* - \mathbb{J}w &= (\alpha_1 - \mathbb{J}_{11}v_1 - \mathbb{J}_{12}\alpha_2, v_2 - \mathbb{J}_{21}v_1 - \mathbb{J}_{22}\alpha_2) \\ &= (-\mathbb{J}_{11}v_1, v_2 - \mathbb{J}_{21}v_1) + (\alpha_1 - \mathbb{J}_{12}\alpha_2, -\mathbb{J}_{22}\alpha_2) \\ &= a^*(v_1, v_2) + b^*(\alpha_1, \alpha_2) \end{aligned}$$

Thus we deduce a fundamental relationship relating representations 2 and 3.

$$w^* - \mathbb{J}w = a^* v + b^* \alpha. \tag{3.8}$$

Hence $(w, w^*) \in$ graph $\mathbb{J}$ if and only if $(v, \alpha) \in \ker(a^* + b^*)$. It is easily verified that $a^*b + b^*a = 0$, and we may express $a^* + b^*$ in a matrix representation as

$$a^* + b^* = \begin{pmatrix} -\mathbb{J}_{11} & 0 & I & -\mathbb{J}_{12} \\ -\mathbb{J}_{21} & I & 0 & -\mathbb{J}_{22} \end{pmatrix}$$

It follows that Rank $(a^* + b^*) =$ Rank $(a, b) = n$. Thus indeed we have displayed L in the form of representation 2, where A, B have been replaced by the special forms a and b, respectively.

The relationship between representation 1 and 2 is fully explained in [**12**] and [**5**]. However, for completeness we briefly discuss the mechanism of moving from representation 2 to representation 1. Given the Dirac structure L represented as in proposition 2, $L = \ker(A^* + B^*)$ define $G : U \to V$, with Range $G = \ker A^*$ and set $E =$ Range $A \subset V^*$. Let $J : E \to E^*$ be a skew symmetric mapping which satisfies

$$A^*JA + B^*A = 0 \tag{3.9}$$

It follows that we may solve the equation $A^*v + B^*\alpha = 0$ to obtain

$$L = \{(v, \alpha); v = J\alpha + \hat{v}, \hat{v} \in \text{Range } G, \alpha \in (\text{Range } G)^{\perp}\} \tag{3.10}$$

Moreover, since $E = (\text{Range } G)^{\perp} \subset V^*$ we may also express L in the form

$$L = \{(v, \alpha); \quad \alpha \in E, \quad v \in V \cong V^{**}, \quad v|_E = J\alpha\}.$$

This is simply the dual version of the representation 1 presented in proposition 1. Clearly we may also solve the equation $A^*v + B^*\alpha = 0$ for α and obtain the representation 1 as presented directly in proposition 1.

We note that the two skew forms J and $\mathbb{J}$ encountered in representations 1 and 3 are very different objects, and related by the equations (3.8) and (3.9).

4. Flows on Vector Spaces Defined by Dirac Structures

DEFINITION 2. *If $L \subset V \oplus V^*$ is a Dirac structure on a vector space V, and H is a function on V, we say that H is admissible if there is a vector field X_H on V such that (X_H, dH) is a section of L, where L is viewed as a subbundle of the tangent bundle of $V \oplus V^*$.*

If H is an admissible function, and (X_H, dH) is a section of L, viewed as above, then X_H defines a flow on V which we shall call a Dirac flow. These flows have been discussed in a more general context of manifolds [**4, 9, 10, 11, 12**], and in particular in a general context for certain nonlinear L-C electrical circuits, and some mechanical systems. Our purpose here is to show how our formulation of Dirac structures is particularly useful for elucidating the nature of general L-C electrical circuits. To explore the flow of X_H corresponding to a Dirac structure L on V, and an admissible function H on V, we must choose a representation for L. Our main contribution in this paper is to exploit the existence of the representation 3. Using the notation in proposition 3 we first interpret the statement that (X_H, dH) is a section of L, noting that the flow of the vector field is generated by the differential equation

$$\dot{v} = X_H(v), \quad v \in V. \tag{4.1}$$

We see that if $(X_H, dH) \in \operatorname{graph} \mathbb{J}$ and (4.1) holds then $X_H = (\dot{v}_1, \dot{v}_2)$, $dH = (d_1H, d_2H)$ and by equation (3.4)

$$\begin{aligned} d_1H &= \mathbb{J}_{11}\dot{v}_1 + \mathbb{J}_{12}d_2H \\ \dot{v}_2 &= \mathbb{J}_{21}\dot{v}_1 + \mathbb{J}_{22}d_2H. \end{aligned} \tag{4.2}$$

If we view these equations in a particular coordinate representation of V,

$$v_1 \triangleq (x_1^1, x_1^2, \dots, x_1^m) = (x_1), \quad v_2 \triangleq (x_2^1, x_2^2, \dots, x_2^{n-m}) = (x_2)$$

then the skew symmetry of $\mathbb{J}$, as defined by equation (3.5), gives the matrix identities

$$\mathbb{J}_{11}^T = -\mathbb{J}_{11}, \quad \mathbb{J}_{22}^T = -\mathbb{J}_{22}, \quad \mathbb{J}_{12}^T = -\mathbb{J}_{21}. \tag{4.3}$$

We now write equations (4.2) in the form

$$\begin{aligned} \frac{\partial H^T}{\partial x_1} &= \mathbb{J}_{11}\dot{x}_1 + \mathbb{J}_{12}\frac{\partial H^T}{\partial x_2} \\ \dot{x}_2 &= \mathbb{J}_{21}\dot{x}_1 + \mathbb{J}_{22}\frac{\partial H^T}{\partial x_2}. \end{aligned} \tag{4.4}$$

Note first that these equations demonstrate that H is constant along the Dirac flow (4.1),

$$\frac{dH}{dt} = \frac{\partial H}{\partial x_1}\dot{x}_1 + \frac{\partial H}{\partial x_2}\dot{x}_2 = \dot{x}_1^T\left(\mathbb{J}_{11}\dot{x}_1 + \mathbb{J}_{12}\frac{\partial H^T}{\partial x_2}\right) + \frac{\partial H}{\partial x_2}\left(\mathbb{J}_{21}\dot{x}_1 + \mathbb{J}_{22}\frac{\partial H^T}{\partial x_2}\right) = 0$$

by use of the identities (4.3). There is a better form for the equations (4.4), which displays their underlying Hamiltonian nature. We employ a change of coordinates in equation (4.4), namely

$$u = x_2 - \mathbb{J}_{21}x_1, \qquad z = x_1$$

and set

$$\tilde{H}(u, z) = H(x_1, x_2) = H(z, u + \mathbb{J}_{21}z).$$

Thus

$$\frac{\partial \tilde{H}}{\partial u} = \frac{\partial H}{\partial x_2}; \qquad \frac{\partial \tilde{H}}{\partial z} = \frac{\partial H}{\partial x_1} + \frac{\partial H}{\partial x_2}\mathbb{J}_{21}.$$

It is now easy to see that we may rewrite the equations (22) in the form:

$$\begin{aligned} &\text{(i)} \quad \dot{u} = \mathbb{J}_{22}\left(\frac{\partial \tilde{H}}{\partial u}\right)^T \\ &\text{(ii)} \quad \mathbb{J}_{11}\dot{z} = \left(\frac{\partial \tilde{H}}{\partial z}\right)^T \end{aligned} \tag{4.5}$$

We may further choose coordinates (u_1, u_2), (z_1, z_2) in which $\mathbb{J}_{22}$ and $\mathbb{J}_{11}$ have the block form

$$\mathbb{J}_{22} = \begin{bmatrix} \hat{\mathbb{J}}_{22} & 0 \\ 0 & 0 \end{bmatrix}, \quad \mathbb{J}_{11} = \begin{bmatrix} \hat{\mathbb{J}}_{11} & 0 \\ 0 & 0 \end{bmatrix},$$

where $\hat{\mathbb{J}}_{22}$ and $\hat{\mathbb{J}}_{11}$ are invertible and skew symmetric. Thus, in these coordinates, equation (23) takes the form

$$(4.6)\qquad \begin{array}{ll} \text{(i)}\quad \dot{u}_1 = \hat{\mathbb{J}}_{22}\left(\dfrac{\partial \tilde{H}}{\partial u_1}\right)^T, & \text{(ii)}\quad \dot{u}_2 = 0 \\ \text{(iii)}\quad \dot{z}_1 = \hat{\mathbb{J}}_{11}^{-1}\left(\dfrac{\partial \tilde{H}}{\partial z_1}\right)^T, & \text{(iv)}\quad \dfrac{\partial \tilde{H}}{\partial z_2} = 0. \end{array}$$

Thus, as stated in the introduction, the flow (4.1) decomposes into a Hamiltonian flow in the (u_1, z_1) coordinates, coordinates u_2 which are constants of motion, and coordinates z_2, fixed by algebraic constraints. In the case of this constant Dirac structure there are no integrability constraints.

5. Application to Nonlinear L-C Circuits

In this section we shall demonstrate that general L-C circuits, with input-output points, may be cast very naturally in the form of flows corresponding to Dirac structures on vector spaces. In particular we will work with the representation 3, while other authors have employed representations 1 and 2.

We consider a network N in which the branches are viewed as the set $N = \Gamma \cup \Sigma$, where Γ is a maximal tree (has no loops) and Σ is the set of links ($\Gamma \cup \{b\}, b \in \Sigma$ has exactly one loop). Let B = number of branches in the network N, M = number of loops and N = number of nodes -1, then $B = N + M$, and Γ has N branches while Σ has M branches.

Let

$$\begin{aligned} I \in \mathbf{R}^M &= \text{vector of mesh current values} \\ V \in \mathbf{R}^N &= \text{vector of node voltage values} \\ i \in \mathbf{R}^B &= \text{vector of branch current values} \\ v \in \mathbf{R}^B &= \text{vector of branch voltage values.} \end{aligned}$$

The network may be described by matrices A and B such that

$$AV = v; \qquad BI = i.$$

Kirchoff's laws are equivalent to $A^T B = 0$. Thus $A^T BI = A^T i = 0$ and $B^T AV = B^T v = 0$. Let $i_\Sigma, v_\Sigma, i_\Gamma, v_\Gamma$ denote the vectors of current and voltage values in branches of Σ and Γ.

Write $A = (A_\Sigma, A_\Gamma)$, $B = (B_\Sigma, B_\Gamma)$ where

$$A_\Sigma V = v_\Sigma, \quad A_\Gamma V = v_\Gamma, \quad B_\Sigma I = i_\Sigma, \quad B_\Gamma I = i_\Gamma.$$

Using standard analysis from network theory

$$(5.1)\qquad P \overset{\Delta}{=} -(A_\Gamma^{-1})^T A_\Sigma^T = B_\Gamma B_\Sigma^{-1} : \mathbf{R}^M \to \mathbf{R}^N$$

and

$$(5.2)\qquad i_\Gamma = P i_\Sigma, \quad v_\Sigma = -P^T v_\Gamma.$$

We employ several further partitions of $i = (i_\Sigma, i_\Gamma)$ and $v = (v_\Sigma, v_\Gamma)$ to reflect branches containing capacitors, inductors and voltage/current sources (input/output ports). The superscripts S, C, L denote input/output, capacitor, inductor branch, respectively. Thus we set

$$
\begin{aligned}
&i = (i^S, i^C, i^L), && v = (v^S, v^C, v^L) \\
&i_\Sigma = (i_\Sigma^S, i_\Sigma^C, i_\Sigma^L), && v_\Sigma = (v_\Sigma^S, v_\Sigma^C, v_\Sigma^L) \qquad (5.3)\\
&i_\Gamma = (i_\Gamma^S, i_\Gamma^C, i_\Gamma^L), && v_\Gamma = (v_\Gamma^S, v_\Gamma^C, v_\Gamma^C).
\end{aligned}
$$

Recognizing the fundamental laws of inductors and capacitors, we introduce charges and fluxes through the relationships

$$
\begin{aligned}
\phi^L &= L(i^L) = (L_1(i_i^L), L_2(i_2^L), \ldots) \\
Q^C &= C(v^C) = (C_1(v_1^C), C_2(v_2^C), \ldots).
\end{aligned}
$$

We assume that the nonlinear inductors and capacitors are such that we may define the following vectors

$$
\begin{aligned}
L^{-1}(\phi^L) &= (L_1^{-1}(\phi_1^L), L_2^{-1}(\phi_2^L), \ldots) \\
C^{-1}(Q^C) &= (C_1^{-1}(Q_1^C), C_2^{-1}(Q_2^C), \ldots).
\end{aligned}
$$

We may therefore write

$$
\begin{aligned}
&i_\Sigma = (i_\Sigma^S, \dot{Q}_\Sigma^C, L_\Sigma^{-1}(\phi_\Sigma^L)), \quad v_\Sigma = (v_\Sigma^S, C_\Sigma^{-1}(Q_\Sigma^C), \dot{\phi}_\Sigma^L) \qquad (5.4)\\
&i_\Gamma = (i_\Gamma^S, \dot{Q}_\Gamma^C, L_\Gamma^{-1}(\phi_\Gamma^L)), \quad v_\Gamma = (v_\Gamma^S, C_\Gamma^{-1}(Q_\Gamma^c), \dot{\phi}_\Gamma^C).
\end{aligned}
$$

We set

$$
\begin{aligned}
x_1 &= (y_1, y_2) = (\phi_\Gamma^L, Q_\Sigma^C), \quad & x_2 &= (y_3, y_4) = (Q_\Gamma^C, \phi_\Sigma^L) \\
e &= (e_1, e_2) = (v_\Sigma^S, i_\Gamma^S), & f &= (f_1, f_2) = (i_\Sigma^S, v_\Gamma^S),
\end{aligned}
$$

and

$$
\begin{aligned}
&H(y_1, y_2, y_3, y_4) = H(x_1, x_2) = \\
&\int^{y_1} L_\Gamma^{-1}(\eta_1)^T(d\eta_1) + \int^{y_2} C_\Sigma^{-1}(\eta_2)^T(d\eta_2) + \int^{y_3} C_\Gamma^{-1}(\eta_3)^T(d\eta_3) \\
&+ \int^{y_4} L_\Sigma^{-1}(\eta_4)^T(d\eta_4).
\end{aligned}
$$

It follows from (5.4) that we may write

$$
\begin{aligned}
&i_\Sigma = (f_1, \dot{y}_2, (\partial H/\partial y_4)^T), \quad v_\Sigma = (e_1, (\partial H/\partial y_2)^T, \dot{y}_4) \\
&i_\Gamma = (e_2, \dot{y}_3, (\partial H/\partial y_1)^T), \quad v_\Gamma = (f_2, (\partial H/\partial y_3)^T, \dot{y}_1).
\end{aligned}
$$

Substituting these into the relationships (18) we obtain a system of equations in the form

$$
\begin{pmatrix} (\partial H/\partial x_1)^T \\ \dot{x}_2 \end{pmatrix} = \begin{pmatrix} J_{11} & J_{12} \\ J_{21} & J_{22} \end{pmatrix} \begin{pmatrix} \dot{x}_1 \\ (\partial H/\partial x_2)^T \end{pmatrix} + \begin{pmatrix} G_1 \\ G_2 \end{pmatrix} f
$$

$$
e = -(G_1^T, G_2^T) \begin{pmatrix} \dot{x}_1 \\ (\partial H/\partial x_2)^T \end{pmatrix} + Kf. \qquad (5.5)
$$

By partitioning the matrix P in (5.1), in a manner consistent with the partitions of the branch current and voltages in (5.3), we see that if

$$
J = \begin{pmatrix} J_{11} & J_{12} \\ J_{21} & J_{22} \end{pmatrix}, \qquad G = \begin{pmatrix} G_1 \\ G_2 \end{pmatrix}
$$

then

$$J = \begin{pmatrix} O & P_{32} & O & P_{33} \\ -P_{32}^T & O & -P_{22}^T & O \\ O & P_{22} & O & P_{23} \\ -P_{33}^T & O & -P_{23}^T & O \end{pmatrix}, \quad G = \begin{pmatrix} P_{31} & O \\ O & -P_{12}^T \\ P_{21} & O \\ O & -P_{13}^T \end{pmatrix}, \tag{5.6}$$

$$K = \begin{pmatrix} O & -P_{11}^T \\ P_{11} & O \end{pmatrix}.$$

We may now write (5.5) in the compact form

$$\begin{pmatrix} (\partial H/\partial x_1)^T \\ \dot{x}_2 \\ e \end{pmatrix} = \begin{pmatrix} J & G \\ -G^T & K \end{pmatrix} \begin{pmatrix} \dot{x}_1 \\ (\partial H/\partial x_2)^T \\ f \end{pmatrix}. \tag{5.7}$$

Note that H, K and $\mathbb{J} = \begin{pmatrix} J & G \\ -G^T & K \end{pmatrix}$ are all skew symmetric. In the context of lossless nonlinear electrical circuits with input-output ports, it is well known that if the topology of the network is such that there are no capacitor loops or inductor cutsets then we may represent the input-output system in a reduced form of equation (5.5) by the equations

$$\begin{aligned} \dot{x}_2 &= J_{22}(\partial H/\partial x_2)^T + G_2 f \\ e &= -G_2^T(\partial H/\partial x_2)^T + Kf. \end{aligned}$$

There are two natural Dirac structures associated with system (5.5). Assuming that

$$\phi_\Gamma^L \in \mathbf{R}^{N_1}, Q_\Sigma^C \in \mathbf{R}^{N_2}, Q_\Gamma^C \in \mathbf{R}^{N_3}, \phi_\Sigma^L \in \mathbf{R}^{N_4}, v_\Gamma^S \in \mathbf{R}^{N_5}, i_\Sigma^S \in \mathbf{R}^{N_6}$$

set

$$V_1 = \mathbf{R}^{N_1} \times \mathbf{R}^{N_2}, \quad V_2 = \mathbf{R}^{N_3} \times \mathbf{R}^{N_4}, \quad U = \mathbf{R}^{N_6} \times \mathbf{R}^{N_5}.$$

Setting $W = V_1 \oplus U \oplus V_2$, $W^* \cong V_1^* \oplus U^* \oplus V_2$, then $\mathbb{J} : W \to W^*$ in (5.7) defines a Dirac structure L on $V = V_1 \oplus U \oplus V_2$. We easily compute using (5.5)

$$\frac{dH}{dt} = \left(\dot{x}_1^T, \frac{\partial H}{\partial x_2}\right)\left(\frac{\partial H^T}{\partial x_1}, \dot{x}_2\right) = \left(\dot{x}_1^T, \frac{\partial H}{\partial x_2}\right) Gf$$

and

$$e^T = -(x_1^T, \partial H/\partial x_2)G - f^T K.$$

Thus,

$$\frac{dH}{dt} + e^T f = 0$$

This generalizes the notion of conservation law for Dirac (Hamiltonian) systems to Dirac (Hamiltonian) systems with inputs and outputs, as in [**5**], [**11**].

Setting $\overline{W} = V_1 \oplus V_2^*$, $\overline{W}^* = V_1^* \oplus V_2$, then $J : \overline{W} \to \overline{W}^*$ defines a Dirac structure $\bar{L}$ on $\bar{V} = V_1 \oplus V_2$. In case of no input-output ports system (5.5) reduces to the form of equation (4.4). We may therefore apply the analysis in section 4 to obtain the canonical coordinates for the underlying Hamiltonian system. Specifically the vectors u and z are given in terms of circuit states as

$$z = x_1 = (\phi_\Gamma^L, Q_\Sigma^C), \quad u = x_2 - J_{21}x_1 = (Q_\Gamma^C - P_{22}Q_\Sigma^C, \phi_\Sigma^L + P_{33}^T \phi_\Gamma^L) \tag{5.8}$$

Clearly, z_1, z_2, u_1 and u_2 are linear combinations of these canonical coordinates. This ability to write out canonical coordinates for general nonlinear L-C electrical circuits is due to the underlying Dirac structure.

We may also apply the analysis in section 4 to system (5.5) in the presence of input-output ports. In terms of the coordinates u and z we may rewrite system (5.5) in the form

$$
\begin{aligned}
\frac{\partial \tilde{H}^T}{\partial z} &= J_{11}\dot{z} + G_1 f \\
\dot{u} &= J_{22}\frac{\partial \tilde{H}^T}{\partial u} + G_2 f \\
e &= -G_1^T\dot{z} - G_2^T\frac{\partial \tilde{H}^T}{\partial u} + Tf
\end{aligned} \tag{5.9}
$$

Now if $G_1 \neq 0$, system (5.9) contains the following subsystems

$$
\begin{aligned}
L_\Gamma^{-1}(\phi_\Gamma^L) &= P_{31}i_\Sigma^S, \quad v_\Sigma^S = -P_{31}^T\dot{\phi}_\Gamma^L \\
C_\Sigma^{-1}(Q_\Sigma^C) &= -P_{12}^T v_\Gamma^S, \quad i_\Gamma^S = P_{12}\dot{Q}_\Sigma^C.
\end{aligned}
$$

Thus system (5.9) involves $\dot{f}$ unless $G_1 = 0$. Hence, if $G_1 = 0$ we may rewrite system (5.9) in z_1, z_2, u_1, u_2 coordinates as

$$\dot{z}_1 = \hat{J}_{11}^{-1}\frac{\partial \tilde{H}^T}{\partial z_1}, \quad \frac{\partial \tilde{H}}{\partial z_2} = 0$$

$$\dot{u}_1 = \hat{J}_{22}\frac{\partial \tilde{H}^T}{\partial u_1} + G_{12}f, \quad \dot{u}_2 = G_{22}f$$

$$e = -G_{12}^T\frac{\partial \tilde{H}^T}{\partial u_1} - G_{22}^T\frac{\partial \tilde{H}}{\partial u_2} + Tf.$$

Again these coordinates are linear combinations of the canonical coordinates in equation (5.8). Thus we have transformed the system into a system of differential equations and constraints generalizing the case without inputs and outputs in section 4.

6. Example

In this section we give an example which demonstrates the structure derived in the previous section. It is based on the circuit in Figure 1.

Write $N = \Gamma \cup \Sigma$; $\Gamma = \{C_2, L_1, C_1, L_3\}$, $\Sigma_1 = \{L_2, C_3\}$.

The equations governing the network are obtained from Kirchoff's laws in the form:

a) $\dot{\phi}_1 + C_1^{-1}(Q_1) + \dot{\phi}_2 + C_2^{-1}(Q_2) = 0$
b) $\dot{\phi}_3 + C_3^{-1}(Q_3) - C_1^{-1}(Q_1) = 0$
c) $\dot{Q}_2 = L_1^{-1}(\phi_1)$
d) $\dot{Q}_2 = L_2^{-1}(\phi_2)$
e) $\dot{Q}_1 + L_3^{-1}(\phi_3) = L_1^{-1}(\phi_1)$
f) $\dot{Q}_3 = L_3^{-1}(\phi_3)$

Writing $H = \Sigma_i \int^{Q_i} C_i^{-1}(\eta)d\eta + \Sigma_i \int^{\phi_i} L_i^{-1}(\eta)d\eta$ and making the identifications

$$y_1 = (\phi_1, \phi_3), \quad y_2 = Q_3, \quad y_3 = (Q_1, Q_2), \quad y_4 = \phi_2$$

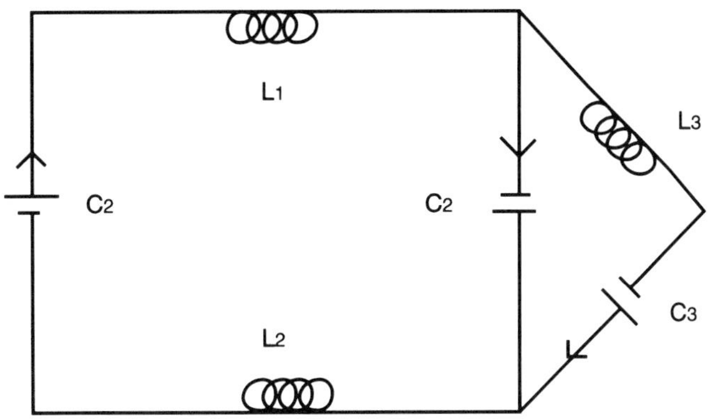

FIGURE 6.1. LC Circuit

we may write the six equations defining the circuit in the form of equations (29) as

$$\begin{pmatrix} \frac{\partial H}{\partial \phi_1} \\ \frac{\partial H}{\partial \phi_3} \\ \frac{\partial H}{\partial Q_3} \\ \dot{Q}_1 \\ \dot{Q}_2 \\ \dot{\phi}_2 \end{pmatrix} = \begin{pmatrix} 0 & 0 & 0 & 0 & 0 & 1 \\ 0 & 0 & 1 & 0 & 0 & 0 \\ 0 & -1 & 0 & 1 & 0 & 0 \\ 0 & 0 & -1 & 0 & 0 & 1 \\ 0 & 0 & 0 & 0 & 0 & 1 \\ -1 & 0 & 0 & -1 & -1 & 0 \end{pmatrix} \begin{pmatrix} \dot{\phi}_1 \\ \dot{\phi}_3 \\ \dot{Q}_3 \\ \frac{\partial H}{\partial Q_1} \\ \frac{\partial H}{\partial Q_2} \\ \frac{\partial H}{\partial \phi_2} \end{pmatrix}.$$

In particular, the canonical coordinates (32) become

$$\begin{pmatrix} u_1 \\ u_2 \end{pmatrix} = y_3 - P_{22} y_2 = \begin{pmatrix} Q_1 \\ Q_2 \end{pmatrix} - \begin{pmatrix} -1 \\ 0 \end{pmatrix} Q_3 = \begin{pmatrix} Q_1 + Q_3 \\ Q_2 \end{pmatrix}$$

$$u_3 = y_4 - P_{33}^T y_1 = \phi_2 + \begin{pmatrix} 1 & 0 \end{pmatrix} \begin{pmatrix} \phi_1 \\ \phi_3 \end{pmatrix} = \phi_2 + \phi_1$$

$$\begin{pmatrix} z_1 \\ z_2 \end{pmatrix} = y_1 = \begin{pmatrix} \phi_1 \\ \phi_3 \end{pmatrix}$$

$$z_3 = y_2 = Q_3$$

and we obtained the decoupled flows

$$\begin{pmatrix} \dot{u}_1 \\ \dot{u}_2 \\ \dot{u}_3 \end{pmatrix} = \begin{pmatrix} 0 & 0 & 1 \\ 0 & 0 & 1 \\ -1 & -1 & 0 \end{pmatrix} \begin{pmatrix} \frac{\partial \tilde{H}}{\partial u_1} \\ \frac{\partial \tilde{H}}{\partial u_2} \\ \frac{\partial \tilde{H}}{\partial u_3} \end{pmatrix}, \begin{pmatrix} \frac{\partial \tilde{H}}{\partial z_1} \\ \frac{\partial \tilde{H}}{\partial z_2} \\ \frac{\partial \tilde{H}}{\partial z_3} \end{pmatrix} = \begin{pmatrix} 0 & 0 & 0 \\ 0 & 0 & 1 \\ 0 & -1 & 0 \end{pmatrix} \begin{pmatrix} \dot{z}_1 \\ \dot{z}_2 \\ \dot{z}_3 \end{pmatrix}.$$

Setting $w_1 = u_1 - u_2$, $w_2 = u_1 + u_2$, $w_3 = u_3$, these equations reduce to ($\tilde{H} = \tilde{H}(w, z)$)

$$\dot{w}_1 = 0, \quad \dot{w}_2 = +\frac{\partial \tilde{H}}{\partial w_3}, \quad \dot{w}_3 = -\frac{\partial \tilde{H}}{\partial w_2}$$

$$\frac{\partial \tilde{H}}{\partial z_1} = 0, \quad \dot{z}_2 = -\frac{\partial \tilde{H}}{\partial z_3}, \quad \dot{z}_3 = +\frac{\partial \tilde{H}}{\partial z_2}.$$

This set of equations is just those expressed in (24). The final set of canonical coordinates is now

$$\begin{aligned} w_1 &= Q_1 + Q_3 - Q_2, & z_1 &= \phi_1 \\ w_2 &= Q_1 + Q_3 + Q_2, & z_2 &= \phi_3 \\ w_3 &= \phi_1 + \phi_2, & z_3 &= Q_3. \end{aligned}$$

We can thus see that our analysis of this example elucidates its generalized Hamiltonian or Dirac structure. As mentioned earlier, the representation of the Dirac structure used here is particularly helpful in finding the canonical coordinates which demonstrate the detailed structure of the system, consisting in general of constants of motion, algebraic constraints, and standard Hamiltonian equations. While it is possible of course to write down its equations by other means, the standard methods cannot capture its Hamiltonian nature. An alternative analysis of LC circuits using the Dirac approach is given in Maschke, van der Schaft and Breedveld [**9**], but see also Dalsmo and van der Schaft [**5**].

7. Conclusions

We have demonstrated here a practical and elegant application of the theory of Dirac or generalized Hamiltonian structures. This theory gives a natural way of simultaneously capturing constraints and constants of motion within the Hamiltonian framework. These ideas are very important for the description of coupled systems of all types, including coupled mechanical systems, as shown in the work of Maschke and van der Schaft [**13**]. We intend to pursue the application of our formalism to other systems such as coupled mechanical systems and in particular to consider this formalism on manifolds rather than just on vector spaces.

References

[1] R. Abraham and J. E. Marsden, *Foundations of Mechanics*, Second Edition, Addison Wesley, (1978).
[2] V. I. Arnold, *Mathematical Methods of Classical Mechanics*, Second Edition, Graduate Texts in Mathematics, Vol. 60, Springer Verlag, (1989).
[3] G.M. Bernstein and M.A. Lieberman, "A Method for Obtaining a Canonical Hamiltonian for Nonlinear L C Circuits," IEEE Trans. Circuits and Systems, Vol. 36, pp. 411–420 (1989).
[4] T. Courant, *Dirac Manifolds*, Trans. American Math. Soc. Vol. 319, pp. 631–661 (1990).
[5] M. Dalsmo and A. J. van der Schaft, "On Representations and Integrability of Mathematical Structures in Energy-Conserving Physical Systems," Memorandum 1350, Faculty of Applied Mathematics, University of Trente (1996).
[6] I. Dorfman, *Dirac Structures and Integrability of Nonlinear Evolution Equations*, Chichester: John Wiley (1993).
[7] H. G. Kwatny, F. M. Massimo and L. Y. Bahor, "The Generalized Lagrange Formulation for Nonlinear RLC Networks," IEEE Trans. Circuits and Systems, Vol. 29, pp. 220–233 (1982).
[8] J. E. Marsden and T. S. Ratiu, *Introduction to Mechanics and Symmetry*, Texts in Applied Mathematics, Vol. 17, Springer Verlag, (1994).
[9] B. M. Maschke, A. J. van der Schaft, and P. C. Breedveld, "An Intrinsic Hamiltonian Formulation of the Dynamics of LC-Circuits," IEEE Trans. Circuits and Systems, Vol. 42, pp. 73–82 (1995).
[10] A. J. van der Schaft and B. M. Maschke, "On the Hamiltonian Formulation of Nonholonomic Mechanical Systems," Rep. Math. Phys. Vol 34, pp. 225–233 (1994).
[11] A. J. van der Schaft and B. M. Maschke, "The Hamiltonian Formulation of Energy Conserving Physical Systems with External Ports," Archiv. fur Elektronik und Uberlragungstechnik, Vol. 49, pp. 362–371 (1995).

[12] A. J. van der Schaft and B. M. Maschke, "Mathematical Modelling of Constrained Hamiltonian Systems," preprints 3rd NOLCOS, Tahoe, CA (1995).
[13] B. M. Maschke and A. J. van der Schaft, "Interconnected Mechanical Systems," in Modelling and Control of Mechanical Systems, (eds. A. Astolfi et al.), Imperial College Press, London, 1997.

DEPARTMENT OF MATHEMATICS, UNIVERSITY OF MICHIGAN, ANN ARBOR, MICHIGAN 49109
E-mail address: `abloch@umich.edu`

CENTER FOR SYSTEMS SCIENCE AND ENGINEERING, ARIZONA STATE UNIVERSITY, TEMPE, AZ 85287.
E-mail address: `peter.crouch@asu.edu`

Proceedings of Symposia in Pure Mathematics
Volume **64**, 1999

NON INTEGRABLE GEODESICS IN SR-MARTINET GEOMETRY

BERNARD BONNARD, MONIQUE CHYBA, AND IVAN KUPKA

ABSTRACT. Consider a sub-Riemannian geometry (U, D, g) where U is a neighborhood of 0 in $I\!R^3$, D is a Martinet type distribution identified to ker ω, ω being the one form : $dz - \frac{y^2}{2}dx$ and g is a metric on D which can be written as $a(q)dx^2 + 2b(q)dxdy + c(q)dy^2$ where $q = (x, y, z)$. In our previous study we prove that g can be written in a normal form where $b = 0$, $a = 1 + \alpha y + ...$, $c = 1 + \beta x + \gamma y + ...$, $\alpha, \beta, \gamma, ...$ being real parameters. Moreover we analyze the case $\beta = 0, ...$ where the geodesics are integrable by quadratures. In this article we complete the analysis by considering the non integrable case : $\alpha = \gamma = 0, ...$, and $\beta \neq 0$. We describe the set of geodesics. This allows to understand the role of the parameter β in the normal form. Moreover we provide a geometric framework to analyze the SR-sphere and wave front in the general case. A consequence is the conjecture that the sphere is not a sub-pfaffian set.

1. INTRODUCTION

In this article, we consider the sub-Riemannian geometry (U, D, g) where U is a small open set of $I\!R^3$ containing 0, D is the distribution ker ω, ω being the one-form $dz - \dfrac{y^2}{2}dx$ where $q = (x, y, z)$ are the coordinates of $I\!R^3$ and g is a metric on D written as $a(q)dx^2 + 2b(q)dxdy + c(q)dy^2$ where a, b, c are germs at 0 of analytic functions. The normal form computed in [1] shows that g can be taken as a sum of squares : $adx^2 + cdy^2$, where $a = 1 + yF(q)$, $c = 1 + G(q), G_{|x=y=0} = 0$. Our study is localized around 0 and the coordinates (x, y, z) have to be gradated as follows : the weight of x, y is one and the weight of z is three. Using this gradation the normal form of order -1 is the *flat case* where $g = dx^2 + dy^2$ and the normal form of order 0 is given by $g = (1+\alpha y)dx^2 + (1+\beta x+\gamma y)dy^2$ and is depending upon 3 real parameters α, β, γ. It is a model to study the *generic SR-Martinet geometry.* In [1] we analyze the sub-Riemannian problem in the flat case ; we compute the conjugate and cut loci, the sphere and the wave front. Also we compute precise estimates concerning the distance function. In particular we identify the *transcendence of the problem* which can be resumed as follows. The trace of the sphere with the Martinet plane $y = 0$ where ω is not a contact form is not *sub-analytic* but belongs to the *sub-pfaffian category.* Also we give a geometric explanation of the phenomenon. The set of normal geodesics project onto the trajectories of an *oscillating pendulum* and the sphere is related to Dulac-Poincaré return mapping evaluated along the separatrix connecting the two saddles. In [3] we complete the analysis by considering the case where $g = (1 + \alpha y)dx^2 + (1 + \gamma y)dy^2$ where the set of geodesics is *integrable by*

Date: October 2, 1997.
1991 *Mathematics Subject Classification.* Primary 49K15, 53B21, 58F10.
Key words and phrases. Sub-Riemannian geometry, Abnormal extremals, Generic Martinet case.

quadratures. We describe the trace of the sphere and the wave front of small radius with the Martinet plane $y = 0$. Also we compute estimates concerning the distance function near the *abnormal geodesics.* In particular we give a clear explanation of the role of the two parameters α, γ in the normal form. The analysis depends strongly upon the integrability property of the set of geodesics and is related to the *oscillating and rotating* trajectories of a pendulum.

This analysis is not generic and the objective of this article is to analyze the role of the remaining parameter β. This parameter is crucial : if β is non zero it breaks the integrability property of the set of the geodesics. The model we need in this case is a *damped pendulum.* This requires additional transcendence to parametrize the sphere because we must deal with a Dulac-Poincaré return mapping evaluated near a *non integrable saddle.*

To introduce precisely our problem, we need some definitions. An *admissible curve* is an absolutely continuous curve $q : [0,T] \mapsto U \subset I\!R^3$ such that $\dot{q}(t) = \dfrac{dq(t)}{dt} \in D(q(t))\backslash\{0\}$ for almost every t. Let $(,)$ be scalar product defined by g. The *length* and the *energy* of an admissible curve q are respectively :

$$L(q) = \int_0^T (\dot{q}(t), \dot{q}(t))^{\frac{1}{2}} dt \ , \ E(q) = \int_0^T (\dot{q}(t), \dot{q}(t)) dt.$$

Let $q_0, q_1 \in U$, a *minimizing curve* joining q_0 to q_1 is an admissible curve of minimal length. To compute the minimizers it is convenient to use the equivalence between the length and the energy minimizing problems and the *maximum principle.* According to this principle, minimizers can be selected among a restricted set of curves as follows.

Let G_1, G_2 be two real analytic vector fields on U such that $D =\text{Span}\{G_1, G_2\}$ and introduce the *control system*

$$\frac{dq(t)}{dt} = \sum_{i=1}^{2} u_i(t) G_i(q(t))$$

where $u(t) = (u_1(t), u_2(t))$ is the *control* associated to q. Let η be a constant equal to 0 or $\frac{1}{2}$ and consider the *Hamilton function* associated to the energy minimizing problem

$$H_\eta(q,p,u) = \langle p, G(q)u \rangle - \eta(G(q)u, G(q)u)$$

where $p = (p_x, p_y, p_z)$ denotes the adjoint vector, (p, η) is non zero, $\langle , \rangle$ is the standard inner product in $I\!R^3$ and $G(q)u$ stands for $\sum_{i=1}^{2} u_i G_i(q)$. A *bi-extremal* is an absolutely continuous curve (q,p) defined on $[0,T]$ such that $t \mapsto (q(t), p(t), u(t))$ is solution for almost all t of the following equations :

$$\frac{dq}{dt} = \frac{\partial H_\eta}{\partial p}(q,p,u) \ , \ \frac{dp}{dt} = -\frac{\partial H_\eta}{\partial q}(q,p,u) \tag{1.1}$$

$$\frac{\partial H_\eta}{\partial u}(q,p,u) = 0. \tag{1.2}$$

Its projection q on $I\!R^3$ is called a *geodesic.* Since the energy and the length minimizing problems are equivalent then according to the maximum principle, a minimizer is a geodesic. When we study the bi-extremals curves, we must distinguish between two cases. If $\eta = 0$, the bi-extremal is called *abnormal* and if $\eta \neq 0$, it is called *normal* and the associated geodesics are respectively called abnormal and normal. A geodesic is called *strictly abnormal* if it is the projection of an abnormal bi-extremal but not a normal one.

In the normal case, equation (1.2) can be solved as follows. Since it is linear with respect to u, the control associated to (q,p) can be computed as an analytic mapping $\hat{u} : (q,p) \mapsto I\!R^2$. By plugging $\hat{u}$ in H_η where $\eta = \frac{1}{2}$ we define an Hamilton function H_n on $U \times I\!R^3$ by setting $H_n(q,p) = H_{\frac{1}{2}}(q,p,\hat{u})$. Then using (1.1) and (1.2), we observe that a normal bi-extremal (q,p) is solution of the following *analytic differential equation*

$$\frac{dq}{dt} = \frac{\partial H_n}{\partial p}(q,p)\ ,\ \frac{dp}{dt} = -\frac{\partial H_n}{\partial q}(q,p). \tag{1.3}$$

Similary, the computation of abnormal bi-extremals is straightforward when $D = \ker\omega$, $\omega = dz - \frac{y^2}{2}dx$. Since $\eta = 0$, the constraint (1.2) takes the form

$$\langle p, G_i(q)\rangle = 0 \qquad \text{for } i = 1,2 \tag{1.4}$$

and if we differentiate twice this equation with respect to time, we get that abnormal geodesics are contained in the set

$$\det(G_1, G_2, [G_1, G_2]) = 0$$

which is the plane $y = 0$ corresponding to the points where ω is not a contact form. Moreover they are *lines* $z = z_0$. The set $y = 0$ which has an important geometric meaning is called the *Martinet plane.*

Assume now that the geodesics are *parametrized by arc length* : $(\dot{q}, \dot{q}) = 1$. We fix the initial point q_0 to 0 and let $q(t,p_0), p(t,p_0)$ be the solution of (1.3) starting at $t = 0$ from $q_0 = 0$ and p_0. It is contained in the level set $H_n = \frac{1}{2}$. The *exponential mapping* is the map

$$\exp_0 : (p_0, t) \mapsto q(t, p_0).$$

The *sub-Riemannian sphere* with radius $r > 0$ is the set $S(0,r)$ of points which are at SR-distance r from 0. The *wave front* of length r is the set $W(0,r)$ of end-points of geodesics with length r, starting from 0. Our study is localized near 0. Hence in the sequel we shall assume r small enough such that we can connect the point 0 using a minimizing curve to each point at SR-distance r from 0. In particular $S(0,r)$ is a subset of $W(0,r)$.

In our previous work [1] we have described the sphere and the wave front in the *flat case* $g = dx^2 + dy^2$. In particular we have proved that the sphere is not sub-analytic but is pfaffian. In [3] we have generalized this analysis to the case where $g = (1+\alpha y)dx^2 + (1+\gamma y)dy^2$ explaning in particular the role of the two parameters α, γ in the normal form. The main object of this article is to describe the set of geodesics in the case where $g = dx^2 + (1+\beta x)dy^2$. Contrarily to the previous situation, the set of geodesics cannot be integrated by quadratures. Hence we shall require more transcendence to compute the geodesics and hence the sphere. Also it will provide a *geometric framework to understand the role of abnormal geodesics in SR-geometry.*

2. Computation of the normal geodesics

Our analysis is localized in a neighborhood U of 0. The normal form computed in [1] for the SR-geometry in the Martinet case is (D,g) with $D = \ker\omega$, $\omega = dz - \frac{y^2}{2}dx$

and $g = a(q)dx^2 + c(q)dy^2$ where a, c are germs of analytic functions of the form $a = 1 + yF(q)$, $c = 1 + G(q)$,$G_{|x=y=0} = 0$, $q = (x, y, z)$. The coordinates x, y, z are gradated as follows : the weight of x,y is 1 and the weight of z is 3. This induces the following gradation on vector fields : the weight of $\frac{\partial}{\partial x}, \frac{\partial}{\partial y}$ is -1 and the weight of $\frac{\partial}{\partial z}$ is -3. We introduce the following two vector fields :

$$G_1 = \frac{\partial}{\partial x} + \frac{y^2}{2}\frac{\partial}{\partial z} \ , \ G_2 = \frac{\partial}{\partial y}$$

such that D =Span$\{G_1, G_2\}$. If (,) designs the scalar product associated to g we have

$$(G_1, G_1) = a \ , \ (G_2, G_2) = c \ , \ (G_1, G_2) = 0.$$

If we set $F_1 = \frac{G_1}{\sqrt{a}}$,$F_2 = \frac{G_2}{\sqrt{c}}$, F_1, F_2 are two orthonormal vector fields such that D =Span$\{F_1, F_2\}$. They are completed by $F_3 = \frac{\partial}{\partial z}$ to form a frame and we can define locally a Riemannian metric on $I\!R^3$ by taking the F_i's as orthonormal vector fields. The restriction of this metric to D is the original SR-metric.

The normal form of *order* p, where $p \geq -1$ is defined by taking all the terms of order $\leq p$. Hence we get :

- Normal form of order -1 : flat case

$$g = dx^2 + dy^2.$$

- Normal form of order 0 :

$$g = (1 + \alpha y)dx^2 + (1 + \beta x + \gamma y)dy^2.$$

The energy minimizing problem is equivalent to the length minimizing problem and can be written using the optimal control framework :

$$\text{(2.1)} \qquad \begin{gathered} \frac{dq(t)}{dt} = \sum_{i=1}^{2} u_i(t) G_i(q(t)), \\ \min_{u(.)} \int_0^T \Big(a(q(t)) u_1^2(t) + c(q(t)) u_2^2(t) \Big) dt. \end{gathered}$$

The Hamiltonian associated to *normal geodesics* is

$$H_{\frac{1}{2}}(q, p, u) = \sum_{i=1}^{2} u_i \langle p, G_i(q) \rangle - \frac{1}{2}(a u_1^2 + c u_2^2)$$

where $p = (p_x, p_y, p_z)$. To compute a geodesic control we must solve : $\frac{\partial H_{\frac{1}{2}}}{\partial u} = 0$ where $u = (u_1, u_2)$ and we get :

$$u_1 = \frac{1}{a}(p_x + p_z \frac{y^2}{2}) \ , \ u_2 = \frac{p_y}{c}$$

and the associated Hamilton function is :

$$H_n(q, p) = \frac{1}{2}(a u_1^2 + c u_2^2) = \frac{1}{2}\Big[\frac{(p_x + p_z \frac{y^2}{2})^2}{a} + \frac{p_y^2}{c}\Big].$$

Basic assumption. From now on we shall assume that H_n is not depending on z. It corresponds to the so-called *isoperimetric situation.* In this case, z is a *cyclic coordinate* and p_z is a *first integral.* Observe that the normal form of order 0 is isoperimetric.

The normal bi-extremals are solutions of the following equations :

$$\begin{aligned}
\dot{x} &= \frac{1}{a}(p_x + p_z\frac{y^2}{2}) \\
\dot{y} &= \frac{p_y}{c} \\
\dot{z} &= \frac{y^2}{2a}(p_x + p_z\frac{y^2}{2}) \\
\dot{p}_x &= \frac{p_y^2 c_x}{2c^2} + \frac{(p_x + p_z\frac{y^2}{2})^2 a_x}{2a^2} \\
\dot{p}_y &= \frac{p_y^2 c_y}{2c^2} + \frac{(p_x + p_z\frac{y^2}{2})^2 a_y}{2a^2} - \frac{(p_x + p_z\frac{y^2}{2})}{a}p_z y \\
\dot{p}_z &= 0.
\end{aligned} \tag{2.2}$$

To compute the normal geodesics it is convenient to introduce : $P_i = \langle p, F_i(q)\rangle$, for $i = 1, 2, 3$ and let P be the vector whose components are (P_1, P_2, P_3). In this representation the Hamiltonian takes the form

$$H_n = \frac{1}{2}(P_1^2 + P_2^2)$$

where $P_1 = \dfrac{p_x + p_z\frac{y^2}{2}}{\sqrt{a}}$, $P_2 = \dfrac{p_y}{\sqrt{c}}$, $P_3 = p_z$ and the normal bi-extremals are solutions of the following equations in the (q, P) coordinates :

$$\begin{aligned}
\dot{x} &= \frac{1}{\sqrt{a}}P_1 \\
\dot{y} &= \frac{1}{\sqrt{c}}P_2 \\
\dot{z} &= \frac{y^2}{2}\frac{P_1}{\sqrt{a}} \\
\dot{P}_1 &= \frac{P_2}{\sqrt{a}\sqrt{c}}[yP_3 - \frac{1}{2}\frac{a_y}{\sqrt{a}}P_1 + \frac{c_x}{2\sqrt{c}}P_2] \\
\dot{P}_2 &= -\frac{P_1}{\sqrt{a}\sqrt{c}}[yP_3 - \frac{1}{2}\frac{a_y}{\sqrt{a}}P_1 + \frac{c_x}{2\sqrt{c}}P_2] \\
\dot{P}_3 &= 0.
\end{aligned} \tag{2.3}$$

If we set : $P_1 = \cos\theta$, $P_2 = \sin\theta$, $P_3 = p_z = \lambda$ we get :

$$\begin{aligned}
&\dot{x} = \frac{\cos\theta}{\sqrt{a}} \qquad \dot{y} = \frac{\sin\theta}{\sqrt{c}} \qquad \dot{z} = \frac{y^2}{2\sqrt{a}}\cos\theta \\
&\dot{\theta} = -\frac{1}{\sqrt{a}\sqrt{c}}(y\lambda - \frac{a_y}{2\sqrt{a}}\cos\theta + \frac{c_x}{2\sqrt{c}}\sin\theta)
\end{aligned} \tag{2.4}$$

which can be expanded using :

$$a = (1 + \alpha y + ...) \;,\; c = (1 + \beta x + \gamma y + ...).$$

If we make the following parametrization : $\sqrt{c}\dfrac{d}{dt} = \dfrac{d}{d\tau}$, we get the equations

$$\frac{dx}{d\tau} = \frac{\sqrt{c}}{\sqrt{a}}\cos\theta \quad \frac{dy}{d\tau} = \sin\theta \quad \frac{dz}{d\tau} = \frac{y^2}{2}\frac{\sqrt{c}}{\sqrt{a}}\cos\theta$$
$$\frac{d\theta}{d\tau} = -\frac{1}{\sqrt{a}}(y\lambda - \frac{a_y}{2\sqrt{a}}\cos\theta + \frac{c_x}{2\sqrt{c}}\sin\theta) \tag{2.5}$$

The following properties are straightforward.

Lemma 2.1. *If $a_x = c_x = 0$, then x, z are cyclic coordinates and the equations whose solutions are the normal bi-extremals can be integrated by quadratures.*

Lemma 2.2. *The abnormal geodesics : $t \mapsto (\pm t, 0, 0)$ are not strictly abnormal if and only if the restriction of a_y to the Martinet plane : $y = 0$ is 0.*

3. Formal expansions - Mathieu equation

To analyze the normal geodesics in a neighborhood of 0, we can blow-up equations (2.3) using the *quasi-homogeneity property of the flat case.* Indeed if we set

$$x = \varepsilon X\ ,\ y = \varepsilon Y\ ,\ z = \varepsilon^3 Z$$
$$P_1 = Q_1\ ,\ P_2 = Q_2\ ,\ P_3 = \frac{Q_3}{\varepsilon^2}$$

and if we use the following parametrization : $ds = \dfrac{dt}{\varepsilon}$, the equations (2.3) can be written

$$\begin{aligned}
\frac{dX}{ds} &= \frac{1}{\sqrt{a}}Q_1\\
\frac{dY}{ds} &= \frac{1}{\sqrt{c}}Q_2\\
\frac{dZ}{ds} &= \frac{Y^2}{2}\frac{Q_1}{\sqrt{a}}\\
\frac{dQ_1}{ds} &= \frac{Q_2}{\sqrt{a}\sqrt{c}}[YQ_3 - \frac{1}{2}\frac{\varepsilon a_y}{\sqrt{a}}Q_1 + \frac{\varepsilon c_x}{2\sqrt{c}}Q_2]\\
\frac{dQ_2}{ds} &= -\frac{Q_1}{\sqrt{a}\sqrt{c}}[YQ_3 - \frac{1}{2}\frac{\varepsilon a_y}{\sqrt{a}}Q_1 + \frac{\varepsilon c_x}{2\sqrt{c}}Q_2]\\
\frac{dQ_3}{ds} &= 0.
\end{aligned} \tag{3.1}$$

If we introduce the angle θ by setting : $Q_1 = \cos\theta, Q_2 = \sin\theta$ we obtain the following equations :

$$\begin{aligned}
X' &= \frac{1}{\sqrt{a}}\cos\theta\\
Y' &= \frac{1}{\sqrt{c}}\sin\theta\\
Z' &= \frac{Y^2}{2}\frac{\cos\theta}{\sqrt{a}}\\
\theta' &= -\frac{1}{\sqrt{a}\sqrt{c}}[YQ_3 - \frac{1}{2}\frac{\varepsilon a_y}{\sqrt{a}}\cos\theta + \frac{\varepsilon c_x}{2\sqrt{c}}\sin\theta]
\end{aligned} \tag{3.2}$$

which must be integrated with $Q_3 = 0$ or $Q_3 = 1$ and $'$ is the derivative with respect to s. The case $Q_3 = 0$ is trivial to analyze and if $Q_3 = 1$, we get $\varepsilon = \frac{1}{\sqrt{\lambda}}$ and

$$\begin{aligned} X' &= \frac{1}{\sqrt{a}}\cos\theta \\ Y' &= \frac{1}{\sqrt{c}}\sin\theta \\ Z' &= \frac{Y^2}{2}\frac{\cos\theta}{\sqrt{a}} \\ \theta' &= -\frac{1}{\sqrt{a}\sqrt{c}}[Y - \frac{1}{2}\frac{\varepsilon a_y}{\sqrt{a}}\cos\theta + \frac{\varepsilon c_x}{2\sqrt{c}}\sin\theta]. \end{aligned} \tag{3.3}$$

We set

$$\begin{aligned} X &= X_0 + \varepsilon X_1 + o(\varepsilon) \\ Y &= Y_0 + \varepsilon Y_1 + o(\varepsilon) \\ Z &= Z_0 + \varepsilon Z_1 + o(\varepsilon) \\ \theta &= \theta_0 + \varepsilon\theta_1 + o(\varepsilon) \end{aligned}$$

and we integrate with the following initial conditions

$$\begin{aligned} X_i(0) = Y_i(0) = Z_i(0) = 0 \\ \theta_0(0) = \theta(0)\ ,\ \theta_1(0) = 0, \end{aligned} \tag{3.4}$$

using the gradated expansions

$$a(x, y) = 1 + \alpha y + ...$$

$$c(x, y) = 1 + \beta x + \gamma y + ...$$

Identifying terms by terms in (3.2) we obtain the following equations

$$\begin{aligned} X_0' &= \cos\theta_0 \\ Y_0' &= \sin\theta_0 \\ Z_0' &= \frac{Y_0^2}{2}\cos\theta_0 \\ \theta_0' &= -Y_0 \end{aligned} \tag{3.5}$$

and

$$\begin{aligned} X_1' &= -\frac{\alpha Y_0}{2}\cos\theta_0 - \theta_1\sin\theta_0 \\ Y_1' &= -\frac{\sin\theta_0}{2}(\beta X_0 + \gamma Y_0) + \theta_1\cos\theta_0 \\ Z_1' &= Y_0 Y_1\cos\theta_0 - \frac{\alpha Y_0^3}{4}\cos\theta_0 - \frac{Y_0^2}{2}\theta_1\sin\theta_0 \\ \theta_1' &= -Y_1 + \frac{1}{2}(\alpha\cos\theta_0 - \beta\sin\theta_0) + \frac{Y_0}{2}[(\alpha+\gamma)Y_0 + \beta X_0]. \end{aligned} \tag{3.6}$$

Equations (3.5) correspond to the *flat case*. The angle θ_0 is a solution of the *oscillating pendulum* ($\theta_0'(0) = 0$)

$$\theta_0'' + \sin\theta_0 = 0. \tag{3.7}$$

The main property of equations (3.6) is the following. We observe that θ_1 is governed by an equation of the form :

$$\theta_1'' + \theta_1\cos\theta_0 = K(X_0, Y_0, Z_0, \theta_0) \tag{3.8}$$

where θ_0 is a periodic motion. It is an equation of the *Mathieu type*, with a right hand side. This equation corresponds to the *variational equation* of our problem. It reflects the *stability property* of the flat case, especially when computing the *wave front.* A better expansion can be computed by taking into account the *secular effects*. Indeed we can expand θ as $\theta = \theta_0(\varepsilon) + \varepsilon\theta_1(\varepsilon) + o(\varepsilon)$ where the *amplitude* and the *phase* of the pendulum is depending on ε, see [10].

4. Phase portraits in the case $g = dx^2 + (1+\beta x)^2 dy^2$.

In this section we analyze the geodesics in the case $g = dx^2 + c(x)dy^2$, $c(x) = (1+\beta x)^2$ to understand the role of the parameter β in the normal form. Hence $\frac{c_x}{2\sqrt{c}} = \beta$ and if we set $\sqrt{c}\frac{d}{dt} = \frac{d}{d\tau}$, we end up with the system :

$$\frac{dy}{d\tau} = \sin\theta, \qquad \frac{d\theta}{d\tau} = -(y\lambda + \beta\sin\theta) \tag{4.1}$$

which can be written as :

$$\theta'' + \lambda\sin\theta + \beta\cos\theta\theta' = 0 \tag{4.2}$$

where θ' is the derivative with respect to τ. We are concerned by the solution for $|\lambda|$ large and if $\lambda >> \beta$ the previous equation can ve written :

$$\frac{d^2\theta}{d\delta^2} + \frac{\lambda}{|\lambda|}\sin\theta + \varepsilon\cos\theta\frac{d\theta}{d\delta} = 0 \tag{4.3}$$

where $\delta = \sqrt{|\lambda|}\tau$ and $\varepsilon = \frac{\beta}{\sqrt{|\lambda|}}$ is a small parameter. It is then a *perturbation of the pendulum equation.*

Lemma 4.1. *Let $\theta : [a,b] \mapsto I\!R$ be a solution of equations (4.2), then we have the following energy-balance formula :*

$$\begin{aligned}\frac{\theta'^2(b) - \theta'^2(a)}{2} + \lambda\beta\int_a^b \sin^2\theta(\tau)d\tau &= \lambda(\cos\theta(b) - \cos\theta(a)) \\ +\frac{\beta^2}{4}(\cos 2\theta(b) - \cos 2\theta(a)) &+ \beta(\sin\theta(a)\theta'(a) - \sin\theta(b)\theta'(b)).\end{aligned} \tag{4.4}$$

Proof. Using (4.2) we have

$$\theta''\theta' + \beta\cos\theta\theta'^2 + \lambda\sin\theta\theta' = 0$$

and integrating between a and b we obtain :

$$\Big[\frac{1}{2}\theta'^2\Big]_a^b + \beta\int_a^b \cos\theta\theta'^2 d\tau - \Big[\lambda\cos\theta\Big]_a^b = 0.$$

Moreover we have :

$$\beta\int_a^b \cos\theta\theta'^2 d\tau = \beta\int_a^b (\sin\theta)'\theta' d\tau$$

and integrating by parts, we obtain :

$$\beta\int_a^b \cos\theta\theta'^2 d\tau = \beta\Big[\sin\theta\theta'\Big]_a^b - \beta\int_a^b \sin\theta\theta'' d\tau.$$

Using (4.2) we have :

$$-\beta\int_a^b \sin\theta\theta'' d\tau = \beta\int_a^b \sin\theta(\lambda\sin\theta + \beta\cos\theta\theta')d\tau$$

$$= \lambda\beta\int_a^b \sin^2\theta d\tau + \frac{\beta^2}{2}\int_a^b \sin 2\theta\theta' d\tau = \lambda\beta\int_a^b \sin^2\theta d\tau - \frac{\beta^2}{4}\Big[\cos 2\theta\Big]_a^b.$$

Hence we have proved the energy-balance formula. □

Properties 4.2. Equation (4.2) :

$$\theta'' + \lambda \sin\theta + \beta \cos\theta\theta' = 0$$

is a *spray* on S^1 endowed with the coordinate $Z = e^{i\theta}$ and the associated system

$$\theta' = v \tag{4.5}$$
$$v' = -\lambda\sin\theta - \beta\cos\theta v$$

defines a vector field $\aleph$ on the *cylinder* $S^1 \times I\!R$. It has *two equilibrium points* $M_1 = (1,0)$ and $M_2 = (-1,0)$ corresponding to $\theta = k\pi$ and $v = 0$. *If* $\lambda >> \beta$, the singular points are the following :

• Near $M_1 = (1,0)$.

The two eigenvalues of the linearized system are

$$\alpha = \frac{-\beta \pm \sqrt{\beta^2 - 4\lambda}}{2}.$$

Hence :

1. *If* $\lambda > 0$: it is a *stable focus* if $\beta > 0$ and an *unstable focus* if $\beta < 0$.
2. *If* $\lambda < 0$: it is a *saddle.*

• Near $M_2 = (-1,0)$.

The two eigenvalues of the linearized system are

$$\alpha = \frac{\beta \pm \sqrt{\beta^2 + 4\lambda}}{2}.$$

Hence :

1. *If* $\lambda > 0$: it is a *saddle.*
2. *If* $\lambda < 0$: it is a *stable focus* if $\beta < 0$ and an *unstable focus* if $\beta > 0$.

Lemma 4.3. *The vector field* $\aleph$ *is complete.*

Proof. Let $\theta :]e_-, e_+[\mapsto I\!R$ be a maximal solution of (4.2) and assume for instance $e_+ < +\infty$. Hence $\theta'(t) \to \infty$ when $t \to e_+$ and for each $M > 0$ there exists $b_0 \subset]c_-, c_+[$ such that

$$|\theta'(b)| \geq 4M \qquad \text{if} \qquad b \geq b_0.$$

In particular if $M \geq 1$ we have

$$\frac{\theta'^2(b)}{2} > M(|\theta'(b)| + 1). \tag{4.6}$$

Let $e_- < a < b < e_+$ and applying the energy-balance formula we get :

$$\frac{\theta'^2(b)}{2} \leq |\lambda\beta|(e_+ - a) + 2|\lambda| + \frac{\beta^2}{2} + |\beta|(|\theta'(a)| + |\theta'(b)|) + \frac{\theta'^2(a)}{2}.$$

Hence if

$$M \geq \max\Big(|\beta|, |\lambda\beta|(e_+ - a) + 2|\lambda| + \frac{\beta^2}{2} + |\beta||\theta'(a)| + \frac{\theta'^2(a)}{2}\Big)$$

we have

$$\frac{\theta'^2(b)}{2} \leq M(|\theta'(b)| + 1). \tag{4.7}$$

This contradicts (4.6). The proof is similar if $e_- = -\infty$. □

Lemma 4.4. *The only periodic trajectories of* $\aleph$ *are the stationary points (1,0) and (-1,0) if* $\lambda >> \beta$.

Proof. Let $\theta : I\!R \to I\!R$ be a solution of (4.2) such that the associated trajectory γ of $\aleph$ on the cylinder is a periodic solution. We must analyze two cases :
• Case a: γ is not homotopic to 0. Then there exists $a, b \in I\!R$, $a < b$ such that $\theta(a) = 0$, $\theta(b) = 2\pi\mathbf{Z}$ and $\theta'(a) = \theta'(b)$. If we apply the energy-balance formula to $\theta : [a, b] \mapsto I\!R$ we get :

$$\lambda\beta \int_a^b \sin^2\theta(\tau)d\tau = 0.$$

Since $\lambda\beta \neq 0$ we obtain $\sin^2\theta(\tau) = 0$, $\forall\ \tau \in [a, b]$. Hence $\theta(\tau) = k\pi$, $k \in \mathbf{Z}$, for each $\tau \in I\!R$.
• Case b: γ is homotopic to 0. We observe that $\aleph$ admits two singular points (1,0),(-1,0) whose index is +1,-1 if $\lambda > 0$ and -1,+1 if $\lambda < 0$. According to Poincaré theorem γ must encircled the singular point of index 1. Assume $\lambda > 0$, hence (1,0) is inside the closed path γ and there exists $a < b$ such that $\theta(a) = \theta(b) = 0$, $\theta'(a) = \theta'(b)$. If we apply again the energy-balance formula to θ we obtain as before : $\int_a^b \sin^2\theta(\tau)d\tau = 0$ and hence $\theta(\tau) = k\pi$. The case $\lambda < 0$ is similar replacing (1,0) by (-1,0). □

Lemma 4.5. *If* $\lambda\beta \neq 0$, *then* $\aleph$ *has no saddle connexions.*

Proof. If $\lambda > 0$, (-1,0) is the only saddle, if $\lambda < 0$ it is the point (1,0). It is sufficient to consider the first case. Assume that $\theta : I\!R \to I\!R$ is a saddle connexion, then $\lim_{\tau\to\pm\infty} e^{i\theta\tau} = -1$, $\lim_{\tau\to\pm\infty} \theta'(\tau) = 0$. Let $a < b$, and applying the energy-balance formula with $a \to -\infty$ and $b \to +\infty$ we get $\int_{-\infty}^{+\infty} \sin^2\theta(\tau)d\tau = 0$ and this is absurd. □

Phase portrait. Using the previous analysis and the Poincaré-Bendixon theorem on the cylinder we can represent the phase portrait in the case $\lambda > 0$, $\beta > 0$, the others cases being similar.

We observe that the system is left invariant by the symmetry $S : (\theta, \theta') \mapsto (-\theta, -\theta')$. Hence we can assume $\theta' \geq 0$. The motion is similar to a *damped pendulum.* Every trajectory other than the separatrix going to the saddle point ultimately winds about the stable focus. Hence the $\omega-$limit set of each trajectory can be either the focus or the saddle point and the $\alpha-$limit set is either empty or the saddle point. The behavior when $t \to -\infty$ is complicated due to the nature of the damping. See figure 1 (resp. figure 2) for the phase portrait on the cylinder (resp. on the phase space).

5. A GEOMETRIC FRAMEWORK TO ANALYZE THE SINGULARITY OF THE SPHERE IN THE ABNORMAL DIRECTION

5.1. **Geometric framework.** In the flat case, the trace of the sphere $S(0, r)$ with the Martinet plane $y = 0$ is obtained by taking the end-points of the normal geodesics of length r at *their first intersections* with the Martinet plane $y = 0$. Moreover their sucessive intersections will define the trace of the wave front $W(0, r)$ with the Martinet plane. Their behaviors in the abnormal directions are obtained

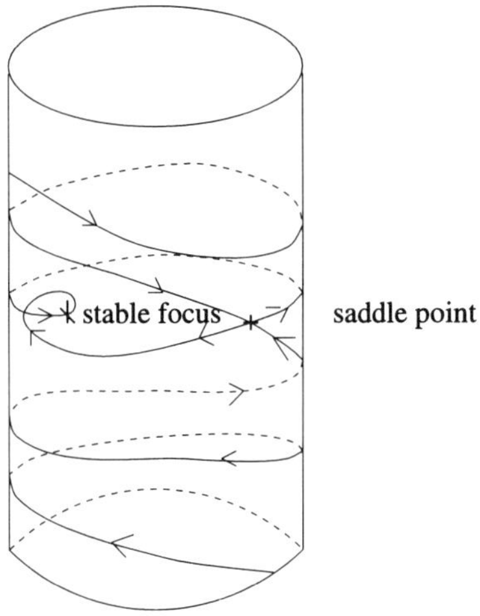

FIGURE 1

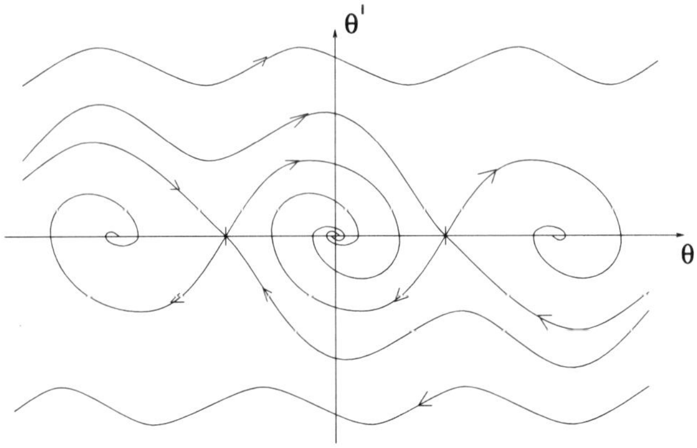

FIGURE 2

by making $\theta(0) \to k\pi$. We represent on figure 3 the behaviors when $\theta(0) \to -\pi$ or 0 and the corresponding trace of the sphere with $y = 0$.

The interesting phenomenon is when $\theta(0) \to -\pi$, the trajectory of the pendulum tends to the separatrix and the trace of the sphere with $y = 0$ is *not subanalytic* but is a *sub-pfaffian set* whose graph can be written as a *converging* power series of

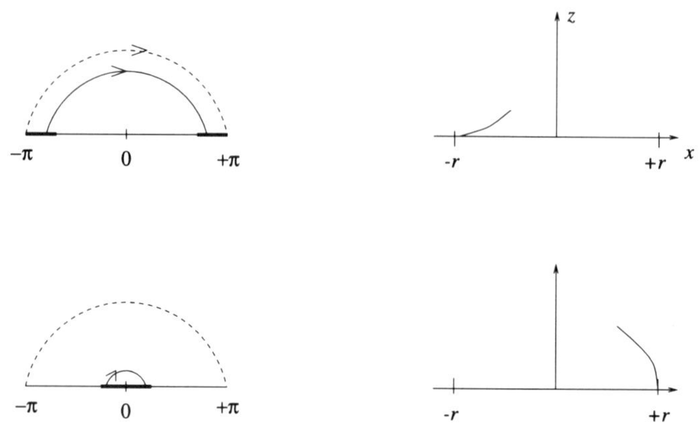

FIGURE 3

the form

$$z = \sum_{k,k'=0}^{+\infty} a_{k,k'} X^k Y^{k'}$$

where $X = \dfrac{x+r}{2r}$, $Y = \frac{e^{-\frac{1}{X}}}{X^2}$, see [1].

Similar expansions can be computed in the so-called *integrable case* where the set of geodesics can be integrated by quadratures. In particular we observe that in this case the saddle of the pendulum admits an *integrating factor.*

When $g = dx^2 + (1 + \beta x)dy^2$ the first intersection of the geodesics with $y = 0$ is constructed in the abnormal direction by taking in the pendulum representation the trajectories such that $\theta(0) \to -\pi$, $\theta'(0) = -\beta \sin\theta(0)$, see figure 4.

Hence we have a similar geometric framework than in the flat case and we have to analyze a return mapping near a separatrix tending to the separatrix of the flat case when $\lambda \to +\infty$. But the analytic situation is different. Indeed the eigenvalues of the saddle will form a *one parameter family* when λ varies. In general a saddle of this family *will not admit* an *integrating factor.* Now from [8], Poincaré-Dulac return mapping near a saddle is sub-pfaffian if and only if the saddle has an integrating factor. Hence we can conjecture.

Conjecture 5.1. *In the generic Martinet case the sphere is not sub-pfaffian.*

5.2. **Stability.** We observe that in the non flat case the geodesics are in general not bounded, even in the neighborhood of the original saddle connection. This phenomenon is a consequence of the non existence of periodic trajectory. In particular the expansion computed in section 3 is only formal. This tells us that the wave front of the flat case is *very unstable.* It is a confirmed by the computation in [3].

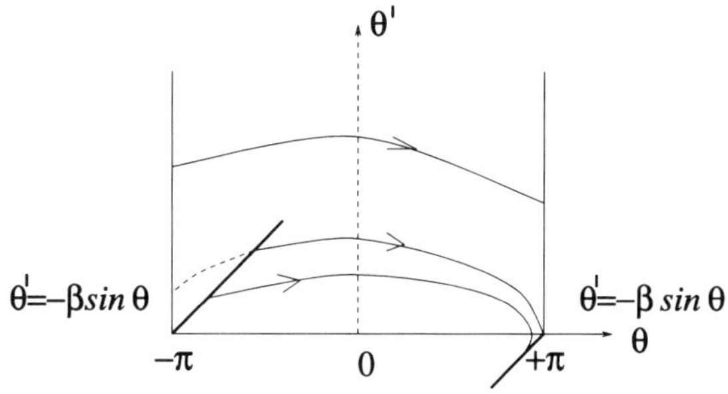

FIGURE 4

6. CONCLUSION

Our study explains the role of the parameter β in the normal form and completes the analysis of [3] were the role of α, γ where analyzed. Also it furnish a general framework to analyze the SR-sphere and wave front. Indeed the non subanalytic sphere is a geometric consequence of the behaviors of the geodesics near a separatrix. The transcendence we need to compute the sphere is connected to Poincaré-Dulac return mapping near a singularity. It is a problem which is analyzed in numerous articles in connection with 16th Hilbert-problem. In this framework the role of the variable z has to be taken into account because in the isoperimetric case there is an additional first integral and it allows to deal with less transcendence than in the generic case.

7. APPENDIX - NUMERICAL SIMULATIONS

On figure 5,6 and 7 we compare the respective curves $t \mapsto x(t), t \mapsto y(t), t \mapsto z(t)$ in the flat case $\beta = 0$ and in the non flat case. On figure 8 we have represented the sphere of radius $r = 0.01$, when $\beta = 0.8$

REFERENCES

1. A. Agrachev. B. Bonnard, M.Chyba, I. Kupka, *Sub-Riemannian sphere in Martinet flat case.*, COCV/ESAIM, Vol. 2, 377-448, 1997.
2. A. A. Andronov, A. A. Vitt, S.E. Khaikin, *Theory of oscillators*, Dover, 1987.
3. B. Bonnard, M.Chyba, *Méthodes géométriques et analytiques pour étudier l'application exponentielle, la sphère et le front d'onde en géométrie sous-riemannienne dans le cas Martinet*, submitted to COCV/ESAIM, 1998.
4. B. Bonnard, M.Chyba, E. Trélat, *Sub-Riemannian geometry : one parameter deformation of the Martinet flat case*, Journal of Dynamical and Control systems, Vol 4, N°1, 59-76, 1998.
5. J. Ecalle, *Introduction aux fonctions analysables et preuve constructive de la conjecture de Dulac*, Actualités Math., Hermann, Paris, 1992.

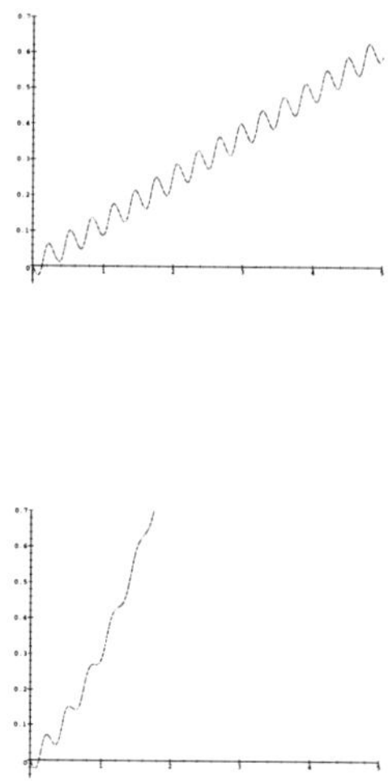

FIGURE 5. curve $t \mapsto x(t)$

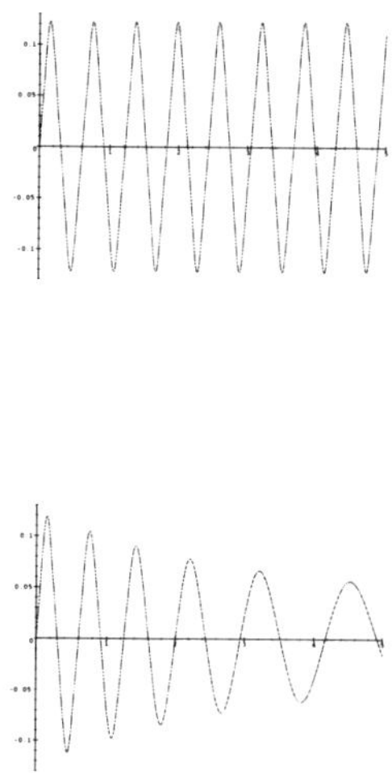

FIGURE 6. curve $t \mapsto y(t)$

6. A. G. Khovanski, *Fewnomials*, Trans. Math. Monographs, Vol. 88, AMS 1991.
7. M. A. Liapounoff, *Problème général de la stabilité du mouvement*, Annals of Math. Studies, N° 17, Princeton University Press, 1947.
8. A. Mourtada, R. Moussu, *Applications de Dulac et applications pfaffiennes*, Bulletin SMF 125, 1-13, 1997.
9. R. Moussu, A. Roche, *Théorie de Khovanskii et problème de Dulac*, Inv. Math 105, 431-441, 1991.

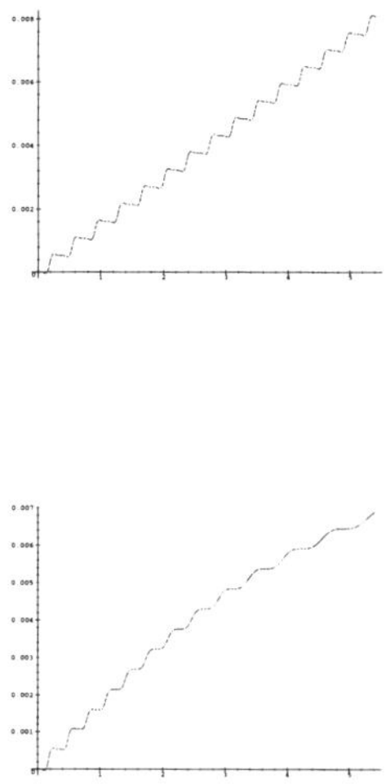

FIGURE 7. curve $t \mapsto z(t)$

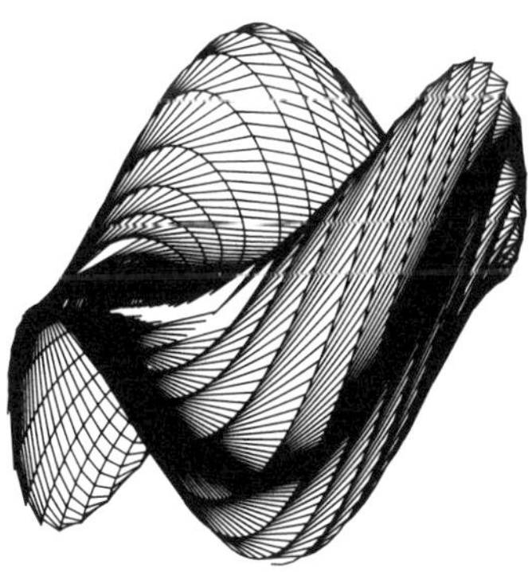

FIGURE 8. sphere

10. R. A. Struble, *Nonlinear differential equations*, McGraw Hill, 1962.

Current address: Université de Bourgogne, Laboratoire de Topologie, UMR 5584 B.P. 400, 21400 Dijon Cedex France
E-mail address: bbonnard@u-bourgogne.fr

Section de Mathématiques, Université de genève, 2-4 rue du Lièvre, Case 240 1211 Genève 24, Suisse
E-mail address: chyba@sunny.unige.ch

Université de Paris VI, 4 place Jussieu, 75232 Paris Cedex 05 France
E-mail address: kupka@math.jussieu.fr

Proceedings of Symposia in Pure Mathematics
Volume **64**, 1999

SINGULAR EXTREMALS OF ORDER 3 AND CHATTERING

BORISOV, V. F.

ABSTRACT. For generic Hamiltonian system with tangential discontinuity, it is proved that in a vicinity of singular solutions of order 3 there exists a bundle with the three–dimensional fibers filled by chattering trajectories, i.e., by trajectories with infinitely many switches at finite time intervals.

The paper deals with optimal control problems and modes of infinitely frequent switches of the control at finite time intervals (chattering). We aim to generalize some results of the Chattering Theory developed in [1] and to extend them from the case of discontinuous Hamiltonian systems with singular solutions of order 2 to the case of the third order singular solutions. The main result is a theorem on existence and inner structure of bundles with the three–dimensional fibers filled by chattering trajectories in a vicinity of singular solutions of order 3. Let us begin with a short introduction and state a problem served as a starting point in our exploration.

1. THE THREE–DIMENSIONAL FULLER'S PROBLEM

Consider the following control problem

Problem 1. *Minimize*

$$\int_0^\infty x_1^2\,dt$$

subject to

$$\dot{x}_1 = x_2, \quad \dot{x}_2 = x_3, \quad \dot{x}_3 = u \in [\alpha, \beta] \qquad (\alpha < 0,\ \beta > 0) \tag{1}$$

under the conditions

$$x_i(0) = x_i^0, \qquad i = 1, 2, 3.$$

1991 *Mathematics Subject Classification.* Primary 49B10, 34C05; Secondary 93C15.

Key words and phrases. Pontryagin's maximum principle, order of singular extremal, chattering.

The author was supported in part by RFBR Grants #96–01–00890, #98–01–00535.

The admissible trajectories $x(t) = (x_1(t), x_2(t), x_3(t))$ are absolutely continuous, the admissible controls $u(t)$ belong to $L_\infty[0, \infty)$.

Though control problems of dimension 3 and higher usually object explicit solutions, a completed optimal synthesis to Problem 1 was designed [2]. It was done due to the presence of a one–parameter homogeneity group which enables to decrease the dimension of the state space by means of transfer to a quotient system in the orbit space. Namely, consider the mapping $g_\lambda : \mathbb{R}^3 \to \mathbb{R}^3$,

$$g_\lambda(x_1, x_2, x_3) = (\lambda^3 x_1, \lambda^2 x_2, \lambda x_3).$$

It can be immediately checked that if $(x(t), u(t))$ is an optimal pair for the initial value $x^0 = x(0)$ then for any $\lambda > 0$ the pair $(x_\lambda(t), u_\lambda(t)) = (g_\lambda(x(t/\lambda)), u(t/\lambda))$ is an optimal solution to Problem 1 for the initial value $x^\lambda = g_\lambda(x^0)$. Let $u = \hat{u}(x)$ be an optimal synthesis to Problem 1. It follows that the dynamic system (1) with $u = \hat{u}(x)$ is homogenous under the action of the group g_λ.

Denote

$$\Gamma(x^0) = \bigcup_{\lambda>0} g_\lambda(x^0)$$

the orbit of the point $x^0 \neq 0$. The orbit space is homeomorphic to the standard two–dimensional unit sphere $S^2 \subset \mathbb{R}^3$. The homogeneity property allows us to follow the motion of a trajectory to (1) considering the motion of the point on S^2 at which $\Gamma(x(t))$ intersects S^2. Thus, we arrive at some dynamic system on S^2 which is specified up to orbital equivalence and is usually called as *quotient system.* Denote the quotient system by $\mathfrak{S}$. The qualitative behavior of its trajectories can be described as follows [2].

There exists a unique closed cycle of $\mathfrak{S}$ which divides S^2 into two the two–dimensional disks – D_1 and D_2. System $\mathfrak{S}$ in D_1 is orbitally equivalent to $\mathfrak{S}$ in D_2 and it is enough to imagine the qualitative behavior of $\mathfrak{S}$ inside any of the disks, say inside D_1.

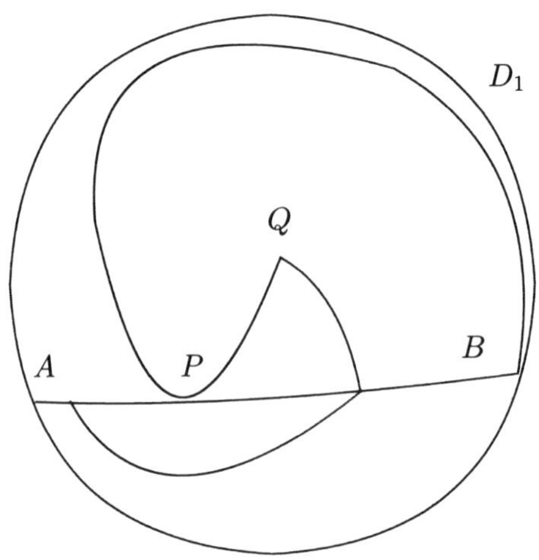

Figure 1

$\mathfrak{S}$ is a discontinuous systems and its switching surface is a one–dimensional curve – the segment AB in Fig. 1. There exists a unique stationary point of $\mathfrak{S}$ in D_1 (the point Q in Fig. 1) and it is the α–limit set of all trajectories of $\mathfrak{S}$ except for the cycle lying in the border of D_1. It is worth noting that there is a unique point on D_1 (point P in Fig. 1) at which trajectories of $\mathfrak{S}$ intersect the switching surface with tangency. The facts are summarized in the following theorem. To state it we need an auxiliary notion.

Let us call the optimal trajectory $x(t)$ as *self–similar* if there exist $\lambda_0 > 0$, $\tau_0 > 0$ such that for all $t \geq 0$

$$x(t + \tau_0) = g_{\lambda_0}(x(t/\lambda_0)).$$

A self–similar solution is called a *bi–constant–ratio* trajectory if the corresponding control $u(t)$ is equal to $\beta > 0$ at $t \in (0, \tilde{\tau})$ and to $\alpha < 0$ at $t \in (\tilde{\tau}, \tau_0)$ for some $\tilde{\tau} \in (0, \tau_0)$. It can be easily seen that if $x(t)$ is a self–similar solution to Problem 1 then for any $\lambda > 0$ the trajectory $g_\lambda(x(t/\lambda))$ is also self–similar with the same parameter λ_0.

Theorem 1 (Borisov, V.F. [2]). (1) *For any x^0 there exists a unique optimal solution $x(t)$ to Problem 1.*

(2) *There are two g_λ–orbits, Γ^α and Γ^β, such that optimal trajectories go along Γ^α and Γ^β with constant control $u(t) \equiv \alpha$ or $u(t) \equiv \beta$ and enter into the origin at finite instants. For the initial values $x^0 \neq 0$ which do not belong $\Gamma^\alpha \bigcup \Gamma^\beta$, the optimal trajectories are chattering and enter into the origin at finite instants after a countable number of switches.*

(3) *The switching points of all optimal solutions fill a two–dimensional surface Π that is homeomorphic to $\mathbb{R}^2$.*

(4) *All self–similar solutions are bi–constant–ratio trajectories, fill a two–dimensional surface Π^s in $\mathbb{R}^3$, and are specified by two switching curves, Γ^1 and Γ^2, that are some g_λ–orbits.*

(5) *While approaching the origin, any chattering solution that is different from a self–similar one reels more and more closely to Π^s in the angle metric of the orbit space. In the reverse time current, the solutions have a finite number of switches and approach (in the metric of the orbit space) one of the curves Γ^α or Γ^β.*

2. Singular solutions to discontinuous Hamiltonian systems

For optimal control problems that are affine in the control $u \in [-1, 1]$, Pontryagin's maximum principle [3] leads to a Hamiltonian system of the following form:

$$\dot{\psi} = -\frac{\partial H}{\partial x}, \quad \dot{x} = \frac{\partial H}{\partial \psi}. \tag{2}$$

Here $H = H_0(\psi, x) + uH_1(x, \psi)$, $u = \operatorname{sgn} H_1(\psi, x)$, $\psi, x \in \mathbb{R}^n$. System (2) is discontinuous at points of the surface $S = \{\psi, x \mid H_1(\psi, x) = 0\}$. The jump of the velocity is equal to $2(-\frac{\partial H_1}{\partial x}, \frac{\partial H_1}{\partial \psi})$ and the vector is orthogonal with the gradient to the surface S. Due to I. Kupka [4], we call (2) as a *Hamiltonian system with tangential discontinuity.* Solutions to (2) are specified in the sense of the definition due to A. F. Filippov [5].

Definition 1. *A solution $(\psi(t), x(t))$ to (2) is called singular on (t_0, t_1) if $H_1(\psi(t), x(t)) \equiv 0$ for all $t \in (t_0, t_1)$.*

Definition 2. *An integer q is called an order of the singular solution $(\psi(t), x(t))$ if*

$$\begin{aligned} \frac{\partial}{\partial u}\frac{d^k}{dt^k}\bigg|_{(2)} H_1 &= 0, \quad k = 1, \dots, 2q-1, \\ \frac{\partial}{\partial u}\frac{d^{2q}}{dt^{2q}}\bigg|_{(2)} H_1 &\neq 0. \end{aligned} \tag{3}$$

The number q is called as *local order* of the singular trajectory if relations (3) take place at points of the trajectory itself and as *intrinsic order* if (3) holds true in an open neighborhood of the trajectory (see [6]–[7]).

Everywhere below we assume that the following *Kelley's necessary optimality condition* [6] is satisfied along any singular trajectory of order q:

$$(-1)^q \frac{\partial}{\partial u}\frac{d^{2q}}{dt^{2q}} H_1 < 0.$$

Assume now that system (2) has a singular solution of the intrinsic order q. Let us show that in this case one can write (2) in a special form in a vicinity of the solution.

Introduce the variable $z_1 = H_1(\psi, x)$. The derivative of z_1 along a solution to (2) is equal to

$$\dot{z}_1 = \mathrm{ad}_{H_0 + uH_1} H_1 = \mathrm{ad}_{H_0} H_1$$

where

$$\mathrm{ad}_g f = \sum_{i=1}^{n} \left(\frac{\partial f}{\partial \psi_i}\frac{\partial g}{\partial x_i} - \frac{\partial f}{\partial x_i}\frac{\partial g}{\partial \psi_i} \right)$$

is the *Poisson bracket* of functions $f(\psi, x)$, $g(\psi, x)$. Furthermore, denoting $z_2 = \mathrm{ad}_{H_0} H_1$ we obtain

$$\dot{z}_2 = \mathrm{ad}^2_{H_0} H_1 + u\, \mathrm{ad}_{H_1} \mathrm{ad}_{H_0} H_1.$$

It follows from Definition 2 that $\mathrm{ad}_{H_1}\mathrm{ad}_{H_0} H_1 \equiv 0$ as a function of (ψ, x). Hence,

$$\dot{z}_2 = \mathrm{ad}^2_{H_0} H_1.$$

If we put $z_i = \mathrm{ad}^i_{H_0} H_1$, $i = 3, \dots, 2q$, then

$$\dot{z}_2 = z_3, \ \dots, \ \dot{z}_{2q-1} = z_{2q}, \quad \dot{z}_{2q} = A(\psi, x) + uB(\psi, x)$$

where $B(\psi, x) \neq 0$ in a vicinity of the singular solution. Kelley's condition implies that $(-1)^q B(\psi, x) < 0$.

Let us assume that functions $z_1, \dots, z_{2q}$ are functionally independent and complete them till a basis of $\mathbb{R}^{2n}$ by some functions $w_1, \dots, w_{2n-2q}$. In these coordinates, system (2) is written as follows:

$$\dot z_1 = z_2, \dots, \dot z_{2q-1} = z_{2q}, \quad \dot z_{2q} = \alpha(z,w) + u\beta(z,w), \\ \dot w = F(z,w,u), \quad u = \operatorname{sgn} z_1. \tag{4}$$

Let us explore the simplest case of system (4) when functions α and β are identically constant:

$$\alpha(z,w) \equiv \alpha_0, \quad \beta(z,w) \equiv \beta_0,$$

and function F does not depend on u, $F = F(z,w)$.

3. Bundles with chattering solutions. Homogeneous case

Let $\alpha = \alpha_0 - |\beta_0| < 0$, $\beta = \alpha_0 + |\beta_0| > 0$. Denote $\Sigma = \{(z,w) \mid z = 0\}$.

Theorem 2. *For any $(0,w) \in \Sigma$, solutions to (4) passing through $(0,w)$ fill a C^0-variety $\mathfrak{N}^+(w)$ that is homeomorphic to $\mathbb{R}^q$. All solutions to (4) emanating from $(0,w)$ fill a q-dimensional variety $\mathfrak{N}^-(w)$. Solutions to (4) that do not belong to $\mathfrak{N}^+(w) \bigcup \mathfrak{N}^-(w)$ leave any small neighborhood of the surface Σ in the direct and the reverse time current.*

Proof. Consider the system

$$\dot z_1 = z_2, \dots, \dot z_{2q-1} = z_{2q}, \quad \dot z_{2q} = \begin{cases} \alpha, & \text{if } (-1)^q z_1 > 0, \\ \beta, & \text{if } (-1)^q z_1 < 0. \end{cases} \tag{5}$$

One can check that up to denotation system (5) is the system of Pontryagin's maximum principle for the following problem:

Problem 2.

$$\int_0^\infty x_1^2(t)\,dt \to \min,$$
$$\dot x_1 = x_2, \ \dots, \ \dot x_{q-1} = x_q, \ \dot x_q = u,$$
$$u \in [\alpha, \beta], \quad x_i(0) = x_i^0, \quad i = 1, \dots, q.$$

Indeed, the Hamiltonian of Problem 2 is equal to

$$H = \psi_1 x_2 + \dots + \psi_{q-1} x_q + \psi_q u - \lambda_0 x_1^2/2,$$

and the conjugate system can be written as follows

$$\dot\psi_1 = \lambda_0 x_1, \ \dot\psi_2 = -\psi_1, \ \dots, \ \dot\psi_q = -\psi_{q-1},$$
$$u = \begin{cases} \alpha, & \text{if } \psi_q < 0, \\ \beta, & \text{if } \psi_q > 0. \end{cases}$$

It is easy to see that $\lambda_0 \neq 0$ and one can put $\lambda_0 = 1$. Putting

$$z_1 = \psi_q, \ z_2 = -\psi_{q-1}, \ z_3 = \psi_{q-2}, \ \dots, \ z_q = (-1)^{q+1}\psi_1,$$
$$z_{q+1} = (-1)^{q+1} x_1, \ z_{q+2} = (-1)^{q+1} x_2, \ \dots, \ z_{2q} = (-1)^{q+1} x_{2q},$$

we obtain (5).

System (5) is homogeneous. Consider the mapping $g_\lambda : \mathbb{R}^{2q} \to \mathbb{R}^{2q}$, $(g_\lambda(z))_i = \lambda^{2q+1-i} z_i$, $i = 1, \dots, 2q$, $\lambda > 0$. For $z(t)$ being a solution to (5), straightforward differentiation yields that $\tilde{z}(t) = g_\lambda(z(t/\lambda))$ is also a solution.

It is proved in [1] that any solution $z(t)$ to (5) such that $\lim\limits_{t\to\infty} z(t) = 0$ specifies an optimal solution to Problem 2 with the corresponding initial values (Kuhn–Tucker Theorem). Conversely, any optimal solution $x(t)$ to Problem 2 and its conjugate trajectory $\psi(t)$ enter into the origin at some finite instant. Since the functional in Problem 2 is strictly convex and the differential system is linear in x and u, for any initial value $x(0)$ there exist a unique optimal solution $x(t)$ and a unique conjugate trajectory $\psi(t)$. This specifies a mapping $\Psi(\cdot) : \mathbb{R}^q \to \mathbb{R}^q$ that transfers $x(0)$ to $\psi(0)$. It is proved in [1] that function $\Psi(\cdot)$ is C^0 in $\mathbb{R}^q$. This implies that solutions to (5) passing through the origin $z = 0$ fill some variety Σ^+ that is a homeomorphic image of $\mathbb{R}^q$ under the mapping $x \mapsto (x, \Psi(x))$. The same assertion holds true for solutions to (5) which are emanating from the origin. To prove that, it is enough to consider extremals of the problem

$$\int_0^T x_1^2 dt \to \min,$$

$$\dot{x}_1 = x_2, \ \dots, \ \dot{x}_{q-1} = x_q, \ \dot{x}_q = u \in [\alpha, \beta],$$
$$x_i(0) = 0, \quad x_i(T) = x_i^0, \quad i = 1, \dots, q.$$

Denote by Σ^- the variety filled by solutions to (5) emanating from the origin. Since any solution to (5) that does not belong to Σ^+ or to Σ^- cannot tend to 0, it follows that the solution leaves any small neighborhood of the origin in the direct and the reverse time current. Let us note that both Σ^+ and Σ^- are invariant under the action of the group g_λ.

Let us prove now that the perturbation of system (5) in the $(2n)$–dimensional extended space by means of adding the equation

$$\dot{w} = F(z, w)$$

has the same structure of solutions at the origin, i.e., for any $w^0 \in \mathbb{R}^{2n-2q}$ solutions to (4) entering the point $(0, w^0)$ fill a q–dimensional variety $\mathfrak{N}^+(w^0)$. Similarly, solutions to (4) emanating from $(0, w^0)$ fill a q–dimensional variety $\mathfrak{N}^-(w^0)$. It is enough to prove the existence of the variety for $\mathfrak{N}^-(w^0)$ since for $\mathfrak{N}^+(w^0)$ the proof is quite similar up to conversion of time current.

Let us consider the variety Σ^- for system (5). For any $z \in \Sigma^-$ denote by $T(z)$ the time instant at which the corresponding solution to (5) emanating from the origin approaches z. It is proved in [1] that function $T(z)$ is continuous. It follows that points of Σ^- can be parametrized by (t, ξ) where t is the time that is necessary for a solution to (4) to reach the point z, and ξ is some $(q-1)$–dimensional parameter on the $(q-1)$–dimensional unit sphere. Thus, any solution to (5) in Σ^- can be written as $z = z(t, \xi)$. Let us substitute the relation into (4):

$$\dot{w} = \tilde{F}(t, w, \xi), \tag{6}$$

where function $\tilde{F}$ is continuous in all arguments and smooth in w. Since solutions of such equations are continuous in the initial values, the solution to (6) with an initial value $w(0) = w^0$ can be written as $w = W(t, w^0, \xi)$ where W is C^0 in the totality of variables. This implies the statement to be proved. □

The structure of solutions inside $\mathfrak{N}^{\pm}$ has not studied much except for the cases $q = 2$ and $q = 3$. If $q = 2$ then both $\mathfrak{N}^{\pm}$ are two–dimensional and the behavior of solutions to (5) inside $\mathfrak{N}^{+}$ is depicted in Fig. 2.

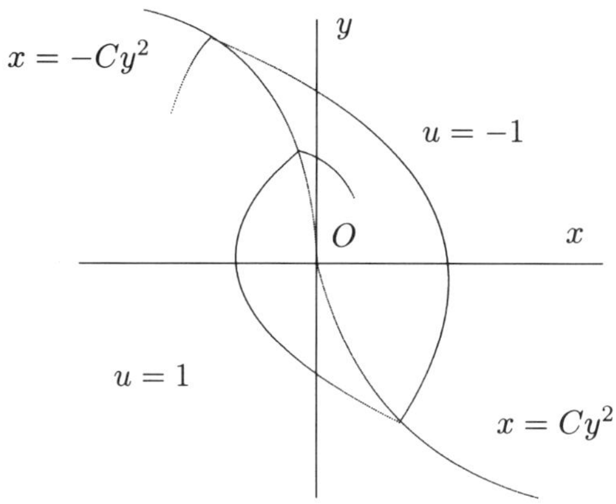

Figure 2

For $q = 3$ the behavior of solutions inside $\mathfrak{N}^{+}$ is described by Theorem 1 above and is depicted (in the orbit space) in Fig. 1. For an arbitrary $q > 3$ it is known only ([8]–[9]) that there exist $[q/2]$ (the integral part of $q/2$) two–dimensional integral varieties inside which the solutions are quite similar to those in Fig. 2. These solutions are self–similar with respect to the action of g_λ and each of the two–dimensional manifolds corresponds to a closed cycle in the orbit space.

Our purpose is to generalize the result of Theorem 1 for $q = 3$ to the case of a perturbed system (4) in which the perturbation terms are small enough with respect to the action of the group g_λ.

4. Bundles with chattering solutions. A perturbed system

Everywhere below we restrict ourselves to the case $q = 3$. Consider the system

$$\begin{gathered}\dot{z}_1 = z_2 + f_1(z, w, u), \quad \dot{z}_2 = z_3 + f_2(z, w, u), \ \dots\ , \ \dot{z}_5 = z_6 + f_5(z, w, u), \\ \dot{z}_6 = \alpha_0 + u\,\beta_0 + f_6(z, w, u), \quad \dot{w} = F(z, w, u), \\ u = \operatorname{sgn} z_1 . \end{gathered} \tag{7}$$

Here $z = (z_1, \dots, z_6) \in \mathbb{R}^6$, $w = (w_1, \dots, w_{2n-6}) \in \mathbb{R}^{2n-6}$. We explore solutions to system (7) in the region $\|z\| \le \epsilon$, $\|w - w^0\| \le \epsilon$ for some $w^0 \in \mathbb{R}^{2n-6}$ and $\epsilon > 0$. Functions f_i, F are assumed to be smooth enough (say, C^∞) and to be small comparatively with z_i in the following sense:

$$\lim_{\lambda\to+0} \sup_{\substack{\|z\|\le\epsilon,\ \|w-w^0\|\le\epsilon \\ u\in[-1,1]}} |f_i(g_\lambda(z),w,u)|/\lambda^{6-i} = 0, \quad i=1,2,\dots,6. \tag{8}$$

We prove now that system (7) has a C^0–bundle with the $(2n-6)$–dimensional base $\Sigma = \{z, w \mid z = 0\}$ and with the three–dimensional fibers $\mathfrak{N}^+(w)$ such that $\mathfrak{N}^+(w)$ is filled by solutions to (7) and the behavior of solutions inside $\mathfrak{N}^+(w)$ is the same as that of optimal solutions to Problem 1. Denote $S = \{z, w \mid z_1 = 0\}$.

Theorem 3. *If $\epsilon > 0$ is small enough then for any w, $\|w - w^0\| \le \epsilon$, there exists a three–dimensional variety $\mathfrak{N}^+(w)$ filled by solutions to (7) with the following properties.*

(1) *There are two smooth curves in $\mathfrak{N}^+(w)$, say L^+ and L^-, such that solutions to (7) with $u \equiv 1$ or $u \equiv -1$ approach the point $(0, w)$ and do not intersect the switching surface S. For any initial value not in $L^\pm$, the solution comes at $(0, w)$ at a finite time instant after a countable number of switches.*
(2) *There exists a two–dimensional integral submanifold $\mathfrak{M}(w) \subset \mathfrak{N}^+(w)$ such that $\mathfrak{M}(w) \cong \mathbb{R}^2$, $\mathfrak{M}(w) \bigcap S \cong \mathbb{R}^1$. The restriction of system (7) to $\mathfrak{M}(w)$ is orbitally equivalent to the restriction of system (5) to $\mathfrak{N}^+$ for $q = 2$. For any initial value different from points of $\mathfrak{M}(w)$, the solution has a finite number of switches in the reverse time current.*
(3) *The totality of switching points fill a two–dimensional surface $\Pi(w)$ that is homeomorphic to $\mathbb{R}^2$.*
(4) *The fibers are continuous in w.*

We will see below that smoothness properties of $\mathfrak{N}^+(w)$ and of the corresponding structural elements of $\mathfrak{N}^+(w)$ could be slightly strengthened. The principal part in the proof is based on Theorem 1 that implies the statement of Theorem 3 for the non–perturbed system. The rest of the paper is devoted to the proof of Theorem 3.

5. Resolution of singularities of a Poincaré mapping at points of surface Σ

The standard method to explore a discontinuous system is to consider the *Poincaré mapping* of the switching surface into itself which associates any initial point on the surface with the point at which the corresponding solution returns to the surface at the first time instant.

Let $y = (z, w) \in \mathbb{R}^{2n}$, and $y(t) = y(t, y^0)$ be a solution to (7) with the initial value $y(0) = y^0$. Consider the mapping

$$\Phi : S \to S$$

such that

$$\Phi(y^0) = y(\tau(y^0), y^0),$$

where $\tau = \tau(y^0)$ be the smallest positive root of the equation

$$z_1(\tau, y^0) = 0.$$

Denote by Φ_0 the mapping Φ in the case $f_i \equiv 0$. Let us write out the mapping Φ_0 in a special coordinates system induced by the action of the group g_λ. Let

$$z_i = \xi_i r^{7-i}, \quad i = 1, \dots, 6,$$

where

$$r = \sqrt[60]{z_1^{10} + z_2^{12} + |z_3|^{15} + z_4^{20} + z_5^{30} + z_6^{60}},$$

and the variables ξ_i are subjected to the restriction

$$\xi_1^{10} + \xi_2^{12} + |\xi_3|^{15} + \xi_4^{20} + \xi_5^{30} + \xi_6^{60} = 1.$$

Since the corresponding variety is homeomorphic to the standard 5–dimensional sphere S^5, one can consider ξ as the coordinates on S^5. Let us write out system (5) in terms of the coordinates r, ξ. Straightforward calculations yield

$$\begin{aligned} \dot{r} &= \frac{1}{60r^{59}}(10z_1^9 z_2 + 12z_2^{11} z_3 + \cdots + 60z_6^{59}(\alpha_0 + u\beta_0)) \\ &= \frac{1}{60}(10\xi_1^9\xi_2 + 12\xi_2^{11}\xi_3 + \cdots + 60\xi_6^{59}(\alpha_0 + u\beta_0)), \end{aligned} \tag{9}$$

i.e., $\dot{r}$ does not depend on r. Denote by $\theta = \theta(\xi, u)$ the right–hand side of the equation (9). Then

$$\begin{gathered} \dot{\xi}_1 = \frac{1}{r}(\xi_2 - 6\xi_1\theta), \quad \dot{\xi}_2 = \frac{1}{r}(\xi_3 - 5\xi_2\theta), \ \dots\ , \ \dot{\xi}_5 = \frac{1}{r}(\xi_6 - 2\xi_5\theta), \\ \dot{\xi}_6 = \frac{1}{r}(\alpha_0 + u\beta_0 - \xi_6\theta). \end{gathered} \tag{10}$$

Any change of length of the state velocity vector without change of its direction preserves the Poincaré mapping. We can multiply therefore the right–hand side of systems (9)–(10) by r. We obtain

$$\dot{r} = r\theta(\xi, u), \quad \dot{\xi} = \Xi(\xi, u), \tag{11}$$

where

$$\begin{gathered} \Xi_i(\xi, u) = \xi_{i+1} - (7 - i)\xi_i\theta(\xi, u), \quad i = 1, \dots, 5, \\ \Xi_6(\xi, u) = \alpha_0 + u\beta_0 - \xi_6\theta(\xi, u). \end{gathered}$$

We need the following technical result.

Lemma 1. *In coordinates (r, ξ, η), system (7) is orbitally equivalent to a system*

$$\begin{gathered} \dot{r} = r\theta(\xi, u) + r^2 R(r, \xi, w, u), \quad \dot{\xi} = \Xi(\xi, u) + r\Xi_1(r, \xi, w, u), \\ \dot{w} = F(0, w, u) + rF_1(r, \xi, w, u), \quad u = \operatorname{sgn} \xi_1, \end{gathered} \tag{12}$$

for R, Ξ_1, F_1 being some smooth functions of (r, ξ, w, u).

Proof. It follows from (8) that

$$f_i(z, w, u) = r^{7-i}\tilde{f}_i(r, \xi, w, u), \quad i = 1, \ldots, 6, \tag{13}$$

where $\tilde{f}_i(r, \xi, w, u)$ are smooth in the totality variables (this is an analog of Weierstrass Preparation Theorem [10]). For details, see [1]. Taking into account (13), the statement is proved by a straightforward differentiation. □

The transfer from (z, w) to (r, ξ, w) can be considered as a blowing–up procedure that resolves a singularity of system (7) at points of the surface Σ. The mapping

$$(z, w) \to (r, \xi, w) \tag{14}$$

is a one–to–one correspondence everywhere in $\mathbb{R}^{2n}$ except for the point $(0, w)$. The change (14) associates the point $(0, w)$ with a 5–dimensional plane $r = 0$ that is parametrized by the values of ξ.

6. Switching surface of $\mathfrak{N}^+(w^1)$ as an invariant manifold of Poincaré mapping. Perturbation of the manifold of self–similar solutions

Consider first the non–perturbed system (7) when $f_i \equiv 0$. Then $R \equiv 0$, $\Xi_1 \equiv 0$, $F_1 \equiv 0$ in (12) and after the resolution of singularities we obtain the system

$$\begin{aligned} \dot{r} &= r\theta(\xi; u), \quad \dot{\xi} = \Xi(\xi, u), \\ \dot{w} &= F(0, w, u), \quad u = \operatorname{sgn} \xi_1. \end{aligned} \tag{15}$$

Let us consider separately the following system on S^5:

$$\dot{\xi} = \Xi(\xi, u), \quad u = \operatorname{sgn} \xi_1. \tag{16}$$

Theorem 1 yields existence of a two–dimensional integral variety $\mathfrak{N}^+ \cong S^2$ on S^5 for (16). Theorem 1 establishes the inner structure of trajectories inside $\mathfrak{N}^+$. In particular, there exists a unique closed cycle in $\mathfrak{N}^+$ corresponding to self–similar solutions of Problem 1 (denote it by Π^s). The switching surface $\Pi = \mathfrak{N}^+ \bigcap \{\xi | \xi_1 = 0\}$ is homeomorphic to the circle S^1 and it intersects Π^s by two points, say A and B. The curve Π is invariant under the corresponding Poincaré mapping Φ_0 of the surface $\{\xi | \xi_1 = 0\}$ and both A and B are the fixed points of the twice repeated Poincaré mapping Φ_0^2. It was proved in [1] that A and B are hyperbolic fixed points of Φ_0^2 and the differential $D\Phi_0^2$ at A and at B has a single real positive eigenvalue less than 1 and four eigenvalues greater than 1 in the absolute value. Invariant Manifold Theorem [11] implies that there exists a one–dimensional manifold in S^5 through A (and through B) which is Φ_0^2–invariant. It can be easily proved that this very curve is a part of the total switching surface Π in S^5.

Extending (16) by adding the equation $\dot{r} = r\theta(\xi, u)$, we obtain that the manifold $\mathfrak{N}^+ \times R_+ \cong \mathbb{R}^3$ is an integral variety of the extended system. Indeed, this a consequence of homogeneity of the non–disturbed system. But having in mind the general non–homogeneous case, let us prove the fact in another way.

Consider the point $(r, \xi) = (0, \xi^0)$ where ξ^0 is the coordinate of point A. The point $(0, \xi^0)$ is a fixed point of the twice–repeated Poincaré mapping for the extended system in (r, ξ)–space. The differential of the mapping at $(0, \xi^0)$ is represented by the matrix

$$\begin{pmatrix} \lambda_0 & 0 & \dots & 0 \\ 0 & & & \\ \vdots & & D\Phi_0^2(\xi^0) & \\ 0 & & & \end{pmatrix}.$$

Spectrum of the matrix consists of the eigenvalues of the matrix $D\Phi_0^2(\xi^0)$ and of a number $\lambda_0 \in (0,1)$. Since $D\Phi_0^2(\xi^0)$ is a hyperbolic matrix (with a single eigenvalue less than 1 in the absolute value) there exists a two–dimensional manifold $\tilde{\mathfrak{N}}$ in (r,ξ)–space in a vicinity of the point A such that $\tilde{\mathfrak{N}}$ is invariant with respect to Φ^2 and the restriction of Φ^2 to $\tilde{\mathfrak{N}}$ is a contracting mapping. In the non–perturbed case $\tilde{\mathfrak{N}}$ is exactly the direct product $\mathfrak{N}^+ \times \mathbb{R}_+$.

Arguing as in the proof of Theorem 2, for the total system (15) we obtain an integral variety $\mathfrak{N}^+ \times R_+ \times \mathbb{R}^{2n-6}$ which is invariant under the flow of the equation (15) and is stratified by the three–dimensional fibers homeomorphic to $\mathbb{R}^3$. Denote the fiber through the point $(0,\xi^0,w^0)$ by $\mathfrak{N}^+(w^0)$.

Let us mention that each $\mathfrak{N}^+(w^0)$ contains a one–dimensional switching curve that is invariant under the twice repeated Poincaré mapping Φ_0^2. Moreover, there exists a two–dimensional Φ_0^2–invariant surface containing the one–dimensional invariant curve such that the restriction of Φ_0^2 on the surface is a contracting mapping. Applying Invariant Manifold Theorem once more, we obtain that a small perturbation of system (15) has also an w^0–family of Φ^2–invariant curves through $(r,\xi,w) = (0,\xi^0,w^0)$ and a family of two–dimensional Φ^2–invariant surfaces in their small vicinity. The detailed proof reduces to designing the corresponding mapping with the fixed points $(0,\xi^0,w^0)$ and with $(0,\Phi_0(\xi^0),w^0)$ and then to straightforward calculating its differential at the fixed points.

Denote the one–dimensional invariant curves by $\Gamma^\pm(w^0)$ and the two–dimensional surfaces in a vicinity of $\Gamma^\pm(w^0)$ by $\Pi^\pm(w^0)$.

7. Backward procedure of reconstruction of the switching surface

Invariant Manifold Theorem allows us to explore the switching surface Π only in a small neighborhood of the curves $\Gamma^\pm(w^0)$. To design the total switching surface inside $\mathfrak{N}^+(w^0)$ we need a special topological procedure.

Let us return again to the non–perturbed system. Apply the mapping $(\Phi_0)^{-1}$ to the part of the switching surface that was obtained as the two–dimensional invariant manifold of Φ_0^2 in a vicinity of $\Gamma^\pm$ (recall that $\Gamma^\pm$ are the switching curves for the family of self–similar solutions).

It can be easily seen that $(\Phi_0)^{-1}(\Pi^+)$ contains Π^- and, conversely, $\Pi^+ \subset (\Phi_0)^{-1}(\Pi^-)$. Repeating the procedure, we obtain two chains of mutually imbedded Φ_0^2–invariant subspaces $\Pi^+ \subset (\Phi_0^{-1})^2\Pi^+ \subset \dots \subset (\Phi_0^{-1})^{2n_0}\Pi^+$ and $\Pi^- \subset (\Phi_0^{-1})^2\Pi^- \subset \dots \subset (\Phi_0^{-1})^{2n_0}\Pi^-$. There are two possibilities for the chain: either it grows infinitely or stops for some n_0 due to appearance of points at which Φ_0^{-1} becomes undefined. It was proved in [1] that the first case is impossible for the non–perturbed system and the second case takes place indeed.

The points at which Φ_0^{-1} is undefined fill a strip I homeomorphic to a direct product of a segment $[0,1]$ with a half-line $\mathbb{R}_+$: $I \cong [0,1] \times \mathbb{R}_+$. Solutions to (15) are transversal to Π at all points of I except for points of a one–dimensional curve (denote it by Γ^0) that corresponds (in the orbit space) to point P in Fig. 1.

At points of Γ^0, solutions to (15) are tangent to the switching surface $\xi_1 = 0$ with the first order of tangency (i.e., $\dot{\xi}_1 = 0$, $\ddot{\xi}_1 \neq 0$).

Finally, there exist two half straightlines $\xi = \xi^{\pm}$ (for $r \geq 0$ being a parameter) such that $\xi^{\pm}$ are the stationary points of the differential systems $\dot{\xi} = \Xi(\xi, \pm 1)$ on S^5 correspondingly. A result of straightforward calculation is the fact that the points $\xi^{\pm}$ are unsteady knots and solutions to (15) with the initial values on I tend to these half–lines in the reverse time current (in the metric of ξ–space).

Let us discuss now what steps of the just designed procedure hold unchanged for the case of small perturbation of system (15).

Since the Poincaré mapping Φ_0 is a diffeomorphizm everywhere in Π except for the line Γ^0, one can take a small neighborhood of Γ^0 such that for r being small enough the Poincaré mapping Φ has the same properties as Φ_0 everywhere outside the neighborhood. In particular, we have the same chain of mutual imbeddings $\Pi^+ \subset (\Phi^{-1})^2\Pi^+ \subset \cdots \subset (\Phi^{-1})^{2n_0}\Pi^+$ (as well as $\Pi^- \subset (\Phi^{-1})^2\Pi^- \subset \cdots \subset (\Phi^{-1})^{2n_0}\Pi^-$) as for Φ_0. It can be easily seen that if $(\Phi^{-1})^{2n_0}\Pi^+$ (or $(\Phi^{-1})^{2n_0}\Pi^-$) intersects the surface $\dot{\xi}_1 = 0$ then (since the points are close to Γ^0) solutions to (12) are tangent to $\xi_1 = 0$ with the first order of tangency ($\ddot{\xi}_1 \neq 0$). It follows that the perturbed system is orbitally equivalent to the non–perturbed one in a vicinity of Γ^0. This implies that the total switching surface Π inside $\mathfrak{N}^+$ is homeomorphic to $\mathbb{R}^2$ and the restriction of Φ to Π is orbitally equivalent to the restriction of the non–perturbed mapping Φ_0 to the corresponding switching surface.

At last, in a vicinity of the curves $\xi = \xi^{\pm}$ the perturbation of system (15) is also orbitally equivalent to the non–perturbed system because unsteady knot is a stable stationary point with respect to small perturbations of the right–hand side of the differential equation. This completes the proof of Theorem 3. □

References

1. Zelikin, M. I., Borisov, V. F. *Theory of Chattering Control with Applications to Astronautics, Robotics, Economics, and Engineering.* Birkhäuser, Boston, 1994.
2. Borisov, V. F. *The design of an optimal synthesis with a chattering mode* (in Russian). Dokl. Akad. Nauk **302** (1989), no. 4, pp. 785–789. English transl. Soviet Math. Dokl. **38** (1989), no. 2, pp. 332–336
3. Pontryagin, L. S., Boltyanskii, V. G., Gamkrelidze, R. V., Mischenko, E. F. The Mathematical Theory of Optimal Processes (in Russian). Nauka, Moscow, 1976. English transl. of 2nd ed. Wiley, 1962, and MacMillan, 1964
4. Kupka, I. *The Ubiquity of Fuller's Phenomenon.* Nonlinear Controllability and Optimal Control, Monograph Textbooks Pure Appl. Math. (ed. by H.Sussman), no. 133, Dekker, N. Y., 1990, 313–350.
5. Filippov, A. F. *Differential equations with discontinuous right–hand side.* Kluwer, Dordrecht, 1988.
6. Kelley, H. J., Kopp, R. E., Moyer, H. G. *Singular extremals.* Topics in Optimization (ed. by G. Leitmann), Acad. Press, N.Y., 1967, pp. 63–103
7. Lewis, R. M. *Definition of order and junction condition in singular control problems.* SIAM J. Control and Optimizat., **18** (1980), no. 1, pp. 21–32.
8. Borisov, V. F. *Number of limit cycles of the quotient system of the n–dimensional Fuller problem* (in Russian). Matematicheskii Sbornik, **187** (1996), no. 12, pp. 3–20. English transl. Sbornik: Mathematics **187**:12 (1997), 1737–1753
9. Borisov, V. F. *On two–dimensional integral varieties of a class of discontinuous Hamiltonian systems* (in Russian). Fundametalnaya i Prikladnaya Matematika (to appear).
10. Malgrange B. *Ideals of differential functions.* Oxford University Press, England, 1966.

11. Hirsh, M. W., Pugh, C. C., Shub, M. *Invariant manifolds.* Springer–Verlag, Berlin, Heidelberg, N.Y., 1977.

DEPARTMENT OF MATHEMATICS, STATE ACADEMY OF CUSTOMARY SERVICES, MOSCOW, RUSSIA

Proceedings of Symposia in Pure Mathematics
Volume **64**, 1999

SOME OPEN PROBLEMS IN CONTROL THEORY

JEAN-MICHEL CORON

ABSTRACT. In this paper we present some open problems on the controllability and stabilizability of nonlinear control system in finite and infinite dimension.

1. SMALL TIME LOCAL CONTROLLABILITY AND CONTINUOUS REACHABILITY

For this open problem we consider the nonlinear control system

$$\dot{x} = f(x, u), \tag{1.1}$$

where $x \in \mathbb{R}^n$ is the state, $u \in \mathbb{R}^m$ is the control. Throughout all this paper we assume that f is a map of class C^∞ from $\mathbb{R}^n \times \mathbb{R}^m$ into $\mathbb{R}^n$ such that

$$f(0, 0) = 0. \tag{1.2}$$

Let us first give two definitions

Definition 1.1. *The control system (1.1) is small time locally controllable if, for all real number $\varepsilon > 0$, there exists a real number $\eta > 0$ such that, for all $x_0 \in B_\eta := \{x \in \mathbb{R}^n; |x| < \eta\}$, there exists a measurable function $u : [0, \varepsilon] \to \mathbb{R}^m$ such that*

$$|u(t)| < \varepsilon, \ \forall t \in [0, T], \tag{1.3}$$

$$(\dot{x} = f(x, u(t)), x(0) = x_0) \Rightarrow (x(\varepsilon) = 0). \tag{1.4}$$

Definition 1.2. *The origin (of $\mathbb{R}^n$) is locally continuously reachable (for (1.1)) in small time if, for all positive real number ϵ, there exist a continuous map u from $\mathbb{R}^n$ into $L^1((0, \epsilon); \mathbb{R}^m)$ and a positive real number η such that*

$$Sup\{|u(a)(t)|; t \in (0, \epsilon)\} \to 0 \ as \ a \to 0,$$

$$(\dot{x} = f(x, u(x(0))(t)), \ |x(0)| < \eta) \Rightarrow x(\epsilon) = 0.$$

1991 Mathematics Subject Classification. Primary 93B05 93C10 93C20 93D15; Secondary 76C05 76D05.

Partially supported by the PRC-GDR "Automatique" of the CNRS and the MST.

Clearly, if the origin is locally continuously reachable in small time, then the the control system is small time locally controllable. Let us notice that, following ideas due to Kawski [25] and Hermes [26], we have proved in [6, Lemma 3.1 and Section 5] that "many" sufficient conditions for small time local controllability imply that the origin is locally continuously reachable in small time. This is in particular the case for the Hermes condition [27] or [39] and its generalization due to Sussmann [43, Theorem 7.3]; this is in fact also the case for the Bianchini and Stefani condition [2, Corollary p. 970], which extends [43, Theorem 7.3]. A natural open problem is

Open Problem 1. *Assume that f is analytic. Does the small time local controllability implies that the origin is locally continuously reachable in small time?*

2. STABILIZABILITY AND SMALL TIME LOCAL CONTROLLABILITY

In this section we use the definitions and notations of Section 1. Moreover the feedback laws considered in this paper are continuous and, unless otherwise specified, are static and stationary It is well known, see e.g. [40, Theorem 7 p. 134], that any linear control system which is controllable can be asymptotically stabilized by means of feedback laws. A natural question is if this result still holds for nonlinear control systems. In 1979 Sussmann has shown that the global version of this result does not hold for nonlinear control systems: in [42] he has given an example of a nonlinear analytic control system which is globally controllable but cannot be globally asymptotically stabilized by means of feedback laws. In 1983 Brockett has shown that the local version also does not hold: in [3] he has given a necessary condition for local asymptotic stabilizability by means of feedback laws which is not implied by local controllability; for example, as pointed in [3], the analytic control system

$$\dot{x}_1 = u_1,\ \dot{x}_2 = u_2,\ \dot{x}_3 = x_1 u_2 - x_2 u_1, \tag{2.5}$$

where the state is $x = (x_1, x_2, x_3) \in \mathbb{R}^3$ and the control $u = (u_1, u_2) \in \mathbb{R}^2$, is locally (and even globally) controllable but does not satisfy the Brockett necessary condition (and therefore cannot be asymptotically stabilized by means of feedback laws).

In 1980 Sontag and Sussmann [41] have proved that any one dimensional state nonlinear control system which is locally (resp. globally) controllable can be locally (resp. globally) asymptotically stabilized by means of **time-varying** feedback laws and Samson has proved in [38] that the nonlinear control system (2.5) can be globally asymptotically stabilized by means of **time-varying** feedback laws. In [5, 6, 10] it is shown that, in many cases, controllability implies stabilizability by means of time-varying feedback laws. One may wonder if this is a general fact. The example given by Sussmann in [42] shows that this is not the case for the global statement (i.e. for global controllability and global stabilizability). In order to state a precise open problem for a local statement, let us give some definitions

Definition 2.1. *Let X be in $C^0(\mathbb{R}^n \times \mathbb{R};\mathbb{R})$. One says that 0 is locally asymptotically stable for $\dot{x} = X(x,t)$ if*
(i) for all $\varepsilon > 0$, there exists $\eta > 0$ such that, for all $\tau \in \mathbb{R}$ and for all $t \geq \tau$,

$$(\dot{x} = X(x,t), |x(\tau)| < \eta) \Rightarrow (|x(t)| < \varepsilon), \tag{2.6}$$

(ii) there exists $\delta > 0$ such that, for all $\varepsilon > 0$, there exists $M > 0$ such that, for all s in $\mathbb{R}$,

$$(\dot{x} = X(x,t) \text{ and } |x(s)| < \delta) \Rightarrow (|x(\tau)| < \varepsilon, \forall \tau > s + M). \tag{2.7}$$

Let us emphasize that since the vector field X is only continuous the Cauchy problem $\dot{x} = X(x,t), x(t_0) = x_0$ where t_0 and x_0 are given may have many maximal solutions. Let us recall that Kurzweil in [28] has shown that, even for vector fields which are only continuous, asymptotic stability is equivalent to the existence of a Lyapunov function.

Before stating our open problem, let us give two new definitions

Definition 2.2. *System (1.1) is locally stabilizable in small time by means of almost smooth periodic time-varying feedback laws if, for any positive real number T, there exist ε in $(0,+\infty)$ and u in $C^0(\mathbb{R}^n \times \mathbb{R};\mathbb{R}^m)$ of class C^∞ on $(\mathbb{R}^n\backslash\{0\}) \times \mathbb{R}$ such that*

$$u(0,t) = 0, \forall t \in \mathbb{R}, \tag{2.8}$$

$$u(x,t+T) = u(x,t), \forall x \in \mathbb{R}^n, \forall t \in \mathbb{R}, \tag{2.9}$$

$$((\dot{x} = f(x,u(x,t)) \text{ and } x(s) = 0) \Rightarrow (x(t) = 0, \forall t \geq s)), \forall s \in \mathbb{R}, \tag{2.10}$$

$$((\dot{x} = f(x,u(x,t)) \text{ and } |x(s)| \leq \varepsilon) \Rightarrow (x(t) = 0, \forall t \geq s+T)), \forall s \in \mathbb{R}. \tag{2.11}$$

Note that (2.9), (2.10), and (2.11) imply that 0 is locally asymptotically stable for $\dot{x} = f(x,u(x,t))$; see Lemma 2.15 in [8] for a proof.

Definition 2.3. [7]. *The strong jet accessibility subspace of (C) at $(\bar{x},\bar{u}) \in \mathbb{R}^n \times \mathbb{R}^m$ is the subspace of $\mathbb{R}^n$, denoted by $a(\bar{x},\bar{u})$, spanned by*

$$\{h(\bar{x}); h \in \{\partial^{|\alpha|} f/\partial u^\alpha(\cdot,\bar{u})\alpha \in \mathbb{N}^m, |\alpha| \geq 1 \cup Br_2(f,\bar{u})\}\} \tag{2.12}$$

where $Br_2(f,\bar{u})$ denotes the set of iterated Lie brackets of at least 2 vector fields in $\{\partial^{|\alpha|} f/\partial u^\alpha(\cdot,\bar{u}); \alpha \in \mathbb{N}^m\}$.

Remark 2.4. *One easily checks that the usual strong accessibility subspace of (C) at $\bar{x}$ (see e.g.* [44, p. 101]*) contains $a(\bar{x},\bar{u})$ for all $\bar{u}$ in $\mathbb{R}^m$ and that, if f is analytic with respect to x and u or is a polynomial with respect to u, these inclusions are all equalities.*

Our last definition is

Definition 2.5. *System (1.1) satisfies the strong jet accessibility rank condition at* $(\bar{x}, \bar{u})$ *if*

$$a(\bar{x}, \bar{u}) = \mathbb{R}^n. \tag{2.13}$$

Remark 2.6. *It follows from Remark 2.4 that if (1.1) satisfies the strong jet accessibility rank condition at* $(\bar{x}, \bar{u})$ *then it satisfies the usual strong accessibility rank condition at* $\bar{x}$ *and the converse holds if* f *is analytic with respect to* x *and* u *or is a polynomial with respect to* u.

Note that, if (1.1) is locally stabilizable in small time by means of almost smooth periodic time-varying feedback laws, then $0 \in \mathbb{R}^n$ is locally continuously reachable in small time for (1.1). The converse holds if $n \notin \{2, 3\}$ and if (1.1) satisfies the strong jet accessibility rank condition at $(0, 0)$, i.e.

Theorem 2.7. *Assume that* 0 *is locally continuously reachable in small time, that (1.1) satisfies the strong jet accessibility rank condition at* $(0, 0)$, *and that*

$$n \notin \{2, 3\}. \tag{2.14}$$

Then (1.1) is locally stabilizable in small time by means of almost smooth periodic time-varying feedback laws.

Theorem 2.7 is proved in [8] when $n \geq 4$ and in [10] when $n = 1$. The open problem of this section is

Open Problem 2. *Does Theorem 2.7 holds if assumption (2.14) is removed?*

3. Partial stabilization

One may wonder if the use of (stationary) feedback laws is sufficient if one wants only a "partial" stabilization. Let us explain what we mean by "partial" stabilization. Let $\phi \in C^\infty(\mathbb{R}^n; \mathbb{R}^p)$ be such that

$$\text{rank } \phi'(0) = p < n, \tag{3.15}$$

$$\phi(0) = 0, \tag{3.16}$$

$$(\phi(x) = 0 \Rightarrow f(x, 0) = 0), \ \forall x \in \mathbb{R}^n. \tag{3.17}$$

Then let us introduce the following definition

Definition 3.1. *System 1.1 is locally* ϕ*-asymptotically stabilizable by means of feedback laws if there exists* $u \in C^0(\mathbb{R}^n; \mathbb{R}^m)$ *such that*

(i) *For all* $\varepsilon > 0$ *there exists* $\eta > 0$ *such that*

$$(\dot{x} = f(x, u(x)), |x(\tau)| < \eta) \Rightarrow (|x(t)| < \varepsilon), \tag{3.18}$$

(ii) *there exists* $\nu > 0$ *such that*

$$(\dot{x} = f(x, u(x)), |x(0)| < \nu) \Rightarrow \lim_{t \to +\infty} h(x(t)) = 0. \tag{3.19}$$

(iii)

$$(\phi(x) = 0 \Rightarrow u(x) = 0), \forall x \in \mathbb{R}^n.$$

Then the open problem of this section is, with Definitions 1.2 and 2.5,

Open Problem 3. *Assume that* 0 *is locally continuously reachable in small time and that (1.1) satisfies the strong jet accessibility rank condition at* $(0,0)$. *Is System 1.1 locally* ϕ*-asymptotically stabilizable by means of feedback laws?*

Remark 3.2. *The same question without the assumption of the strong jet accessibility rank condition at* $(0,0)$ *is also open.*

4. Dynamic feedback laws

Again we use the same notations and assumptions on f as in Section 1. Let us recall that the feedback laws considered in this paper are continuous and, unless otherwise specified, are static and stationary. In this section all the feedback law considered are stationary. Let us start by recalling the following necessary condition for stabilizability due to Brockett [3]

Theorem 4.1. *If the control system (1.1) can be locally asymptotically stabilized by means of feedback laws, then the image of any neighborhood of* $(0,0) \in \mathbb{R}^n \times \mathbb{R}^m$ *is a neighborhood of* $0 \in \mathbb{R}^n$.

In Zabczyk [48] has observed that, from a theorem due to Krasnosel'skiĭ [32, 33], one can deduce the following stronger necessary condition, that we will call the index condition,

Theorem 4.2. *If the control system (1.1) can be locally asymptotically stabilized by means of feedback laws, then there exists* $u \in C^0(\mathbb{R}^n;\mathbb{R}^m)$ *vanishing at* 0 *such that* $f(x,u(x)) \neq 0$ *for* x *small enough but not* 0 *and the index of* $x \to f(x,u(x))$ *at* 0 *is equal to* $(-1)^n$.

For a definition of the index, see, for example, [33, p. 9]. In order to state the open problem of this section, let us introduce a definition

Definition 4.3. *The control system (1.1) can be locally asymptotically stabilized by means of dynamic feedback laws if, for some integer* $p \in \mathbb{N}$, *the control system*

$$\dot{x} = f(x,u),\ \dot{y} = v \in \mathbb{R}^p, \tag{4.20}$$

where the control is $(u,v) \in \mathbb{R}^m \times \mathbb{R}^p$ *and the state is* $(x,y) \in \mathbb{R}^n \times \mathbb{R}^p$ *is locally asymptotically stabilizable by means of feedback laws. By convention, when* $p = 0$, *system (4.20) is just system (1.1).*

Clearly, if control system (1.1) can be locally asymptotically stabilized by means of a feedback law, it can be locally asymptotically stabilized by means of dynamic feedback laws. But it is proved in [15] that the converse does not all. Moreover the example given in [15] shows that the index condition is not necessary for local asymptotic stabilizability by means of dynamic feedback laws. Clearly the Brockett necessary condition is still necessary local asymptotic stabilizability by means of dynamic feedback laws. But this condition turns out to be not sufficient for local asymptotic stabilizability by means of dynamic feedback laws even if one assumes that 0 is locally continuously reachable in small time and that the system is analytic. In [4] we have proposed a slightly stronger necessary condition; we have

Theorem 4.4. *Assume that control system (1.1) can be locally asymptotically stabilized by means of dynamic feedback laws. Then, for any positive and small enough* ϵ,

$$f_\star\left(\sigma_{n-1}\left(\{(x,u);\ |\,x\,|+|\,u\,|<\epsilon, f(x,u)\neq 0\}\right)\right) = \sigma_{n-1}(\mathbb{R}^n\setminus\{0\})\,(=\mathbb{Z}), \tag{4.21}$$

where $\sigma_{n-1}(A)$ *denotes the stable homotopy group of order* $(n-1)$*–for a definition of stable homotopy group, see e.g.* [45].

Let us point out that the index condition implies (4.21). Moreover (4.21) implies that a "dynamic extension" of (1.1) satifies the index condition if the system is analytic. More precisely, one has

Theorem 4.5. [9, Section 2]*Assume that* f *is analytic (or continuous and subanalytic). Assume that (4.21) is satisfied. Then, if* $p\geq 2n+1$, *control system (4.20) satisfies the index condition.*

The open section of this section is

Open Problem 4. *Let us assume that* f *is analytic, satisfies (4.21) and that* 0 *is locally continuously reachable in small time. Is control system (1.1) locally asymptotically stabilizable by means of dynamic feedback laws?*

A natural guess is that, unfortunately, a positive answer is unlikely to be true. A possible candidate for a negative answer is the control system, with $n=3$ and $m=1$,

$$\dot{x}_1 = x_3^2(x_1-x_2),\ \dot{x}_2 = x_3^2(x_2-x_3),\ \dot{x}_3 = u.$$

5. Controllability of the viscous Burger equation

The viscous Burger equation on the space interval $\{x;\ x\in[0,1]\}$ and the interval of time $\{t;\ t\in[0,T]\}$ is

$$\frac{\partial y}{\partial t}-\frac{\partial^2 y}{\partial x^2}+y\frac{\partial y}{\partial x}=0 \text{ in } [0,1]\times[0,T]. \tag{5.22}$$

One can considers the two following natural cases of boundary controls

(i) The control is at $x=0$ only: one makes no restriction on $y(0,t)$ and one requires

$$y(1,t)=0,\ \forall t\in[0,T], \tag{5.23}$$

(ii) The control is at $x = 0$ and $x = 1$ and so one makes no restriction on $y(1, t)$ and no restriction on $y(0, t)$.

In case (i) it has been shown by Fursikov and Imanuvilov in [18] that, for $T > 0$ given, one does not have approximate controllability in $L^2(0, 1)$, that is one can find y_0 in $C^\infty([0, 1])$, y_1 in $C^\infty([0, 1])$ and $\epsilon \in (0, +\infty)$ such that, for any solution $y \in C^\infty([0, 1] \times [0, T])$ of (5.22) and of (5.23) such that

$$y(x, 0) = y_0(x), \ \forall x \in [0, 1],$$

one has

$$\mid y(\cdot, T) - y_1 \mid_{L^2(0,1)} > \epsilon.$$

In fact they have proved the following stronger statement

Theorem 5.1. [18, Section 2] *There exists a positive constant C such that, for any solution $y \in C^\infty([0, 1] \times [0, T])$ of (5.22) and of (5.23) one has*

$$\frac{d}{dt} \int_0^1 x^6 y_+^4(x, t) dx < C, \tag{5.24}$$

where $y_+(x, t) = max(y(x, t), 0)$.

Another obstruction for approximate controllability has been given by Diaz in [16].

Of course, due to smoothing property of the viscous Burger equation, one cannot expect to have in general $y(\cdot, T) = y_1$. This a general phenomenon for parabolic equations. For these equations, as introduced by Fursikov and Imanuvilov in [18, 19], the natural definition for exact controllability is not passing from a state (y_0) to another state (y_1) but passing from one solution to another solution. This definition of controllability, in the framework of finite-dimensional control systems, was introduced by Willems [46]. In [18] Fursikov and Imanuvilov have proved exact local controllability, even in case (i). More precisely they have proved proved the following theorem

Theorem 5.2. [18]. *Let $T > 0$. Let $\hat{y}_1 \in C^\infty([0, 1] \times [0, T])$ be a solution of (5.22) and of (5.23) on $[0, 1] \times [0, T]$ and on $[0, T]$ respectively. Then there exists $\eta > 0$ such that, for any solution $\hat{y}_0 \in C^\infty([0, 1] \times [0, T]$ of (5.22) and of (5.23) such that*

$$\mid \hat{y}_0(\cdot, 0) - \hat{y}_1(\cdot,) \mid_{C^1([0,1])} < \eta,$$

there exits a solution $y \in C^\infty([0, 1] \times [0, T])$ of (5.22) and of (5.23) such that

$$y(x, 0) = \hat{y}_0(x, 0), \ \forall x \in [0, 1],$$

$$y(x, T) = \hat{y}_1(x, T), \ \forall x \in [0, 1].$$

Let us also point out that Theorem 5.1 does not hold if (5.23) is not required for y. So it is tempting to ask if, in case (ii) -i.e. when the control is on the full boundary of $[0, 1]$-, one has global approximate controllability or even global exact controllability in an arbitrary time. One may ask for example the following question

Open Problem 5. *Let $T > 0$. Let $\hat{y}_0$ and $\hat{y}_1$ be two solutions of (5.22) on $[0,1] \times [0,T]$ of class C^∞. Does there exist a solution $y \in C^\infty([0,1] \times [0,T])$ of (5.22) such that*

$$y(x,0) = \hat{y}_0(x,0),\ \forall x \in [0,1],$$
$$y(x,T) = \hat{y}_1(x,T),\ \forall x \in [0,1]?$$

Let us end this section by mentionning the paper [1] by F. Ancona and A. Marson and the paper [29] by T. Horsin which both deal with the controllability of the non-viscous Burger equation, i.e. equation (5.22) without the the dissipative term $\partial^2 y/\partial t^2$. In particular, in [29], Horsin presents a "return method" (see [5, 12, 13] and Section 7 below) to describe the attainable sets on an interval of the non-viscous Burger equation by means of variations of the domain. This might be also useful for the open problem of this section.

6. Blow-up and controllability of a non-linear heat equation

For simplicity, let us restrict ourself to the following simple nonlinear heat equation

$$\frac{\partial y}{\partial t} - \frac{\partial^2 y}{\partial x^2} - y^3 = 0, \tag{6.25}$$

where t is the time and $x \in [0,1]$. It is well known that for numerous initial Cauchy data $y_0 \in C^\infty([0,1])$ satisfying

$$y_0(0) = y_0(1) = 0 \tag{6.26}$$

there is no solution $y \in C^\infty([0,1] \times [0,+\infty))$ of (6.25) on $\{(x,t); x \in [0,1],\ t \in [0,+\infty)\}$ satisfying the initial condition

$$y(x,0) = y_0(x),\ \forall x \in [0,1] \tag{6.27}$$

and the Dirichlet boundary condition

$$y(0,t) = y(1,t) = 0,\ \forall t \in [0,+\infty). \tag{6.28}$$

This is due to a blow-up phenomenon. More precisely, for numerous $y_0 \in C^\infty([0,1])$ satisfying (6.26), there exists $T \in (0,+\infty)$ such that the unique solution to the Cauchy problem (6.25), (6.26) with the boundary condition (6.28) is defined on $[0,1] \times [0,T)$ and satisfied

$$\lim_{t \to T} \mid y(\cdot,t) \mid_{L^\infty(0,1)} = +\infty.$$

One may ask, if acting on the boundary of [0,1], on can avoid the blow-up phenomenon, that is

Open Problem 6. *Let $y_0 \in C^\infty([0,1])$ be given. Does there exists $y \in C^\infty([0,1] \times [0,+\infty))$ satisfying the heat equation (6.25) and the initial condition (6.27)?*

One could be even more "ambitious" and ask if the nonlinear heat equation is exactly controllable in an arbitrary time, i.e. ask the following question

Open Problem 7. *Let $T > 0$. Let $\hat{y}_0 \in C^\infty([0,1] \times [0,T])$ and $\hat{y}_1 \in C^\infty([0,1] \times [0,T])$ be two solutions of the nonlinear heat equation (6.25) on $[0,1] \times [0,+\infty)$ be given. Does there exists a solution $y \in C^\infty([0,1] \times [0,T])$ of the nonlinear heat equation (6.25) on $[0,1] \times [0,T]$ such that*

$$y(x,0) = \hat{y}_0(x,0),\ \forall x \in [0,1],$$

$$y(x,t) = \hat{y}_1(x,t),\ \forall (x,t) \in [0,1] \times [0,T]?$$

Let us point out that A. bamberger has proved that, if in (6.25) one replaces $-y^3$ by $+y^3$, the answer to this question is no: he has proved that with the + sign the heat equation is not even approximatively controllable; see [24]. For weaker nonlinearity and related questions let us mention the papers [17, 18, 30, 47] and the references therein.

7. Controllability of incompressible fluids

Let $l \in \{2,3\}$ and let Ω be a bounded nonempty connected open subset of $\mathbb{R}^l$ of class C^∞. Let $\Gamma^\#$ be an open subset of $\Gamma := \partial\Omega$ and let $\Omega^\#$ be an open subset of Ω. We assume that

$$\Gamma^\# \cup \Omega^\# \neq \emptyset, \tag{7.29}$$

$$\overline{\Omega^\#} \cap \Gamma = \emptyset. \tag{7.30}$$

We denote by n the outward unit normal vector field on Γ. The set $\Gamma^\#$ is the part of the boundary and $\Omega^\#$ is the part of the domain Ω on which the controls acts. The fluid that we consider is incompressible so that the velocity field y satisfies

$$\text{div } y = 0.$$

On the part of the boundary $\Gamma \backslash \Gamma^\#$ where there is no control the fluid does not cross the boundary: it satisfies

$$y \cdot n = 0 \text{ on } \Gamma \backslash \Gamma^\#. \tag{7.31}$$

Let $\nu \geq 0$ be the viscosity of the fluid. When $\nu \geq 0$ the fluid is called "perfect"; it then satisfies the Euler eqution. When $\nu > 0$ the fluid is viscous; it then satisfies the Navier-Stokes equation. When the fluid is viscous it satisfies on $\Gamma \backslash \Gamma^\#$, besides (7.31), some extra conditions which are the controls. For the moment being, let us just call by **BC** all the boundary conditions (*including* (7.31)) satisfied by the fluid on $\Gamma \backslash \Gamma^\#$.

Let us introduce the following definition

Definition 7.1. *A trajectory of the Navier Stokes ($\nu > 0$) or Euler ($\nu = 0$) control system on the interval of time $[0,T]$ is a map $y : \overline{\Omega} \times [0,T] \to \mathbb{R}^l$ of class C^∞ such that, for some function $p : \overline{\Omega} \times [0,T] \to \mathbb{R}$ of class C^∞,*

$$\frac{\partial y}{\partial t} - \nu \Delta y + (y \cdot \nabla) y + \nabla p = 0 \text{ in } (\overline{\Omega} \backslash \Omega^\#) \times [0,T], \tag{7.32}$$

$$\text{div } y = 0 \text{ in } \overline{\Omega} \times [0,T], \tag{7.33}$$

and, for any t in [0, T],

(7.34) $\quad y(\cdot, t)$ *satisfies the boundary conditions BC on* $\Gamma\backslash\Gamma^{\#}$.

Note that (7.32) to (7.36) have many solutions. (even up to arbitrary depending only on time added to p). In order to have uniqueness, up to arbitrary depending only on time added to p, one needs to add extra conditions. These extra conditions are the controls.

Let us point out that, when $\nu = 0$, (7.32) is simply the Euler equation (for incompressible fluids)

$$\frac{\partial y}{\partial t} + (y \cdot \nabla)y + \nabla p = 0 \text{ in } (\overline{\Omega}\backslash\Omega^{\#}) \times [0, T]. \tag{7.35}$$

When $\nu > 0$, (7.32) is the Navier Stokes equations. Our final definition is

Definition 7.2. *The Euler (resp. Navier-Stokes) control system is controllable if , given* $T > 0$ *and two trajectories* $\hat{y}_0$ *and* $\hat{y}_1$ *two trajectories of the Euler (resp. Navier-Stokes) control system on* $[0, T]$ *, there exists a trajectory* y *of the Euler (resp. Navier-Stokes) control system on* $[0, T]$ *such that*

$$y(x, 0) = \hat{y}_0(x, 0), \, \forall x \in \overline{\Omega}, \tag{7.36}$$

$$y(\cdot, T) = \hat{y}_1(x, T), \, \forall x \in \overline{\Omega}. \tag{7.37}$$

For the Euler control system one has

Theorem 7.3. *The Euler control system is controllable if and only if* $\Gamma^{\#}$ *meets any connected component of* Γ

Theorem 7.3 has been proved in

- [11] when Ω is simply-connected and $l = 2$,
- [12] when Ω is multi-connected and $l = 2$,
- [22] when Ω is contractible and $l = 3$,
- [23] when Ω is not contractible and $l = 3$.

The strategy of the proof of Theorem 7.3 relies on a method described in [11], [12], and [13] under the name of "return method" and introduced in [5] for a stabilization problem. Roughly speaking it consists in looking for $(\bar{y}, \bar{p})$ such that (7.33), (7.35), (7.36) and (7.37) hold, with $y = \bar{y}, p = \bar{p}$, $\hat{y}_0 = \hat{y}_1 = 0$ and such that the linearized control system around $\bar{y}$ is controllable under the assumptions of Theorem 7.3. With such a $(\bar{y}, \bar{p})$ one may hope that there exists (y, p)- close to $(\bar{y}, \bar{p})$- satisfying the required conditions, at least if y_0 and y_1 are "small". Finally, by using some scaling argument, one can deduce from the existence of (y, p) when y_0 and y_1 are "small" the existence of (y, p) even if y_0 and y_1 are not "small".

Let us emphasize that one cannot take $(\bar{y}, \bar{p}) = (0, 0)$. Indeed, with such a choice of $(\bar{y}, \bar{p})$, (7.33), (7.36), (7.37), (7.35) hold, with $y = \bar{y}, p = \bar{p}$, $y_0 = y_1 = 0$, but the linearized control system around $\bar{y} = 0$ is not at all controllable. Indeed this linearized control system is

$$\text{div } z = 0 \text{ in } \overline{\Omega} \times [0, T], \tag{7.38}$$

$$\frac{\partial z}{\partial t} + \nabla \pi = 0 \text{ in } \overline{\Omega} \times [0,T], \tag{7.39}$$

$$z(x,t).n(x) = 0, \ \forall (x,t) \in (\Gamma \backslash \Gamma^{\#}) \times [0,T].$$

Taking the curl of (7.39), one gets using (7.38)

$$\frac{\partial \text{curl } z}{\partial t} = 0,$$

which clearly shows that the linearized control system is not controllable. So one needs to consider other $(\bar{y}, \bar{p})$. Let us briefly explain how one constructs "good" $(\bar{y}, \bar{p})$ when $l = 2$ and Ω is simply connected. In such a case one easily checks the existence of a harmonic function θ in $C^\infty(\overline{\Omega})$ such that

$$\nabla \theta(x) \neq 0, \ \forall x \in \overline{\Omega},$$

$$\frac{\partial \theta}{\partial n} = 0 \text{ on } \Gamma \backslash \Gamma^{\#}.$$

Let $\alpha \in C^\infty(0,T)$ vanishing 0 and T. Let

$$(\bar{y}, \bar{p})(x,t) = (\alpha(t) \nabla \theta(x), -\alpha'(t)\theta(x) - \frac{1}{2}\alpha^2(t) \mid \nabla \theta(x) \mid^2).$$

Then (7.33), (7.35), (7.36) and (7.37) hold, with $y = \bar{y}, p = \bar{p}$, $\hat{y}_0 = \hat{y}_1 = 0$. Moreover using arguments relying on an extension method analogous to the one introduced by Russell in [37] one can see that the linearized control system around $\bar{y}$ is controllable.

The open problem of this section, which has been raised by J.-L. Lions in [34, 35], is

Open Problem 8. *Are the Navier-Stokes equations controllable?*

Let us end this section by giving briefly some known results related to this open problem. The known results depend on the boundary conditions BC for the Navier-Stokes equations. Three types of conditions are considered

- Stokes condition,
- Navier slip boundary condition,
- curl condition.

The Stokes condition is the well known no-slip boundary condtion

$$y = 0 \text{ on } \Gamma \backslash \Gamma^{\#}, \tag{7.40}$$

which of course implies (7.31)

The Navier slip boundary condition [36] imposes, condition (7.31), which is always assumed, and

$$\overline{\sigma} y \cdot \tau + (1 - \overline{\sigma}) n^i \left(\frac{\partial y^i}{\partial x^j} + \frac{\partial y^j}{\partial x^i} \right) \tau^j = 0 \text{ on } \Gamma \backslash \Gamma^{\#} \tag{7.41}$$

where $\overline{\sigma}$ is a constant in $[0,1)$, $n = (n^1, \dots, n^l)$ and $\tau = (\tau^1, \dots, \tau^l)$ is any tangent vector field on the boundary Γ. In (7.41) we also have used the usual summation convention. Note that the Stokes condition (7.40) corresponds to the case $\overline{\sigma} = 1$, which we do not include in the Navier slip boundary condition.

Finally the curl condition is considered in dimension 2 ($l = 2$). This condition is, condition (7.31) which is always assumed, and

$$\text{curl } y = 0 \text{ on } \Gamma \backslash \Gamma^{\#}. \tag{7.42}$$

The known results we want to recall are

(i) Local controllability for $l = 2$ and and the curl condition. By local controllability one means that there exists a trajectory of the Navier-Stokes control system connecting two given trajectories $\hat{y}_0$ and $\hat{y}_1$ of the Navier-Stokes control system if $\hat{y}_0$ and $\hat{y}_1$ are close together. This result is due to Fursikov and Imanuvilov [19, 20, 21].

(ii) Local controllability for any l and the Stokes condition. This result is due to O. Imanuvilov [31].

(iii) Approximate controllability for the Navier boundary condition when $l = 2$. This result is proved in [13]. It's proof relies on the return method.

(iv) Controllability if $\Gamma^{\#} = \Gamma$ and $l = 2$. (Of course when $\Gamma^{\#} = \Gamma$ one does not to precise the boundary conditions considered.) This result is proved in [14] using the methods of [13, 19, 20, 21].

REFERENCES

[1] F. Ancona and A. Marson, On the Attainable Set for Scalar Nonlinear Conservation Laws with Boundary Control, to appear in *SIAM J. of control.*

[2] R.M. Bianchini and G. Stefani, Sufficient conditions for local controllability, Proc. 25th Conference on Decision and Control, Athens (1986) pp. 967-970.

[3] R.W. Brockett, Asymptotic stability and feedback stabilization, in: *Differential Geometric Control Theory*, R.W. Brockett, R.S. Millman et H.J. Sussmann eds, Progress in Math., 27 (1983) pp. 181-191, Birkhäuser, Basel-Boston.

[4] J.-M. Coron, A necessary condition for feedback stabilisation, *Systems and Control Letters*, 14 (1990) pp. 227-232.

[5] J.-M. Coron, Global asymptotic stabilization for controllable systems without drift, *Math. Control Signals Systems*, 5 (1992) p. 295-312.

[6] J.-M. Coron, Links between local controllability and local continuous stabilization, in: *IFAC Nonlinear Control Systems Design*, M. Fliess ed., (1992) pp. 165-171.

[7] J.-M. Coron, Linearized controlled systems and applications to smooth stabilization, *SIAM J. Control and Optimization*, 32 (1994) pp. 358-386.

[8] J.-M. Coron, Stabilization in finite time of locally controllable systems by means of continuous time-varying feedback laws, *SIAM J. Control and Optimization*, 33 (1995) pp. 804-833.

[9] J.-M. Coron, Relations entre commandabilité et stabilisation non linéaire, in: *Nonlinear partial differential equations and their applications, Collège de France Seminar* vol. 11, H. Brezis and J.-L Lions ed., Research Notes in Math. Pitman, Boston, (1994) pp. 68-86.

[10] J.-M. Coron, Stabilizing time-varying feedback, in: *IFAC Nonlinear Control Systems Design*, Tahoe city, July 1995.

[11] J.-M. Coron, Contrôlabilité exacte frontière de l'équation d'Euler des fluides parfaits incompressibles bidimensionnels, *C. R. Acad. Sci. Paris*, 317, Série 1 (1993) pp. 271-276.

[12] J.-M. Coron, On the controllability of 2-D incompressible perfect fluids, *J. Math. Pures & Appliquées*, 75 (1996) pp. 155-188.

[13] J.-M. Coron, On the controllability of the 2-D incompressible Navier-Stokes equations with the Navier slip boundary conditions, *ESAIM: Control, Optimization, and Calculus of Variations, www.emath.fr/cocv/*, 1 (1996) pp. 35-75.

[14] J-M. Coron and A. Fursikov, Global exact controllabilty of the 2D Navier-Stokes equations on a manifold without boundary, preprint, *Russian Journal of Mathematical Physics*, 4, (1996) pp. 429-448.

[15] J.-M. Coron and L. Praly, Adding an integrator for the stabilization problem, *Systems and Control Letters*, 17 (1991) pp. 89-104.

[16] J.I. Diaz, Sobre la controlabilidad approximada de problemas no lineales disipativos, in Proceedings of *Jornadas Hispano-Francesas sobre Control de Sistemas Distribuidos*, A. Valle ed., (1991) pp. 41-48.

[17] C. Fabre, J.-P. Puel and E. Zuazua, Approximate controllability for semilinear heat equation, *Proc. Royal Soc. Edinbourgh*, 125A (1995), pp. 31-61.

[18] A. Fursikov and O. Yu. Immanuvinov, On controllability of certain systems simulating a fluid flow, in: *Flow Control*, IMA vol. in Math. and its Appl. , M.D. Gunzburger ed., Springer Verlag, New York, 68 (1994) pp. 149-184.

[19] A. Fursikov and O. Yu. Imanuvilov, Local exact controllability of the Navier-Stokes equations, *C. R. Acad. Sci. Paris*, 323 (1996) pp. 275-280.

[20] A. Fursikov and O. Yu. Imanuvilov, On exact boundary zero controllability of the two-dimensional Navier-Stokes equations *Acta Appl. Math.* 36 (1994) pp. 1-10.

[21] A. Fursikov and O. Yu. Imanuvilov, Local exact controllability for $2-D$ Navier-Stokes equations, *Sbornik Math.*, 187 (1996) (in Russian).

[22] O. Glass, Contrôlabilité exacte frontière de l'équation d'Euler des fluides parfaits incompressibles en dimension 3, *C.R. Acad. Sci. Paris*, 325, Série I, (1997) pp. 987-992.

[23] O. Glass, Exact boundary controllability of 3-D Euler equation, preprint, 1998.

[24] J. Henry, Étude de la contrôlabilité de certaines équations paraboliques, Thèse d'État de l'université Paris 6, (1978).

[25] M. Kawski, High-order small time local controllability, in: *Nonlinear Controllability and Optimal Control*, H.J. Sussmann ed., Monographs and Textbooks in Pure and Applied Mathematics, 113, M. Dekker, Inc., New York, 1990, p. 431-467.

[26] H. Hermes, On the synthesis of a stabilizing feedback control via Lie algebraic methods, *SIAM J. on Control and Optimization*, 18 (1980) pp. 352-361.

[27] H. Hermes, Control systems which generate decomposable Lie algebras, *J. Differential Equations*, 44 (1982) pp. 166-187.

[28] J. Kurzweil, On the inversion of Lyapunov's second theorem on stability of motion, *Ann. Math. Soc. Trans.* Ser.2, 24 (1956) p. 19-77.

[29] Th. Horsin, On the controllability of the Burger equation, preprint Univervité de Versailles, 1997, *ESAIM: Control, Optimization, and Calculus of Variations, www.emath.fr/cocv/*, 3 (1998), pp. 83-95.

[30] O. Yu Imanuvilov, Thesis, Moscow 1991 (in Russian) and Exact boundary controllability of the parabolic equation, *Russian Math. Surveys*, 48 (1993), pp. 211-212.

[31] O. Yu. Imanuvilov, On exact controllability for Navier Stokes equations,*ESAIM: Control, Optimisation and Calculus of Variations, www.emath.fr/cocv/*, 3 (1998), pp. 97-131.

[32] M.A. Krasnosel'skiĭ, The operator translation along the trajectories of differential equations, *Trans. Math. Monographs*, 19 (1968).

[33] M.A. Krasnosel'skiĭ and P.P. Zabreiko, *Geometric Methods in Nonlinear Analysis,* Springer-Verlag, Berlin, 1983.

[34] J.-L. Lions, Are there connections between turbulence and controllability?, in: *9th INRIA International Conference*, Antibes, June 12-15, 1990.

[35] J.-L. Lions, Exact controllability for distributed systems. Some trends and some problems, in: *Applied and Industrial Mathematics*, R. Spigler ed., Kluwer Academic Publishers, Dodrecht, Boston, London, (1991) pp. 59-84.

[36] C.L.M.H. Navier, Sur les lois du mouvement des fluides, *Mem. Acad. R. Sci. Inst. France*, 6 (1823) pp. 389-440.

[37] D.L. Russell, Exact boundary value controllability theorems for wave and heat processes in star-complemented regions, in: *Differential Games and Control Theory,* Roxin, Liu et Sternberg eds., Marcel Dekker, New York, (1974) pp. 291-319.

[38] C. Samson, Velocity and torque feedback control of a nonholonomic cart, in: *Advanced Robot Control, Proceedings of the International Workshop on Nonlinear and Adaptative Control: Issues in Robotics*, Grenoble, France, November 21-23, 1990 (C. Canudas de Wit, ed.), Lecture Notes in Control and Information Sciences, vol. 162, pp. 125-151, Springer-Verlag, Berlin Heidelberg New York, 1991.

[39] H.J. Sussmann, Lie brackets and local controllability : a sufficient condition for scalar-input systems,*SIAM J. on Control and Optimization*, 21 (1983) pp. 686-713.

[40] E.D. Sontag, *Mathematical Control Theory: Deterministic Finite Dimensional Systems,* Text in Applied Mathematics 6, Springer-Verlag, New York Berlin Heidelberg, 1990.

[41] E.D. Sontag and H.J. Sussmann, Remarks on continuous feedback, *IEEE CDC*, Albuquerque, 2 (1980) p. 916-921.

[42] H.J. Sussmann, Subanalytic sets and feedback control, *J. Differential Equations*, 31 (1979) pp. 31-52.

[43] H.J. Sussmann, A general theorem on local controllability, *SIAM J. on Control and Optimization*, 25 (1987) pp. 158-194.

[44] H.J. Sussmann and V. Jurdjevic, Controllability of nonlinear systems, *J. Differential Equations*, 12 (1972) pp. 95-116.

[45] G.W. Whitehead, *Elements of Homotopy Theory,* Springer, Berlin-Heidelberg-New York, (1978).

[46] J.C. Willems, Paradigms and puzzles in the theory of dynamical systems, *IEEE Trans. Autom. Control*, 36 (1991) pp. 259-294

[47] E. Zuazua, Exact controllability for semilinear wave equatios in one space dimension, *Ann. Inst. Henri Poincaré, Nonlinear Analysis*, 10 (1993), p. 109-129.

[48] J. Zabczyk, Some comments on stabilizability, *Applied Mathematics and Optimization,* 19 (1989) pp. 1-9.

Université de Paris-Sud, Analyse Numérique et EDP, Bâtiment 425, 91405 Orsay Cedex, France. E-mail: `Jean-Michel.Coron@math.u-psud.fr`.

Proceedings of Symposia in Pure Mathematics
Volume **64**, 1999

QUADRATIC ORDER CONDITIONS OF A LOCAL MINIMUM FOR SINGULAR EXTREMALS IN A GENERAL OPTIMAL CONTROL PROBLEM

A. V. DMITRUK

Abstract

We consider the class of optimal control problems, linear in the control, with control bounded by linear inequalities, and with terminal equality and inequality constraints. Both the control and state variables are multidimensional, and the examined control is totally singular. For such problems we suggest quadratic-order necessary and sufficient conditions for a weak and a so-called Pontryagin minimum, the last being a minimum of an intermediate type between classic weak and strong minima. Necessary conditions transform into sufficient ones only by strengthening an inequality, what is similar to conditions in the classical analysis and calculus of variations (adjoint pairs of conditions).

Key words: singular extremal, weak minimum, Pontryagin minimum, strong minimum, quadratic order of estimation, necessary and sufficient conditions, critical cone, second variations, Legendre conditions.

1 Statement of the problem and preliminaries

Let us consider the control system:

$$\dot{x} = f_0(x,t) + F(x,t)u = f_0(x,t) + \sum_{i=1}^{k} u_i f_i(x,t), \tag{1}$$

where all the functions $f_0, \dots, f_k$ are defined in an open set $\mathcal{D}_{x,t}$ in $R^n \times R$ and take values in R^n, they are assumed to be C^2-smooth in (x,t). Here $f_i(x,t)$, $i = 1, \dots, k$, are the columns of the matrix $F(x,t)$.

We consider this system on a fixed time interval $[t_0, t_1]$. Denote $x_0 = x(t_0)$, $x_1 = x(t_1)$, and $p = (x_0, x_1)$, and consider the following optimal control problem A:

$$J = \varphi_0(p) \longrightarrow \min, \tag{2}$$

1991 *Mathematics Subject Classification.* Primary 49K15.
This research was supported by the Russian Foundation for Basic Research under grants 97-01-00135 and 96-15-96072.

$$\varphi_i(p) \le 0, \quad i = 1, \dots, \nu, \tag{3}$$

$$K(p) = 0, \tag{4}$$

$$u(t) \in U, \tag{5}$$

$$(x(t), t) \in \mathcal{D}_x, \qquad p \in \mathcal{D}_p, \tag{6}$$

where $\varphi_0, \dots, \varphi_\nu$, K are C^2-smooth functions (K is multidimensional), defined in an open set $\mathcal{D}_p$ in R^{2n}, and U is a closed convex solid set in R^k. The sets $\mathcal{D}_{x,t}$ and $\mathcal{D}_p$ are usually just implied, but not indicated explicitly.

Without loss of generality we may put $t_0 = 0$ and $t_1 = T$, denoting then $x(t_1) = x_T$. Note a well known important property of system (1) (see e.g. [41, 33]).

Theorem 1.1. *Let $u_m \in L^k_\infty[0,T]$, and x_m be the solution of (1) for u_m with an initial condition $x_m(0) = a_m$. Let $u_m \overset{\text{weak-*}}{\longrightarrow} u_0 \in L^k_\infty[0,T]$, $a_m \to a_0$, and x be the solution of (1) for u_0 with the initial condition $x_0(0) = a_0$.*

Then $x_m \rightrightarrows x_0$ (uniformly).

In Problem A we will seek a minimum among all absolutely continuous n-vector functions $x(t)$ and bounded measurable k-vector functions $u(t)$. We do not study here the problem of existence of minimum. At least we can say, that if U is a convex compactum, x_0 is fixed, every solution $x(t)$ of system (1) entirely lies in $\mathcal{D}_{x,t}$, and its endpoints (x_0, x_T), satisfying (3), (4), belong to $\mathcal{D}_p$, then the existence follows from the known Filippov's lemma [42], which, in its turn, follows from the theorem of Alaoglu and Theorem 1.1.

Introduce the space $W = AC^n \times L^k_\infty[0,T]$ with elements $w = (x, u)$, and equip it with the norm $||w|| = |x(0)| + ||\dot{x}||_1 + ||u||_\infty$. (Here and throughout the paper $||\cdot||_p$ stands for the norm in the Banach space $L^k_p[0,T]$, and $||\cdot||_C$ stands for the norm in the space of continuous n-vector functions $C^n[0,T]$.) One can easily see, that for solutions of (1) the convergence of a sequence w_m to a point w_0 in the norm of W is equivalent to its convergence to w_0 in the norm $||w||_\infty = ||x||_C + ||u||_\infty$, hence a local minimum w.r.t. the norm of W is a common weak minimum. As usual, by a strong minimum we call a minimum w.r.t. the seminorm $||w||' = ||x||_C$, the u being free.

Now we introduce also another type of minimum, intermediate between the weak and strong minima. Let $\hat{w} = (\hat{x}, \hat{u})$ be an admissible pair.

Definition 1.1. We say that $\hat{w}$ is a Pontryagin minimum point in Problem A, if for all N there exists an $\varepsilon > 0$ such that $\hat{w}$ is a minimum point in Problem A on the set

$$||x - \hat{x}||_C < \varepsilon, \quad ||u - \hat{u}||_1 < \varepsilon, \quad ||u - \hat{u}||_\infty \le N. \tag{7}$$

In other words, there can exist no sequence $w_m = (x_m, u_m)$ such that

$$||x_m - \hat{x}||_C \to 0, \quad ||u_m - \hat{u}||_1 \to 0, \quad ||u_m - \hat{u}||_\infty \le O(1), \tag{8}$$

all constraints are satisfied, and for all m $J(p_m) < J(\hat{p})$.

Sequences that satisfy (8), we call *Pontryagin sequences,* converging to $\hat{w}$. The set of all such sequences we denote by $\Pi(\hat{w})$. The differences $\delta w_m = w_m - \hat{w}$ are Pontryagin sequences, converging to zero; we call them Pontryagin variations. Thus, the Pontryagin minimum (briefly, the Π-minimum) is the minimum in the class of all Pontryagin sequences (or, respectively, in the class of all Pontryagin variations). This type of minimum was introduced by A.Ja.Dubovitskii and A.A.Milyutin as a natural extension of minimum in the class of uniformly small and needle-type variations of the control. The importance of this notion is that the Pontryagin Maximum Principle (MP) is a necessary condition of the first order for the Pontryagin minimum, i.e. for the minimum in the class of Pontryagin variations. (It is said often that the MP is a necessary condition for the *strong* minimum, but this assertion is weaker than the above one.) Moreover, Dubovitskii and Milyutin showed [3], [10], that the fulfillment of MP for a trajectory is *equivalent* to its stationarity in the class of all possible Pontryagin variations. As we told already, the following relations for these three types of minimum are true:

$$\text{weak min} \;<\; \text{Pontryagin min} \;<\; \text{strong min},$$

and both the "inequalities" are strict: it is not difficult to provide corresponding counterexamples, see e.g. [33]. In particular case, when the admissible control set U is bounded, the Π-minimum is in fact the minimum w.r.t. the norm $||w||_1 = ||x||_C + ||u||_1$, and we call it the L_1-minimum w.r.t. the control.

The Π-minimum for Problem A is, generally, rather far from the strong minimum (unlike the case of problems with the general control system, nonlinear in the control, where these two types of minimum are rather close to each other; see [6], [7]). However, note here a special case, when these two types of minimum are equivalent: it is when U is bounded, and $(\forall\, t)\; \hat{u}(t)$ is an extreme point of U. (The set of all extreme points of U is denoted by $ex\, U$.)

Proposition 1.1. *Let U be an arbitrary convex compactum, and $\hat{u}(t) \in ex\, U$ a.e. Suppose also that* $\operatorname{rank} F(\hat{x}(t), t) = k \quad \forall t$, *i.e. the vectors $f_i(\hat{x}(t), t)$, $i = 1, \dots, k$ are linearly independent. Then the Π -minimum is equivalent to the strong minimum.*

The proof readily follows from the next

Lemma 1.1. *Let $\hat{x}, \hat{u}$ satisfy equation (1), $x_m \Rightarrow \hat{x}$, $u_m \in U$, U is a convex compactum, and $\hat{u}(t) \in exU$ a.e. Then $||u_m - \hat{u}||_1 \longrightarrow 0$.*

The proof follows from another two lemmas.

Lemma 1.2. *Under the first four conditions of Lemma 1.1, $u_m \overset{\text{weak-*}}{\longrightarrow} \hat{u}$ (as elements of L_∞ w.r.t. L_1).*

Proof. Since U is bounded, then due to the theorem of Alaoglu, passing if necessary to a subsequence, one can consider $u_m(t) \overset{\text{weak-*}}{\longrightarrow} u_*(t)$. Since U is closed and convex, $u_*(t) \in U$ a.e. Let x_* be the solution of system (1) for this u_* with the initial condition $x_*(0) = \hat{x}(0)$. By Theorem 1.1 $x_m \Rightarrow x_*$, whence due to the uniqueness of the limit we have $x_* = \hat{x}$, thus $\hat{x}$ is the solution of (1) for u_* with the initial value $\hat{x}(0)$. But then, since the vectors $f_i(\hat{x}(t), t)$, $i = 1, ..., k$ are linearly independent, the value of control is uniquely determined by $\hat{x}(t)$, and so $u_*(t) = \hat{u}(t)$ a.e. Thus, $u_m \overset{\text{weak-*}}{\longrightarrow} \hat{u}$ (for the chosen subsequence; hence, for any subsequence of the initial sequence there is a sub-subsequence with this property, hence it is true for the whole initial sequence).

Lemma 1.3. *Let U be a convex compactum, $u_m(t) \in U$, and $u_m(t) \overset{\text{weak-*}}{\longrightarrow} \hat{u}(t)$, where $(\forall t)$ $\hat{u}(t) \in ex\, U$. Then $||u_m - \hat{u}||_1 \longrightarrow 0$.*

Proof. Let $p(t)$, $|p(t)| = 1$ $\forall t$, be a supporting vector for U at the point $\hat{u}(t)$, i.e.

$$(p(t), U) \leq (p(t), \hat{u}(t)). \tag{9}$$

(Such a $p(t)$ does exist by the measurable selection theorem [43].) Since $\int p(t)(u_m - \hat{u})dt \to 0$, then from (9), from the extremality of $\hat{u}(t)$ and the boundedness of U one can show that $\int |u_m - \hat{u}|dt \to 0$. To understand this effect, one may consider, for example, $U = \{u \in R^2 |\ u_1^2 \leq u_2\}$, $\hat{u} = (0,0)$, $p(t) = (0,-1)$. If $\int p u_m dt = \int u_{2,m} dt \to 0$, then $\int u_{1,m}^2 dt \to 0$, whence obviously $\int |u_{1,m}|dt \to 0$ too, i.e. $\int (|u_{1,m}| + |u_{2,m}|)dt \to 0$, q.e.d. The general case involves a bit more technical details, see [41].

2 Maximum Principle and singular controls

Let $\hat{w} = (\hat{x}, \hat{u})$ be a Π-minimum point in Problem A. Then it satisfies the Pontryagin Maximum Principle (MP), which says, that there exist Lagrange multipliers $\alpha = (\alpha_0, \ldots, \alpha_\nu) \geq 0$, $\beta \in R^{d(K)}$, and a Lipschitz $n-$vector function $\psi(t)$ (the adjoint, or costate variable), the collection of which we denote by $\lambda = (\alpha, \beta, \psi)$, and which

generate the terminal Lagrange function $l[\lambda](p) = \alpha \cdot \varphi(p) + \beta \cdot K(p)$, where $\varphi = (\varphi_0, \dots, \varphi_\nu)$, and the Pontryagin function
$H[\lambda](x,u,t) = \psi[f_0(x,t) + F(x,t)u] = (\psi, f_0(x)) + \sum u_i(\psi, f_i(x))$,
such that the following conditions hold:

a) normalization condition: $|\alpha| + |\beta| = 1$ (here $|\cdot|$ is an arbitrary norm in the finite-dimensional space),

b) complementary slackness: $\alpha_i \varphi_i(\hat{p}) = 0, \quad i = 1, ..., \nu$,
(in the sequel we assume, without loss of generality, that $\varphi_i(\hat{p}) = 0, \quad \forall\, i = 1, ..., \nu$, i.e. all indices i are active, and we denote $I = \{0, 1, ..., \nu\}$),

c) adjoint, or costate equations: $\dot{\psi} = -H_x[\lambda], \qquad \dot{H}[\lambda] = H_t[\lambda]$,

d) transversality conditions: $\psi(0) = l'_{x_0}[\lambda], \quad \psi(T) = -l'_{x_T}[\lambda]$,

e) maximality condition: for all t

$$\max_{u \in U} H[\lambda](\hat{x}(t), u, t) = H[\lambda](\hat{x}(t), \hat{u}(t), t). \tag{10}$$

The set of all $\lambda = (\alpha, \beta, \psi)$, satisfying conditions (a)–(e) for the trajectory $\hat{w}$, we denote by $\Lambda(\hat{w})$, or, having in mind that the trajectory $\hat{w}$ will be the same throughout the paper, simply by Λ. Obviously, Λ is a finite-dimensional compact set, and generally it may consist of more than a single point. (In particular case, when Λ consists of a single point, we will write $\Lambda = \{\cdot\}$.) We say that the trajectory $\hat{w}$ is stationary, or extremal, if $\Lambda(\hat{w})$ is nonempty. The MP guarantees that if $\hat{w}$ is a Π-minimum point in Problem A, then the set $\Lambda(\hat{w})$ is nonempty, i.e. $\hat{w}$ is an extremal.

It is well known, however, that the MP is only a necessary, but not a sufficient condition (like any other first order necessary condition for general nonconvex problems): its fulfillment does not guarantee even a weak minimum at $\hat{w}$. Especially it is true for our Problem A . Because of this, many authors made investigations on higher order (mostly "second order") conditions of a local minimum (for some particular statements of Problem A) since early 1960-s; we just mention here Kelley, Kopp, Moyer, Bryson, Robbins, Goh, Vapnyarsky, Speyer, Jacobson, Bell, McDanell, Powers, Gabasov, Kirillova, Krener, Agrachiov, Gamkrelidze, Milyutin, Dmitruk, Knobloch, Zelikin, Gurman, Dykhta, Lamnabhi-Lagarrigue, Stefani, Sarychev, and others, see [4, 12], [17] – [39], and references therein. These more fine conditions require a more fine specification of examined extremals. There are two essentially different classes of extremals: singular and nonsingular extremals. The first one of them includes, as the most pronounced case, the class of totally singular extremals.

Definition 2.1. An extremal $\hat{w}$ is called *totally singular,* if for any $\lambda \in \Lambda(\hat{w})$, and for any t the Pontryagin function $H[\lambda](\hat{x}(t), u, t)$ does not depend on $u \in U$, i.e. takes the same value $H[\lambda](\hat{x}(t), \hat{u}(t), t)$ for all $u \in U$.

Since H is linear in u, and U has nonempty interior, this means simply, that for all t

$$H_u[\lambda](\hat{x}, \hat{u}, t) = \psi(t)F(\hat{x}(t), t) = 0, \tag{11}$$

or, equivalently,

$$(\psi(t), f_i(\hat{x}(t), t)) = 0 \quad \forall\, i = 1, ...k.$$

Remark 2.1. Perhaps, it would be more proper to say in this case, that the constraint $u \in U$ (or the set U) is totally singular for the extremal $\hat{w}$ (rather than the extremal $\hat{w}$ itself is totally singular), because U does not actually enter the MP at all.

Remark 2.2. Note that if U is totally singular, then any other U', containing U, is totally singular as well, and the set $\Lambda(\hat{w})$ is one and the same for all such sets U', and coincides with the set $\Lambda(\hat{w})$ for $U = R^k$, i.e. for Problem A with unconstrained control. However, if one takes U' containing $\hat{u}(t)$, but not containing U, it may happen to exist $\lambda \in \Lambda(\hat{w}, U')$, for which (11) fails.

If there exists at least one $\lambda \in \Lambda$, which does not satisfy (11), then $\hat{w}$ is not totally singular. The most distinct case is when there exists $\lambda \in \Lambda$, for which $\forall t$ $H[\lambda](\hat{x}(t), u, t)$ attains its maximum over U only at the single point $\hat{u}(t)$. In this case we call $\hat{w}$ *strictly nonsingular.* In the intermediate case, when $\hat{w}$ is neither totally singular, nor strictly nonsingular, it is of a mixed type. This case practically has not been studied in the framework of higher order conditions of a local minimum. (Only conditions at junction times of singular and nonsingular subarcs have been analyzed until now, see e.g. [40].) The main efforts have been spent to investigate the two principal cases of totally singular and strictly nonsingular extremals. The higher order conditions for these two cases turned out to be essentially different. IN THIS PAPER WE CONSIDER ONLY THE TOTALLY SINGULAR CASE. Note that in the classical calculus of variations (CCV) this case had not been studied, because all considerations in CCV were made under the assumption of strong Legendre condition: $-H_{uu}[\lambda] \geq \text{const} > 0$, whereas for totally singular extremals this coefficient is identically zero.

Now let $\hat{w}$ be a totally singular extremal. We assume that the control $\hat{u}(t)$ is continuous. In view of equations (11), this is rather a mild requirement, because in a generic case it is possible, by differentiating these equations two times in t, to express $\hat{u}(t)$ in terms of $\psi(t)$ and $\hat{x}(t)$ (see e.g. [17, 19, 20]), whence $\hat{u}(t)$ is in fact Lipschitzian.

Thus, we have taken two assumptions about the examined extremal:

Assumption 1. $\hat{w} = (\hat{x}, \hat{u})$ is a totally singular extremal.

Assumption 2. The control $\hat{u}(t)$ is continuous.

Now we impose an assumption on the character of contact of the control $\hat{u}(t)$ with the boundary ∂U of the admissible control set U.

Assumption 3. U is a closed polyhedral set (may be unbounded), and there exists a face U_0 of U, such that $\forall t \quad \hat{u}(t) \in \operatorname{reint} U_0$. (By *reint* we denote the relative interior of the convex set.)

In particular case, $U_0 = U$ is allowed, which means that $\forall t \;\; \hat{u}(t) \in \operatorname{int} U$.
If U_0 is a proper face of U, then $\hat{u}(t) \in \partial U$, and we call it the simplest case of boundary control.

Assumption 3 can be weakened to the following one.

Assumption 3'. $U = M \cap D$, where M satisfies Assumption 3, and D is an arbitrary convex set, such that $\forall t \;\; \hat{u}(t) \in \operatorname{int} D$.

3 Quadratic order of estimation

As was said already, having got first order conditions for a local minimum, it is natural to move to study "second order" conditions. But here we have to define preciser, what will be meant by "second order" conditions. In finite-dimensional problems it is quite clear: they are conditions of the order $|\delta x|^2$, i.e. when all considerations are made modulo $o(|\delta x|^2)$. These conditions possess the following important properties: a) both necessary and sufficient conditions withstand perturbations of all data functions in the problem within $o(|\delta x|^2)$,
b) the sufficient condition consists of a lower bound of second variations by $|\delta x|^2$,
c) if a point satisfies the necessary condition, one can make an arbitrary small perturbation of the data functions in the C^2-norm, after which this point satisfies the sufficient condition.

However, in infinite-dimensional problems, such as CCV and optimal control problems, the situation is not so simple, and it is not clear a'priori, what is the "second order". If one undertake the straightforward generalization of the finite-dimensional ideology, i.e. perform all considerations w.r.t. $\gamma_0(\delta w) = ||\delta w||^2$, $\delta w = (\delta x, \delta u)$, then unfortunately one would never obtain the fulfillment of sufficient conditions (apart from the case of Hilbert space, which is not characteristic neither for CCV, nor for optimal control): it is well known, that a continuous quadratic form can be bounded from below by the square of the norm only if the space is isomorphic to a Hilbert space. The functional γ_0 can be said to be "too rough" to estimate second variations in these classes of problems. (Note by the way, that in CCV the term "second order conditions" was never used. In those times the authors said "conditions, based on the second variation", see e.g. [1]. Obviously, they understood that the term "second order conditions" is not a proper one for problems in functional spaces.)

Thus, the question arises: *what a functional should be taken instead of* $\gamma_0(\delta w)$, in such a way that, roughly speaking, it would preserve the abovementioned properties of γ_0 ? The answer depends on the class of problems being considered, and may also depend on the type of examined trajectory. It was discovered, due to a series of deep works by A.A.Milyutin and his scientific school [5] – [9], [31] – [38], that for any properly defined class of extremal problems there should exist an estimating functional $\gamma(\delta w)$, positive outside of zero (which we call an order of estimation, or simply an order), characteristic to this class. (See precise definitions and details in [5].) For general optimal control problems, nonlinear in the control (e.g. for CCV), in case when $\hat{u}(t)$ is continuous, such an order is:

$$\gamma_{clas}(\delta w) = |\delta x(0)| + \int |\delta u(t)|^2 dt$$

(here and throughout the paper all integrals without limits are taken over the whole interval $[0,T]$). The corresponding necessary and sufficient conditions (in particular, those in CCV) possess the above properties a) – c) [5]. However, it does not suit to our Problem A: from the very outset one can say, that, since the control comes just linearly in the problem, the second variations in this problem will never be bounded from below by this $\gamma_{clas}(\delta w)$, because they never contain the term with δu^2.

It turns out that the proper quadratic functional of estimation for Problem A is:

$$\gamma(\delta w) = |\, \delta x(0)\, |^2 + |\, \delta y(T)\, |^2 + \int |\, \delta y(t)\, |^2 \, dt, \tag{12}$$

where

$$\delta \dot{y} = \delta u, \qquad \delta y(0) = 0.$$

Note that the control variation δu does not come as such in the quadratic order (12); it comes only through the variation of a new state variable y, where $\dot{y} = u,\ y(0) = 0$. (This relation between y and u will be preserved throughout the paper). Here we give conditions of this order γ.

4 The second and third variations, and the critical cone

For any $\lambda \in \Lambda$ consider the corresponding Lagrange function

$$\Phi[\lambda](w) = \; l[\lambda](p) + \int ((\psi, \dot{x}) - H[\lambda](x,u,t))\,dt$$

(recall that $l[\lambda]$ and $H[\lambda]$ were defined in the beginning of Sec.2), and consider also the half of its second variation at $\hat{w}$ - the quadratic functional

$$\Omega[\lambda](\bar{w}) = \frac{1}{2}(l''[\lambda]\bar{p},\bar{p}) - \int \Big((\frac{1}{2} H_{xx}[\lambda]\,\bar{x},\bar{x}) + (\bar{x}, H_{xu}[\lambda]\,\bar{u}) \Big)\,dt. \tag{13}$$

Define the matrices $A(t) = f_0'(\hat{x}(t),t) + F'(\hat{x}(t),t)\hat{u}$, $\quad B(t) = F(\hat{x}(t),t)$, and the tensor $R(t) = F'(\hat{x}(t),t)$ (by f_i' we denote the derivative of f_i w.r.t. x) in such a way that for $x = \hat{x} + \delta x$ and $u = \hat{u} + \delta u$ equation (1) takes the form:

$$\delta\dot{x} = A(t)\delta x + B(t)\delta u + (R(t)\delta x, \delta u) \; + \; \text{higher order terms.} \tag{14}$$

Denote by $\mathcal{K}$ the so-called critical cone for the problem (1)–(4) with the free control. It consists of all variations $\bar{w} = (\bar{x},\bar{u})$ in W, such that

$$\varphi'(0)\bar{p} \le 0, \qquad K'(0)\bar{p} = 0, \tag{15}$$

and

$$\dot{\bar{x}} = A(t)\bar{x} + B(t)\bar{u}. \tag{16}$$

The cone of critical variations for the "full" Problem A is $\mathcal{K} \cap \mathcal{N}$, where $\mathcal{N} = \{\bar{w} |\; \bar{u}(t) \in N \; a.e.\}$, and $N = con(U - \hat{u}(t)) = con(M - \hat{u}(t))$ is the local (pointwise) tangent cone for the polyhedron M at the point $\hat{u}(t)$. (By $con\, M$ we denote the conical hull $\bigcup \{\alpha M \,|\, \alpha > 0\}$ of the set M.) Due to Assumption 3 (or 3') N does not depend on t.

We also introduce the cubic functional

$$\rho[\lambda](\bar{w}) = \int \Big[-(\frac{1}{2} H_{uxx}[\lambda]\bar{x},\bar{x},\bar{u}) + (H_{xu}[\lambda]\bar{y}, (R(t)\bar{x},\bar{u})) \Big]\,dt. \tag{17}$$

It is one sixth of the third variation of Lagrange function at $\hat{w}$ (in case when all $f_i(x,t)$ are C^3–smooth in x) on equation (1) to within $o(\gamma)$ on Pontryagin sequences, see [32, 33] for details.

It is convenient to consider functionals (13) and (17) in slightly transformed variables. Namely, there is a convenient change of variables - the so-called Goh transformation [18, 19]: $(\bar{x},\bar{u}) \longmapsto (\bar{\xi},\bar{y},\bar{u})$, where $\bar{\xi} = \bar{x} - B\bar{y}$,

$$\dot{\bar{y}} = \bar{u}, \qquad \bar{y}(0) = 0, \tag{18}$$

and hence

$$\dot{\bar{\xi}} = A(t)\bar{\xi} + B_1(t)\bar{y}, \qquad B_1 = AB - \dot{B}. \tag{19}$$

This transformation is convenient because, instead of the state variable $\bar{x}$, related to $\bar{u}$ by a general linear system (16), there are now two state variables $\bar{\xi}, \bar{y}$, such that $\bar{u}$ does not come into equation (19) for $\bar{\xi}$, and comes, in the simplest way, only into equation (18) for $\bar{y}$.

In these new variables, the second variation (13) takes the form:

$$\begin{aligned} \Omega[\lambda](\bar{\xi}, \bar{y}, \bar{u}) &= g[\lambda](\bar{\xi}_0, \bar{\xi}_T, \bar{y}_T) + \\ + \int ((D[\lambda](t)\bar{\xi}, \bar{\xi}) &+ (P[\lambda](t)\bar{\xi}, \bar{y}) + (Q[\lambda](t)\bar{y}, \bar{y}) + (V[\lambda](t)\bar{y}, \bar{u}))dt, \end{aligned} \tag{20}$$

where $g[\lambda]$ is a terminal quadratic form, $Q[\lambda](t)$ is a symmetric and $V[\lambda](t)$ is a skew-symmetric Lipschitz matrices. The term $(G[\lambda]\bar{\xi}, \bar{u})$ has been taken by parts in view of (19) and (18).

Putting in (17) $\bar{x} = \bar{\xi} + B\bar{y}$, reduce $\rho[\lambda]$ in the new variables to the form: $\rho[\lambda] = \theta[\lambda] + \eta[\lambda]$, where

$$\theta[\lambda](\bar{\xi}, \bar{y}, \bar{u}) = \int ((T_1[\lambda](t)\bar{\xi}, \bar{\xi}, \bar{u}) + (T_2[\lambda](t)\bar{\xi}, \bar{y}, \bar{u}))dt,$$

$$\eta[\lambda](\bar{\xi}, \bar{y}, \bar{u}) = \int (\mathcal{E}[\lambda](t)\bar{y}, \bar{y}, \bar{u}))dt, \tag{21}$$

the tensor $\mathcal{E}[\lambda]$ is obtained from (17) by substituting $B\bar{y}$ instead of $\bar{x}$. It can be easily shown (see e.g. [33]) that on any Pontryagin sequence we have $\theta[\lambda] = o(\gamma)$, so the only essential term in $\rho[\lambda]$ is $\eta[\lambda]$.

The cone $\mathcal{K}$ in the new variables is given by equations (18), (19) and terminal relations (15), in which one should put $\bar{p} = (\bar{x}_0 = \bar{\xi}_0,\ \bar{x}_T = \bar{\xi}_T + B_1(T)\bar{y}_T)$.

At last, introduce some notations concerning the local cone N. Let H_2 be the maximal subspace in N, and H_1 be its complement, i.e. $R^k = H_1 \oplus H_2$. Then $N_1 = N \cap H_1$ is a pointed cone in the subspace H_1, and $N = N_1 \oplus H_2$. If one consider H_1, H_2 to be coordinate subspaces, then any $u \in N$ can be uniquely represented in the form $u = (u_1, u_2)$, where $u_1 \in N_1$, $u_2 \in H_2$.

Now we are ready to formulate conditions of the order γ for the presence of weak and Π- minima at $\hat{w}$.

5 Conditions for a weak minimum

We begin with a general description of the quadratic order conditions. For both weak and Pontryagin minimum these conditions are of the same form. To be precise, for any $a \in R$ we define a subset $M_a \subset \Lambda$, nonincreasing while a increases, and define the corresponding functional

$$\Omega[M_a](\bar{w}) = \sup_{\lambda \in M_a} \Omega[\lambda](\bar{w}) \tag{22}$$

(the sup over $\emptyset$, as usually, being equal to $+\infty$), and then conditions of minimality are as follows:

necessary condition:

$$\Omega[M_0](\bar{w}) \geq 0 \qquad \forall \bar{w} \in \mathcal{K} \cap \mathcal{N},$$

sufficient condition: $\exists\, a > 0$, such that

$$\Omega[M_a](\bar{w}) \geq a\gamma(\bar{w}) \qquad \forall \bar{w} \in \mathcal{K} \cap \mathcal{N}.$$

The difference between conditions for a weak minimum and conditions for a Π-minimum is only in the definition of the set M_a. This set consists of all $\lambda \in \Lambda$, for which the Lagrange function satisfies some special pointwise conditions. (Any pointwise conditions we call *conditions of Legendre type*). For a $\Pi-$minimum these pointwise conditions are more restrictive, hence the set M_a is smaller, and the corresponding necessary and sufficient conditions are stronger.

Let us now formulate these pointwise conditions. We begin with a weak minimum. Here one can take as M_a the entire Λ, but it is possible to choose a more narrow set, thus giving more strong necessary conditions, and more simple sufficient ones.

For any $a \in R$ denote by $G_a(\Lambda)$ the set of all $\lambda \in \Lambda$, such that the corresponding second variation $\Omega[\lambda]$ satisfies the following three conditions: $\forall t \in [0, T]$, $\forall\, \bar{h} \in H_1$, and $\forall\, \bar{u}, \bar{v} \in H_2$

$$\begin{aligned} &i)\ \ (V[\lambda](t)\bar{u}, \bar{h}) = 0, \\ &ii)\ \ (V[\lambda](t)\bar{u}, \bar{v}) = 0, \\ &iii)\ \ (Q[\lambda](t)\bar{u}, \bar{u}) \geq a|\bar{u}|^2. \end{aligned} \tag{23}$$

We introduce also the set $G_+(\Lambda) = \bigcup_{a>0} G_a(\Lambda)$.

Remark 5.1. For the case when $\Lambda = \{\cdot\}$, and $\hat{u}(t) \in \operatorname{int} U$ (i.e. $H_2 = R^k$), conditions *(ii)* and *(iii)* with $a = 0$ were obtained by B.S.Goh [18] as necessary conditions for a weak minimum. Note that if *(ii)* holds, the quadratic form (20) does not contain the control u, and then, taking into account (19), the variable y can be regarded as *a new control,* whence condition *(iii)* is just the classical

Legendre condition w.r.t. this new control y. Condition *(i)* is due to the author [38].

Before the formulation of the conditions for a weak minimum, let us recall the following

Definition 5.1. We say that the equality constraints (1) and (4) satisfy Lyusternik condition at $\hat{w}$, or that they are mutually nondegenerate at $\hat{w}$, if the operator

$$g: \quad w = (x, u) \longmapsto (\dot{x} - f_0(x,t) - F(x,t)u, \ K(x_0, x_T)),$$

mapping W to $Z = L_1^n[0,T] \times R^{\dim K}$, has a surjective derivative at $\hat{w}$, i.e. $g'(\hat{w})W = Z$.

Theorem 5.1. *i) Let $\hat{w}$ be a weak minimum point in Problem A. If $U_0 \neq U$, i.e. $\hat{u}(t) \in \partial U$, suppose also that the equality constraints (1) and (4) satisfy Lyusternik condition at $\hat{w}$. Then $G_0(\Lambda)$ is nonempty, and*

$$\Omega[G_0(\Lambda)](\bar{w}) \geq 0 \qquad \forall \bar{w} \in \mathcal{K} \cap \mathcal{N}. \tag{24}$$

ii) Let $G_0(\Lambda)$ be nonempty, and for some $a > 0$

$$\Omega[G_0(\Lambda)](\bar{w}) \geq a\gamma(\bar{w}) \qquad \forall \bar{w} \in \mathcal{K} \cap \mathcal{N}. \tag{25}$$

Then $\hat{w}$ is a strict weak minimum point in Problem A.

iii) From (25) it follows, that $G_a(\Lambda)$ is nonempty, and

$$\Omega[G_a(\Lambda)](\bar{w}) \geq a\gamma(\bar{w}) \qquad \forall \bar{w} \in \mathcal{K} \cap \mathcal{N}. \tag{26}$$

The proof is based on a general theory of higher order conditions for a local minimum in extremum problems with constraints [5]. For the case $U_0 = U$, i.e. $\hat{u} \in \operatorname{int} U$, it is given in [4], [31], for the case $\hat{u} \in \partial U$ it is due to A.A.Milyutin, see [37]. Note that the nondegeneracy of the equality constraints is required only in the necessary conditions, and only in the case of boundary control.

Definition 5.2. We say that $\hat{w}$ is a weak γ-minimum point in Problem A, if condition (25) or (26) is fulfilled for some $a > 0$.

We also give two equivalent formulations of these conditions. Recall that U satisfies Assumption 3$'$. Let the polyhedron $M \subset R^k$ be given by the inequalities:

$$(a_s, u) + b_s \leq 0, \qquad s = 1, ..., \hat{s}. \tag{27}$$

Denote by $S_0 = S_0(\hat{u})$ the set of all active indices s for $\hat{u}(t)$, i.e. such s, that the inequality turns into equality. Again due to Assumption 3$'$ it does not depend on t. Note by the way, that the cone $N = con(U - \hat{u})$ is given by the inequalities $(a_s, \bar{u}) \leq 0, \quad s \in S_0$.

Now let us introduce the so-called violation function:

$$\sigma(w) = \sum_{i \in I} \varphi_i^+(p) + |K(p)| +$$

$$+ \int |\dot{x} - f(x,t) - F(x,t)u| dt + \sum_{s \in S_0} \operatorname{ess\,sup} \left[(a_s, u(t)) + b_s \right]^+.$$

(We use the conventional notation $z^+ = \max\{0, z\}$.)

Theorem 5.2. *i) Conditions (25) and(26) are equivalent to the following one: there exists a neighborhood of $\hat{w}$ in W, in which the violation function has the lower bound:*

$$\sigma(w) \geq a' \gamma(w - \hat{w}), \qquad a' > 0. \tag{28}$$

ii) Suppose that equality constraints (1) and (4) satisfy Lyusternik condition at $\hat{w}$. Then conditions (25) and (26) are equivalent to the following one: $\exists\, a' > 0$, such that inequality (28) holds for any w from a neighborhood of $\hat{w}$ in W, satisfying equations (1) and (4) with $u \in U$. (Note that for such w the σ reduces to the first term only).

Condition *(i)* is proved in the general theory [5], the proof of *(ii)* is given in Appendix.

Let us dwell for a moment on condition (28). It obviously implies that $\hat{w}$ is a strict weak minimum point, thus it is a sufficient condition for a weak minimum. The equivalence of (25) and (26) to (28) has, at first look, rather an abstract character. However, it turned out to be quite an effective tool of investigation. (It will be essentially used in a forthcoming paper, devoted to application of the general conditions to abnormal sub-Riemannian geodesics). It worth noting also, that condition (28) has a "nonvariational" character: it does not include neither critical variations, nor Lagrange multipliers; it includes just the functions from the statement of Problem A as they are, not having been undergone any differentiations or approximations.

Checking conditions (24) and (25), i.e. checking the functional $\Omega[G_0(\Lambda)](\bar{w})$ (the maximum of a family of quadratic forms) for the nonnegative or positive definiteness on $\mathcal{K} \cap \mathcal{N}$, is in itself rather a nontrivial problem. It is known, that for a *single quadratic form* the question of its nonnegative or positive definiteness on a *subspace* can be solved by means of the classical Jacobi theory, which gives an answer in terms of the absence of conjugate (or focal) points. For a quadratic form on a *cone*, the more so for a *maximum of quadratic forms*, this question is much more difficult, because here the Euler-Jacobi equation is no longer linear (the difference of two solutions is not a solution). However, for this nonclassical situation there is an analogue of Jacobi theory [14], [15], [16], which establishes the equivalence of the positive definiteness of Ω to the absence of conjugate (focal) points in [0,T]. But for the nonnegativity of Ω, generally, only the one-way implication is true: $\Omega \geq 0$ implies that there are no conjugate points in (0,T). In the case of a single quadratic form, being considered on a polyhedral cone (hence the local cone must be absent: $N = R^k$, only terminal inequalities are allowed), the reverse implication is true as well [16], and so the Jacobi theory for this case is in a sense complete. In the general case there are counterexamples to this reverse implication [14], [15].

6 Sufficient conditions for small time intervals

There is an important case, when the above question of checking the quadratic functional for the positive definiteness has a simple solution. It is the case when the length of time interval is small enough. Recall that in CCV the positive definiteness of a quadratic functional on small time intervals is guaranteed by the strengthened Legendre condition, provided that at least one of the endpoints in the initial problem is fixed, i.e. one of the endpoint variations equals to zero ($\bar{x}(0) = 0$ or $\bar{x}(T) = 0$). A similar fact holds true for a quadratic functional of the form (20) on an arbitrary cone $\mathcal{K}$, contained in the subspace given by equations (19), (18).

6.1 A general theorem

For any interval $\Delta = [t_0, t_1] \subset [0, T]$ let us consider a quadratic functional of the type

$$\Omega(\bar{\xi}, \bar{y}, \bar{u}) = g(\bar{\xi}_0, \bar{\xi}_1, \bar{y}_1) + \tag{29}$$
$$+ \int_\Delta ((D(t)\bar{\xi}, \bar{\xi}) + (P(t)\bar{\xi}, \bar{y}) + (Q(t)\bar{y}, \bar{y}) + (V(t)\bar{y}, \bar{u}))\, dt,$$

where g is a terminal quadratic form, and matrices D, P, Q, V are defined and uniformly bounded on $[0, T]$. We also consider the subspace $\mathcal{L}_\Delta$, consisting of all $\bar{w} = (\bar{\xi}, \bar{y}, \bar{u})$ satisfying (19) and (18) a.e. on Δ, and the cone $\mathcal{N} = \{\bar{w} \mid \bar{u}(t) \in N \ a.e. \text{ on } \Delta\}$.

Theorem 6.1. *Suppose that* $\exists\, C,\ \delta_0 > 0$ *such that* $\forall \Delta,\ |\Delta| \leq \delta_0$, *the following estimates hold on* $\mathcal{L}_\Delta$*:*

$$||\bar{\xi}||_\infty \leq C||\bar{y}||_1, \qquad |\bar{y}(t_1)| \leq C||\bar{y}||_1, \tag{30}$$

and let $N = N_1 \oplus H_2$, *where* N_1 *is a pointed cone in a subspace* H_1, *and* $H_1 \oplus H_2 = R^k$. *Suppose further that quadratic form (29) satisfies on [0, T] the above three conditions (23) with* $a > 0$. *Then there exists* $\delta > 0$, *depending on the numbers* C, δ_0, a, *and on the coefficients of* Ω, *such that if* $\Delta \subset [0, T]$ *and* $|\Delta| \le \delta$, *then*

$$\Omega(\bar{w}) \ge \frac{a}{2}\gamma(\bar{w}) \qquad \forall \bar{w} \in \mathcal{L}_\Delta \cap \mathcal{N}_\Delta.$$

This theorem in fact claims, that under conditions (23) with $a > 0$, for all small enough Δ, the main term in Ω is $\int (Q_{22}\bar{y}_2, \bar{y}_2)dt$.

Proof. (We omit the bars over the variables.) Take $u = (u_1, u_2)$, $u_1 \in H_1$, $u_2 \in H_2$ on Δ, and $y = (y_1, y_2)$, $\dot{y} = u$, $y(t_0) = 0$. Like in the classical situation, we will evaluate all the terms in Ω, taking into account that Δ is small. Recall the known estimate for the functions on an interval Δ with $|\Delta| \le \delta$:

$$||y||_1 \le \sqrt{\delta} \cdot ||y||_2. \tag{31}$$

Denote $\gamma_1(y) = \int_\Delta y_1^2 dt$, $\gamma_2(y) = \int_\Delta y_2^2 dt$, and let us compare $\gamma_2(y)$ with our "full"

$$\gamma(w) = |\ \xi(t_0)\ |^2 + |\ y(t_1)\ |^2 + \int_\Delta (y_1^2 + y_2^2)dt,$$

when $w = (\xi, y, u) \in \mathcal{L}_\Delta$. Since N_1 is a pointed cone, and $\dot{y} = u \in N_1$, $y(t_0) = 0$, then y_1 is monotone nondecreasing w.r.t. N_1, hence for some constant b

$$||y_1||_\infty \le b|y_1(t_1)|\ , \tag{32}$$

and due to (30), (31)

$$||y_1||_\infty \le bC||y||_1 \le bC\sqrt{\delta}||y||_2,$$

therefore $\gamma_1(y) \le \delta||y_1||_\infty^2 \le b^2C^2\delta^2||y||_2^2 = (b^2C^2\delta^2)(\gamma_1 + \gamma_2)$, which yields that for δ small enough $\gamma_1(y) \le 2(b^2C^2\delta^2)\gamma_2(y)$.

From here, (30) and (31) it follows that $\gamma(y)$ is of the same order as $\gamma_2(y)$; moreover, for $\delta > 0$ small enough

$$\gamma_2(y) \le \gamma(y) \le \frac{3}{2}\gamma_2(y). \tag{33}$$

With (30) and (31) this gives

$$||\xi||_\infty \le \text{const} \cdot \sqrt{\delta}\sqrt{\gamma_2(y)}, \qquad |y(t_1)| \le \text{const} \cdot \sqrt{\delta}\sqrt{\gamma_2(y)}, \tag{34}$$

whence the outside term in Ω is $\le$ const$\cdot\delta \cdot \gamma_2(y)$.

From (34) it follows also that the integrand terms in Ω, containing ξ, have the same estimate. Due to (32) and (34) we get $||y_1||_\infty \le \text{const}\cdot\sqrt{\delta}\sqrt{\gamma_2(y)}$, whence the integrand terms, containing (y_1, y_1) and (y_1, y_2), are $\le$ const$\cdot\delta \cdot \gamma_2(y)$ too. Recall, that due to (23) we have

$$\int_\Delta (Q_{22}y_2, y_2)\, dt \geq a \cdot \gamma_2(y). \tag{35}$$

Thus, it remains to evaluate only the integrand term $(V_{11}y_1, u_1)$. Since $u_1 \in N_1$, and N_1 is a pointed cone, there exists a vector $l \in R^{k_1}$, such that $|u_1| \leq$ const (l, u_1) $\forall u_1 \in N_1$ (actually, here one may take any $l \in \text{int } N_1^*$). Then

$$\int_\Delta |(V_{11}y_1, u_1)|\, dt \leq \int_\Delta |V_{11}| \cdot |y_1| \cdot |u_1|\, dt \leq \text{const} \cdot ||V||_\infty \cdot ||y_1||_\infty \cdot \int_\Delta (l, u_1)\, dt \leq$$

$$\leq \text{const} \cdot ||y_1||_\infty \cdot (l, y_1(t_1)) \leq \text{const} \cdot |y_1(t_1)|^2$$

(the last inequality is due to (32)), and taking into account (34), we continue $\leq$ const$\cdot\delta \cdot \gamma_2(y)$.

Thus, $\Omega = \int (Q_{22}y_2, y_2)dt + \mu(\xi, y, u)$, where $|\mu| \leq \text{const}\cdot\delta \cdot \gamma_2(y)$. If δ is small enough, then $|\mu| \leq \frac{1}{4} a\gamma_2(y)$, and in view of (33) and (35):

$$\Omega \geq \frac{3}{4} a\, \gamma_2(y) \geq a \frac{3}{4}\frac{2}{3} \gamma(y) \geq \frac{a}{2} \gamma(y), \qquad \text{q.e.d.}$$

6.2 A special case

Consider here an important special case, in which the above assumptions about Ω and estimates (30) do hold. Let system (1) be of the form:

$$\dot{x} = z f_0(x) + \sum_{i=1}^{k} u_i f_i(x), \qquad \dot{z} = 0, \tag{36}$$

where $x \in R^n$ and $z \in R^1$ are the state variables, $\mathcal{D}_{x,z,t} = \mathcal{D}_x \times R^1 \times [0,T]$, the vectors $f_i(x)$, $i = 0, 1, ..., k$ are C^2-smooth and linearly independent at any point $x \in \mathcal{D}_x$. Take a trajectory $\hat{w} = (\hat{x}(t),\ \hat{z} = 1,\ \hat{u} = 0)$, $t \in [0,T]$, and for any $\Delta = [t_0, t_1] \subset [0,T]$ let us consider the Problem A_Δ, associated with this trajectory, with the terminal constraints $x(t_0) = \hat{x}(t_0)$, $x(t_1) = \hat{x}(t_1)$, with $z(t_0)$ and $z(t_1)$ being free, the control u being free too, and with the cost functional $J = z(t_0) \longrightarrow \min$. Begin with a simple

Proposition 6.1. *If $\hat{w}$ satisfies MP in Problem $A_{[0,T]}$, then it satisfies MP in Problem A_Δ for any $\Delta \subset [0,T]$.*

Proof. The MP for Problem A_Δ means that there exists a nonzero n-vector function $\psi_x(t)$ (the costate variable, corresponding to x), such that for $H = \psi_x(z f_0(x) + \sum u_i f_i(x))$ the following relations hold on Δ along $\hat{w}(t)$:

$$\dot{\psi}_x = -H_x, \qquad H_u = 0, \qquad H = \text{const} \geq 0.$$

(The costate variable ψ_z, corresponding to z, is a scalar linear function with $\psi_z(t_1) = 0$ and $\dot{\psi}_z = -H$, so it is determined by ψ_x.) If such ψ_x exists for $[0, T]$, it obviously fits for any $\Delta \subset [0, T]$.

Since there are no inequality constraints in our problem, the critical cone here is, in fact, the subspace $\mathcal{L}_\Delta$, consisting of all $\bar{w} = (\bar{x}, \bar{z}, \bar{u})$, satisfying the linear equations:

$$\dot{\bar{x}} = \bar{z} f_0(\hat{x}) + f_0'(\hat{x})\bar{x} + \sum \bar{u}_i f_i(\hat{x}), \tag{37}$$
$$\dot{\bar{z}} = 0, \qquad \bar{x}(t_0) = \bar{x}(t_1) = 0.$$

Passing to the Goh variables $(\bar{z}, \bar{\xi}, \bar{y}, \bar{u})$, where

$$\bar{x} = \bar{\xi} + \sum \bar{y}_i f_i(\hat{x}),$$

$$\dot{\bar{y}}_i = \bar{u}_i, \quad \bar{y}_i(t_0) = 0, \quad i = 1, ..., k, \tag{38}$$

we get instead of (37):

$$\dot{\bar{\xi}} = \bar{z} f_0 + f_0' \bar{\xi} + \sum \bar{y}_j [f_0, f_j], \qquad \bar{\xi}(t_0) = 0 \tag{39}$$

(here $[f, g] = f'g - g'f$ are Lie brackets), and the terminal equality $\bar{x}(t_1) = 0$ becomes:

$$\bar{\xi}(t_1) + \sum \bar{y}_i(t_1) f_i(\hat{x}(t_1)) = 0, \tag{40}$$

so that now we may consider $\mathcal{L}_\Delta$ as the set of all $(\bar{z}, \bar{\xi}, \bar{y}, \bar{u})$, satisfying (39), (40) and (38). Since the vectors $f_i(\hat{x}(t))$ are linearly independent, and continuous in t, there exists a number C_0 (independent of t_1), such that (40) implies

$$|\bar{y}(t_1)| \le C_0 \left|\bar{\xi}(t_1)\right| . \tag{41}$$

Besides, (39) obviously implies that for some C_1 (independent of Δ)

$$||\bar{\xi}||_\infty \le C_1 \left(|\bar{z}| \cdot |\Delta| + \int_\Delta |\bar{y}| \, dt \right). \tag{42}$$

Let us show that $\exists\, C_2$, such that

$$|\bar{z}| \le C_2 \frac{1}{|\Delta|} ||\bar{y}||_1 \quad \text{on } \mathcal{L}_\Delta. \tag{43}$$

In order to do this, consider the space $W_\Delta = R \times L_1^k(\Delta) \times R^k$ with elements $\bar{w} = (\bar{z}, \bar{y}, \bar{h})$, and consider the subspace $\mathcal{M}_\Delta \subset W_\Delta$, defined by the equality

$$\bar{\xi}(t_1) + \sum_{i=1}^{k} \bar{h}_i \hat{f}_i(t_1) = 0, \tag{44}$$

where $\bar{\xi}$ is the solution to (39) for $\bar{z}, \bar{y}$, and for brevity we write $\hat{f}_i(t) = f_i(\hat{x}(t))$. There is a natural injection $\mathcal{L}_\Delta \longrightarrow \mathcal{M}_\Delta$, $(\bar{z}, \bar{\xi}, \bar{y}, \bar{u}) \longmapsto (\bar{z}, \bar{y}, \bar{h})$ (we just ignore relations (38) and put $\bar{h} = \bar{y}(t_1)$), so that $\mathcal{M}_\Delta$ can be considered as an extension of $\mathcal{L}_\Delta$. We will show that estimate (43) holds on $\mathcal{M}_\Delta$.

First we establish the following abstract fact.

Lemma 6.1. *Let Y and H be Banach spaces, $W = R \times Y \times H$, and $\mathcal{M}$ be a subspace in W. Suppose that $\exists\, \rho > 0$ such that for any $y \in Y$, $||y|| < \rho$, for any $h \in H$ the point $(1, y, h) \notin \mathcal{M}$. Then $\forall w = (z, y, h) \in \mathcal{M} \quad |z| \le \frac{1}{\rho}||y||$.*

Proof. Take any $w = (z, y, h) \in \mathcal{M}$. If $z = 0$, the desired estimate is fulfilled trivially. If $z \ne 0$, then $w' = (1, y/z, h/z) \in \mathcal{M}$, and hence $||y/z|| \ge \rho$ (otherwise $w' \notin \mathcal{M}$), which gives $\rho|z| \le ||y||$, q.e.d.

Let us return to our space W_Δ. Here $Y = Y_\Delta = L_1^k(\Delta)$, $H = R^k$, and $\mathcal{M}$ is the above $\mathcal{M}_\Delta$. In the space Y_Δ we will consider the norm $||y|| = \frac{1}{|\Delta|}||y||_1$, and we have to check the premise of Lemma 6.1.

First, we show that $\forall \bar{h}$ the vector $\bar{w} = (\bar{z} = 1,\ \bar{y} = 0,\ \bar{h}) \notin \mathcal{M}_\Delta$. To prove this, we have only to check that (44) does not hold. Note that for $\bar{z} = 1$, $\bar{y} = 0$ equation (39) reads: $\dot{\bar{\xi}} = f_0 + f_0'\bar{\xi}$, $\bar{\xi}(t_0) = 0$, and it has the explicit solution $\bar{\xi}(t) = (t - t_0) f_0(\hat{x}(t))$ (this was noticed by A.A.Milyutin), hence its terminal value on Δ is $\bar{\xi}(t_1) = (t_1 - t_0) f_0(\hat{x}(t_1))$. Since $\hat{f}_i(t)$, $i = 0, 1, \ldots, k$ are linearly independent $\forall t \in [0, T]$, then for any $\bar{h} \in R^k$

$$|\Delta| \cdot \hat{f}_0(t_1) + \sum_{i=1}^{k} \bar{h}_i \hat{f}_i(t_1) \ne 0, \tag{45}$$

and so $\bar{w} = (1, 0, \bar{h}) \notin \mathcal{M}_\Delta$.

Now let us check this condition for small nonzero $\bar{y}$-s.

Lemma 6.2. *$\exists\, \rho > 0$ such that $\forall \Delta$, for any $\bar{y} \in Y_\Delta$, $||\bar{y}|| \le \rho$, for any $\bar{h} \in R^k$, the point $\bar{w} = (1, \bar{y}, \bar{h}) \notin \mathcal{M}_\Delta$.*

Proof. Denote $L(t) = \mathrm{Lin}\ \{\hat{f}_i(t),\ i = 1, \ldots, k\}$. (We write *Lin* for the linear hull of the set.) Equality (44) means that $\bar{\xi}(t_1) \in L(t_1)$, and we have to show that this is not so. Since the vectors $\hat{f}_i(t)$, $i = 0, 1, \ldots, k$ are linearly independent $\forall t \in [0, T]$ and continuous, $\exists\, \varepsilon > 0$ such that

$$\forall t \in [0, T] \qquad B_\varepsilon(\hat{f}_0(t)) \cap L(t) = \emptyset. \tag{46}$$

Given this ε, $\exists\, \rho > 0$, such that $\forall \Delta$, if $||\bar{y}||_1 \le |\Delta| \cdot \rho$, then for $\bar{\xi}(t_1)$, corresponding to $\bar{z} = 1$ and this $\bar{y}$, we have

$$\left|\bar{\xi}(t_1) - |\Delta| \cdot \hat{f}_0(t_1)\right| \le \mathrm{const} \cdot \int_\Delta |\bar{y}|\, d\tau \le \mathrm{const} \cdot |\Delta| \cdot \rho \le |\Delta| \cdot \varepsilon,$$

i.e. $\bar{\xi}(t_1) \in B_{|\Delta|\varepsilon}(|\Delta| \cdot \hat{f}_0(t_1)) = |\Delta| \cdot B_\varepsilon(\hat{f}_0(t_1))$. In view of (46), for any Δ and any $\bar{y} \in Y_\Delta$, such that $||\bar{y}|| = \frac{1}{|\Delta|}||\bar{y}||_1 \le \rho$, we have $\bar{\xi}(t_1) \notin L(t_1)$, and hence $\forall \bar{h} \in R^k$ the point $\bar{w} = (1, \bar{y}, \bar{h}) \notin \mathcal{M}_\Delta$. Lemma 6.2 is proved.

Thus, the premise of Lemma 6.1 is fulfilled with a $\rho > 0$ independent of Δ. By this Lemma we get the desired estimate

$$|\bar{z}| \le C_2 \frac{1}{|\Delta|} ||\bar{y}||_1 \qquad \text{on } \mathcal{M}_\Delta \supset \mathcal{L}_\Delta \quad \text{with } C_2 = 1/\rho.$$

Finally, (42) and (43) give $||\bar{\xi}||_\infty \le \text{const}\, ||\bar{y}||_1$, and in view of (41) we get $|\bar{y}(t_1)| \le \text{const}\, ||\bar{y}||_1$ on $\mathcal{L}_\Delta$, thus estimates (30) hold, and so Theorem 6.1 applies to the considered special case.

Remark 6.1. Estimates (30) are fulfilled also for the critical subspace, corresponding to the system

$$\dot{x} = z\left(f_0(x) + \sum_{i=1}^{k} u_i f_i(x)\right), \qquad \dot{z} = 0, \tag{47}$$

at the same reference trajectory $\hat{w} = (\hat{x}(t),\ \hat{z} = 1,\ \hat{u} = 0)$, because its linearization has the same form (37) (the difference between systems (36) and (47) is only in their Ω-s).

7 Conditions for a Π-minimum

Let us now pass to conditions for a $\Pi-$minimum. As was said already, they differ from the conditions for a weak minimum by an additional pointwise condition. (Recall that we regard any pointwise condition as a condition of Legendre type.) This new condition of Legendre type involves the third variation of Lagrange function, or, more precisely, the cubic functional (17), and also the admissible control set U. Recall that we denote by U_0 the minimal face of U, containing $\hat{u}(t)$, and that due to Assumption 3$'$ this face is the same for all t. For any $\lambda \in \Lambda$ and any fixed $t_* \in [0, T]$ define the functional

$$L[\lambda, t_*](\bar{y}) = \int_0^1 ((Q[\lambda](t_*)\bar{y}, \bar{y}) + (\mathcal{E}[\lambda](t_*)\bar{y}, \bar{y}, \bar{u}))d\tau, \tag{48}$$

which is to be considered for all absolutely continuous functions $\bar{y}(\tau)$ on $[0, 1]$, such that $\bar{y}(0) = \bar{y}(1) = 0$ (we will call such functions cycles). Here $Q[\lambda]$ is the matrix from the second variation (20) of Lagrange function, and $\mathcal{E}[\lambda]$ is the tensor from the third variation (21), both frozen at the point t_*.

Definition 7.1. We say that λ satisfies the Π–Legendre condition with a parameter $a \in R$, if for every t_*, and every cycle $\bar{y}(\tau)$, such that

$$\dot{\bar{y}}(\tau) = \bar{u}(\tau) \in U - \hat{u}(t_*), \tag{49}$$

the following inequality holds:

$$L[\lambda, t_*](\bar{y}) \geq a \int_0^1 (\bar{y}, \bar{y})\, d\tau. \tag{50}$$

(Relation (49) means that the control set is also frozen at the point t_*.)
Denote by $E_a(\Lambda)$ the set of all $\lambda \in G_a(\Lambda)$ satisfying this condition.

Theorem 7.1. *i) Let $\hat{w}$ be a Π-minimum point in Problem A. If $U_0 \neq U$, i.e. $\hat{u}(t) \in \partial U$, suppose also that the equality constraints (1) and (4) satisfy Lyusternik condition at $\hat{w}$. Then $E_0(\Lambda)$ is nonempty, and*

$$\Omega[E_0(\Lambda)](\bar{w}) \geq 0 \qquad \forall \bar{w} \in \mathcal{K} \cap \mathcal{N}. \tag{51}$$

ii) Let for some $a > 0$ the set $E_a(\Lambda)$ be nonempty, and

$$\Omega[E_a(\Lambda)](\bar{w}) \geq a\gamma(\bar{w}) \qquad \forall \bar{w} \in \mathcal{K} \cap \mathcal{N}. \tag{52}$$

Then $\hat{w}$ is a strict Π-minimum point in Problem A.

The proof again is based on the general theory [5]. For the case $U_0 = U$, i.e. $\hat{u} \in \operatorname{int} U$, it is given in [33], [34] (see also [39]); for the case $\hat{u} \in \partial U$ it is given in [37].

Consider inequality (52). Theorem 6.1 readily implies the following property for small intervals.

Lemma 7.1. *Consider $\Omega[\lambda]$ and $\eta[\lambda]$ on any interval $\Delta \subset [0, T]$, and suppose that for all small enough Δ estimates (30) hold. If for some $a > 0$ the set $E_a(\Lambda)$ is nonempty, then for all small enough Δ inequality (52) is fulfilled with $a/2$.*

Proof. If $\lambda_0 \in E_a(\Lambda)$, then by definition $\lambda_0 \in G_a(\Lambda)$, i.e. (23) is fulfilled with this $a > 0$. By Theorem 6.1 for all small enough Δ we have $\Omega[\lambda_0](\bar{w}) \geq \frac{a}{2}\gamma(\bar{w}) \quad \forall \bar{w} \in \mathcal{K}_\Delta \cap \mathcal{N}_\Delta$, and since $E_{a/2}(\Lambda) \supset E_a(\Lambda) \ni \lambda_0$, we get $\Omega[E_{a/2}(\Lambda)](\bar{w}) \geq \Omega[\lambda_0](\bar{w}) \geq \frac{a}{2}\gamma(\bar{w}) \quad \forall \bar{w} \in \mathcal{K}_\Delta \cap \mathcal{N}_\Delta$, q.e.d.

For large intervals, as was said already, in order to check (52) one should use a theory of Jacobi type (see e.g. [15, 16]).

Definition 7.2. We say that $\hat{w}$ is a $\Pi - \gamma$-minimum point in Problem A, if condition (52) is fulfilled for some $a > 0$.

As before, we give two equivalent formulations of this condition. Denote by $\Pi(U, \hat{w})$ the set of all Pontryagin sequences, converging to $\hat{w}$, such that

$$\text{ess sup } dist\,(u(t), U) \to 0.$$

Theorem 7.2. *i) Condition (52) with $a > 0$ is equivalent to the following one: there exists $a' > 0$, such that for any sequence $\{w_m\} \in \Pi(U, \hat{w})$, for all large enough m, the violation function has the lower bound:*

$$\sigma(w_m) \geq a'\gamma(w_m - \hat{w})\,. \tag{53}$$

ii) Suppose that equality constraints (1) and (4) satisfy Lyusternik condition at $\hat{w}$. Then condition (52) with $a > 0$ is equivalent to the following one: there exists $a' > 0$, such that if $\{w_m\}$ is a Pontryagin sequence, converging to $\hat{w}$, and satisfying equations (1) and (4) with $u_m \in U$, then inequality (53) holds for all large enough m. (Note that for such w_m the σ reduces to the first term only.)

Assertion *(i)* is proved in the general theory [5], the proof of *(ii)* is given in Appendix. Before going further, let us make some remarks about the Π–Legendre condition (50).

Remark 7.1. From (49) and $\bar{y}(0) = \bar{y}(1) = 0$ it follows obviously that actually $\bar{u}(\tau) \in U_0 - \hat{u}(t_*) \subset H_2$, and $\bar{u}_1 = \bar{y}_1 = 0$, hence, for every cycle $\bar{y}$ only the second component $\bar{y}_2$ may be nonzero. Therefore, we can put in (49) $\bar{u}(\tau) \in U_0 - \hat{u}(t_*)$, and so condition (50) in fact concerns the minimal face U_0 of U, containing $\hat{u}(t_*)$ (and hence containing all $\hat{u}(t)$).

Remark 7.2. Condition (50) is invariant w.r.t. the interval of integration: taking $u'(\tau) = \bar{u}(s\tau)$, $y'(\tau) = \bar{y}(s\tau)/s$, by an easy calculation we get the same inequality (50) for interval $[0, 1/s]$, so we may choose to consider $[0, 1]$.

Remark 7.3. Condition (50), though being seemingly of an integral form, has a pointwise character, i.e. it can be verified at each point t_* separately.

To verify condition (50), one have in fact to solve an auxiliary extremal problem: to find the maximal a, for which this inequality holds true for any cycle satisfying (49). This is rather an unusual and difficult problem. At present we can solve this problem completely for three cases of the control set U: a) the whole space, b) an arbitrary stripe of codimension 1, c) an arbitrary ellipse in the plane (see [12, 38]). For a general case some theoretical results are obtained in [13].

Here, in view of our further purposes, we consider only the most simple case (a), in which u is unconstrained. Define the differential 1-form

$$\omega[\lambda](t_*) = (\mathcal{E}[\lambda](t_*)y, y, dy) = \sum_{ijs} \mathcal{E}_{ijs}[\lambda](t_*)y^i y^j dy^s. \tag{54}$$

Theorem 7.3 [32, 33]. *Condition (50) holds for a given* a *iff* $Q[\lambda](t_*) \geq a$ *(Goh condition for a weak minimum), and* $\omega[\lambda](t_*)$ *is closed, i.e.*

$$d\omega[\lambda](t_*) = \sum_{ijs} \mathcal{E}_{ijs}[\lambda](t_*)(y^i dy^j + y^j dy^i) \wedge dy^s = 0, \tag{55}$$

the differential is taken w.r.t. y, *and* λ, t_* *are just parameters.*

Thus, in this case condition (50) decomposes onto conditions concerning the quadratic and cubic parts separately, the last being an additional optimality condition (55) of equality type.

Similar conditions take place in the case when $U = U_1 \oplus H_2$, $R^n = H_1 \oplus H_2$, and U_1 is a polyhedron in H_1, having $\hat{u}_1(t) \equiv \hat{u}_1$ as a vertex. (Here the above face $U_0 = H_2$.) In view of Remark 7.1, condition (50) here concerns only the cycles lying in H_2, hence (50) means that on the subspace H_2 both $Q[\lambda](t_*) \geq a$ and (55) must hold.

Note also that in this case the following property holds [31, 15].

Theorem 7.4. *Suppose that* $E_0(\Lambda)$ *is nonempty, and for some* $a > 0$

$$\Omega[E_0(\Lambda)](\bar{w}) \geq a\gamma(\bar{w}) \qquad \forall \bar{w} \in \mathcal{K} \cap \mathcal{N}.$$

Then $E_a(\Lambda)$ *is nonempty (with the same* a*), and still*

$$\Omega[E_a(\Lambda)](\bar{w}) \geq a\gamma(\bar{w}) \qquad \forall \bar{w} \in \mathcal{K} \cap \mathcal{N}.$$

The reverse implication holds trivially due to inclusion $E_a \subset E_0$. In the general case, when the face U_0 is not a subspace, Theorem 7.4 fails to hold.

Let us now consider another case, when the minimal face U_0 is "small", and establish a relation between conditions $Q[\lambda](t_*) \geq a$ and (50) in this case. As before, we have $R^n = H_1 \oplus H_2$, $H_2 = \mathrm{Lin}\,(U_0 - \hat{u}(t))$, $N = \mathrm{con}\,(U - \hat{u}(t)) = N_1 \oplus H_2$, $N_1 = N \cap H_1$.

Lemma 7.2. *(i) Let for* t_* *condition (50) holds. Then* $Q[\lambda](t_*) \geq a$ *on* H_2.
(ii) Suppose that $Q[\lambda](t_*) \geq a$ *on* H_2. *Then* $\forall a' < a$ $\exists \delta > 0$, *such that if* U_0 *is contained in the* δ*-neighbourhood of* $\hat{u}(t_*)$, *then (50) is fulfilled with* a'. *The number* δ *depends only on* $|a - a'|$ *and* $||\mathcal{E}[\lambda](t_*)||$.

Proof. Assertion *(i)* follows from the fact that one can take as a cycle a function $\bar{y}(\tau)$ passing with a constant velocity along an arbitrary vector $v \in H_2$ and back to zero. On such a cycle any point $\bar{y}$ is passed forward and back with the same velocity, hence the cubic term in (48) obviously vanishes, and the quadratic term reduces to $(Q[\lambda](t_*)v, v)$, which should be $\geq a(v, v)$.

Assertion *(ii)* follows from the simple estimate:

$$\int |(\mathcal{E}[\lambda](t_*)\bar{y}, \bar{y}, \bar{u})| d\tau \leq ||\mathcal{E}[\lambda](t_*)|| \cdot ||\bar{u}||_\infty \int (\bar{y}, \bar{y}) d\tau.$$

If $||\bar{u}||_\infty \leq \delta$, and $\delta||\mathcal{E}|| \leq |a - a'|$, then the cubic term, being added to the quadratic term, can worsen the constant a not lower than to a'.

Assertion *(ii)* means, that the smaller in size is the face U_0 containing $\hat{u}(t)$, the smaller contribution in $L[\lambda]$ gives the cubic term. If $Q[\lambda](t_*) \geq a > 0$ on H_2, and the face U_0 is small enough in size, then (50) is also fulfilled, perhaps with a smaller, still positive a'. Note that here the size of the entire U does not matter, it may be arbitrarily large, only the size of the face U_0 is important. This observation will be used in the next section.

8 Sufficient conditions for a strong minimum in case of a strictly convex set U

Let us now consider the case, when the control set U in Problem A is a strictly convex compactum with a smooth boundary ∂U, and $\hat{u}(t) \equiv \hat{u} \in \partial U$ on $[0, T]$. Since in this case $\partial U = ex\, U$, then, supposing also that $\forall t \quad \operatorname{rank} F(\hat{x}(t), t) = k$, by Proposition 1.1 we have that the Π-minimum implies here the strong minimum (the reverse implication always holds trivially), and so any sufficient conditions for the Π-minimum are at the same time sufficient conditions for the strong minimum. In view of this property, we are interested to obtain sufficient conditions for the Π-minimum in this case.

Remark 8.1. All considerations in this section are in fact valid for a more general case, when U is a convex compact set (not necessarily strictly convex), $\hat{u}(t) \equiv \hat{u} \in \partial U$, there is a unique outward normal $\eta \in R^k$ for U at $\hat{u}$, and $\operatorname{Argmax}(\eta, U) = \{\hat{u}\}$, i.e. $\hat{u}$ is a unique maximum point of the linear functional η on U. For example, if the set U is given by the inequality $\varphi(u) \leq 1$, where φ is a smooth strictly convex sublinear functional, these assumptions obviously are fulfilled for any $\hat{u} \in \partial U$.

Let us try to apply the abovestated theory to this case. Formally, this case does not fit in this theory, because U does not satisfy Assumption 3′ (it is not a polyhedron in a vicinity of $\hat{u}$). However, this theory applies, if we take the following approach. Since we are interested in *sufficient* conditions, then instead of U we may consider any polyhedron U' containing U, and obtain sufficient conditions for Problem A with the control set U' : obviously, these conditions will be as well sufficient for Problem A with the set U. What is important is that the choice of the polyhedron U' is in our disposal.

Let us fix any $\varepsilon > 0$, and take as U' an arbitrary polyhedron containing U, such that $\hat{u} \in \partial U'$, and the corresponding face U_0' (containing $\hat{u}$) has diameter $\leq \varepsilon$. Due to the smoothness of ∂U, $\dim U_0' = k-1$, $\hat{u} \in \operatorname{reint} U_0'$, and the affine hull $\operatorname{Aff} U_0'$ is a (unique) tangent hyperplane for U at $\hat{u}$.

Remark 8.2. There are plenty of ways to construct such a polyhedron. Here we point out two of them.

1) Take any $\delta > 0$, and pick a finite δ-net $\Theta \subset \partial U$ containing $\hat{u}$ as an element. For each $u \in \Theta$ take any support hyperplane for U at u, and consider the corresponding closed halfspace $P(u)$ containing U. Then $U' = \bigcap_{u \in \Theta} P(u)$ satisfies the above requirements with some $\varepsilon(\delta)$ tending to zero as $\delta \to 0$.

2) Let P be an arbitrary polyhedron containing U, such that $\hat{u} \in \partial P$, let η be the outward normal to U (and then to P) at $\hat{u}$, and L be the tangent hyperplane at $\hat{u}$. Choose an arbitrary polyhedral norm in R^k, and for any $\varepsilon > 0$ denote by $B_\varepsilon(\hat{u})$ the ε-ball around $\hat{u}$.

Given an $\varepsilon > 0$, denote by K_δ, $\delta > 0$ the cone generated by $B_\varepsilon(\hat{u}) \cap L$ with vertex at $\hat{u} + \delta\eta$. Since $L \cap U = \{\hat{u}\}$, then $K_\delta \supset U$ for some $\delta > 0$. (The existence of such $\delta > 0$ is obvious for any 2-dimensional cutset of U containing η, and then, by virtue of the arguments of compactness, for the entire U.)

Take, for the certainty, δ be the maximal of those δ-s, and denote it by $\delta(\varepsilon)$. Then the set $U' = P \cap K_{\delta(\varepsilon)}$ will do. (For small $\varepsilon > 0$, its face containing $\hat{u}$, is $U_0' = B_\varepsilon(\hat{u}) \cap L$, which diameter is $\leq \varepsilon$.)

Chosen any method of constructing the polyhedron U', we will consider that $\forall \varepsilon > 0$ we have a polyhedron U_ε with the above properties. Note that generally $\bigcap_{\varepsilon > 0} U_\varepsilon \neq U$ (e.g. in the above second method), and in fact we do not need that U_ε would be close to U; we only need that the size of the face U_ε^0 in U_ε (corresponding to the normal η) would tend to zero. We do not provide here pictures, since the reader can easily draw them himself.

Problem A with the control set U_ε (instead of U) will be denoted by Problem A_ε. We want to obtain sufficient conditions for Π-minimum in Problem A_ε. But now, since U_ε is a polyhedron, we can use the above theory, and formulate sufficient conditions for $\Pi-\gamma$-minimum in Problem A_ε. In order to do this, we need to inspect all objects of this theory for Problem A_ε.

As was noted in Remark 2.2 (and it is very important for us), due to the singularity of $\hat{w}$, the set $\Lambda(\hat{w})$ (and hence the families $\Omega[\lambda]$ and $\rho[\lambda]$) do not depend on the choice of the control set U_ε, since the last contains the initial set U. Besides, for all U_ε the tangent (at $\hat{u}$) cone N is one and the same: $N = \{\bar{u} \mid (\eta, u) \le 0\}$, where η is the outward normal to U at $\hat{u}$ (which is assumed to be unique), the maximal subspace in N is $H_2 = \{\bar{u} \mid (\eta, u) = 0\}$, its complement $H_1 = R\eta$, and $N_1 = -R_+\eta$.

All this implies, that $\forall a \in R$ the set $G_a(\Lambda)$ is one and the same for all Problems A_ε, $\varepsilon > 0$. It consists of all $\lambda \in \Lambda$, such that quadratic form $\Omega[\lambda]$ satisfies conditions:

$$i)\ \ V[\lambda](t) = 0 \tag{56}$$

(here both conditions *(i)* and *(ii)* of (23) are taken into account, as well as the fact that $\dim H_1 = 1$), and

$$ii)\ \ Q[\lambda](t) \ge a \quad \text{on } H_2. \tag{57}$$

As to the set $E_a(\Lambda)$, involving in its definition the face of U_ε containing $\hat{u}$, here the situation is as follows. Denote this set for Problem A_ε by $E_a(\Lambda, U_\varepsilon)$. Let us fix an $a > 0$, and take a $\lambda \in G_a(\Lambda)$. Due to Lemma 7.2, $\exists\, \varepsilon_0 > 0$, such that $\lambda \in E_{a/2}(\Lambda, U_\varepsilon)$ for all $\varepsilon \le \varepsilon_0$. The number ε_0 depends only on a and $||\mathcal{E}[\lambda](t)||_\infty$. Since the tensor $\mathcal{E}[\lambda](t)$ is linear in λ, and the set Λ is bounded, then $\max\{||\mathcal{E}[\lambda]||_\infty,\ \lambda \in \Lambda\}$ is a finite number, hence we can consider that ε_0 depends only on a, and does not depend on λ.

Thus, $\forall \varepsilon \le \varepsilon_0(a)$ we have: if $\lambda \in G_a(\Lambda)$, then $\lambda \in E_{a/2}(\Lambda, U_\varepsilon)$, i.e.

$$G_a(\Lambda) \subset E_{a/2}(\Lambda, U_\varepsilon). \tag{58}$$

The critical cone $\mathcal{K}$ for all Problems A_ε with the free control (i.e. without taking into account the constraint $u \in U_\varepsilon$) is one and the same by its definition, because all these problems differ only in the sets U_ε.

All this obviously implies, that if the inequality

$$\Omega[G_a(\Lambda)](\bar{w}) \ge a\gamma(\bar{w}) \qquad \forall \bar{w} \in \mathcal{K} \cap \mathcal{N} \tag{59}$$

holds, then the inequality

$$\Omega[E_{a/2}(\Lambda, U_\varepsilon)](\bar{w}) \ge a\gamma(\bar{w}) \qquad \forall \bar{w} \in \mathcal{K} \cap \mathcal{N} \tag{60}$$

holds too (and the more so the last one holds, if in its right hand side $a\gamma$ is replaced by $a\gamma/2$). Recall now, that by Theorem 5.1 inequality (59) is a sufficient condition for the weak γ-minimum in Problem A_ε, and by Theorem 7.1 inequality (60) is a sufficient condition for $\Pi - \gamma$-minimum in Problem A_ε. Thus, we have got the following chain of implications:

weak $\gamma-$minimum in Problem A_ε $\Longrightarrow$ $\Pi-\gamma-$*minimum in Problem A_ε* $\Longrightarrow$
strict $\Pi-$minimum in Problem A_ε $\Longrightarrow$
(due to inclusion $U_\varepsilon \supset U$) strict $\Pi-$minimum in Problem A $\Longrightarrow$
(due to Proposition 1.1) strict strong minimum in Problem A.

Finally, in order to complete the picture, we add one more step. Consider the halfspace $U_* = \{u \mid (\eta, u) \leq (\eta, \hat{u})\}$ containing U, bounded by the tangent hyperplane to U at $\hat{u}$. Problem A with U_* we denote by A_*. Obviously, for this problem the local cone is still the same as for Problems A_ε (i.e. N). Therefore, $G_a(\Lambda, U_*) = G_a(\Lambda)$ for all a, whence the weak $\gamma-$minimum in Problem A_* simply coincides with the weak $\gamma-$minimum in any Problem A_ε, $\varepsilon > 0$. Note by the way, that on the level of weak minimum not only $\gamma-$sufficient conditions in all these problems coincide: in fact, the problems themselves do not differ one from another, because in a neighborhood of $\hat{u}$ (depending on ε) the sets U_* and U_ε simply coincide.

These arguments and the above chain of implications yield the following sufficient conditions for the strong minimum in Problem A.

Theorem 8.1. *Suppose that $\hat{w}$ satisfies $\gamma-$sufficient conditions for the weak minimum in Problem A_*, i.e. the set $G_0(\Lambda)$ is nonempty, and for some $a > 0$*

$$\Omega[G_0(\Lambda)](\bar{w}) \geq a\gamma(\bar{w}) \qquad \forall \bar{w} \in \mathcal{K} \cap \mathcal{N}. \tag{61}$$

Then $\hat{w}$ is a strict strong minimum point in the initial Problem A.

Moreover, for some $\varepsilon, C > 0$, on the set $\|x - \hat{x}\|_\infty < \varepsilon$, for all $w = (x, u)$, $u \in U$, satisfying equation (1), the following inequality holds:

$$\sum_{i \in I} \varphi_i^+(p) + |K(p)| \geq C\gamma(w - \hat{w}).$$

Proof. By Theorem 5.1 inequality (61) is equivalent to (26) = (59), and, as was shown, this implies (60) for some $\varepsilon > 0$, which means that $\hat{w}$ is a $\Pi-\gamma$-minimum point in Problem A_ε, and which is a sufficient condition for the strict Π-minimum in Problem A_ε, and hence for the strict Π-minimum in Problem A. In our case the last is equivalent to the strict strong minimum, so the first assertion is proved.

Now let us pass to the second assertion. Due to Theorem 7.2, from (60) it follows that $\exists C > 0$, such that for any sequence $w_m = (x_m, u_m)$, $u_m \in U$, satisfying (1) and such that

$$\|x_m - \hat{x}\|_\infty \to 0, \qquad \|u_m - \hat{u}\|_1 \to 0, \tag{62}$$

we have

$$\sigma(w_m) = \sum_{i \in I} \varphi_i^+(p_m) + |K(p_m)| \geq C \cdot \gamma(w_m - \hat{w}). \tag{63}$$

Due to Lemma 1.1, the first condition in (62) implies (for our U, and $\hat{u} \in \partial U$) the second one, so the last one can be omitted. Then the resulting assertion obviously implies (in fact is equivalent to) the required second assertion of the theorem.

From here and Theorem 6.1 we obtain the following theorem for small time intervals. Consider a special case of Problem A as in Sec. 6.2, for system (47) with the constraint $u \in U$, and with the cost $J = z(t_0) \to \min$. (This corresponds to the problem of time-optimality for system (47) with $z = 1$.)

Theorem 8.2. *Suppose that $\exists\, \delta_0 > 0$, such that $\forall \Delta$, $|\Delta| \le \delta_0$, the set $G_+(\Lambda(A_*(\Delta)))$ is nonempty, i.e. there exists a costate function $\psi(t)$, satisfying on Δ the costate equation and conditions (56) and (57) with an $a > 0$. Then there exists $\delta_1 > 0$, such that $\forall \Delta$, $|\Delta| \le \delta_1$ trajectory $\hat{w}_\Delta$ is a strict weak minimum point in Problem $A_*(\Delta)$, and a strict global minimum point in Problem A_Δ.*

Note that here we do not require that the set $G_+(\Lambda(A_*[0,T]))$ should be nonempty, or even that the Lagrange set $\Lambda(A_*[0,T])$ itself should be nonempty.

In the proof we will need the following three lemmas.

Lemma 8.1. *Suppose that $\exists\, \delta_0 > 0$, such that $\forall\, \Delta_0 \subset [0,T]$, $|\Delta_0| \le \delta_0$, $\exists\, \delta > 0$, such that if $\Delta \subset \Delta_0$, $|\Delta| \le \delta$, then some Property $P(\Delta)$ holds. Then $\exists\, \delta_1 > 0$ such that if $\Delta \subset [0,T]$ and $|\Delta| \le \delta_1$, then $P(\Delta)$ holds.*

The proof follows from simple arguments of compactness.

Lemma 8.2. *Suppose for some $\Delta_0 \subset [0,T]$ the $\hat{w}$ is a strict strong minimum point in Problem A_{Δ_0}. Then $\exists\, \delta, \varepsilon > 0$, such that if $\Delta \subset \Delta_0$, $|\Delta| \le \delta$, a point $w = (z,x,u)$ is admissible in Problem A_Δ, and $|z-1| < \varepsilon$, then $z > 1$.*

Proof. The strict strong minimum in Problem A_{Δ_0} means that $\exists\, \varepsilon_0 > 0$, such that if $w = (z,x,u)$ is admissible in Problem A_{Δ_0}, and $|z-1| < \varepsilon_0$, $||x-\hat{x}||_\infty < \varepsilon_0$ on Δ_0, then $z > 1$.

Consider the operator $L_\infty(\Delta_0) \times L_1(\Delta_0) \to C(\Delta_0)$, mapping $(z,u) \mapsto x$ according to system (47) with the fixed value $x = \hat{x}$ at the left end of Δ_0. Since this operator is continuous at $(\hat{z} = 1,\ \hat{u} = 0)$, and maps it to $\hat{x}$, $\exists\, \varepsilon_1 > 0$, such that if $||z(t)-1||_\infty < \varepsilon_1$, $||u||_1 < \varepsilon_1$, then $||x-\hat{x}||_\infty < \varepsilon_0$ on Δ_0. Take this $\varepsilon_1 \le \varepsilon_0$. We claim that the following Property F holds:

$\exists\, \varepsilon_2 > 0$, such that if $(z(t),x,u)$ satisfies (47) on Δ_0 with $u \in U$, $x = \hat{x}$ at $\partial\,\Delta_0$, $1-\varepsilon_2 \le z(t) \le 1$, and $||u||_1 < \varepsilon_2$, then there exists a triple (z',x',u'), admissible in Problem A_{Δ_0}, such that $1-\varepsilon_0 \le z' \le 1$, and $||x'-\hat{x}||_\infty < \varepsilon_0$.

Indeed, let us reparametrize the given x, u, putting

$$q = \frac{1}{|\Delta_0|} \int_{\Delta_0} z(t)\,dt, \qquad \text{and} \quad d\tau = \frac{1}{q} z(t)\,dt,$$

and define $z' = q$, $x'(\tau) = x(t(\tau))$, $u'(\tau) = u(t(\tau))$. Obviously, the new triple (z', x', u') still satisfies (47), and since z' is constant, this triple is admissible in Problem A_{Δ_0}; besides, since z' is the mean value of $z(t)$, we have $1-\varepsilon_2 \le z' \le 1$. If ε_2 is small enough, then $|z'-1| < \varepsilon_1 \le \varepsilon_0$, and $||u'||_1 < \varepsilon_1$, and hence, due to the continuity of the above operator, $||x' - \hat{x}||_\infty < \varepsilon_0$. Property F is proved.

Now, since U is bounded, $\exists\, \delta_2 > 0$, such that if $|\Delta| < \delta_2$, then $||u||_1 < \varepsilon_2$. We claim that for these δ_2 and ε_2 the assertion of Lemma holds.

Take any $\Delta \subset \Delta_0$, $|\Delta| \le \delta_2$, and let a point $w = (z, x, u)$ be admissible in Problem A_Δ with $|z - 1| < \varepsilon_2$. We have to show that $z > 1$. Suppose that $z \le 1$. By the definition of δ_2 we have $||u||_1 < \varepsilon_2$. Expand $x(t)$ on the entire Δ_0, putting $x = \hat{x}$ outside of Δ (this is possible, because $x = \hat{x}$ at both ends of Δ), and expand $u(t)$, putting $u = 0$ outside of Δ, and define $z(t) = z$ on Δ and $z(t) = 1$ outside of Δ. The triple $(z(t), x(t), u(t))$ satisfies the conditions of Property F, and hence there exists a triple (z', x', u'), admissible in Problem A_{Δ_0}, such that $1 - \varepsilon_0 \le z' \le 1$, $||x' - \hat{x}||_\infty < \varepsilon_0$ and $z' \le 1$. But this contradicts the strict strong minimum at $\hat{w}$ in Problem A_{Δ_0}, and hence $z > 1$. Lemma 8.2 is proved.

Lemma 8.3. *Suppose that for some $\Delta_0 \subset [0,T]$ the $\hat{w}$ is a strict strong minimum point in Problem A_{Δ_0}. Then $\exists\, \delta > 0$, such that if $\Delta \subset \Delta_0$, $|\Delta| \le \delta$, then $\hat{w}$ is a strict global minimum point in Problem A_{Δ_0}.*

Proof. By Lemma 8.2 we have the following Property B: $\exists\, \delta, \varepsilon > 0$, *such that if* $\Delta \subset \Delta_0$, $|\Delta| \le \delta$, *a point* $w = (z, x, u)$ *is admissible in Problem* A_Δ, *and* $|z - 1| < \varepsilon$, *then* $z > 1$. Fix this $\delta > 0$, and let us show that we can get rid of the requirement $|z - 1| \le \varepsilon$, i.e. that the following Property C holds: *if* $|\Delta| \le \delta$, *and the point* $w = (z, x, u)$ *is admissible in Problem* A_Δ, *then* $z > 1$.

Indeed, if it is not true, then $\exists\, \Delta$, $|\Delta| \le \delta$, and $w = (z, x, u)$, admissible in Problem A_Δ, such that $z \le 1$. Taking a linear change of time: $d\tau = z \cdot dt$, we get $x'(\tau)$ on an interval Δ', with $|\Delta'| = z \cdot |\Delta| \le \delta$, satisfying equation (47) with $z = 1$, which contradicts Property B. Thus, Property C is proved, and it is just the strict global minimum in Problem A_Δ for any $\Delta \subset [0, T]$ with $|\Delta| \le \delta$, q.e.d.

Proof of Theorem 8.2. Take any $\Delta_0 \subset [0,T]$, $|\Delta_0| \le \delta_0$, and a $\lambda \in G_+(\Lambda(A_*(\Delta_0)))$. By definition λ satisfies conditions (56) and (57) on Δ_0 with some $a = a(\Delta_0) > 0$. Due to Theorem 6.1, $\exists\,\delta > 0$ such that if $\Delta \subset \Delta_0$, and $|\Delta| \le \delta$, then

$$\Omega[\lambda](\bar{w}) \ge \frac{a}{2}\gamma(\bar{w}) \qquad \forall\,\bar{w} \in \mathcal{L}_\Delta \cap \mathcal{N}_\Delta\,,$$

i.e. $\hat{w}$ is a weak γ-minimum point in Problem $A_*(\Delta)$. By Theorem 8.1 $\hat{w}$ is a strict strong minimum point in Problem A_Δ. From here by Lemma 8.1 we get that $\exists\,\delta_1 > 0$, such that if $\Delta \subset [0,T]$ and $|\Delta| \le \delta_1$, then $\hat{w}$ is a strict strong minimum point in Problem A_Δ. For this δ_1 we have by Lemma 8.3, that $\forall\,\Delta \subset [0,T]$, $|\Delta| \le \delta_1$, $\exists\,\delta > 0$ such that if $\Delta' \subset \Delta$ and $|\Delta'| \le \delta$, then $\hat{w}$ is a strict global minimum point in Problem $A_{\Delta'}$. Using again Lemma 8.1, we obtain that $\exists\,\delta_2 > 0$, such that if $\Delta \subset [0,T]$ and $|\Delta| \le \delta_2$, then $\hat{w}$ is a strict global minimum point in Problem A_Δ. Theorem 8.2 is proved.

9 Acknowledgments

The author is grateful to Prof. A.A.Milyutin for numerous discussions and valuable suggestions. The author thanks the American Mathematical Society for the invitation to the 1997 Summer Research Institute at the University of Colorado, Boulder, where these results were presented, and the anonymous referee for useful comments.

APPENDIX

Here we prove assertions *(ii)* of Theorems 5.2 and 7.2. Begin with some known abstract facts. In a Banach space X consider a system of linear equalities and inequalities:

$$\Phi_i(x) \le 0, \;\; i \in I, \qquad Gx = 0, \tag{64}$$

where $\forall\, i$ Φ_i is a sublinear functional, such that the set $\Phi_i(x) < 0$ is nonempty, and G is a linear operator from X onto a Banach space Y. In [2] A.Ja.Dubovitskii and A.A.Milyutin proved that the system with strict inequalities

$$\Phi_i(x) < 0, \;\; i \in I, \qquad Gx = 0,$$

has no solution if and only if there exist $\alpha_i \ge 0$, $x_i^* \in \partial\Phi_i$, $i \in I$, and $y^* \in Y$, such that $\sum \alpha_i + ||y^*|| = 1$, and

$$\sum \alpha_i x_i^* + y^* G = 0. \tag{65}$$

(Here $\partial\Phi_i$ is the subdifferential in the sense of convex analysis.) We denote by Λ the set of all tuples $\lambda = (\{(\alpha_i, x_i^*),\, i \in I\},\, y^*)$, satisfying these conditions. Suppose that Λ is nonempty.

Definition. We say that an index i_0 (or the constraint $\Phi_{i_0}(x) \le 0$) is singular in system (64), if $\forall\,\lambda \in \Lambda$ the multiplier $\alpha_{i_0} = 0$. Otherwise we say that i_0 is nonsingular.

The following relation between this "dual" and a "primal" characterizations of singular constraints holds.

Lemma A1. *a) Suppose that* $\exists\, x$ *satisfying (64), such that* $\Phi_{i_0}(x) < 0$. *Then the index* i_0 *is singular.*

b) Let I_0 *be the set of all singular indices, and suppose that* $\forall\, j \notin I_0$ *the functional* Φ_j *is linear. Then* $\exists\, x$ *satisfying (64), such that* $\forall\, i \in I_0 \quad \Phi_i(x) < 0$.

Proof. *a)* Suppose that i_0 is nonsingular, i.e. $\exists\,\lambda \in \Lambda$ with $\alpha_{i_0} > 0$. From $x_i^* \in \partial\Phi_i$ we have $(x_i^*, x) \le \Phi_i(x) \le 0 \;\; \forall\, i$, and <0 for i_0, whence this x violates equation (65), a contradiction.

b) Denote by $I_1 = I \setminus I_0$ the set of all nonsingular indices, and suppose there is no such x, i.e. the cones $K_i = \{x \mid \Phi_i(x) < 0\}$, $i \in I_0$, and $C = \{x \mid \Phi_j(x) \le 0,\ \forall\, j \in I_1,\ Gx = 0\}$ do not intersect. By the Dubovitskii–Milyutin theorem [2] there exists a nontrivial collection of $p_i \in K_i^*$ and $q \in C^*$, such that $\sum p_i + q = 0$.

For each $i \in I_0$, since the set $\Phi_i(x) < 0$ is nonempty, we have $p_i = -\alpha_i x_i^*$, where $\alpha_i \ge 0$ and $x_i^* \in \partial\Phi_i$. The collection of α_i is nontrivial, because otherwise all $p_i = 0$, and then also $q = 0$, a contradiction. Now, since for each $j \in I_1$ we assume $\Phi_j(x) = (l_j, x)$, where $l_j \in X^*$, then by the Farkas lemma $q = -\sum \alpha_j l_j - y^* G$, where $\alpha_j \ge 0$ and $y^* \in Y$. Summarizing these expressions for p_i and q, we get

$$\sum_{I_0} \alpha_i\, x_i^* + \sum_{I_1} \alpha_j\, l_j + y^* G = 0.$$

The collection $(\alpha_i,\, \alpha_j,\, y^*)$, properly normalized, belongs to Λ, and since $\sum_{I_0} \alpha_i > 0$, there is an $i_0 \in I_0$, such that $\alpha_{i_0} > 0$. This contradicts the singularity of i_0. Lemma A1 is proved.

Let us return to our Problem A in the space W. First we note that for some $\varepsilon > 0$, $\forall\, t$, in the ε-neighborhood of $\hat{u}(t)$ the set U coincides with the polyhedron M, given by inequalities (27), and hence, for some (may be smaller) $\varepsilon > 0$, $\forall\, t$, in the ball $B_\varepsilon(0)$ the set $U - \hat{u}(t)$ simply coincides with its tangent cone $N = \{\bar{u} \mid (a_s, \bar{u}) \le 0,\ s \in S_0\}$. Since U has nonempty interior, so does N. Put, for convenience in notation, $\hat{w} = 0$; then in a neighborhood of zero $U = N$.
Recall that we denote $\mathcal{N} = \{\bar{w}(t) = (\bar{x}(t), \bar{u}(t)) \in W \mid \bar{u}(t) \in N \text{ a.e. on } [0,T]\}$.

Consider the following system of linear equalities and strict inequalities in W:

$$\varphi_i'(0)\bar{p} < 0, \quad i \in I, \quad K'(0)\bar{p} = 0, \quad \dot{\bar{x}} = A\bar{x} + B\bar{u}, \quad \bar{u} \in \operatorname{int}\mathcal{N}.$$

The last inequality can be written as

$$\Phi(\bar{u}) = \min_{s \in S_0} \operatorname{ess\,sup}\,(a_s, \bar{u}(t)) < 0,$$

where obviously Φ is a sublinear functional. The stationarity of $\hat{w}$ means that this system has no solution, which by the Dubovitskii–Milyutin theorem is equivalent to the existence of Lagrange multipliers $\lambda = (\alpha_i,\ \beta \in R^{dim K},\ \psi \in L_\infty,\ \mu_s \in L_1)$, such that $\alpha_i \geq 0\ \ \forall i \in I,\ \ \mu_s(t) \geq 0$ a.e. on $[0,T]\ \ \forall s \in S_0$, and the Euler–Lagrange equation holds: for all $\bar{w} \in W$

$$\sum \alpha_i \varphi_i' \bar{p} + K' \bar{p} + \int \psi\,(\dot{\bar{x}} - A\bar{x} - B\bar{u})\,dt + \sum \int \mu_s\,\bar{u}\,dt = 0$$

For Problem A this condition is equivalent to MP.

The total singularity of $\hat{w}$ (i.e. its singularity w.r.t. all U) is equivalent to that all $\mu_s(t) = 0$ for all such λ-s, and by Lemma A1 the last property is equivalent to the existence of $\bar{w}_* \in W$, such that

$$\varphi_i'(0)\,\bar{p}_* < 0, \qquad K'(0)\,\bar{p}_* = 0, \qquad \dot{\bar{x}}_* = A\bar{x}_* + B\bar{u}_*, \tag{66}$$

$$\bar{u}_* \in \operatorname{int}\mathcal{N}, \quad \text{i.e. } \exists\,\varepsilon_* > 0 \text{ such that } B_{\varepsilon_*}(\bar{u}_*(t)) \subset N \quad \text{a.e. on } [0,T].$$

(Recall that system (66) defines the critical cone $\mathcal{K}$.) We can take $||\bar{w}_*|| \leq 1$. These $\bar{w}_*$ and the corresponding $\varepsilon_* > 0$ will be used in what follows.

Let us establish a simple property of the cone N in R^k. Define the function $\sigma_N(u) = \sum_{s \in S_0} (a_s, u)^+$, and pick any $\bar{u} \in \operatorname{int} N$ and $\varepsilon > 0$, such that $B_\varepsilon(\bar{u}) \subset N$.

Lemma A2. *For any ρ there exists $C = C(N, \varepsilon, \rho)$, such that $\forall\,\delta > 0$, $\forall\,u \in R^k$ with $\sigma_N(u) \leq \delta$, the point $u + \delta C \bar{u}$ lies in the $\delta\rho$-interior of N, i.e.*

$$u + \delta C\bar{u} + B_{\delta\rho}(0) \subset N.$$

Proof. In view of homogeneity of this property, we can consider only the case when $\delta = 1$, i.e. to prove that for some C, if $\sigma_N(u) \leq 1$, then $(u + C\bar{u}) + B_\rho(0) \subset N$.

Obviously, $\exists\,C_0 = C_0(N, \varepsilon)$ such that for all these u we have $u + C_0\bar{u} \in N$. Then, $\exists\,C_1 = C_1(N, \varepsilon, \rho)$ such that for all $u' \in N$ the point $u' + C_1\bar{u}$ lies in the ρ-interior of N. Taking $C = C_0 + C_1$, we get the required property.

Now we recall the known estimate of the distance to the level set of an operator g, mapping a Banach space X to another Banach space Y (see e.g. [11]). Suppose, g

is strictly differentiable at a point $x_0 \in X$ (e.g. continuously Frechet differentiable at x_0), $g(x_0) = 0$, and denote $\mathcal{M} = \{x \mid g(x) = 0\}$. If $g'(x_0)$ is onto (i.e. g satisfies Lyusternik condition at x_0), then there exist a number L and a neighborhood $\mathcal{B}(x_0)$, such that

$$\forall\, x \in \mathcal{B}(x_0) \qquad \text{dist}\,(x, \mathcal{M}) \le L\,||g(x)||. \tag{67}$$

Finally, we prove an estimate for the increment of the function

$$\sigma_0(w) = \sum \varphi_i^+(p) + |\,K(p)\,| + \int |\,\dot{x} - f_0(x) - F(x)u\,|\,dt, \tag{68}$$

which is crucial for our purposes.

Lemma A3. *Let* $w_n \to 0$, $\bar{w} \in \mathcal{K}$, *and* $\varepsilon_n \to 0$. *Then* $\sigma_0(w_n + \varepsilon_n \bar{w}) \le \sigma_0(w_n) + o\,(\varepsilon_n)$.

Proof. Consider the increment of the first term in (68) (we omit i):

$$\varphi^+(p_n + \varepsilon_n \bar{p}) = [\,\varphi(p_n) + \varphi'(p_n) \cdot \varepsilon_n \bar{p} + o\,(\varepsilon_n)\,]^+ \le$$

(because $(\cdot)^+$ is a sublinear functional)

$$\le \varphi^+(p_n) + \varepsilon_n\,[\,\varphi'(p_n)\,\bar{p}\,]^+ + o(\varepsilon_n). \tag{69}$$

Since $\varphi'(p_n) \to \varphi'(0)$, and $\varphi'(0)\,\bar{p} \le 0$ (because $\bar{w} \in \mathcal{K}$), the second term in (69) is $o(\varepsilon_n)$, whence for the increment of the first term in (68) we obtain the desired estimate. Similar arguments work for other two terms in (68). Lemma A3 is proved.

Proof of Theorem 5.2 (ii). In view of assertion *(i),* we have to show, that if $\exists\, a > 0$ and a neighborhood $\mathcal{B}(\hat{w})$, such that $\forall\, w \in \mathcal{B}(\hat{w})$ satisfying (1), (4) with $u \in N$, the inequality $\sigma(w) \ge a\gamma(w)$ holds, then $\exists\, a' > 0$ and a neighborhood $\mathcal{B}'(\hat{w})$, in which the inequality $\sigma(w) \ge a'\gamma(w)$ holds. (The reverse implication is trivial.) We prove the contrary-reverse implication, i.e. if $\exists\, w_n \to 0$, such that $\sigma(w_n) = o(\gamma(w_n))$, then $\exists\, w_n' \to 0$ satisfying (1), (4) with $u_n' \in N$, such that $\sigma(w_n') = o(\gamma(w_n'))$.

Thus, suppose that $\exists\, w_n \to 0$ with $\sigma(w_n) \le \alpha_n\,\gamma(w_n)$, where $\alpha_n \to 0$. In particular, we have $\sigma_N(u_n) \le \alpha_n\,\gamma(w_n)$. Denote by L the constant from the Lyusternik estimate (67) for the operator g, corresponding to equality constraints (1) and (4) (see Def. 5.1).

By Lemma A2, taking $\bar{u} = \bar{u}_*(t)$, $\varepsilon = \varepsilon_*$ (uniform for a.a. t), and $\rho = 2L$, we get $\delta_n = \alpha_n\,\gamma(w_n)$, and $w_n' = w_n + C\,\delta_n\,\bar{w}_*$, such that

$$B_{\rho\delta_n}(u_n') \subset N. \tag{70}$$

In particular, $u'_n \in N$, and hence $\sigma(w'_n) = \sigma_0(w'_n)$. By Lemma A3 we have

$$\sigma_0(w'_n) \leq \sigma_0(w_n) + o(\alpha_n \gamma(w_n)) \leq 2\,\alpha_n\,\gamma(w_n)$$

for large enough n. The point w'_n violates equalities (1) and (4) not more than by $\sigma_0(w'_n)$. Using estimate (67), we get a point w''_n satisfying these equalities, such that $||w''_n - w'_n|| \leq L \cdot 2\,\alpha_n\,\gamma(w_n) = \rho\,\alpha_n\,\gamma(w_n)$. In particular, $|\,u''_n(t) - u'_n(t)\,| \leq \rho\,\alpha_n\,\gamma(w_n)$, and due to (70) we have $u''_n(t) \in N$. Since the function σ_0 is Lipschitzian (at least in a neighborhood of zero), then

$$\sigma(w''_n) = \sigma_0(w''_n) \leq \sigma_0(w'_n) + \text{const}\,||w''_n - w'_n|| \leq$$

$$\leq 2\,\alpha_n\,\gamma(w_n) + \text{const} \cdot 2L\,\alpha_n\,\gamma(w_n) = o(\gamma(w_n)).$$

Finally, since $||w''_n - w_n|| \leq o(\,\gamma(w_n))$, we obviously have $\gamma(w''_n) \sim \gamma(w_n)$, so we may write $\sigma(w''_n) \leq o(\,\gamma(w''_n))$. Thus, we get the desired sequence $w''_n \to 0$, satisfying (1), (4) and the last estimate, with $u''_n(t) \in N$, q.e.d.

Proof of Theorem 7.2 (ii) is similar to the previous one. We have to show, that if there exists a sequence $\{w_n\} \in \Pi(U, \hat{w})$, such that $\sigma(w_n) = o(\gamma(w_n))$, then there exists a sequence $\{w'_n\} \in \Pi(U, \hat{w})$ satisfying equalities (1), (4) with $u'_n \in N$, such that $\sigma(w'_n) = o(\gamma(w'_n))$. The only two points in the previous proof, where we used that $||w_n|| \to 0$, were those, when we used Lyusternik estimate (67) and the Lipschitz continuity of σ_0 in a neighborhood of zero. The validity of these properties in a neighborhood of zero is not sufficient now, because Pontryagin sequences do not, generally, converge to zero in the norm of W. However, these properties are valid not only in a neighborhood of zero, but on a more broad set $||w||_1 = |x(0)| + ||\dot{x}||_1 + ||u||_1 < \varepsilon$, for some $\varepsilon > 0$. For estimate (67) this was proved in [33, Part II], [34], while the Lipschitz continuity of σ_0 on this set (w.r.t. $||w||$, or even w.r.t. $||w||_1$) is rather obvious. Thus, in both these points the previous proof remains valid, and so Theorem 7.2 *(ii)* is proved.

References

[1] G.A.Bliss, *Lectures on the calculus of variations,* Chicago, 1946.

[2] A.Ja.Dubovitskii, A.A.Milyutin, *Extremal problems in the presence of constraints,* USSR Comput. Math. and Math. Physics, 1965, v. 5, No. 3.

[3] A.Ja.Dubovitskii, A.A.Milyutin, *Theory of the Maximum Principle,* "Metody teorii ekstremal'nyh zadach v ekonomike" ("Application of the theory of extremal problems to economics") (V.L.Levin ed.), Moscow, "Nauka", CEMI, 1981, p. 6–47 (in Russian).

[4] A.A.Milyutin, *Quadratic conditions of an extremum in smooth problems with a finite-dimensional image,* – ibid., p. 138–177 (in Russian).

[5] E.S.Levitin, A.A.Milyutin, N.P.Osmolovskii. *Conditions of higher–order for a local minimum in extremal problems with constraints,* Russian Math. Surveys, 1978, v. 33, No. 6, p. 97–168.

[6] N.P.Osmolovskii. *Necessary and sufficient conditions of a high order for a Pontryagin and a bounded-strong minima in an optimal control problem,* Soviet Phys. Doklady, 1988, v. 33, No. 12, p. 883–885.

[7] N.P.Osmolovskii, *Quadratic conditions for nonsingular extremals in optimal control (a theory),* Russian Journal of Mathematical Physics, Wiley, 1994, v. 2, No. 4, 487–516.

[8] V.A.Dubovitskii, *Necessary and sufficient conditions of a Pontryagin minimum in an optimal control problem with singular regimes,* Russian Math. Surveys, 1982, v. 37, No. 6.

[9] E.S.Levitin, *Perturbation theory in the mathematical programming and its applications,* Moscow, Nauka, 1992.

[10] A.A.Milyutin, *Maximum Principle in the general optimal control problem,* Moscow, "Phasis", 1998, to appear (in Russian).

[11] A.V.Dmitruk, A.A.Milyutin, N.P.Osmolovskii, *Lyusternik's theorem and the theory of extremum,* Russian Math. Surveys, 1980, v. 35, No. 6, p. 11–51.

[12] A.V.Dmitruk, A.A.Milyutin, *New Legendre type conditions for optimal control problems, linear in the control,* Proc. of the European Control Conference, Brussels, 1997.

[13] A.A.Milyutin, *On a duality formula for multidimensional vector fields,* Russian Journal of Mathematical Physics, Wiley, 1995, v. 3, No. 1, p. 81–112.

[14] A.V.Dmitruk, *The Euler-Jacobi equation in the calculus of variations,* Math. Notes of the Acad. Sci. USSR, 1976, v. 20, p. 1032–1038.

[15] A.V.Dmitruk, *Jacobi type conditions for the problem of Bolza with inequalities,* ibid., 1984, v. 35, p. 427–435.

[16] A.V.Dmitruk, *A condition of Jacobi type for a quadratic form to be nonnegative on a finite-faced cone,* Mathematics of the USSR, Izvestija, 1982, v. 18, No. 3, p. 525–535.

[17] H.J.Kelley, R.E.Kopp, H.G.Moyer, *Singular extremals,* Topics in Optimization (ed. G.Leitman), Acad. Press, New York–London, 1967, p. 63–101.

[18] B.S.Goh, *Necessary conditions for singular extremals involving multiple control variables,* SIAM J. on Control, 1966, v. 4, No. 4, p. 716–731.

[19] R.Gabasov, F.M.Kirillova, *Singular optimal controls,* Nauka, Moscow, 1973.

[20] D.J.Bell, D.H.Jacobson, *Singular Optimal Control Problems,* Academic Press, NY, 1975.

[21] H.W.Knobloch, *Higher order necessary conditions in optimal control theory,* Lecture Notes in Control and Inf. Sciences, v. 34, 1981.

[22] A.J.Krener, *The high order maximal principle and its application to singular extremals,* SIAM J. on Control, 1977, v. 15, No. 2, p. 256–293.

[23] A.A.Agrachiov, R.V.Gamkrelidze, *Second order optimality principle for a time-optimal problem,* Math. USSR, Sbornik, 1976, v. 100, No. 4.

[24] A.V.Sarychev, *High-order necessary conditions of optimality for nonlinear control systems,* Systems & Control Letters, 1991, v. 16, p. 369–378.

[25] M.I.Zelikin, *An optimality condition for singular trajectories in the problem of minimization of a curvilinear integral,* Soviet Math. Doklady, 1982, v. 267, No. 3.

[26] F.Lamnabhi-Lagarrigue, G.Stefani, *Singular optimal control problems: on the necessary conditions of optimality,* SIAM J. on Control & Optim., 1990, v. 28, No. 4, p. 823–840.

[27] J.L.Speyer, D.H.Jacobson. *Necessary and sufficient conditions of optimality for singular control problems. A transformation approach,* J. Math. Analysis and Appl., 1971, v. 33, No. 1, p. 163–186.

[28] I.B.Vapnyarskii, *Existence theorem in the problem of Bolza, and necessary conditions of sliding and singular regimes,* USSR Comput. Math. and Math. Physics, 1967, v. 7, No. 2.

[29] V.I.Gurman, *Singular optimal control problems,* Moscow, Nauka, 1977.

[30] V.A.Dykhta, *Conditions of a local minimum for singular regimes in systems with linear control,* Automatics and Remote Control, 1981, No. 12.

[31] A.V.Dmitruk, *Quadratic order conditions of a weak minimum for singular regimes in optimal control problems,* Soviet Math. Doklady, 1977, v. 18, No. 2.

[32] A.V.Dmitruk, *Quadratic conditions for a Pontryagin minimum in an optimal control problem, linear in the control, with a constraint on the control,* Soviet Math. Doklady, 1983, v. 28, No. 2, p. 364–368.

[33] A.V.Dmitruk, *Quadratic conditions for a Pontryagin minimum in an optimal control problem, linear in the control,* Mathematics of the USSR, Izvestija, 1987, v. 28, No. 2, p. 275–303, and 1988, v. 31, No. 1, p. 121–141.

[34] A.V.Dmitruk, *Fine theorems on weakening equality constraints in optimal control problems, linear in control,* Siberian Math. Journal, 1990, v. 31, No. 2, p. 37–51.

[35] A.V.Dmitruk, *Second order necessary and sufficient conditions of a Pontryagin minimum for singular regimes,* Lecture Notes in Control and Inf. Sciences, 1992, v. 180, p. 334–343.

[36] A.V.Dmitruk, *Second order optimality conditions for singular extremals,* in Computational Optimal Control (R.Bulirsch and D.Kraft eds.), Intern. Ser. on Numer. Mathematics, Birkhaueser, Basel, v. 115 (1994), p. 71–81.

[37] A.V.Dmitruk, *Quadratic order conditions of a Pontryagin minimum for singular extremals in optimal control problems,* Dr.Sci. Dissertation, St. Petersburg University, 1994.

[38] A.V.Dmitruk, *Second Order Necessary and Sufficient Conditions of a Pontryagin Minimum for Singular Boundary Extremals,* Proc. of the Third Intern. Congress on Industrial and Appl. Math., Hamburg, 1995: ZAMM, Issue 3, p. 411–412.

[39] A.V.Dmitruk, *Quadratic order conditions of a local minimum for abnormal extremals,* Nonlinear Analysis, Theory, Methods & Appl., 1997, v. 30, No. 4, p. 2439–2448.

[40] D.J.W.Ruxton, D.Bell, *Junction times in singular optimal control,* Appl. Math. Comput., 1995, v. 70, No. 2-3, p. 143–154.

[41] A.Visintin, *Strong convergence results related to strict convexity,* Commun. Partial Diff. Equations, 1984, v. 9, No. 5, p. 439–466.

[42] A.F.Filippov, *On some topics in the theory of optimal regulation,* Vestnik Moskovskogo Universiteta, Ser. mat. mech. phys., 1959, No. 2 (in Russian).

[43] Ch.Castaing, *Sur les multi-applications measurables,* Rev. Francaise Inf. Rech. Oper., 1967, v. 1, p. 91–126.

Central Economics and Mathematics Institute of the Russian Academy of Sciences, Russia 117418, Moscow, Nakhimovskii prospekt, 47

e-mail: dmitruk@member.ams.org

Proceedings of Symposia in Pure Mathematics
Volume **64**, 1999

Some Solvable Infinite Time Horizon Stochastic Control Problems in Hyperbolic Three Space

T. E. Duncan

January 1, 1994 and, in revised form, June 22, 1994

ABSTRACT. An infinite time horizon, discounted cost stochastic control problem and an ergodic cost stochastic control problem are formulated and explicitly solved in a real hyperbolic space of dimension three. The problems are to control a Brownian motion in a real hyperbolic three space by a drift vector field so that the resulting process remains close to the origin. Real hyperbolic three space is a symmetric space and the cost function is invariant under the associated maximal compact subgroup. This invariance is exploited to find explicitly an optimal control.

1. Introduction

While many results are available for the existence of optimal controls and the existence and the uniqueness of the solution of the Hamilton-Jacobi-Bellman equation for stochastic control problems, there are relatively few examples where an optimal control is expressed explicitly as a function of the state. This is especially true for the infinite time horizon stochastic control problems. Some examples are the well known linear quadratic Gaussian problem and some portfolio and portfolio/consumption problems (e.g., [2, 11, 12]). In this paper some examples are given with a discounted cost and an ergodic cost (an average cost per unit time) where the controlled diffusion evolves in a hyperbolic space. The problem is to control a Brownian motion in a real hyperbolic three space, $\mathbb{H}^3(\mathbb{R})$, by a drift vector field. Since $\mathbb{H}^3(\mathbb{R})$ is the symmetric space $SL(2,\mathbb{C})/SU(2)$ and the cost functionals are $SU(2)$ invariant, this invariance plays a basic role in the solutions of the control problems. This work generalizes some finte time horizon problems in [3, 5].

1991 *Mathematics Subject Classification.* Primary 93E20, 58G32; Secondary 53C35.
Key words and phrases. Infinite time horizon stochastic control, solvable stochastic control, discounted cost stochastic control, ergodic cost stochastic control, hyperbolic three space..
Research supported by NSF Grant DMS 9623439.

2. Preliminaries and Main Results

A real hyperbolic three space, $\mathbb{H}^3(\mathbb{R})$, is a noncompact symmetric space of rank one that can be described as

$$\mathbb{H}^3(\mathbb{R}) = SL(2,\mathbb{C})/SU(2) \tag{1}$$

The unit ball model for $\mathbb{H}^3(\mathbb{R})$ is

$$B_1(0) = \{y \in \mathbb{R}^3 : |y| < 1\} \tag{2}$$

where $|\cdot|$ is the canonical norm in $\mathbb{R}^3$. The unit ball $B_1(0)$ with the Riemannian structure

$$ds^2 = 4(1-|y|^2)^{-2}(dy_1^2 + dy_2^2 + dy_3^2) \tag{3}$$

has constant sectional curvature -1. Since any space of constant negative curvature is globally trivial, the geodesic polar coordinates for this model of $\mathbb{H}^3(\mathbb{R})$ are a useful coordinate system. These coordinates are defined by the map

$$\mathrm{Exp}_0 Y \mapsto (r, \theta_1, \theta_2) \tag{4}$$

where $Y \in T_0\mathbb{H}^3(\mathbb{R})$, $r = |Y|_0$, $|\cdot|_0$ is given by the Riemannian metric at 0, and (θ_1, θ_2) are coordinates of the unit vector $Y/|Y|_0$ for $Y \neq 0$. In these coordinates the Riemannian structure is

$$ds^2 = dr^2 + (\sinh r)^2 d\sigma^2 \tag{5}$$

where $d\sigma^2$ is the Riemannian structure on the unit sphere in $T_0\mathbb{H}^3(\mathbb{R})$ (e.g., [8]).

The controlled diffusion has the infinitesimal generator

$$\frac{1}{2}\Delta_{\mathbb{H}^3(\mathbb{R})} + u\frac{\partial}{\partial r} \tag{6}$$

so that the diffusion is an $\mathbb{H}^3(\mathbb{R})$-valued Brownian motion plus a drift vector field that is the control. The discounted cost function is

$$J_d(U) = E_{|Y(0)|_0} \int_0^\infty e^{-\rho t}\left[a \sinh^2 \frac{|Y(t)|_0}{2} + \left(\cosh^2 \frac{|Y(t)|_0}{2}\right) U^2(t)\right] dt \tag{7}$$

and the ergodic cost is

$$J_e(U) = \limsup_{T\to\infty} \frac{1}{T} E \int_0^T \left[a \sinh^2 \frac{|Y(t)|_0}{2} + \left(\cosh^2 \frac{|Y(t)|_0}{2}\right) U^2(t)\right] dt \tag{8}$$

where $\rho > 0$, $a > 0$, and $(Y(t), t \geq 0)$ is the controlled diffusion in geodesic polar coordinates with the infinitesimal generator (6). For notational simplicity the dependence of Y on U is suppressed in (7,8).

Since the cost functionals with the control fixed are invariant under the action of the maximal compact subgroup $SU(2)$ and the Laplacian is invariant under $SU(2)$, it suffices to consider the controlled diffusion projected to the positive Weyl chamber, $\{r : r \geq 0\}$. So the infinitesimal generator (6) becomes

$$\frac{1}{2}\hat{\Delta}_{\mathbb{H}^3(\mathbb{R})} + u\frac{\partial}{\partial r} = \frac{1}{2}\frac{\partial^2}{\partial r^2} + \coth r \frac{\partial}{\partial r} + u\frac{\partial}{\partial r} \tag{9}$$

and the associated controlled diffusion is

$$
\begin{aligned}
dX(t) &= (\coth X(t) + U(t))dt + dB(t) \\
X(0) &= |Y(0)|_0.
\end{aligned}
\tag{10}
$$

The differential operator $\hat{\Delta}_{\mathbb{H}^3(\mathbb{R})}$ is called the *radial part* of the Laplace-Beltrami operator $\Delta_{\mathbb{H}^3(\mathbb{R})}$ [8].

The family of (Markovian) admissible controls is all Borel measurable functions of the controlled diffusion such that the associated stochastic differential equation has one and only one strong solution and the associated costs are finite. By the $SU(2)$ invariance of the cost functions the family of admissible controls are

$$
\begin{aligned}
\mathcal{U} = \{u : \mathbb{R}_+ \to \mathbb{R} \mid\ & u \text{ is Borel measurable and } u(t) = \tilde{u}(X(t)) \text{ in (10) has} \\
& \text{one and only one strong solution and the cost is finite}\}.
\end{aligned}
\tag{11}
$$

Initially it is necessary to verify that the stochastic control problems (6,7) and (6,8) are well posed, that is, there is at least one admissible control that gives a finite value to (7) and (8).

LEMMA. *Let* $(\tilde{X}(t), t \geq 0)$ *be the diffusion process on* $\mathbb{R}_+$ *with the control* $U^1(X(t)) = -bX(t)$ *for* $t \geq 0$ *where* $b > 0$. *Then for* $\rho > 0$

$$
E_{\tilde{X}(0)} \int_0^\infty e^{-\rho t} \left[a \sinh^2 \frac{\tilde{X}(t)}{2} + \left(\cosh^2 \frac{\tilde{X}(t)}{2} \right) b^2 \tilde{X}^2(t) \right] dt < \infty \tag{12}
$$

and

$$
\limsup_{T \to \infty} \frac{1}{T} E_{\tilde{X}(0)} \int_0^T \left[a \sinh^2 \frac{\tilde{X}(t)}{2} + \left(\cosh^2 \frac{\tilde{X}(t)}{2} \right) b^2 \tilde{X}^2(t) \right] dt < \infty \tag{13}
$$

PROOF. The process $(\tilde{X}(t), t \geq 0)$ satisfies the stochastic differential equation

$$
d\tilde{X}(t) = (\coth \tilde{X}(t) + U^1(t))dt + dB(t). \tag{14}
$$

By a comparison theorem for stochastic differential equations (p. 437 [9]) it suffices for the verification of (12,13) to consider the stochastic differential equation

$$
dZ(t) = \left(\frac{\alpha}{Z(t)} - \beta Z(t) \right) dt + dB(t) \tag{15}
$$

where $\alpha > 0$ and $\beta > 0$. It can be shown [6] that a solution of (14) or (15) with a positive initial condition does not hit the origin with probability 1 so that both of these equations have unique strong (pathwise) solutions.

It is shown that

$$
\sup_{t \geq 0} E[\exp(\gamma Z(t))] < \infty
$$

for each $\gamma > 0$ by computing the moments of $(Z(t), t \geq 0)$. Fix $p \in \mathbb{N}$. A differential equation and a differential inequality are obtained for $EZ^p(t)$. The differential equation can be obtained by applying Itô's formula to $(Z^p(t), t \geq 0)$ and taking the expectation. The stochastic integral term is a martingale by the integrability form the Picard iteration construction of the solution or using a family of stopping terms. Alternatively, Dynkin's formula can be obtained directly from the (local) integrability of the generator acting on the function $f(x) = x^p$. The differential equation is

$$\frac{d}{dt}EZ^p(t) = \left(\alpha p + \frac{1}{2}p(p-1)\right) EZ^{p-2}(t) - \beta p EZ^p(t).$$

Using Hölder's inequality the following differential inequality is obtained

$$\frac{d}{dt}EZ^p(t) \leq \left(\alpha p + \frac{1}{2}p(p-1)\right) (EZ^p(t))^{(p-2)/p} - \beta p EZ^p(t).$$

Let $v(t) = EZ^p(t)$ so the inequality (17) can be written as

$$\frac{dv}{dt} \leq \left(\alpha p + \frac{1}{2}p(p-1)\right) v^{(p-2)/2} - \beta p v. \tag{18}$$

For a suitable $K_1 > 0$ on the set $\Gamma = \{x : K_1 p^{p/2} < x\}$ the right hand side of the inequality (18) is negative so for $t \geq T$ where T does not depend on p it follows that

$$v(t) \leq K_2 p^{p/2}$$

or equivalently

$$EZ^p(t) \leq K_2 p^{p/2}.$$

where K_2 does not depend on p. Thus,

$$\begin{aligned} E\exp[\gamma Z(t)] &\leq \sum_{k=0}^{\infty} \frac{\gamma^k}{k!} EZ^p(t) \\ &\leq \sum_{k=0}^{\infty} \frac{\gamma^k}{k!} K_2 k^{k/2} \leq K < \infty \end{aligned}$$

where K is a constant that does not depend on t. Thus,

$$E\exp[\gamma \tilde{X}(t)] \leq K$$

and (12,13) are satisfied. □

The solution of the infinite time horizon, discounted expected cost control problem is given now.

THEOREM 1. *The stochastic optimal control problem given by* (6, 7) *has an admissible optimal control* U^* *given by*

$$U^*(x) = -\frac{1}{2}g \tanh \frac{x}{2} \tag{19}$$

where g *is the real positive root of*

$$\frac{1}{4}g^2 + g\left(\rho - \frac{3}{2}\right) - a = 0 \tag{20}$$

and the optimal cost is

$$W(x) = g \sinh^2 \frac{x}{2} + h \tag{21}$$

where

$$h = \frac{3g}{4\rho} \tag{22}$$

and $X(0) = x$.

PROOF. The Hamilton-Jacobi-Bellman (HJB) equation for an infinite time horizon, discounted expected cost [7, 9] is

$$0 = \min_{v \in U} [A^v W(x) - \rho W(x) + L(x, v)] \tag{23}$$

where A^v is the infinitesimal generator of the controlled diffusion with the control v and the cost functional is

$$J(u) = E_x \int_0^\infty \mathrm{e}^{-\rho s} L(x(s), u(s)) ds.$$

It is assumed that

$$E_x \int_0^\infty \mathrm{e}^{-\rho s} |L(x(s), u(s))| ds < \infty$$

and W is sufficiently smooth to define a "classical" solution.

Applying (23) to the control problem (6,7) it is clear that the integrand of the cost functional (7) as a function of the state with the control fixed is invariant under the action of the maximal compact subgroup $SU(2)$, so the HJB equation (23) can be simplified by using the radial part of the infinitesimal generator (9) as

$$0 = \min_{v \in U} \left[\frac{1}{2} \frac{\partial^2 W}{\partial r^2} + \coth r \frac{\partial W}{\partial r} + v \frac{\partial W}{\partial r} - \rho W + a \sinh^2 \frac{r}{2} + v^2 \cosh^2 \frac{r}{2} \right]. \tag{24}$$

Performing the minimization in (24) it is clear that the candidate for an optimal control is

$$U^*(r) = \frac{-1}{2 \cosh^2 \frac{r}{2}} \frac{\partial W}{\partial r}. \tag{25}$$

Assume a solution W of (24) as

$$W(r) = g \sinh^2 \frac{r}{2} + h. \tag{26}$$

To apply a verification theorem (p. 147 [7]) it is necessary to verify that for each admissible control

$$\liminf_{T \to \infty} W(X(T)) \leq 0. \tag{27}$$

For all functions β such that $|\beta(x)| \leq K e^{-x/2}$ it follows from the finiteness of (7) that the Dynkin formula can be applied. Since W satisfies this growth condition, it follows by the finiteness of the cost functional (7) for each admissible control to be considered that (27) is satisfied.

It is useful to recall that

$$\sinh^2 \frac{r}{2} = \frac{1}{2}(\cosh r - 1) \tag{28}$$

and

$$\frac{1}{2} \sinh r = \sinh \frac{r}{2} \cosh \frac{r}{2}. \tag{29}$$

Substitute (25) and (26) in (24) and use (28,29) to obtain

$$\begin{aligned} 0 &= \frac{1}{4}g\left(1+2\sinh^2\frac{r}{2}\right)+\frac{1}{2}g\left(1+2\sinh^2\frac{r}{2}\right)-\frac{1}{2}g^2\sinh^2\frac{r}{2} \\ &\quad -\rho\left(g\sinh^2\frac{r}{2}+h\right)+a\sinh^2\frac{r}{2}+\frac{1}{4}g^2\sinh^2\frac{r}{2} \\ &= \sinh^2\frac{r}{2}\left(\frac{1}{2}g+1g-\frac{1}{4}g^2-\rho g+a\right)+\frac{1}{4}g+\frac{1}{2}g-\rho h. \end{aligned} \tag{30}$$

Let g be the positive real root of

$$\frac{1}{4}g^2+g\left(\rho-\frac{3}{2}\right)-a=0 \tag{31}$$

and

$$h=\frac{3g}{4\rho}. \tag{32}$$

Then (26) is a classical (smooth) solution to (24). Since the solution of the stochastic differential equation (10) with $U \equiv 0$ does not hit the origin it is elementary to verify that there is a strong solution (10) with the (optimal) control (25) and this solution does not hit the origin almost surely. □

The solution of the ergodic cost control problem (6,8) is given now.

THEOREM 2. *The stochastic optimal control problem given by* (6, 8) *has an admissible optimal control* U^* *given by*

$$U^*(x)=-\frac{1}{2}g\tanh\frac{x}{2} \tag{33}$$

where g *is the real positive root of*

$$\frac{1}{4}g^2-\frac{3}{2}g-a=0 \tag{34}$$

and the optimal cost W^* *is*

$$W^*=\frac{3}{4}g. \tag{35}$$

PROOF. Since (13) is satisfied, the ergodic control problem is well posed. The HJB equation for an ergodic control problem [1, 9] is

$$0=\min_{v\in U}[A^vV(x)-\gamma+L(x,v)] \tag{36}$$

where the cost is

$$C(u)=\limsup_{T\to\infty}\frac{1}{T}\int_0^T L(X(s),U(s))ds. \tag{37}$$

A^v is the infinitesimal genenerator of the controlled diffusion and γ is the optimal cost.

Since there is an inherent degeneracy or ambiguity in (36) because it is necessary to solve for both V and γ, a direct construction of an optimal control and the optimal cost is made. The construction uses the controls and the optimal costs from a family of finite time horizon control problems. This family of finite time horizon control problems arises essentially from the cost function (8) by fixing T. More precisely, for $T > 0$ consider the finite time horizon control problem with the system (10) and the cost function

$$C_T(U) = E_{X(0)} \int_0^T \left[a \sinh^2 \frac{|Y(t)|_0}{2} + \left(\cosh^2 \frac{|Y(t)|_0}{2} \right) U^2(t) \right] dt. \tag{38}$$

The HJB equation for the control problem (6,38) is

$$0 = \frac{\partial W_T}{\partial s} + \min_{v \in \mathbb{R}} \left[\frac{1}{2} \frac{\partial^2 W_T}{\partial r^2} + \coth r \frac{\partial W_T}{\partial r} + v \frac{\partial W_T}{\partial r} + a \sinh^2 \frac{r}{2} + v^2 \cosh^2 \frac{r}{2} \right]. \tag{39}$$

Performing the minimization in (39), it follows that the candidate for an optimal control is

$$U_T^*(s,r) = \frac{-1}{2 \cosh^2 \frac{r}{2}} \frac{\partial W_T}{\partial r}. \tag{40}$$

Assume a solution W_T of (39) as

$$\begin{aligned} W_T(s,r) &= g_T(s) \sinh^2 \frac{r}{2} + h_T(s) \\ W_T(T,r) &\equiv 0. \end{aligned} \tag{41}$$

The validity of a verification theorem (p. 141 [7]) follows as in the proof of Theorem 1.

Substitute (41) in (39) to obtain

$$\begin{aligned} 0 &= g_T' \sinh^2 \frac{r}{2} + h_T' + \frac{1}{4} g_T \left(1 + 2\sinh^2 \frac{r}{2}\right) + \frac{1}{2} g_T \left(1 + 2\sinh^2 \frac{r}{2}\right) \\ &\quad - \frac{1}{2} g_T^2 \sinh^2 \frac{r}{2} + a \sinh^2 \frac{r}{2} + \frac{1}{4} g_T^2 \sinh^2 \frac{r}{2} \\ &= \sinh^2 \frac{r}{2} \left(g_T' + \frac{1}{2} g_T + 1 g_T - \frac{1}{4} g_T^2 + a \right) + h_T' + \frac{3}{4} g_T. \end{aligned} \tag{42}$$

Let

$$\begin{aligned} g_T' + \frac{3}{2} g_T - \frac{1}{4} g_T^2 + a &= 0 \\ g_T(T) &= 0 \end{aligned} \tag{43}$$

and

$$\begin{aligned} h_T' + \frac{3}{4} g_T &= 0 \\ h_T(T) &= 0. \end{aligned} \tag{44}$$

Thus W_T given by (41) satisfies (39).

It is clear that the family of optimal controls (U_T^*, $T > 0$) converges (at least pointwise) to

$$U^*(r) = -\frac{g}{2} \tanh \frac{r}{2} \tag{45}$$

where g satisfies

$$\frac{1}{4} g^2 - \frac{3}{2} g - a = 0.$$

It is simple to verify by contradiction that U^* is an optimal control for (6,8).

The optimal cost using the cost function

$$\tilde{C}_T(U) = \frac{1}{T} E_{X(0)} \int_0^T \left[a \sinh^2 \frac{|Y(t)|_0}{2} + \left(\cosh^2 \frac{|Y(t)|_0}{2} \right) U^2(t) \right] dt \tag{46}$$

is

$$\begin{aligned} \tilde{W}_T(0,x) &= \frac{1}{T} W_T(0,x) = \frac{1}{T} \left(g_T(0) \sinh^2 \frac{x}{2} + h_T(0) \right) \\ &= \frac{1}{T} \left(g_T(0) \sinh^2 \frac{x}{2} + \int_0^T \frac{3}{4} g_T dt \right). \end{aligned} \tag{47}$$

It easily follows from (47) that

$$\lim_{T \to \infty} \tilde{W}_T(0,x) = \frac{3}{4} g. \tag{48}$$

□

Discounted expected cost and ergodic cost control problems analogous to (6,7) and (6,8) can be solved in other noncompact symmetric spaces. For other real hyperbolic spaces, $\mathbb{H}^n(\mathbb{R})$, $n \geq 2$, the approach is identical because only the coefficient of $\coth r$ in the radial part of the Laplace-Beltrami operator changes. The results in Theorems 1 and 2 can be extended to all rank one symmetric spaces using the results in [4]. Other cost functions and higher rank symmetric spaces can be employed using the methods in [6].

References

[1] A. Bensoussan and J. Frehse, *On Bellman equations of ergodic control in* $\mathbb{R}^N$, J. Reine Angew. Math. **429** (1992), 125–160.

[2] M. H. A. Davis and A. R. Norman, *Portfolio selection with transaction costs*, Math. Operators Res. **15** (1990), 676–713.

[3] T. E. Duncan, *A solvable stochastic control problem in hyperbolic three space*, System Control Lett. **8** (1987), 435–439.

[4] T. E. Duncan, *Some solvable stochastic control problems in noncompact symmetric spaces of rank one*, Stochastics Stochastic Repts. **35** (1991), 129–142.

[5] T. E. Duncan, *A solvable stochastic control problem in real hyperbolic three space. II*, Ulam Quart. **1** (1992), 13–18.

[6] T. E. Duncan and H. Upmeier, *Stochastic control problems and spherical functions on symmetric spaces*, Trans. Amer. Math. Soc. **347** (1995), 1083–1130.

[7] W. H. Fleming and H. M. Soner, *Controlled Markov Processes and Viscosity Solutions*, Springer-Verlag, 1993.

[8] S. Helgason, *Groups and Geometric Analysis*, Academic Press, 1984.

[9] N. Ikeda and S. Watanabe, *Differential Equations and Differential Processes*, North Holland, 1989.

[10] H. J. Kushner and P. G. Dupuis, *Numerical Methods for Stochastic Control Problems in Continuous Time*, Springer-Verlag, 1992.

[11] R. C. Merton, *Optimal consumption rules in a continuous-time model*, J. Economic Theory **3** (1971), 373–413.

[12] M. Taksar, M. J. Klass, and D. Assaf, *A diffusion model for optimal portfolio selection in the presence of brokerage fees*, Math. Operations Res. **13** (1988), 277–294.

Department of Mathematics, University of Kansas, Lawrence, KS 66045
E-mail address: duncan@math.ukans.edu

Proceedings of Symposia in Pure Mathematics
Volume **64**, 1999

Geometric Control and Lie Semigroup Theory

Jimmie D. Lawson

ABSTRACT. In this paper we survey aspects of geometric control theory on noncompact Lie groups and its connections with Lie semigroup theory.

1. Introduction

An interesting class of geometric control problems is comprised of those problems that can be interpreted as problems on Lie groups. There are significant advantages to interpreting problems in this framework whenever possible; these advantages include the availability of a rich arsenal of highly developed theoretical tools from Lie group and Lie algebra theory. In addition, Lie groups typically appear as matrix groups and one has available the concrete computational methods and tools of linear algebra and matrix theory. We refer to [**15**], where one can find a variety of control problems that can be interpreted in a Lie group context.

The central aim of this paper is to sketch the contours of the recently emerging Lie theory of semigroups and its connections with geometric control theory. On the one hand the methods of geometric control theory have proved quite useful in the development of the theory. Conversely the types of semigroups studied have a number of close connections with geometric control problems that arise on noncompact Lie groups. The theory also provides a rich source of examples and some theoretical tools for various aspects of geometric control.

2. Matrix Lie Groups

A *matrix Lie group* is a closed subgroup G of $\mathrm{GL}(n,\mathbb{F})$ where $\mathbb{F}=\mathbb{R}$ or $\mathbb{F}=\mathbb{C}$. It is then known to be an embedded analytic submanifold of $\mathrm{GL}(n,\mathbb{F})$, where the latter has the manifold structure inherited from being an open subset of $\mathbb{F}^{n^2}$. Furthermore the multiplication function is polynomial and inversion can be defined via rational functions, so multiplication and inversion are analytic mappings.

As usual, we identify the tangent bundle $T\big(\mathrm{GL}(n,\mathbb{F})\big)$ with $\mathrm{GL}(n,\mathbb{F})\times \mathrm{g}\ell(n,\mathbb{F})$, where $\mathrm{g}\ell(n,\mathbb{F})$ denotes all $n\times n$ matrices over $\mathbb{F}$. The tangent bundle for G is then an appropriate subbundle. Let e denote the identity matrix. Then the tangent space at e is given by

$$T_eG=\{\dot\alpha(0)\colon \alpha\colon(-\epsilon,\epsilon)\to G \text{ is a smooth curve with } \alpha(0)=e\}.$$

1991 *Mathematics Subject Classification.* 93B27, 22A15.

The author gratefully acknowledges the support of the National Science Foundation Grant DMS9623724.

It turns out that the tangent space is closed with respect to the commutator product of matrices,

$$A, B \in T_eG \Rightarrow [A, B] := AB - BA \in T_eG,$$

and hence T_eG is a Lie algebra with respect to the commutator product. We typically denote T_eG endowed with this structure of a Lie algebra by $\mathfrak{g}$ and call it the *Lie algebra* of the Lie group G. It follows directly from the facts that left and right translations ($x \mapsto gx$ and $x \mapsto xg$) by $g \in G$ are linear on $\mathfrak{g}\ell(n, \mathbb{F})$ and their restrictions to G are diffeomorphisms that $T_gG = g\mathfrak{g} = \mathfrak{g}g$ for all $g \in G$.

Each $A \in \mathfrak{g}$ gives rise to an analytic right-invariant vector field $\vec{A}$ on G defined by left multiplication by A: $\vec{A}(g) = Ag$ for all $g \in G$. All right-invariant vector fields arise in this way, they are closed under the usual Lie bracket of vector fields ($[X, Y]f = Y(Xf) - X(Yf)$), and the mapping $A \mapsto \vec{A}$ from $\mathfrak{g}$ to the Lie algebra of right-invariant vector fields on G is an isomorphism of Lie algebras with inverse $X \mapsto X(e)$. Similarly $g \mapsto gA$ defines the Lie algebra of left-invariant vector fields on G and in this case the identification with $\mathfrak{g}$ is a Lie algebra anti-isomorphism.

The exponential function $\exp : \mathfrak{g}\ell(n, \mathbb{F}) \to \mathrm{GL}(n, \mathbb{F})$ is defined by

$$\exp(A) = I + A + \frac{1}{2!}A^2 + \frac{1}{3!}A^3 + \dots.$$

Its restriction always carries $\mathfrak{g}$ into G for any matrix Lie group G contained in $\mathrm{GL}(n, \mathbb{F})$. We list some of the basic properties of the exponential map $\exp : \mathfrak{g} \to G$ for an arbitrary matrix Lie group G.

1. A one-parameter group in G (a continuous homomorphism $\sigma : (\mathbb{R}, +) \to (G, \cdot)$) is uniquely determined by a member of $\mathfrak{g}$, i.e., the one-parameter groups all arise as differentiable mappings of the form

$$t \mapsto \exp(tA) : \mathbb{R} \to G$$

for some $A \in \mathfrak{g}$. Thus $\mathfrak{g}$ organizes the one-parameter groups into a Lie algebra.

2. The Lie algebra $\mathfrak{g}$ of G can be alternately characterized by

$$\mathfrak{g} = \{A \in \mathfrak{g}\ell(n, \mathbb{F}) : \ \forall t \in \mathbb{R}, \ \exp(tA) \in G\}.$$

3. The flow corresponding to a right invariant vector field $\vec{A}$ is given by $\Phi(t, g) = \exp(tA)g$.
4. The exponential map has derivative at 0 given by $d_0 \exp = \mathrm{id}_{\mathfrak{g}}$ (where $T\mathfrak{g} = \mathfrak{g} \times \mathfrak{g}$), and hence by the Inverse Function Theorem exp is locally a diffeomorphism at 0 with local inverse "log."
5. For $A, B \in \mathfrak{g}$, $A + B = \lim_{n\to\infty} n \log\left(\exp(\frac{1}{n}A) \cdot \exp(\frac{1}{n}B)\right)$;
$[A, B] = \lim_{n\to\infty} n^2 \log\left(\exp(\frac{1}{n}A)\exp(\frac{1}{n}B)\exp(-\frac{1}{n}A)\exp(-\frac{1}{n}B)\right)$.

3. Semigroups via Control Systems on Lie Groups

In the study of linear time-varying systems $\dot{x}(t) = A(t)x(t)$ on $\mathbb{R}^n$, an important role is played by the *transition matrix* function Φ, the solution of the corresponding matrix differential equation. We can interpret Φ as a solution of a differential equation on a matrix Lie group, where the vector fields involved are right invariant. Many of the control-theoretic problems that arise on Lie groups (or can be reinterpreted as problems on Lie groups) also involve vector fields that are right invariant. We seek a framework for the study of geometric control problems on Lie groups that is general enough to encompass the variety of problems involving right invariant vector fields that typically arise. This leads to the consideration of what we call the *basic group control equation.*

Let G denote a Lie group with Lie algebra $\mathfrak{g}$; one may think matrix Lie groups although the results of this section are valid for more general Lie groups. We henceforth freely identify $\mathfrak{g}$ both with the set of right invariant vector fields on G and with T_eG, the tangent space at e. For $X \in \mathfrak{g}$ and $g \in G$, let $X(g) \in T_gG$ denote the value of the vector field X at g.

Let Ω be a subset of $\mathfrak{g}$. We consider the control system given by the *Basic Group Control Equation* (BGCS) on G:

$$\text{(BGCE)} \qquad \dot{g}(t) = \vec{U}(t)\,(g(t)) = U(t)g(t),$$

where $U(\cdot)$ is called a *steering* or *control* function and belongs to the class $\mathcal{U}(\Omega)$ of measurable functions from $\mathbb{R}^+ = [0, \infty)$ into Ω which are locally bounded, that is, bounded on every finite subinterval. (A measurable set in $\mathfrak{g}$ is one whose image is measurable in the euclidean space which is vector space isomorphic to $\mathfrak{g}$; equivalently the set is measurable with respect to Haar measure on $(\mathfrak{g}, +)$.) A solution of (BGCE), called an *Ω-trajectory* or simply a *trajectory*, is an absolutely continuous function $x(\cdot)$ from $\mathbb{R}^+$ into G such that the equation (BGCE) holds a.e. Control systems such as the one just described are called *right invariant*, since right translates of solutions of (BGCE) are again solutions.

EXAMPLE 3.1. (Bilinear Systems) Let $\Delta \subseteq \mathbb{R}^m$ and let $A_0, A_1, \dots, A_m \in g\ell(n, \mathbb{R})$, the Lie algebra of $n \times n$-matrices with the commutator Lie product. The control system on the corresponding Lie group $\mathrm{GL}(n, \mathbb{R})$ given by

$$\text{(BIL)} \qquad \dot{g}(t) = \big(A_0 + \sum_{i=1}^{m} u_i(t)A_i\big)g(t), \text{ where } u(t) := \big(u_1(t), \dots, u_m(t)\big) \in \Delta$$

is called a *bilinear* control system with control set Δ. Note that the steering function $u(\cdot)$ can alternately be interpreted as a steering function into

$$\Omega := \{A_0 + \sum_{i=1}^{m} u_i A_i \colon (u_1, \dots, u_m) \in \Delta\} \subseteq \mathrm{gl}(n, \mathbb{R}),$$

and can thus be interpreted as a BGCE system. Conversely the system (BGCE) may be interpreted as a special case of a bilinear system by letting $A_0 - 0$, picking a basis $A_1, \dots, A_m$ of $\mathfrak{g}$, and considering all steering functions $U = (u_1, \dots, u_m)$ such that each component is locally bounded and measurable and $\sum_{i=1}^m u_i(t)A_i \in \Omega$ for each t. Note in this case that $U \colon [0, \infty) \to \Omega$ is measurable and locally bounded if and only if each of its coordinate functions $u_i(\cdot)$ with respect to A_i is locally bounded and measurable.

REMARK 3.2. The preceding observations concerning interpreting a BGCE system as a bilinear system make it possible to apply the theory of bilinear systems, or more generally of systems affine in control, to BGCE systems.

We give specific references for the following results, where they are typically found stated in the context of bilinear systems. A good general reference is [**15**], particularly Chapter 3.

For any right invariant vector field X, the solution of (BGCE) with initial value $x(t_0) = g$ and with constant control function $U(t) \equiv X$ is given by $x(t) = \exp\big((t - t_0)X\big)g$; by induction one has the following result.

PROPOSITION 3.3. *Let* $0 = t_0 < t_1 < \cdots < t_n = T$ *and suppose that* $U : [0,T] \to \mathfrak{g}$ *is a piecewise constant function given by* $U(t) = X_i \in \mathfrak{g}$ *for* $t_{i-i} < t < t_i$. *Then the solution* $x(\cdot)$ *to (BGCE) with initial value* $x(0) = g \in G$ *is continuous and piecewise differentiable and is given by*

$$x(t) = \exp\big((t - t_{i-1})X_i\big) \exp\big((t_{i-i} - t_{i-2})X_{i-1}\big) \cdots \exp(t_1 X_1) g,$$

where $t_{i-i} \le t < t_i$, $i = 1, \ldots, n$.

LEMMA 3.4. *Let* $U : \mathbb{R}^+ \to \Omega \subseteq \mathfrak{g}$ *be locally bounded and measurable. Then there exists a unique solution* $x(\cdot)$ *on* G *of the initial value problem*

$$\dot{x}(t) = U(t)\big(x(t)\big), \quad x(0) = g.$$

This solution is globally defined in the sense that $x(t)$ *is defined for all* $t \ge 0$.

PROOF. By passing to a chart, one obtains a system of ordinary differential equations satisfying the Carathéodory conditions; hence there exist unique local solutions. The homogeneity of G is used for extending local solutions to global solutions. See Lemma 2.1 of [**18**] for details. □

If U is a locally bounded measurable steering function, we denote the solution $x(\cdot)$ of (BGCE) which satisfies $x(0) = g$ by $\pi(g, \cdot, U)$, i.e., $\pi(g, 0, U) = g$ and $\pi(g, t, U) = x(t)$ for all $t \ge 0$. If $h = \pi(g, t, U)$ for some $t \ge 0$, we say that U *steers* g *to* h *in* t *units of time.* If there exists $U \in \mathcal{U}(\Omega)$ such that $h = \pi(g, t, U)$, then we say that h *is attainable from* g *at time* t *for the system* Ω. The set of all such elements attainable from g at time t is denoted $A(g, t, \Omega)$. We also employ the notation

$$\mathbf{A}(g, \le T, \Omega) = \bigcup_{0 \le t \le T} A(g, t, \Omega),$$

$$\mathbf{A}(g, \Omega) = \bigcup_{0 \le t < \infty} A(g, t, \Omega).$$

The set $\mathbf{A}(g, \Omega)$ is called the *attainability set* or *reachable set* from g. We use the subscript "pc" to denote the appropriately modified versions of the preceding notions for the case that only piecewise constant controls into Ω are used. For example, the set of elements attainable from g at time t using only piecewise constant controls into Ω is denoted $A_{pc}(g, t, \Omega)$.

Note that the equality $\mathbf{A}(g, \Omega) = \mathbf{A}(e, \Omega)g$ and analogs for other attainability sets follow from the right invariance of the control system. Thus attainability sets at the identity determine the attainability sets at other points, and we can restrict our attention primarily to the former. The following proposition, which can be deduced from Proposition 3.3, provides the first link between infinitesimally generated semigroups and attainability sets.

PROPOSITION 3.5. *Let* $\Omega \subseteq \mathfrak{g}$. *The set* $A_{pc}(e, T, \Omega)$ *consists of all* $g \in G$ *that can be written as a product*

$$g = \exp(\tau_1 X_1) \exp(\tau_2 X_2) \cdots \exp(\tau_n X_n),$$

where each $\tau_i \ge 0$, $\sum_{i=1}^n \tau_i = T$ *and* $X_1, \ldots, X_n \in \Omega$. *The set* $\mathbf{A}_{pc}(e, \Omega)$ *is equal to the subsemigroup of* G *generated by the set* $\exp(\mathbb{R}^+\Omega)$.

From the right invariance of solutions of (BGCE), one verifies directly for any $T_1, T_2 \geq 0$ that

$$\mathbf{A}(e, \leq T_1, \Omega)\mathbf{A}(e, \leq T_2, \Omega) \subseteq \mathbf{A}(e, \leq T_1 + T_2, \Omega).$$

This containment yields the following proposition.

PROPOSITION 3.6. *For $\Omega \subseteq \mathfrak{g}$, the set $\mathbf{A}(e, \Omega)$ is a subsemigroup of G.*

Proposition 3.6 establishes a crucial link between geometric control theory and Lie semigroup theory. CHARLES LOEWNER recognized as early as the nineteen-fifties that the attainable set of the basic group control equation with initial value the identity (and apparently with class of controls the piecewise continuous functions) formed a semigroup (although, of course, the language of control theory was not avaiable to him at that time). He refered to such subsemigroups of Lie groups as "infinitesimally generated semigroups" and launched their study in a series of papers [**26**], [**27**], [**28**]. That the attainability set from the identity forms a semigroup was also noted by R. BROCKETT and others in the early development of geometric control theory (see [**2**] and [**11**], for example).

We fix some inner product $\langle \cdot, \cdot \rangle$ on $\mathfrak{g}$ making it a Hilbert space (in the matrix Lie group setting, one can choose the inner product to be the trace of AB^* for $A, B \in \mathfrak{g}$, where B^* denotes the conjugate transpose). We fix some interval $[0, T]$ and consider the Hilbert space of measurable functions $f\colon [0, T] \to \mathfrak{g}$ which are square integrable, i.e., $\int_0^T \langle f(t), f(t) \rangle \, dt$ exists and is finite. (As customary, functions that are equal a.e. are identified.) The topology of weak convergence on a set of steering functions is then by definition the topology of weak convergence on this Hilbert space restricted to the given set.

Let $\mathcal{U}(T)$ denote the set of bounded measurable functions (hence square integrable) from $[0, T]$ into $\mathfrak{g}$ endowed with the weak topology. The following lemma is Lemma 4.3 of [**18**], although the proof is originally due to SUSSMANN [**36**].

LEMMA 3.7. *The mapping $(U, t) \mapsto \pi(g, t, U)\colon \mathcal{U}(T) \times [0, T] \to G$ is continuous for each $g \in G$ and each $T > 0$, where $\mathcal{U}(T)$ is given the topology of weak convergence.*

The preceding lemma has a number of applications to control. We list some of these. For proofs, see [**18**], [**36**], [**3**].

COROLLARY 3.8. *Let $\Omega \subseteq \mathfrak{g}$ be compact and convex. Then $A(e, T, \Omega)$ and $\mathbf{A}(e, \leq T, \Omega)$ are compact.*

COROLLARY 3.9. *Suppose that Ω is compact and convex. If M is a closed set and $M \cap \mathbf{A}(g, \Omega) \neq \emptyset$, then there exists a time optimal control function steering g into M.*

LEMMA 3.10. *Let $\emptyset \neq \Omega \subseteq \mathfrak{g}$ and let K be its closed convex hull. Then for $T > 0$, the set of piecewise constant functions from $[0, T]$ into Ω is weakly dense in $\mathcal{U}(T, K)$.*

This lemma leads to the "chattering principle" in the Lie group context.

PROPOSITION 3.11. *Let $\emptyset \neq \Omega \subseteq \mathfrak{g}$ and let K be its closed convex hull. Then $A_{pc}(e, T, \Omega)$ is dense in $A(e, T, K)$. Hence the semigroup $\mathbf{A}_{pc}(e, \Omega)$ is dense in the semigroup $\mathbf{A}(e, K)$.*

It is a standard result of geometric control theory that certain sets of accessibility have non-empty interior if and only if Ω satisfies the "full rank" condition, i.e., it generates (as a Lie subalgebra) $\mathfrak{g}$ (see, for example, Chapter 3 of [**15**]).

THEOREM 3.12. *For $\Omega \subseteq \mathfrak{g}$, the Lie algebra of G, the set $\mathbf{A}(e, \leq T, \Omega)$ has dense interior for each $T > 0$ if and only if Ω generates $\mathfrak{g}$.*

4. Lie's Fundamental Theorems Revisited

After laying the foundations of the theory of Lie groups, LIE singled out certain foundational results for the theory and grouped them into three "fundamental theorems." A central ingredient of these theorems was the correpondence between finite-dimensional Lie groups and their Lie algebras. Every Lie group uniquely determined a Lie algebra and conversely every Lie algebra determined a local Lie group. We turn to an analogous program for semigroups.

DEFINITION 4.1. A subset S of a Lie group G is called a *semigroup* resp. *local semigroup* if (i) $e \in S$ and (ii) $S \cdot S \subseteq S$ resp. $(S \cap U) \cdot (S \cap U) \subseteq S$ for some U open containing e.

We have identified the set of the tangent vectors at the identity with the Lie algebra of G. We follow a similar procedure for (local) semigroups.

DEFINITION 4.2. Let G be a Lie group, $\exp : \mathfrak{g} \to G$. Let S be a subsemigroup of G. The *subtangent set* of S is

$$\begin{aligned} \mathbf{L}(S): &= \left\{\dot{\alpha}(0) : \alpha : \mathbb{R} \to G,\, \alpha(0) = e,\, \alpha(\mathbb{R}^+) \subseteq S\right\} \\ &= \{X \in \mathfrak{g} : \forall t \geq 0,\, \exp(tX) \in S\}. \end{aligned}$$

The definition for a local semigroup S is similar except that we only require membership in S for $t \in [0, \epsilon)$ for some $\epsilon > 0$.

It is not obvious that the two sets on the right are equal, but needs to be shown. The proof uses the fact that S is a (local) semigroup and the fifth property of the exponential function at the end of Section 2. The second version of the definition is analogous to the second property of the exponential function.

For any $A \in \mathfrak{g}$, the linear operator $\operatorname{ad} A$ defined by $\operatorname{ad} A(B) = [A, B]$ is a derivation of the Lie algebra $\mathfrak{g}$, called an *inner derivation.* Applying the exponential function to this linear operator, one obtains the Lie algebra automorphism $e^{\operatorname{ad} A}$ of $\mathfrak{g}$, called an *inner automorphism* of $\mathfrak{g}$. For the case of matrix Lie groups, it agrees with conjugation by $\exp A$, i.e., for all $X \in \mathfrak{g}$, $e^{\operatorname{ad} A}(X) = (\exp A) X (\exp A)^{-1}$.

DEFINITION 4.3. Let $\mathfrak{g}$ be a (finite dimensional) real or complex Lie algebra. A subset $\mathbf{W}$ of $\mathfrak{g}$ is called a *closed wedge* if $\mathbf{W}$ is closed under vector addition, scalar multiplication by non-negative scalars, and is topological closed. We use the terminology "closed wedge" instead of the more customary "closed convex cone" to emphasize the role played by the largest subspace $\mathbf{W} \cap -\mathbf{W}$ contained in $\mathbf{W}$, called the *edge* of $\mathbf{W}$. If $\mathbf{W}$ is a closed wedge in $\mathfrak{g}$, and the edge $\mathbf{W} \cap -\mathbf{W}$ is a Lie subalgebra of $\mathfrak{g}$, then $\mathbf{W}$ is called a *Lie wedge* or a *Lie-Loewner wedge* if $e^{\operatorname{ad} X}(\mathbf{W}) \subseteq \mathbf{W}$ for each $X \in \mathbf{W} \cap -\mathbf{W}$, the edge of $\mathbf{W}$.

The next theorems are analogs for semigroups of the heart of LIE'S Fundamental Theorems.

THEOREM 4.4. *If $S \subseteq G$ is a closed subsemigroup containing e, then the subtangent set $\mathbf{L}(S)$ is a Lie-Loewner wedge.*

One would expect that $\mathbf{L}(S)$ should approximate S in some appropriate way near e. We give one result along this line (see Chapter IV of [**9**]). Given closed wedges $\mathbf{W}$ and $\mathbf{W}'$, we say that $\mathbf{W}'$ *surrounds* $\mathbf{W}$ if $\mathbf{W} \subseteq \mathbf{W}'$ and $\mathbf{W} \setminus (\mathbf{W} \cap -\mathbf{W})$ is contained in the interior of $\mathbf{W}'$.

PROPOSITION 4.5. *Let S be a (local) closed semigroup and let $\mathbf{W}$ be a closed wedge which surrounds $\mathbf{L}(S)$. Then for all sufficiently small open sets U containing e, $S \cap U \subseteq \exp(\mathbf{W} \cap (\log U))$.*

It was LIE'S fundamental insight to introduce Lie algebras as the "infinitesimal generators" of Lie groups and to develop the correspondence between subgroups of the Lie group and subalgebras of the corresponding Lie algebra. LOEWNER had the vision and did the pioneering work of extending this program to semigroups in a series of papers beginning in the 1950's (see [**26**], [**27**], and [**28**]) and introduced an early version of the Lie-Loewner wedge as the appropriate (sub)tangent object in the Lie algebra. The reverse problem of constructing a semigroup structure that corresponds with a given Lie-Loewner wedge is a much more difficult problem and a general solution was not given until that by OL'SHANSKIĬ [**31**] and independently by HILGERT and HOFMANN [**8**]. A detailed treatment may be found in Chapter IV of [**9**].

THEOREM 4.6. *Let $\mathbf{W}$ be a Lie-Loewner wedge in a finite dimensional Lie algebra $\mathfrak{g}$ with Lie group G. Then there exists a local semigroup $S \subseteq G$ with $\mathbf{L}(S) = \mathbf{W}$.*

Just as in general one obtains only local solutions to differential equations, so one obtains here only a "local semigroup" solution for a given Lie-Loewner wedge. Indeed in LIE'S original version of the Fundamental Theorems, he obtained only a local solution. It was only through the later work of É. CARTAN that the existence of a corresponding global Lie group was established. In what may at first glance appear surprising this result does not extend to semigroups. At least there exist Lie-Loewner wedges inside Lie algebras for which there are no (global) semigroups with the given wedge as subtangent wedge inside any Lie group with the given Lie algebra as its Lie algebra. Whether there exists in the abstract some semigroup which would be a good candidate to wear the name "Lie semigroup" and have the wedge in question as its subtangent wedge remains an unanswered question.

The problem of determing whether a given Lie-Loewner wedge in a Lie algebra is the subtangent wedge of some semigroup in the corresponding Lie group has received considerable attention and some theory does now exist for attempting to answer this question. In general the question of whether a Lie-Loewner wedge is *global* in a Lie group, i.e., is the subtangent wedge of some semigroup in a corresponding Lie group, is quite difficult because it involves computing, or at least approximating, the corresponding reachable semigroup (the reachable set from the identity) in the Lie group. In some cases techniques of control theory are useful in attacking this problem. We have seen previously that members of the attainable set are reachable via time-optimal controls into a compact convex set generating the Lie-Loewner wedge. Thus the Pontryagin Maximality Principle can sometimes be successfully exploited to obtain a description of the reachable semigroup. We refer the reader to [**29**] for further details and some examples of this approach.

DEFINITION 4.7. A subsemigroup S of a Lie group G is *infinitesimally generated* if the one-parameter semigroups contained in S generate a dense subsemigroup of S. If S is a closed subsemigroup, this is equivalent to saying that

$$S = \overline{\langle \exp \mathbf{L}(S) \rangle_{\mathrm{sgp}}},$$

the closure of the subsemigroup generated by the exponential image of the subtangent wedge of S.

EXAMPLE 4.8. (Totally Positive Matrices) Let S denote the semigroup of invertible $n \times n$ totally positive matrices (matrices for which all subdeterminants are non-negative). Then $\mathbf{L}(S)$ consists of all non-negative linear combinations of matrices with one non-zero entry of 1 somewhere on the main diagonal and Jacobi extreme intensity matrices (which have two non-zero entries, an entry of 1 somewhere on the diagonal and an entry of -1 immediately adjacent to the 1 in the same row). LOEWNER **[25]** has observed that any element of S is a finite product of exponentials of elements in $\mathbf{L}(S)$, and hence that S is infinitesimally generated.

EXAMPLE 4.9. (Stochastic and Embeddable Matrices) The set S of invertible $n \times n$ stochastic matrices (matrices with all entries non-negative and all row sums equal to 1) forms a closed subsemigroup of $\mathrm{GL}(n, \mathbb{R})$. Then the subtangent wedge $\mathbf{W} := \mathbf{L}(S)$ is the set of $n \times n$ intensity matrices (all row sums are 0 and every non-diagonal entry is non-negative). The "embedding problem" is the problem of characterizing those matrices which are members of the infinitesimally generated attainability semigroup $\mathbf{A}\big(I, \mathbf{W}\big) = \overline{\langle \exp \mathbf{W} \rangle_{\mathrm{sgp}}}$, in this case a proper subsemigroup of the original S; such matrices are called *embeddable.* (The original formulation of this problem was to characterize those matrices appearing in a Markov chain consisiting of a two-parameter continuous family of stochastic matrices satisfying the Chapman-Kolmogorov equations; Goodman **[7]** showed the equivalence to the formulation given here.) This problem remains open at this time, but has been solved in the 3×3 case by Johansen and Ramsey **[14]**.

The *Lie theory of semigroups* denotes primarily the theory of closed infinitesimally generated subsemigroups of Lie groups. These are the ones for which the data of the Lie algebra are sufficient for the reconstruction of the semigroup. Their study combines the tools of classical Lie theory with appropriate new techniques imported from other disciplines such as convex analysis, geometric control theory, and the theory of ordered topological spaces. Major treatments of the theory can be found in the monographs **[9]**, **[10]**, and **[13]**.

5. Attainable Sets and Local Semigroups

Let G be a Lie group with Lie algebra $\mathfrak{g}$, and let $\|\cdot\|$ be a norm on $\mathfrak{g}$. We equip each tangent space T_gG with a norm so that the derivative of right translation by g is an isometric isomorphism from $\mathfrak{g}$ to T_gG. Let $\Omega \subseteq \mathfrak{g}$. For any bounded function $U\colon [0, T] \to \Omega$, we define the length of the trajectory $x(\cdot)$ satisfying (BGCE) by

$$L\big(x(\cdot)\big) := \int_0^T \|\dot{x}(t)\|\, dt = \int_0^T \|U(t)\|\, dt.$$

We can think of arc length as a cost function on controls, and consider corresponding reachable sets:

$$\mathbf{R}(e, \leq L, \Omega) := \{g \in G \quad : \quad \exists \text{ an } \Omega\text{-trajectory } x(\cdot)\colon [0,T] \to G \text{ of length } \leq L \text{ such that } x(0) = e,\ x(T) = g\}.$$

Assume now that $\Omega = \mathbb{R}^+\Omega$. If we reparameterize a trajectory $x(\cdot)$ by arc length, then we obtain a trajectory of the same length in which the tangent vector is of unit length a.e., but still in Ω since $\mathbb{R}^+\Omega = \Omega$. For the arc length parametrization the length is just the length L of the interval $[0, L]$ over which the trajectory is defined. In this case L agrees with the time to travel the trajectory. Thus if we set $\Omega_1 := \{X \in \Omega\colon \|X\| = 1\}$, we have that $A(e, T, \Omega_1)$ is equal to the set of all endpoints of trajectories of length T which begin at e.

Again employing arc length reparameterizations, one can show directly that $A(e, T, \Omega_1 \cup \{0\}) = \mathbf{A}(e, \leq T, \Omega_1)$. By Proposition 3.11 the closure of $A(e, T, \Omega_1)$ is equal to $A(e, T, K)$, where K is the closed convex hull of $\Omega_1 \cup \{0\}$. Note that K is also the closed convex hull of the intersection of Ω and the unit ball. We summarize these results.

LEMMA 5.1. *Let $\|\cdot\|$ be a norm on $\mathfrak{g}$, let $\Omega = \mathbb{R}^+\Omega \subseteq \mathfrak{g}$, and let $g \in G$. The following are equivalent:*

1. *g lies on an Ω-trajectory of length T (resp. of length $\leq T$);*
2. *g lies on an Ω_1-trajectory of length T (resp. of length $\leq T$), where Ω_1 consists of all members of Ω of length 1;*
3. *$g \in A(e, T, \Omega_1)$ (resp. $g \in \mathbf{A}(e, \leq T, \Omega_1) = A(e, T, \Omega_1 \cup \{0\})$).*

If K denotes the closed convex hull of $\Omega_1 \cup \{0\}$, then

$$\mathbf{R}(e, \leq T, \Omega)^- = A(e, T, \Omega_1 \cup \{0\})^- = A(e, T, K) = \mathbf{A}(e, \leq T, \Omega_1)^- = \mathbf{A}(e, \leq T, K)$$

where A^- denotes the closure of A.

Now let $\mathbf{W}$ be a Lie wedge in $\mathfrak{g}$, and let U be an open set containing e in G. We let $S(\mathbf{W}, U)$ be the smallest local subsemigroup which is closed in U and contains $(\exp \mathbf{W}) \cap U$. Then by LIE'S Fundamental Theorem for Semigroups, if U is chosen small enough, $S(\mathbf{W}, U)$ has subtangent wedge $\mathbf{L}\big(S(\mathbf{W}, U)\big) = \mathbf{W}$. Indeed as $S(\mathbf{W}, U)$ varies over small U, the sets all belong to the same germ in the sense that given any two, their intersections with some open set around e are equal. This theorem can be stated and proved in a geometric control theoretic setting.

THEOREM 5.2. *Let G be a Lie group with Lie algebra $\mathfrak{g}$ endowed with a norm $\|\cdot\|$. Let $\Omega = \mathbb{R}^+\Omega \subseteq \mathfrak{g}$ have closed convex hull $\mathbf{W}$ a Lie wedge. Let K be the intersection of $\mathbf{W}$ with the closed unit ball. Then there exists $T_0 > 0$ such that if $0 < T < T_0$, the sets*

$$\mathbf{R}(e, \leq T, \Omega)^- = A(e, T, \Omega_1 \cup \{0\})^- = A(e, T, K) = \mathbf{A}(e, \leq T, \Omega_1)^- = \mathbf{A}(e, \leq T, K)$$

are all closed local subsemigroups with tangent wedge $\mathbf{W}$ and all belong to the same germ as the sets $S(\mathbf{W}, U)$ for small U.

This theorem has been proved for $\mathbf{R}(e, \leq T, \Omega)^-$ for a specially chosen norm on $\mathfrak{g}$ (see Chapter IV of [**9**], where the proof has been translated back into the Lie algebra). However it is easy to see that if a different norm is chosen on $\mathfrak{g}$, then since the two norms are equivalent, the sets $\mathbf{R}(e, \leq T, \Omega)^-$ for the two norms and various T are filtered inside each other. Thus if for one of the norms the collection of sets

$\mathbf{R}(e, \leq T, \Omega)^-$, $T > 0$ has the property that the sets belong to a certain germ for small T, then that property will persist for the other collection.

The main idea of the proof of Theorem 5.2 that yields both the germ property and the local semigroup property is that members of the attainability set that are close to e can be reached by trajectories that are relatively small in length. These ideas have been further developed and amplified in [**12**].

The union $S = \bigcup_{T>0} \mathbf{R}(e, \leq T, \Omega)$ will always form a semigroup. For a Lie-Loewner wedge $\mathbf{W}$ and for large values of T the sets $\mathbf{R}(e, \leq T, \mathbf{W})$ may exhibit varying behavior.
(i) It may be the case that all $\mathbf{R}(e, \leq T, \mathbf{W})$ continue to belong to the same germ as the collection for small T; in this case the semigroup S will be a global semigroup with Lie-Loewner wedge $\mathbf{W}$. We say that $\mathbf{W}$ is *global* in G. It is not difficult to show that there exists a closed semigroup in G with Lie-Loewner wedge $\mathbf{W}$ if and only if this case obtains.
(ii) At the other extreme, it may be the case that S is all of G. This happens if and only if the system (BGCE) is a controllable system for the control set $\mathbf{W}$. In this case we say that $\mathbf{W}$ is *controllable.*
(iii) It may happen that neither of the two preceding cases occur. In this case S will have a strictly larger Lie-Loewner wedge that $\mathbf{W}$, but one that is not all of $\mathfrak{g}$.

6. Lie Saturates and Semigroups

In practice one finds that it is generally fairly straightfoward to compute the Lie-Loewner wedge for a given closed semigroup in a Lie group G (a semigroup analog of differentiation). However the reverse problem of computing the closed semigroup (which is equal to the closure of the attainability set from e) generated by a Lie wedge or a more general set Ω is generally a much more difficult problem. Even the somewhat less ambitious problems of trying to decide whether the set Ω defines a controllable system or whether a Lie wedge is global for a given Lie group G can be quite difficult. In this and the following sections we sketch some of the main approaches that have been developed to problems of this type.

DEFINITION 6.1. Let $\Omega \subset \mathfrak{g}$ be a non-empty subset of the Lie algebra $\mathfrak{g}$ of a connected Lie group G. Then there exists a largest subset $\mathcal{L}\mathrm{S}(\Omega) \subseteq \mathfrak{g}$, called the *Lie saturate* of Ω, such that the closures of the attainability sets for $\mathcal{L}\mathrm{S}(\Omega)$ and Ω are equal.

The introduction of the notion of Lie saturate and its application to questions of controllability originated in the work of V. JURDJEVIC and I. KUPKA [**16**],[**17**]. In the setting of Lie group theory, the notion is very closely linked to that of a Lie-Loewner wedge as will emerge in what follows.

Since $\mathbf{A}(g, \Omega) = \mathbf{A}(e, \Omega)g$, $\mathcal{L}\mathrm{S}(\Omega)$ has an alternate characterization as the largest subset of $\mathfrak{g}$ such that the closures of the attainability sets at e are equal. The next proposition collects together some basic connections between notions of geometric control and those of Lie semigroup theory.

PROPOSITION 6.2. *Let G be a connected Lie group with Lie algebra $\mathfrak{g}$. The following items are equivalent for a non-empty closed subset S of G:*

1. *S is an infinitesimally generated subsemigroup;*
2. *$S = \langle \exp \mathbf{L}(S) \rangle_{sgp}^-$, the closure of the subsemigroup generated by $\exp \mathbf{L}(S)$;*
3. *$S = \langle \exp \mathbb{R}^+ \Omega \rangle_{sgp}^-$ for some $\Omega \subseteq \mathfrak{g}$;*

4. *S is the closure of $\mathbf{A}(e,\Omega)$ for some $\Omega \subseteq \mathfrak{g}$;*
5. *S is the closure of $\mathbf{A}\big(e,\mathbf{L}(S)\big)$.*

For any S satisfying these conditions, $\mathbf{L}(S)$ is the largest Ω which satisfies condition (3) or (4). Hence for any $\Omega \subseteq \mathfrak{g}$, we have that its Lie saturate is the Lie-Loewner wedge of $\mathbf{A}(e,\Omega)^-$ and is the smallest Lie-Loewner wedge containing Ω that is global in G.

The setting for genuine Lie semigroup theory is of necessity within those Lie groups which are not compact. In the compact case the theory collapses, since compact subsemigroups of groups must be subgroups (see e.g., Chapter V, Section 0 of [**9**]).

PROPOSITION 6.3. *Let G be a Lie group. If S is a subsemigroup such that S^- is compact, then S^- is a compact group. If the interior of S meets a compact subgroup K, then the interior intersects K in a compact subgroup of K which is open in K.*

The preceding results provide an approach for attempting to find the Lie saturate of a set Ω and hence for attempting to determine whether the system (BGCE) determined by Ω on a connected Lie group G is controllable. *We take as a standing hypothesis that Ω generates $\mathfrak{g}$* (otherwise we would work in the analytic subgroup with Lie algebra generated by Ω). Then by Theorem 3.12 the semigroup $S := \mathbf{A}(e,\Omega)$ has dense interior. It is easy to see that if a subsemigroup S of G is dense in G and has non-empty interior, then $S = G$. It follows that either S^- is proper in G or $S = G$ and the system determined by Ω is controllable. We have thus shown the following.

LEMMA 6.4. *Assume Ω generates $\mathfrak{g}$, the Lie algebra of a connected Lie group G. The following are equivalent:*

1. *the system determined by Ω is controllable;*
2. *the semigroup $\mathbf{A}(e,\Omega)$ is all of G;*
3. *the semigroup $\mathbf{A}(e,\Omega)^-$ is all of G;*
4. *the Lie saturate of Ω is all of $\mathfrak{g}$.*

Thus to show that the system corresponding to Ω is controllable, one can attempt to show that its Lie saturate is everything. This determination typically involves the following steps.

Step 1. Check that Ω generates $\mathfrak{g}$ as a subalgebra. This is a necessary condition for controllability.

Step 2. Form the smallest closed wedge (i.e., closed convex cone) containing Ω.

Step 3. If $\mathbb{R}X$ is contained in the subset generated in Step 2, apply $e^{\operatorname{ad} X} = \operatorname{Ad}(\exp X)$ to the set generated in Step 2. Take the union of such sets. The fact that the Lie saturate is a Lie wedge and hence closed with respect to the operations of Steps 2 and 3 insures that these steps keep one within the Lie saturate of Ω.

Step 4. If the ray $\mathbb{R}^+X$ appears in Ω or one of the preceding constructions and if $\exp(\mathbb{R}^+X)$ has compact closure in G, then add all of $\mathbb{R}X$. Proposition 6.3 insures that $\exp(\mathbb{R}X)$ will be contained in $\mathbf{A}(e,\Omega)$.

Step 5. Keep repeating Steps 2-4 until it is not possible to add anything further. If at any stage all of $\mathfrak{g}$ is obtained, then the system determined by Ω is controllable.

The preceding algorithm need not necessarily produce the Lie saturate of Ω; it may stabilize before reaching that far. However it does always produce a Lie-Loewner wedge that sits between Ω and the Lie saturate of Ω.

These steps have been cleverly and skillfully applied to the study of the control equation

$$\dot{g}(t) = (A + u(t)B)g(t),$$

where A and B are constant right invariant vector fields on G, to deduce necessary conditions for controllability for the case that G is semisimple. The program was initiated by Jurdjevic and Kupka in [**16**], [**17**] and extended by work of El Assoudi, Gauthier, and Kupka [**6**]. Since this work has been adequately surveyed in [**15**] and [**19**], we do not trace these developments further.

7. Maximal Semigroups

The study of maximal subsemigroups of Lie groups has provided another fruitful approach to the study of controllability. A subsemigroup S of a group G is a *maximal* (*infinitesimally generated*) semigroup if S is not a subgroup and if T is a(n) (infinitesimally generated) subsemigroup which properly contains S, then $T = G$. The following observation is useful.

LEMMA 7.1. *Let G be a connected Lie group and let $S \neq G$ be a(n) (infinitesimally generated) subsemigroup with non-empty interior. Then S is contained in a closed (infinitesimally generated) maximal subsemigroup. In particular, if Ω generates $\mathfrak{g}$ but the system determined by Ω is not controllable, then $\mathbf{A}(e, \Omega)$ is contained in a closed (infinitesimally generated) maximal subsemigroup.*

PROOF. Let M be the union of a maximal chain of proper subsemigroups containing S. If M were to meet the inverse of the interior of S, then so would some T in the chain. But then e would be contained in the interior of T, and hence T would equal G since G is connected. Essentially the same proof works for infinitesimally generated semigroups. The second part follows from the first and 3.12. □

For those Lie groups G for which one can determine all maximal infinitesimally generated subsemigroups and their Lie-Loewner tangent wedges, one has available a test of whether a subset Ω of $\mathfrak{g}$ is controllable. This will hold, in particular, if and only if Ω is not contained in any of the tangent wedges of the maximal infinitesimally generated subsemigroups. In the solvable case, the latter have been determined [**20**].

THEOREM 7.2. *Let G be a connected solvable Lie group. If S is a maximal closed subsemigroup of G, then S is infinitesimally generated and $\mathbf{L}(S)$ is a half-space such that $\mathfrak{h} := \mathbf{L}(S) \cap -\mathbf{L}(S)$ is a subalgebra. Conversely, if G is simply connected, any such half-space determines an infinitesimally generated closed maximal subsemigroup.*

A simple illustration of the preceding theorem is the nonabelian Lie group of dimension 2. In this case every half-space of the Lie algebra determines a maximal subsemigroup of the group and all the closed maximal subsemigroups arise in this way. A survey of further developments in the study of controllability of systems on solvable groups may be found in [**33**] of this volume.

Via Proposition 6.3 one can extend the preceding theorem to Lie groups that are compact modulo their radical.

THEOREM 7.3. *Let G be a connected Lie group such that $G/\mathrm{Rad}\,G$ is compact, where $\mathrm{Rad}\,G$ denotes the maximal solvable normal subgroup of G. If S is a maximal closed subsemigroup of G, then S is infinitesimally generated and $\mathbf{L}(S)$ is a half-space such that $\mathfrak{h} := \mathbf{L}(S) \cap -\mathbf{L}(S)$ is a subalgebra. Conversely, if G is simply connected, any such half-space determines an infinitesimally generated closed maximal subsemigroup.*

This result significantly generalizes earlier work giving conditions for controllabilty involving actions of compact groups on vector spaces [**1**].

COROLLARY 7.4. *Let G be a semidirect product of a compact group K and a vector group V on which K acts linearly without fixed points (except 0). Then the system induced by a subset Ω in $\mathfrak{g}$ is controllable if and only if Ω generates $\mathfrak{g}$ (as a Lie algebra).*

Only scattered results exist concerning the identification of maximal subsemigroups of semisimple Lie groups and more generally of Lie groups with a non-trivial proper radical. These primarily concern a fundamental type of semigroup called an *Ol'shanskiĭ semigroup* (see [**22**] for a survey of this class of semigroups). These are subsemigroups of a semisimple Lie group G that admit a factorization of the form $H\exp(\mathbf{C})$, where H is a group fixed by some involution on G and $\mathbf{C}$ is an $\mathrm{Ad}(H)$-invariant hyperbolic cone in the (-1)-eigenspace of the induced involution at the Lie algebra level. If the cone $\mathbf{C}$ is a maximal $\mathrm{Ad}\,H$-invariant cone, then the corresponding Ol'shanskiĭ semigroup turns out to be a maximal semigroup (see [**23**] or [**10**] for a proof for complex Lie groups; the results have recently been extended to Ol'shanskiĭ semigroups in more general involutive Lie groups by A. Neumann and G. Ólafsson [**30**]). L. San Martin [**35**] also has obtained some results on maximality, particularly for rank 1 real Lie groups, using the machinery of control sets, which are defined and discussed briefly in the next section.

8. Control Sets and Semigroups

In recent years the study of the asymptotic behavior of dynamical systems has been extended to control trajectories in control systems. Many of the usual concepts in dynamical systems such as limit sets, sets of recurrence and chain-recurrence, topological transitivity and Morse decompositions can be carried over to the control setting (see [**4**] for a comprehensive reference). A main concept in the theory is that of a control set, a region of the state space where the control semigroup acts transitively. This notion of a control set can be abstracted to subsemigroups of Lie groups acting on homogeneous spaces.

DEFINITION 8.1. Let G be a Lie group, H a closed subgroup, and S a subsemigroup with non-empty interior. A *control set* for S in G/H is a subset D with non-empty interior such that $D \subseteq (Sx)^-$ for all $x \in D$ and D is maximal with respect to this property. The control set is called *invariant* if further $S \cdot D \subseteq D$.

Given any semisimple connected Lie group G, there is associated with G an important class of compact coset spaces that have been widely studied called its *flag manifolds.* For the case that $G = \mathrm{SL}(n, \mathbb{R})$, its maximal flag manifold consists of all chains of length n of subspaces of $\mathbb{R}^n$ and its minimal flag submanifolds consist of the various Grassmanians of k-dimensional subspaces of $\mathbb{R}^n$. In general the flag manifolds are the homogeneous spaces that arise by dividing out a parabolic subgroup. For the case that G has finite center, the understanding of the control sets of S on the flag manifolds of G yields much information about the structure of S itself. For example, in the context of controllable systems, L. San Martin has shown that the question of controllability reduces down to a flag manifold question [**34**].

THEOREM 8.2. *Let S be a subsemigroup with non-empty interior in a connected semisimple Lie group G with finite center. If B is a maximal flag manifold of G, then S has a unique invariant control set in G. This control set is proper in B if and only if $S \neq G$.*

For a more detailed survey of semigroups and their control sets we refer to [**35**]. For a survey of connections between semigroups and the dynamics and asymptotics of control systems, see [**5**] and the references cited there.

References

[1] B. Bonnard, V. Jurjevic, I. Kupka, G. Sallet, Transitivity of families of invariant vector fields on semi-direct products of Lie groups, *Trans. Am. Math. Soc.* **271** (1982), 525–535.

[2] R. Brockett, Lie algebras and Lie groups in control theory, in: Geometric Methods in Systems Theory, Reidel, 1973, 43–82.

[3] I. Chon and J. D. Lawson, Attainable sets and one-parameter semigroups of sets, *Glasgow Math. J.* **33** (1991), 187–201.

[4] F. Colonius and W. Kliemann, *Dynamics and Control*, Birkhäuser, Boston, in press.

[5] F. Colonius, W. Kliemann, and L. San Martin, Asymptotic problems–from control systems to semigroups, *Positivity in Lie Theory: Open Problems*, Editors: Hilgert, Lawson, Vinberg, de Gruyter, Berlin, in press.

[6] R. El Assoudi, J. P. Gauthier, I. Kupka, Controllability of right invariant systems on semi-simple Lie groups, *Ann. Inst. H. Poincaré*, in press.

[7] G. Goodman, An intrinsic time for non-stationary finite Markov chains, *Z. Wahrscheinlichkeitstheorie verw. Gebiete* **16** (1970), 165–180.

[8] J. Hilgert and K. H. Hofmann, On Sophus Lie's Fundamental Theorem, *Journal of Functional Analysis* **67** (1986), 209–216.

[9] J. Hilgert, K. H. Hofmann, and J. D. Lawson, *Lie Groups, Convex Cones, and Semigroups*, Oxford Press, Oxford, 1989.

[10] J. Hilgert and K.-H. Neeb, *Lie Semigroups and their Applications*, Springer Lecture Notes in Mathematics **1552**, Springer, 1993.

[11] R. Hirschorn, Topological semigroups, sets of generators and controllability, *Duke J. Math.* **40** (1973), 937–947.

[12] K. H. Hofmann and J. D. Lawson, Local semigroups in Lie group and locally reachable sets, *Rocky Mountain J. Math.* **20** (1990), 717–735.

[13] K. H. Hofmann, J. D. Lawson, and E. B. Vinberg, Editors, *Semigroups in Algebra, Geometry and Analysis*, de Gruyter, Berlin, 1995.

[14] S. Johansen and F. Ramsey, A bang-bang representation for 3×3 embeddable stochastic matrices, *Z. Wahrscheinlichkeitstheorie verw. Gebiete* **47** (1979), 107–118.

[15] V. Jurdjevic, *Geometric Control Theory*, Cambridge Press, Cambridge, 1997.

[16] V. Jurdjevic and I. Kupka, Control systems subordinated to a group action: accessibility, *J. Diff. Eq.* **39** (1981), 186–211.

[17] V. Jurdjevic and I. Kupka, Control systems on semi-simple Lie groups and their homogeneous spaces, *Ann. Inst. Fourier* **31** (1981), 151–179.

[18] V. Jurdjevic and H. Sussmann, Control systems on Lie groups, *J. Diff. Eq.* **12** (1972), 313–329.

[19] I. Kupka, Applications of semigroups to geometric control theory, in: *The Analytic and Topological Theory of Semigroups*, Editors: Hofmann, Lawson, Vinberg, de Gruyter, Berlin, 1990, 337–345.

[20] J. D. Lawson, Maximal subsemigroups of Lie groups that are total, *Proceedings of the Edinburgh Math. Soc.* **87** (1987), 479–501.

[21] —, Ordered manifolds, invariant cone fields, and semigroups, *Forum Math.* **1** (1989), 273–308.

[22] —, Semigroups of Ol'shanskii type, in: *Semigroups in Algebra, Geometry, and Analysis*, Editors: Hofmann, Lawson, Vinberg, de Gruyter, Berlin, 1995, 121–157.

[23] —, Maximality of Ol'shanskiĭ semigroups, *Journal of Lie Theory* **4** (1995), 17–35.

[24] —, Semigroups in Möbius and Lorentzian geometry, to appear *Geom. Dedicata.*

[25] C. Loewner, *On totally positive matrices*, Math. Zeit. **63** (1955), 338–340.

[26] —, On some transformation semigroups, *J. Rat. Mech. and Anal.* **5** (1956), 791–804.

[27] —, On some transformation semigroups invariant under Euclidean or non-Euclidean isometries, *J. Math. and Mech.* **8** (1959), 393-409.

[28] —, On semigroups in analysis and geometry, *Bulletin A.M.S.* **70** (1964), 1–15.

[29] D. Mittenhuber, Applications of the maximum principle to problems in Lie semigroups, in: *Semigroups in Algebra, Geometry, and Analysis*, Editors: Hofmann, Lawson, Vinberg, de Gruyter, Berlin, 1995, 313–338.

[30] A. Neumann and G. Ólafsson, Minimal and maximal semigroups related to causal symmetric spaces, to appear *Semigroup Forum.*

[31] G. I. Ol'shanskii, Convex cones in symmetric Lie algebras, Lie semigroups, and invariant causal structures on pseudo-Riemannian symmetric spaces, *Sov. Math. Dokl.* **26** (1982), 97–101.

[32] —, Invariant cones in Lie algebras, Lie semigroups, and the holomorphic discrete series, *Funct. Anal. and Appl.* **15** (1982), 275–285.

[33] Y. L. Sachkov, Controllability of right-invariant systems on solvable Lie groups: a survey and some new results, this Volume.

[34] L. San Martin, Invariant control sets on flag manifolds, *Math. Control Signals Systems* **6** (1993), 41–61.

[35] L. San Martin, Control sets and semigroups in semisimple Lie groups, in: *Semigroups in Algebra, Geometry, and Analysis*, Editors: Hofmann, Lawson, Vinberg, de Gruyter, Berlin, 1995, 275–291.

[36] H. Sussmann, The bang-bang problem for certain control systems in $GL(n,\mathbb{R})$, *SIAM J. Control* **10** (1972), 470–476.

Department of Mathematics, Louisiana State University, Baton Rouge, Louisiana 70803

E-mail address: `lawson@math.lsu.edu`

Proceedings of Symposia in Pure Mathematics
Volume **64**, 1999

Analysis of Abnormal Extremals in Optimal Control[1]

Urszula Ledzewicz[2] and Heinz Schättler[3]

ABSTRACT: Abnormal extremals are analyzed for an optimal control problem with fixed terminal point. Second-order approximations are used to determine tangent directions to the dynamics in the case when the linearization is not completely controllable. Based on these approximations necessary conditions for optimality of abnormal trajectories are given which are nontrivial in the sense that the multiplier associated with the objective is nonzero. The procedure is illustrated for the case of bang-bang extremals for a control-affine system.

1. Introduction

If one formulates the optimal control problem for ordinary differential equations with terminal constraints in the framework of optimization theory, then the dynamical equations together with their respective boundary conditions define an operator F which describes an equality constraint Q. Necessary conditions for optimality are typically derived using tangent approximations to this set Q. If the Frèchet derivative $F'(x_*)$ at a reference point x_* is onto (in other words the operator F is *regular* at x_* or the *surjectivity condition* holds), then the classical Lyusternik theorem describes the tangent space to Q at x_* as the kernel of $F'(x_*)$. If $F'(x_*)$ is not onto, but $\operatorname{Im} F'(x_*)$ is closed, then the associated necessary conditions for optimality can be satisfied trivially by choosing a multiplier which annihilates this image. This generates so-called *abnormal extremals.* While the nomenclature, which has its origin in Calculus of Variations, suggests that abnormal processes are an aberration, for optimal control problems this is not the case. The phenomenon is quite general, can be observed in a multitude of problems in optimal control and abnormal extremals cannot be excluded from optimality apriori. For instance, there exist optimal abnormal trajectories for the problem of stabilizing

---

[1] *1991 AMS Subject Classification.* Primary 49K15, 49K27, 49L05; Secondary 93B29, 93C10

[2] supported in part by NSF Grant DMS-9622967, SIUE Research Scholar Award and 1997 Summer Fellowship

[3] supported in part by NSF Grant DMS-9503356

the harmonic oscillator time-optimally in minimum time (see, for instance, [13]), a simple time-invariant linear system. Locally length minimizing abnormal geodesics in Sub-Riemannian manifolds have been studied from an optimal control perspective by Liu and Sussmann [22, 23] and by Agrachev and Sarychev [1, 2] and it has been found that abnormal minimizers are ubiquitous rather than exceptional.

Since the conventional necessary conditions for optimality provide conditions which degenerate for abnormal processes, it becomes necessary to develop theories which are tailored to abnormal processes. This has been a quite active area of research recently. A general method to overcome the difficulty of abnormality has been introduced by Milyutin in [24] known as "weakening equality constraints" and was investigated further by Dmitruk in [9]. This method applies to general extremum problems where the degenerate part of the operator F which describes equality constraints has a finite-dimensional image. In papers by Avakov [5, 6] and by Arutyunov [3, 4] both necessary and sufficient conditions for optimality of abnormal extremals are given based on quadratic approximations. Izmailov [12] also considers quadratic approximations, but for problems with inequality constraints. In our research we follow the approach initiated by Avakov [5, 6] which relies on an extension of the Lyusternik theorem. Rather than simply considering the Fréchet derivative $F'(x_*)$ if the operator F is not regular at a point x_*, the operator F is analyzed further using higher order derivatives. Replacing linear approximations by polynomial approximations of degree p we obtained a generalization of the classical Lyusternik Theorem which describes the structure of p-order tangent directions to the equality constraint $Q = \{x \in X : F(x) = 0\}$ in a Banach space also for non-regular operators [19]. This result has then been applied to derive generalized first- and second-order necessary conditions for optimality for general optimization problems in Banach spaces [15, 20]. Based on these results both a local and global version of an extended p-order Maximum Principle have been formulated [17, 18].

In this paper we analyze second-order approximations for abnormal extremals in the case when the terminal manifold is a single point. Specifically, we formulate necessary and sufficient conditions under which second-order tangent directions exist and show how these approximations can be used to analyze the optimality of abnormal extremals. In Section 2 we briefly recall some results on second-order tangent approximations for non-regular equality constraints. In Section 3 we describe and analyze abnormal extremals from the point of view of their relations with non-regular equality constraints and analyze the corresponding second-order approximations. These constructions are then formulated for the case of bang-bang extremals for a control-affine system in Section 4. In Section 5 we give necessary conditions for the optimality of abnormal extremals which are based on these quadratic approximations.

2. Second-order tangent vectors.

We briefly recall some earlier results [15] on high-order approximations to equality constraints given by a non-regular operator. In this paper we only pursue quadratic approximations. The notation is simpler for this case and the basic ideas of the approximation come out clearly without being obstructed by technicalities. However, these results were also obtained for arbitrary orders p of approximation based on a p-order generalization of the Lyusternik theorem given in [19]. In the case of quadratic approximations also results of Avakov, who originally obtained the relevant results for the case $p = 2$ [6], could be used instead.

Let X and Y be Banach spaces, $F : X \to Y$ an operator and $Q = \{x \in X : F(x) = F(x_*)\}$. Assuming that $F : X \to Y$ is three times continuously Fréchet differentiable in a neighborhood of x_* we have the Taylor-expansion of F along a curve $\gamma(\varepsilon) = x_* + \sum_{i=1}^{3} \varepsilon^i h_i$ as

$$\begin{aligned} F(x_* + \sum_{i=1}^{3} \varepsilon^i h_i) &= F(x_*) + \varepsilon F'(x_*)h_1 \\ &\quad + \varepsilon^2 \left(F'(x_*)h_2 + \frac{1}{2}F''(x_*)(h_1, h_1) \right) + \\ &\quad + \varepsilon^3 \left(F'(x_*)h_3 + F''(x_*)(h_1, h_2) + \frac{1}{6}F'''(x_*)(h_1, h_1, h_1) \right) \\ &\quad + \widetilde{r}(\varepsilon) \end{aligned} \tag{1}$$

where $\widetilde{r}(\varepsilon)$ denotes a function of order $o(\varepsilon^3)$ as $\varepsilon \to 0$. If F is regular at x_*, then the classical Lyusternik theorem describes the tangent space to Q at x_*, $T_{x_*}Q$, as $\ker F'(x_*)$. Intuitively, it is clear why this is correct: if more generally $x_* + \sum_{i=1}^{p} \varepsilon^i h_i$ is an approximating curve, then the coefficient arising at ε^p will be of the form $F'(x_*)h_p$ plus terms involving higher order derivatives of F at x_* acting on the vectors $h_1, \dots, h_{p-1}$. If $F'(x_*)$ is onto, then no matter what these terms are, they can be cancelled using h_p. If, however, $F'(x_*)$ is no longer surjective, not all of $\ker F'(x_*)$ may lie in $T_{x_*}Q$. For instance, if $F'(x_*)h_1 = 0$, but $F''(x_*)(h_1, h_1) \notin \operatorname{Im} F'(x_*)$, then no choice of h_2 can make the quadratic term vanish and thus $h_1 \notin T_{x_*}Q$. Even if $F''(x_*)(h_1, h_1) \in \operatorname{Im} F'(x_*)$, $h_1 \in T_{x_*}Q$ is not guaranteed, but whether this holds depends on higher order derivatives of F in x_*. For, in this case, we can certainly make the quadratic term vanish, but if $F'''(x_*)(h_1, h_1, h_1)$ cannot be compensated by choosing simultaneously h_2 and h_3, then the cubic term cannot be made zero. If, however, the first and second derivatives combine to a surjective operator, then very much like in the classical Lyusternik theorem higher order terms can always be cancelled. This is the *motivation* behind the definition of 2-regularity below which formalizes these comments. The definition of p-regularity and the analogous results for p-order approximating sequences are given in [19].

Definition 1 *We say the operator F is 2-regular at x_* in direction of $h_1 \in X$ if*

(A1) *$F : X \to Y$ is three times continuously Fréchet differentiable in a neighborhood of x_*;*

(A2) *the subspace $\operatorname{Im} F'(x_*) = Y_1$ is closed;*

(A3) *the map $\mathcal{G}_2 = \mathcal{G}_2[F](x_*; h_1)$, $X \to Z \doteq Y_1 \times (Y/Y_1) = \operatorname{Im} F'(x_*) \times \operatorname{coker} F'(x_*)$, given by*

$$v \mapsto \mathcal{G}_2(v) = (F'(x_*)v, \pi F''(x_*)(h_1, v)), \tag{2}$$

where the $\pi : Y \to Y/Y_1$ denotes the canonical projection into the quotient space, is onto.

By construction the first coordinate of $\mathcal{G}_2$ is onto $\operatorname{Im} F'(x_*)$. Given any $k \in \operatorname{Im} F'(x_*)$, any two solutions of $F'(x_*)v = k$ differ by elements in the kernel of $F'(x_*)$, $\ker F'(x_*)$. Therefore $\mathcal{G}_2 = \mathcal{G}_2(h_1)$ is onto if and only if the linear mapping

$$\begin{aligned} \mathcal{H}_2(h_1) \quad : \quad & \ker F'(x_*) \to \operatorname{coker} F'(x_*), \\ & v \mapsto \pi F''(x_*)(h_1; v) \end{aligned} \tag{3}$$

is onto. (For, given $k \in \operatorname{Im} F'(x_*)$ and $[\ell] \in \operatorname{coker} F'(x_*)$, choose v so that $F'(x_*)v = k$. If $\mathcal{H}_2(h_1)$ is onto, then there exists a $w \in \ker F'(x_*)$ such that $\pi F''(x_*)(h_1; w) = [\ell] - \pi F''(x_*)(h_1; v)$ an thus $\mathcal{G}_2(v + w) = (k, [\ell])$. Conversely, if $\mathcal{G}_2$ is onto, given any $[\ell] \in \operatorname{coker} F'(x_*)$ there exists a v such that $\mathcal{G}_2(v) = (0, [\ell])$. Hence $v \in \ker F'(x_*)$ and $\mathcal{H}_2(h_1)v = [\ell]$.)

There is a close relation between 2-regularity and the intrinsic *Hessian* of the operator F at x_*, which we denote by $\mathcal{H}F(x_*)$ to avoid ambiguities with the notation $F''(x_*)$. The Hessian is the quadratic mapping

$$\begin{aligned} \mathcal{H}F(x_*) \quad : \quad & \ker F'(x_*) \to \operatorname{coker} F'(x_*), \\ & h \mapsto \mu F''(x_*)(h, h) \qquad \text{where } \mu \text{ ranges over } \left(\operatorname{Im} F'(x_*)\right)^{\perp}. \end{aligned} \tag{4}$$

In this notation, it follows from our motivation above that if $h \in \ker F'(x_*)$, then $h \in \ker \mathcal{H}F(x_*)$ is a necessary condition for the existence of quadratic approximations along h or, equivalently, for h to be a tangent direction to Q. If the operator F is 2-regular at x_* in direction of h then this condition is also sufficient. Based on a high-order generalization of the Lyusternik theorem [19] we can give a precise description of quadratic (and higher order) approximating sequences for a 2-regular operator. The following definition is standard (see, for instance [7]):

Definition 2 *Let $(h_1, h_2) \in X^2$ and set $x(\varepsilon) \doteq x_* + \varepsilon h_1 + \varepsilon^2 h_2$. We call a pair (h_1, h_2) a second-order approximating sequence to a set $S \subseteq X$ at $x_* \in \operatorname{Clos} S$, respectively we call $x : \varepsilon \longmapsto x(\varepsilon)$, a second-order approximating curve, if there exist an $\varepsilon_0 > 0$ and a function r defined on $[0, \varepsilon_0]$ with values in X, $r : [0, \varepsilon_0] \to X$, with the property that for $\varepsilon > 0$*

$$x(\varepsilon) + r(\varepsilon) = x_* + \varepsilon h_1 + \varepsilon^2 h_2 + r(\varepsilon) \in S \tag{5}$$

and

$$\lim_{\varepsilon \to 0} \frac{||r(\varepsilon)||}{\varepsilon^2} = 0. \tag{6}$$

Using the standard Landau notation we say that the function r is of order $o(\varepsilon^2)$ as $\varepsilon \to 0$. Also, we call a second-order approximating sequence/curve *second-order feasible* if S is an inequality constraint respectively *second-order tangent* if S is an equality constraint. More general versions of this definition exist for nonsmooth systems (for instance, [25]), but for our purpose the above definition suffices.

Definition 3 *We call h_2 a second-order tangent vector to an equality constraint Q at x_* in direction of $h_1 \in X$ if $(h_1, h_2) \in X^2$ is a second-order approximating sequence to the set Q at $x_* \in Q$. The collection of all second-order tangent vectors to Q at x_* in direction of h_1 will be called the second-order tangent set to Q at x_* in direction of h_1 and will be denoted by $TS^{(2)}(Q; x_*, h_1)$.*

Then, for the case $p = 2$, the generalized Lyusternik theorem in [19] can be summarized as follows:

Proposition 1 *[15, 19] Suppose $h \in \ker F'(x_*)$ and the operator F is 2-regular at x_* in direction of h. Then $TS^{(2)}(Q; x_*, h)$ is nonempty if and only if*

$$F''(x_*)(h,h) \in \operatorname{Im} F'(x_*) \tag{7}$$

or, equivalently, if

$$h \in \ker \mathcal{H}F(x_*). \tag{8}$$

In this case $TS^{(2)}(Q; x_*, h)$ consists of all $v \in X$ which are solutions to the equations

$$F'(x_*)v + \frac{1}{2}F''(x_*)(h,h) = 0 \tag{9}$$

$$F''(x_*)(h,v) + \frac{1}{6}F'''(x_*)(h,h,h) \in \operatorname{Im} F'(x_*) \tag{10}$$

Geometrically, second-order tangent sets are closed affine subspaces of X through a point $\mathcal{R}_1(x_*; h) \in Z$ determined by remainder terms and generated by the kernel of $\mathcal{G}_2$, $\ker \mathcal{G}_2$. This also is the geometric structure for general p-order approximations [19].

3. Nonregular operators in optimal control problems

Let $W_{11}^n(0,T)$ denote the space of $\mathbb{R}^n$-valued absolutely continuous functions and let $\overline{W}_{11}^n(0,T)$ denote the subspace of functions $x \in W_{11}^n(0,T)$ which satisfy $x(0) = 0$. We consider the problem Σ to minimize the functional

$$I(x,u) = \int_0^T L(x(t), u(t))dt \tag{11}$$

over all $(x,u) \in \overline{W}_{11}^n(0,T) \times L_\infty^m(0,T)$ subject to the constraints

$$\dot{x}(t) = f(x(t), u(t)), \quad x(0) = x_0, \quad x(T) = q, \tag{12}$$

$$u(\cdot) \in \mathcal{U} = \{u(\cdot) \in L_\infty^m(0,T) : u(t) \in U \text{ for } t \in [0,T]\}. \tag{13}$$

The important restriction for the following discussions is that we only consider the optimal control problem to a fixed *terminal point* $q \in \mathbb{R}^n$. The terminal time T is free and the control set U is an arbitrary subset of $\mathbb{R}^m$. We make the following **regularity assumptions** on the data: $L : \mathbb{R}^n \times \mathbb{R}^m \to \mathbb{R}$, $f : \mathbb{R}^n \times \mathbb{R}^m \to \mathbb{R}^n$ and for every $u \in U$ the functions $L(\cdot, u)$ and $f(\cdot, u)$ are C^∞ in x; for every $x \in \mathbb{R}^n$ the functions $L(x, \cdot)$ and $f(x, \cdot)$ and all partial derivatives with respect to x are continuous in (x,u) on $\mathbb{R}^n \times U$.

For this problem of steering the system optimally into a terminal point *nonregular operators arise if the linearization of the system around a reference trajectory is not completely controllable.* Let $\Gamma = (x_*, u_*)$ be a trajectory defined over the compact interval $[0,T]$ and let

$$\Sigma_L : \qquad \dot{h} = A(t)h + b(t)z, \qquad h(0) = 0 \tag{14}$$

be the time-varying linear system with system matrix $A(t) = f_x(x_*(t), u_*(t))$ given by the dynamics of the variational equation in x along Γ and single-input control term $b(t) = f(x_*(t), u_*(t))$ given by the dynamics along the reference trajectory.

Here and in the following partial derivatives are denoted by subscripts. It is a well-known result about controllability of linear systems that Σ_L is *not completely controllable* if and only if there exists a nontrivial solution λ to the adjoint equation

$$\dot{\lambda} = -\lambda f_x(x_*(t), u_*(t)) = -\lambda A(t), \tag{15}$$

such that $\lambda(t)b(t) \equiv 0$ on $[0,T]$. (Note that we write λ as a row vector.) This is easily seen: Let R denote the reachable set for Σ_L at time T. For any trajectory h of Σ_L corresponding to control z, we have

$$\lambda(T)h(T) = \int_0^T \dot{\lambda}(t)h(t) + \lambda(t)\dot{h}(t)dt = \int_0^T \lambda(t)b(t)z(t)dt. \tag{16}$$

Thus, if R is not all of $\mathbb{R}^n$, then there exists a vector $\overline{\lambda}\perp R$ and the corresponding solution λ of (15) satisfies $\lambda(t)b(t) \equiv 0$ on $[0,T]$. Conversely, for any solution of (15) which satisfies $\lambda(t)b(t) \equiv 0$ on $[0,T]$ we have $\lambda(T)\perp R$. In particular, Σ_L is not completely controllable if and only if there exists a nontrivial solution λ to the adjoint equation such that

$$\lambda(T)\perp R. \tag{17}$$

Our main interest here is to *connect the fact that Σ_L is not completely controllable along the reference trajectory Γ to the nonregularity of an operator related to the dynamics with initial and terminal condition.* Define the operator

$$F : \overline{W}^n_{11}(0,T) \times L^1_\infty(0,T) \to \overline{W}^n_{11}(0,T) \times \mathbb{R}^n$$

by

$$(x,w) \longmapsto (x(\cdot) - \int_0^{(\cdot)} w(t)f(x(t),u_*(t))dt,\ x(T) - q). \tag{18}$$

Note that the control has been fixed to be u_*, the reference control. The extra variable w allows for reparametrizations of the reference trajectory which corresponds to $w \equiv 1$. The Fréchet-derivative of F at the point $(x_*,1)$ is the linear operator

$$F'(x_*,1) : \overline{W}^n_{11}(0,T) \times L^1_\infty(0,T) \to \overline{W}^n_{11}(0,T) \times \mathbb{R}^n$$

given by

$$(h,z) \longmapsto (h(\cdot) - \int_0^{(\cdot)} (A(t)h(t) + b(t)z(t))dt,\ h(T)). \tag{19}$$

Let $\Psi^\Gamma(t,s)$ denote the fundamental matrix of the equation $\dot{h} = A(t)h$, i.e.

$$\frac{\partial}{\partial t}\Psi^\Gamma(t,s) = A(t)\Psi^\Gamma(t,s), \qquad \Psi^\Gamma(s,s) = Id. \tag{20}$$

Then the image of $F'(x_*,1)$ can be expressed as

Lemma 1

$$\operatorname{Im} F'(x_*,1) = \{(a, \int_0^T \Psi^\Gamma(T,s)\dot{a}(s)ds + R) : \quad a \in \overline{W}^n_{11}(0,T)\}. \tag{21}$$

In particular, $\operatorname{Im} F'(x_*,1)$ *is closed and of finite codimension given by the codimension of* R.

Proof: Given any $a \in \overline{W}^n_{11}(0,T)$, the unique solution $h \in \overline{W}^n_{11}(0,T)$ to

$$h(t) - \int_0^t A(s)h(s)ds = a(t) = \int_0^t \dot{a}(s)ds$$

is given by $\int_0^t \Psi^\Gamma(t,s)\dot{a}(s)ds$. We can still superimpose $h(T)$ with any solution to the linear equation

$$h(t) - \int_0^t (A(s)h(s) + b(s)z(s))\, ds = 0.$$

This adds the reachable set R in the terminal condition without changing the first component verifying (21). Hence $\operatorname{Im} F'(x_*, 1)$ is the direct sum of a closed subspace isomorphic to $\overline{W}^n_{11}(0,T)$ and the finite-dimensional subspace $\{0\} \times R$. Thus $\operatorname{Im} F'(x_*, 1)$ is closed and the codimension is given by the codimension of R.□

Thus we have shown the following Proposition:

Proposition 2 *Let $\Gamma = (x_*, u_*)$ be a trajectory of the system Σ defined over the compact interval $[0,T]$. The following are equivalent:*

(i) *The operator F defined in (18) is not regular at $(x_*, 1)$ (i.e. the Frèchet-derivative $F'(x_*, 1)$ is not onto).*

(ii) *The single-input linear time-varying system Σ_L is not completely controllable, i.e. $\dim R < n$.*

(iii) *There exists a nontrivial solution λ of the equation $\dot{\lambda} = -\lambda A(t)$ which satisfies $\lambda(t)b(t) \equiv 0$ on $[0,T]$*

(iv) *There exists a nontrivial solution λ of the equation $\dot{\lambda} = -\lambda A(t)$ which satisfies $\lambda(T) \perp R$.*

If the trajectory Γ is optimal, then by the Maximum Principle there exist a constant $\lambda_0 \geq 0$ and a solution λ to the adjoint equation (15) which do not vanish together such that the control u_* minimizes the Hamiltonian

$$H = \lambda_0 L(x,u) + \lambda f(x,u)$$

over the control set U along the trajectory x_* with minimum value 0. Extremals for which $\lambda_0 = 0$ are called *abnormal.* If Γ is abnormal, then the multiplier λ of the Maximum Principle therefore satisfies the condition $\lambda(t)b(t) \equiv 0$ and thus the operator F is nonregular at $(x_*, 1)$. In this case the codimension of R is called the order of abnormality by Hestenes in [11] or in more modern terminology the *corank of the abnormal extremal* by Agrachev and Sarychev [2].

We now specialize the results of Section 2 for the operator F defined in (18) along a trajectory Γ for which the equality constraint

$$Q = \{(x,z) \in \overline{W}^n_{11}(0,T) \times L^1_\infty(0,T) : F(x,z) = F(x_*, 1)\} \tag{22}$$

is non-regular at $(x_*, 1)$. By Lemma 1 condition (A2) is satisfied and we need to analyze in which directions $(h,z) \in F'(x_*, 1)$ the operator F is 2-regular. Note that $\ker F'(x_*, 1)$ consists of all pairs $(h,z) \in \overline{W}^n_{11}(0,T) \times L^1_\infty(0,T)$ where h is the trajectory of Σ_L corresponding to control z with terminal condition $h(T)=0$. Next we determine in which directions $(h,z)\in\ker F'(x_*,1)$ the operator F is 2-regular at $(x_*,1)$. This requires to analyze $\pi F''(x_*,1)(h,z;\eta,\xi)$ for $(\eta,\xi)\in \overline{W}^n_{11}(0,T)\times L^1_\infty(0,T)$ where π denotes the projection onto coker $F'(x_*,1)$. We have

$$F'(x_*,1)(\eta,\xi)(t)=\left(\eta(t)-\int_0^t A(s)\eta(s)+b(s)\xi(s)ds,\eta(T)\right)$$

and

$$\begin{aligned}F''(x_*,1)(h,z;\eta,\xi)(t) &= \Big(-\int_0^t f_{xx}(x_*(s),u_*(s))(h(s),\eta(s))\\ &\qquad +f_x(x_*(s),u_*(s))(\eta(s)z(s)+h(s)\xi(s))ds,0\Big)\end{aligned}$$

Since $\operatorname{Im}F'(x_*,1)=\{(a,\int_0^T\Psi^\Gamma(T,s)\dot a(s)ds+R): a\in\overline{W}^n_{11}(0,T)\}$, it follows that

$$\pi F''(x_*,1)(h,z;\eta,\xi)=(0,\widehat{\pi}\widehat{a}(T))$$

where

$$\widehat{a}(T)=\int_0^T\Psi^\Gamma(T,s)\left[f_{xx}(x_*,u_*)(h,\eta)+f_x(x_*,u_*)(\eta z+h\xi)\right]ds$$

and $\widehat{\pi}:\mathbb{R}^n\to\mathbb{R}^n$ denotes the orthogonal projection onto the complement of the reachable set, $R^\perp$. In view of (3) we therefore have:

Proposition 3 *The operator*

$$\begin{aligned}F \quad &: \quad \overline{W}^n_{11}(0,T)\times L^1_\infty(0,T)\to\overline{W}^n_{11}(0,T)\times\mathbb{R}^n\\ (x,z) \quad &\longmapsto \quad (x(\cdot)-\int_0^{(\cdot)}z(t)f(x(t),u_*(t))dt,\ x(T)-q),\end{aligned}$$

is 2-regular at $(x_*,1)$ in direction of $(h,z)\in\ker F'(x_*,1)$ if and only if the linear operator

$$\begin{aligned}\mathcal{H}_2(h,z) \quad &: \quad \ker F'(x_*,1)\to\operatorname{co}\ker F'(x_*,1)\simeq R^\perp\subset\mathbb{R}^n,\\ (\eta,\xi) \quad &\longmapsto \quad \widehat{\pi}\int_0^T\Psi^\Gamma(T,s)\left[f_{xx}(x_*,u_*)(h,\eta)+f_x(x_*,u_*)(\eta z+h\xi)\right]ds\end{aligned}$$

is onto .

For example, if the codimension of R is one, i.e. if Γ is a corank 1 abnormal extremal, then F is 2-regular at $(x_*,1)$ in direction of $(h,z)\in\ker F'(x_*,1)$ if and only if there exists a pair $(\eta,\xi)\in\ker F'(x_*,1)$ such that

$$\lambda(T)\int_0^T\Psi^\Gamma(T,s)\left[f_{xx}(x_*,u_*)(h,\eta)+f_x(x_*,u_*)(\eta z+h\xi)\right]ds\neq 0. \tag{23}$$

where $\lambda(T)$ spans $R^\perp$. But $\widetilde{\lambda}(s)=\lambda(T)\Psi^\Gamma(T,s)$ is a scalar multiple of the adjoint multiplier λ of the Maximum Principle and thus (23) is equivalent to

$$\int_0^T\lambda(s)\left[f_{xx}(x_*,u_*)(h,\eta)+f_x(x_*,u_*)(\eta z+h\xi)\right]ds\neq 0. \tag{24}$$

Since $\pi F''(x_*,1)(h,z;\eta,\xi)=0$ if and only if $\widehat{a}(T)\in R$, we also have the following result by Proposition 1.

Proposition 4 *Suppose $(h,z) \in \ker F'(x_*,1)$ and F is 2-regular at $(x_*,1)$ in direction of (h,z). Then the second-order tangent set $TS^{(2)}(Q;(x_*,1);(h,z))$ is non-empty if and only if $(h,z) \in \ker \mathcal{H}F(x_*,1)$. Equivalently, if and only if*

$$\int_0^T \Psi^\Gamma(T,s)\,[f_{xx}(x_*,u_*)(h,h) + 2f_x(x_*,u_*)hz]\,ds \in R \tag{25}$$

or if

$$\int_0^T \lambda(s)\,[f_{xx}(x_*,u_*)(h,h) + 2f_x(x_*,u_*)hz]\,ds = 0 \tag{26}$$

for every solution λ of the adjoint equation $\dot{\lambda} = -\lambda f_x(x_*(t),u_*(t))$ which satisfies $\lambda(T) \perp R$.

4. Abnormal bang-bang extremals

We now spezialize further to the problem of time optimal control to a terminal point q for a system which is affine in the control of the form

$$\Sigma: \qquad \dot{x}= f(x) + ug(x), \qquad x \in \mathbb{R}^n, \qquad |u| \leq 1, \tag{27}$$

where f and g are C^∞ (respectively sufficiently smooth) vector fields.

As example we consider bang-bang trajectories with $(n-1)$ switchings, but the theory formulated above can be applied to any kind of trajectory. Naturally, the actual calculations will be different depending on the structure of the specific extremal. We chose bang-bang trajectories to illustrate the theory since some of the calculations can be carried out easily thus providing geometric insight. Furthermore, for the time-optimal control problem for nonlinear control-affine systems an optimal synthesis typically includes bang-bang trajectories as well as more general concatenations of bang-bang trajectories with singular arcs. For instance, if q is an equilibrium point of f, then in small dimensions ($n \leq 4$) it is known that time-optimal controls are bang-bang or concatenations of bang and singular arcs provided that the Lie-bracket configuration of the system at q is non-degenerate (e.g. [30, 27, 26].) In the codimension 0 case there are at most $(n-1)$ switchings if the initial condition is sufficiently close to q [8, 14, 29]. In cases of positive codimension typically longer concatenations occur [30, 26, 28].

If a trajectory $\Gamma = (x_*,u_*)$ defined over $[0,T]$ is time optimal, then the conditions of the Maximum Principle for this problem state that there exist a constant $\lambda_0 \geq 0$ and an absolutely continuous function $\lambda : [0,T] \to (\mathbb{R}^n)^*$, called adjoint state, which do not vanish simultaneously, such that

$$\dot{\lambda}= -\lambda(Df(x_*) + u_* Dg(x_*)), \tag{28}$$

$$u_*(t) = \begin{cases} +1 & \text{if} \quad \langle \lambda(t), g(x_*(t))\rangle < 0 \\ -1 & \text{if} \quad \langle \lambda(t), g(x_*(t))\rangle > 0 \end{cases} \tag{29}$$

and

$$\lambda_0 + \langle \lambda(t), f(x_*(t)) + u_*(t)g(x_*(t))\rangle \equiv 0. \tag{30}$$

The function

$$H(\lambda_0,\lambda,x_*,u_*) = \lambda_0 + \langle \lambda, f(x) + ug(x)\rangle \tag{31}$$

is called the *Hamiltonian* function for the time-optimal control problem. The expression

$$\varphi(t) = \langle \lambda(t), g(x_*(t)) \rangle \tag{32}$$

which determines the optimal control is called the *switching function.* Also we denote the vector fields which correspond to the constant controls $u \equiv -1$ and $u \equiv 1$ by $X = f - g$ and $Y = f + g$ respectively.

We consider a bang-bang extremal Γ with $(n-1)$ switchings at times $0 < t_1 < \dots < t_{n-1} < T$ and switching points $p_i = x_*(t_i)$, $i = 1, \dots, n-1$. Also set $t_0 = 0$, $t_n = T$, $p = p_0$ and $q = p_n$. Without loss of generality suppose $u \equiv -1$ on $[0, t_1]$ and then the control switches between $u \equiv -1$ and $u \equiv +1$. Denote the times along the trajectories by s_i, $i = 1, \dots, n-1$, so that $s_1 = t_1$, $s_2 = t_2 - t_1$ etc. and $s_n = T - t_{n-1}$. It follows from (29) that we have

$$\langle \lambda(t_i), g(x_*(t_i)) \rangle = 0 \tag{33}$$

at every switching time t_i, $i = 1, \dots, n-1$. We now combine these conditions by evaluating them all at one point. This will be achieved by moving both the vectors and covectors along the flow of the reference trajectory Γ to one point along the trajectory, for instance a specific switching point or just the initial or terminal point. We move all the vectors to the terminal point q of the trajectory Γ. Tangent vectors $v \in T_{x_*(s)}$ are moved along the flow of Γ by integrating the variational equation of the system along Γ with initial condition v at time s. Thus, if $\Psi^\Gamma(t,s)$ denotes the fundamental matrix of the variational equation (here $A(t) = Df(x_*(t)) + u_*(t)Dg(x_*(t))$ and D denotes the Jacobian matrix), then the vector v moved along the flow of Γ to the point $x_*(t)$ is given by $\Psi^\Gamma(t,s)v$. Specifically, moving the vectors $g(p_i)$ to the terminal point q generates the vectors $\Psi^\Gamma(T, t_i)g(p_i)$, $i = 1, \dots, n-1$. In the notation we use Γ as a symbol to indicate that the variational equation is taken along X respectively Y depending on the specific domains. Analogously, covectors $\eta \in T^*_{x_*(s)}$ are moved along the flow of Γ by integrating the covariational equation of the system along Γ with initial condition η at time s. But the covariational equation is precisely the adjoint equation (28) of the Maximum Principle. Therefore, if we want to move $\lambda(s)$ to time t, the result is simply $\lambda(t)$. Since (28) is the adjoint equation to the variational equation, it also follows for any vector $v \in T_{x_*(s)}$ that the function

$$t \mapsto \langle \lambda(t), \Psi^\Gamma(t,s)v \rangle \tag{34}$$

is constant. Thus in particular

$$\langle \lambda(T), \Psi^\Gamma(T, t_i)g(p_i) \rangle = 0 \qquad \text{for } i = 1, \dots, n-1. \tag{35}$$

The extremal Γ is abnormal if $\lambda_0 = 0$. Combining $\lambda_0 = 0$ with (30) and (33), we also get at the switching points that

$$\langle \lambda(t_i), f(p_i) \rangle = 0 \qquad \text{for} \qquad i = 1, \dots, n-1. \tag{36}$$

Equivalently, Γ is abnormal if and only if $\lambda(t_i)$ vanishes against both $X(p_i)$ and $Y(p_i)$ for every switching. If the control $u \equiv 1$ is used on the interval (t_i, t_{i+1}), then noticing that $X(p_i) = Y(p_i) - 2g(p_i)$ it also follows that

$$\Psi^\Gamma(t_{i+1}, t_i)X(p_i) = \Psi^\Gamma(t_{i+1}, t_i)(Y(p_i) - 2g(p_i)) = Y(p_{i+1}) - 2\Psi^\Gamma(t_{i+1}, t_i)g(p_i).$$

Moving the vector $Y(p_i)$ with the flow of Y to the point p_{i+1} gives $Y(p_{i+1})$. Thus the second condition in (36) is redundant. Altogether, (33) and (36) therefore generate at most n linearly independent conditions which we can describe by adding the vector $\Gamma(q)$, i.e. $X(q)$ respectively $Y(q)$ depending on whether the control is $u \equiv -1$ or $u \equiv 1$ on $[t_{n-1}, T]$, to the vectors $\Psi^\Gamma(T, t_i)g(p_i)$, $i = 1, \ldots, n-1$. Hence we have

Proposition 5 *If Γ is abnormal, and we denote by S the linear hull of the vectors*

$$\left\{\Gamma(q),\ \Psi^\Gamma(T, t_i)g(p_i),\ i = 1, \ldots, n-1\right\}, \tag{37}$$

then $\dim S < n$.

This set S is precisely the reachable set at time T of the linearization Σ_L around Γ.

Proposition 6 *The reachable set R for Σ_L at time T is given by*

$$R = lin\ span\left\{\Gamma(q),\ \Psi^\Gamma(T, t_i)\Gamma(p_i),\ i = 1, \ldots, n-1\right\} = S. \tag{38}$$

Here $\Gamma(p_i) = X(p_i)$ if $u \equiv -1$ on (t_{i-1}, t_i) and $\Gamma(p_i) = Y(p_i)$ if $u \equiv 1$ on (t_{i-1}, t_i).

Proof. It follows from the variation of parameters formula that

$$h(T) = \int_0^T \Psi^\Gamma(T, s)b(s)z(s)ds.$$

Separating the intervals on which the control is constant we have

$$h(T) = \sum_{i=1}^n \Psi^\Gamma(T, t_i) \int_{t_{i-1}}^{t_i} \Psi^\Gamma(t_i, s)b(s)z(s)ds. \tag{39}$$

But for $s \in [t_{i-1}, t_i]$, $b(s) = X(x_*(s))$ or $b(s) = Y(x_*(s))$ and this vector is moved forward along the flow of the same vector field. Therefore

$$\Psi^\Gamma(t_i, s)b(s) = \Psi^\Gamma(t_i, s)\Gamma(x_*(s)) = \Gamma(x_*(t_i)) = b(t_i) \tag{40}$$

and thus

$$h(T) - \sum_{i=1}^n \Psi^\Gamma(T, t_i)\Gamma(p_i)\left(\int_{t_{i-1}}^{t_i} z(s)ds\right). \tag{41}$$

Hence the reachable set of Σ_L at time T is spanned by the vectors

$$R = linspan\{\Psi^\Gamma(T, t_i)\Gamma(p_i) : i = 1, \ldots, n\} \tag{42}$$

where $\Gamma(p_i) = X(p_i)$ if $u \equiv -1$ on (t_{i-1}, t_i) and $\Gamma(p_i) = Y(p_i)$ if $u \equiv 1$ on (t_{i-1}, t_i). But this is the space S. For, moving all vectors back to the last switching point p_{n-1}, for $i = n$ and $i = n-1$ we obtain the vectors $X(p_{n-1})$ and $Y(p_{n-1})$ or equivalently $f(p_{n-1})$ and $g(p_{n-1})$. (For instance, if $u \equiv -1$ on (t_{n-1}, T), then $\Gamma(q) = X(q)$ is moved back by its own flow to give $X(p_{n-1})$ and $\Psi^\Gamma(T, t_{n-1})\Gamma(p_{n-1})$ of course gives $\Gamma(p_{n-1}) = Y(p_{n-1})$ because of the switching.) Similarly, moving the vectors $\Psi^\Gamma(T, t_i)\Gamma(p_i)$ and $\Psi^\Gamma(T, t_{i+1})\Gamma(p_{i+1})$back to the switching point p_i, gives the vectors $X(p_i)$ and $Y(p_i)$. But these are precisely the vectors which span S. □

Indeed, it follows from the last argument that we can add arbitrary vectors $\Psi^\Gamma(T,t_i)\left(f(p_i)+ug(p_i)\right)$ for $|u|\leq 1$ to the spanning set of R. Furthermore, adjoint vectors λ which make Γ an abnormal extremal have terminal conditions orthogonal to the reachable set for Σ_L.

Also $\ker F'(x_*,1)$ can easily be expressed in terms of the vectors from R. We have already seen that

$$h(T)=\sum_{i=1}^{n}\Psi^\Gamma(T,t_i)\Gamma(p_i)\left(\int_{t_{i-1}}^{t_i} z(s)ds\right)$$

and therefore, if

$$a_i=\int_{t_{i-1}}^{t_i} z(s)ds, \tag{43}$$

then $(h,z)\in\ker F'(x_*,1)$ if and only if

$$\sum_{i=1}^{n}\Psi^\Gamma(T,t_i)\Gamma(p_i)a_i=0. \tag{44}$$

Nontrivial solutions exist by the linear dependence of these vectors and the dimension of the solution space is equal to the codimension of R.

5. On the optimality of abnormal extremals

We now return to the more general system Σ and apply the results of sections 2 and 3 to investigate the optimality of an abnormal extremal Γ. We *assume* as given a pair $(h,z)\in\ker F'(x_*,1)\cap\ker\mathcal{H}F(x_*,1)$ for which F is 2-regular at $(x_*,1)$ in direction of (h,z). Since the compatability condition (25) is satisfied there exist vectors $(\eta,\xi)\in\overline{W}^n_{11}(0,T)\times L^1_\infty(0,T)$ and a function $r:[0,\varepsilon_0]\to\overline{W}^n_{11}(0,T)\times L^1_\infty(0,T)$ which is of order $o(\varepsilon^2)$ as $\varepsilon\to 0$ such that

$$F\left((x_*,1)+\varepsilon(h,z)+\varepsilon^2(\eta,\xi)+(r_1(\varepsilon),r_2(\varepsilon)\right)=0.$$

The corresponding input-trajectory pair $\Gamma_\epsilon=(x_\epsilon,u_\epsilon)$ for the system Σ consists of the reference control u_* reparametrized in time according to the invertible transformation $\tau:[0,T]\to[0,\tau(T)]$ defined by

$$\tau(t)=\int_0^t(1+\varepsilon z(s)+\varepsilon^2\xi(s)+r_2(\varepsilon)(s))ds. \tag{45}$$

and corresponding trajectory x_ϵ which is obtained from the curve

$$x_*(t)+\varepsilon h(t)+\varepsilon^2\xi(t)+r_1(\varepsilon)(t)$$

by reparametrization.

For instance, for the time-optimal control problem and the case of the bang-bang trajectory considered in Section 4 the new trajectory is also bang-bang with the same switching order, but the switchings occur at times $\tau(t_i)$, $i=1,\dots,n$. Since $(\eta,\xi)\in TS^{(2)}(Q;(x_*,1);(h,z))$, the new trajectory Γ_ϵ also steers p into q, but the total time is given by

$$T_\epsilon=\tau(T)=T+\varepsilon\int_0^T z(s)ds+\varepsilon^2\int_0^T\xi(s)ds+\rho(\varepsilon) \tag{46}$$

where the function $\rho : [0, \varepsilon_0] \to \mathbb{R}$ is of order $o(\varepsilon^2)$ as $\varepsilon \to 0$. In particular, if $\int_0^T z(s)ds < 0$, then Γ is not time-optimal. If $\int_0^T z(s)ds > 0$, we can change (h, z) into $-(h, z)$ without violating the assumptions that F is 2-regular at $(x_*, 1)$ in direction of $-(h, z)$ or the compatability condition (25). Hence Γ will not be optimal if $\int_0^T z(s)ds \neq 0$.

This argument is valid in general: the cost for the new trajectory Γ_ϵ is given by

$$\begin{aligned} J_\epsilon &= \int_0^T L(x_* + \varepsilon h + o(\varepsilon), u_*)(1 + \varepsilon z + o(\varepsilon))dt \\ &= \int_0^T L(x_*, u_*)dt + \varepsilon \int_0^T \left[L(x_*, u_*)z + L_x(x_*, u_*)h\right] dt + o(\varepsilon). \end{aligned} \tag{47}$$

If $\int_0^T (L(x_*, u_*)z + L_x(x_*, u_*)h)dt \neq 0$, then, as before, by changing the sign of (h, z) we may assume this quantity is negative and thus the reference trajectory will not be optimal.

Proposition 7 *Let $(h, z) \in \ker F'(x_*, 1) \cap \ker \mathcal{H}F(x_*, 1)$ and suppose F is 2-regular at $(x_*, 1)$ in direction of (h, z). If Γ is optimal, then*

$$\int_0^T (L(x_*, u_*)z + L_x(x_*, u_*)h)dt = 0. \tag{48}$$

Hence, if Γ is optimal, then (h, z) is a critical direction for the objective and we also need to consider second-order (or even higher-order) approximations for the cost-functional. A necessary condition for optimality will be that there exist no second-order directions of decrease among the second-order tangent vectors. The theorem below gives a characterization of this non-empty intersection property in terms of multipliers of the corresponding dual spaces. But besides the equality constraint and the objective considered so far, Theorem 1 also contains variations of the time parametrization. Indeed, crucial for the derivation of the extended minimum condition (52) in Theorem 1 below is to allow singular reparametrizations, i.e. $w \geq 0$ while w vanishes on sets of positive measure, and to insert variations on the zero-set. Details of this argument are given in [16]. The construction is analogous to Dubovitskii's proof of the Maximum principle from its local version [10]. The important aspect of the construction is that it gives non-trivial conditions in the sense that we can normalize the multiplier at the objective to be 1. This is a consequence of the fact that F is 2-regular at $(x_*, 1)$ in direction of (h, z).

Theorem 1 (Extended second-order Maximum Principle for abnormal extremals) *Suppose the abnormal extremal $\Gamma = (x_*, u_*)$ defined on $[0, T]$ is optimal for the system Σ and suppose for $(h, z) \in \ker F'(x_*, 1) \cap \ker \mathcal{H}F(x_*, 1)$ the map F is 2-regular at $(x_*, 1)$ in direction of (h, z). Then there exist functions $\lambda(\cdot) = \lambda(h, z)(\cdot)$ and $\psi(\cdot) = \psi(h, z)(\cdot)$ from $[0, T]$ into $(\mathbb{R}^n)^*$ such that*

$$\dot{\lambda} = -\lambda f_x(x_*, u_*), \qquad \lambda f(x_*, u_*) = 0 \tag{49}$$

$$\dot{\psi} = -L_x(x_*, u_*) - \psi f_x(x_*, u_*) - \lambda \left[f_{xx}(x_*, u_*)h + z f_x(x_*, u_*)\right], \tag{50}$$

and the following minimum condition is satisfied along (x_*, u_*): for every $u \in [-1, 1]$ and a.e. $t \in [0, 1]$ we have

$$0 \equiv L_x(x_*, u_*) + \psi f(x_*, u_*) + \lambda f_x(x_*, u_*)h, \tag{51}$$

$$0 \leq L_x(x_*, u) + \psi f(x_*, u) + \lambda f_x(x_*, u)h, \tag{52}$$

Except for the additional claim that the multiplier at the objective is nonzero (which occurs as the 1 multiplying the Lagrangian terms), Theorem 1 is a special case of [16, Thm 5.1] for the system Σ. The nontriviality statement is new and its proof follows by carefully tracing the structure of the functionals from the dual cones to second-order approximating cones in the proof of [16, Thm. 5.1]. This proof will be given in general for p-regular operators elsewhere [21].

In this paper we include a motivation respectively explanation of the conditions of the result. In the theorem we consider possible expansions $x(\varepsilon) = x_* + \varepsilon h + \cdots$ of trajectories for reparametrizations $w_* z(\varepsilon)$ where $w_* \geq 0$ is fixed and $z(\varepsilon) = 1 + \varepsilon z + \cdots$. Then quadratic terms from the second-order tangent set to Q at x_* in direction (h, z) are added. Formally, if we set

$$\lambda_0(\varepsilon) = 0 + 1 \cdot \varepsilon + \cdots \quad \text{and} \quad \lambda(\varepsilon) = \lambda + \varepsilon\psi + \cdots, \tag{53}$$

then equations (49) and (50) precisely state that the adjoint equation of the Pontryagin maximum principle is satisfied by these multipliers modulo terms of order $o(\varepsilon)$ as $\varepsilon \to 0$. For,

$$\begin{aligned}
\dot{\lambda}(\varepsilon) &= -\lambda_0(\varepsilon) L_x(x(\varepsilon), u_*) - \lambda(\varepsilon) f_x(x(\varepsilon), u_*) \\
&= -\varepsilon L_x(x_*, u_*) \\
&\quad -(\lambda + \varepsilon\psi)\left[f_x(x_*, u_*) + \varepsilon\left(f_{xx}(x_*, u_*)h + f_x(x_*, u_*)z\right)\right] + o(\varepsilon) \\
&= -\lambda f_x(x_*, u_*) \\
&\quad -\varepsilon\left[L_x(x_*, u_*) + \psi f_x(x_*, u_*) + \lambda\left(f_{xx}(x_*, u_*)h + f_x(x_*, u_*)z\right)\right] + o(\varepsilon).
\end{aligned} \tag{54}$$

Note that we need to allow also for the change due to reparametrization which brings in the contribution from z. Formally, we could simply consider the auxiliary dynamics

$$\widetilde{f}(t, x, z) = z f(x, u_*(t))$$

and take the expansion in ε. Similarly, the Hamiltonian expands as

$$\begin{aligned}
H(\varepsilon) &= \lambda_0(\varepsilon) L(x(\varepsilon), u_*) + \lambda(\varepsilon) f(x(\varepsilon), u_*) \\
&= \varepsilon L(x_* + \varepsilon h + o(\varepsilon), u_*) + (\lambda + \varepsilon\psi + o(\varepsilon)) f(x_* + \varepsilon h + o(\varepsilon), u_*) \\
&= \lambda f(x_*, u_*) + \varepsilon\left[L(x_*, u_*) + \psi f(x_*, u_*) + \lambda f_x(x_*, u_*)h\right] + o(\varepsilon).
\end{aligned} \tag{55}$$

Since Γ is abnormal the zero-order term vanishes identically and thus, for $H(\varepsilon)$ to satisfy the condition $H \equiv 0$, we must have (51). The 2-regularity of F in direction of (h, z) guarantees that the quadratic term in the expansion of $\lambda_0(\varepsilon)$ does not vanish and this creates the 1 in this equation.

Thus, formally the multipliers in Theorem 1 provide a first-order expansion in ε for the adjoint variable corresponding to the optimal solution $\Gamma = (x_*, u_*)$. This expansion becomes of interest precisely for abnormal extremals when the leading terms generate conditions which vanish identically. These expansions exist provided the abnormal extremal Γ is not isolated in the sense that no second-order tangent directions exist. Note, however, that this is not the same as the rigidity conditions investigated in [2]. For instance, an abnormal bang-bang extremal Γ may be rigid when the controls are given the L^1_∞-topology, but there may still exist variations in the switching times (the L^1_∞-distance between these controls is 1) which can be used to exclude the optimality of Γ. In our construction of second-order approximating cones such changes in the switching times and more general variations are generated through the singular time-transformation along Γ. In this respect, the conditions which we obtain in our constructions are for strong minimality of abnormal extremals.

6. Conclusion

In this paper we described the geometric content of recently developed constructions which analyze high-order approximations for equality constraints near non-regular points. We illustrated some of the specific calculations in the analysis of abnormal bang-bang trajectories.

Also, the arguments and constructions presented here are not restricted to *abnormal* extremals, but they apply whenever non-regular equality constraints are present. In optimal control theory, another incidence of this occurs for *singular* extremals. However, for singular extremals the operator F which becomes non-regular is different. Specifically, variations in the control variable are made and the variational equation is taken in both x and u in this case. Then the relevant necessary conditions are given by the so-called local or weak version of the extended maximum principle in [16]. This will be pursued elsewhere [21].

References

[1] Agrachev, A.A. and Sarychev, A.V., *On abnormal extremals for Lagrange variational problems*, J. of Math. Syst., Est. and Con., **5** (1995), 127-130

[2] Agrachev, A.A. and Sarychev, A.V., *Abnormal sub-Riemannian geodesics: Morse index and rigidity*, Ann. Inst. Henri Poincaré, **13**, (1996), 635-690

[3] Arutyunov, A.V., *Optimality conditions in abnormal extremal problems*, Systems Control Lett., **27**, (1996), 279-284

[4] Arutyunov, A.V., *Second-order conditions in extremal problems. The abnormal points*, Trans. Am. Math. Soc., to appear

[5] Avakov, E.R., *Necessary conditions for a minimum for nonregular problems in Banach spaces. Maximum principle for abnormal problems of optimal control*, Trudy Mat. Inst. AN. SSSR, **185**, (1988), 3–29, [in Russian]

[6] Avakov, E.R., Necessary extremum conditions for smooth abnormal problems with equality and inequality-type constraints, *Math. Zametki,* **45**, (1989), 3–11.

[7] Ben-Tal, A., and Zowe, J., *A unified theory of first and second-order conditions for extremum problems in topological vector spaces*, Math. Prog. Study, **19**, (1982), 39-76

[8] Bressan, A., *The generic local time-optimal stabilizing controls in dimension 3*, SIAM J. Contr. and Optimiz., **24**, (1986), 177-190

[9] Dmitruk, A.V., *Quadratic order conditions of a local minimum for abnormal extremals*, Proceedings of the Second World Congress of Nonlinear Analysts, Part 4, (Athens, 1996), Nonlinear Anal. **30**, (1998), 2439-2448

[10] Girsanov, I.V., *Lectures on Mathematical Theory of Extremum Problems*, Springer, Berlin, Germany, 1972

[11] Hestenes, M.R., *Calculus of Variations and Optimal Control Theory*, Krieger, New York, NY, 1966

[12] Izmailov, A.F., *Optimality conditions for degenerate extremum problems with inequality-type constraints*, Comp. Maths. Math. Phys., **34**, (1994), pp. 723-736

[13] Jurdjevic, V., *Geometric Control Theory*, Cambridge University Press, Cambridge, United Kingdom, (1997)

[14] Krener, A.J., and Schättler, H., *The structure of small-time reachable sets in low dimensions*, SIAM J. Contr. and Optimiz., **27**, (1989), 120-147

[15] Ledzewicz, U. and Schättler, H., *Second-order conditions for extremum problems with nonregular equality constraints*, J. Opt. Th. and Appl. (JOTA), **86**, (1995), 113-144,

[16] Ledzewicz, U. and Schättler, H., *An extended maximum principle*, Nonlinear Analysis, **29**, (1997), 159-183,

[17] Ledzewicz, U., and Schättler, H., *A high-order generalization of the Lyusternik theorem and its Application to Optimal Control problems,* in: Dynamical Systems and Differential Equations II (W. Chen and S. Hu, Eds.), (1998), 45-59

[18] Ledzewicz, U., and Schättler, H., *High order extended maximum principles for optimal control problems with non-regular constraints*, in: Optimal Control: Theory, Algorithms and Applications, (W.W. Hager and P.M. Pardalos, Eds.), Kluwer Academic Publishers, (1998), to appear

[19] Ledzewicz, U. and Schättler, H., *A high-order generalization of the Lyusternik theorem*, Nonlinear Analysis, to appear

[20] Ledzewicz, U. and Schättler, H., *High-order approximations and generalized necessary conditions for optimality*, SIAM J. Contr. and Optimiz., to appear

[21] Ledzewicz, U. and Schättler, H., *A high-order generalized local Maximum Principle*, preprint

[22] Liu, W.S., and Sussmann, H.J., *Abnormal Sub-Riemannian minimizers*, in: Differential Equations, Dynamical Systems and Control Science, (K.D. Elworthy, W.N. Everitt and E.B.Lee,Eds.), Marcel Dekker, New York, (1993), 705-716

[23] Liu, W.S., and Sussmann, H.J., *Shortest paths for Sub-Riemannian metrics on rank-2 distributions*, Mem. Amer. Math. Soc., **118**, (1995)

[24] Milyutin, A.A., *Quadratic conditions of an extremum in smooth problems with a finite-dimensional image*, in: Methods of the Theory of Extremal Problems in Economics, Nauka, Moscow, Russia, 91981), 138-177, [in Russian]

[25] Páles, Zs., and Zeidan, V., *Nonsmooth optimum problems with constraints*, SIAM J. Contr. and Optimiz., **32**, (1994), 1476-1502

[26] Piccoli, B., *Classification of generic singularities for the planar time-optimal synthesis*, SIAM J. Contr. and Optimiz., **34**, (1996), 1914-1946

[27] Schättler, H., *The local structure of time-optimal trajectories in dimension 3 under generic conditions*, SIAM J. Contr. and Optimiz., **26**, (1988), 899-918

[28] Schättler, H., and Jankovic, M., *A synthesis of time-optimal controls in the presence of saturated singular arcs*, Forum Math., **5**, (1993), 203-241

[29] Shin, Ch. E., *On the structure of time-optimal stabilizing controls in* $\mathbb{R}^4$, Bollettino U.M.I., **7**, (1995), 299-320

[30] Sussmann, H.J., *The structure of time-optimal trajectories for single-input systems in the plane: the general real-analytic case*, SIAM J. Contr. and Optimiz., **25**, (1987), 868-904

Department of Mathematics and Statistics, Southern Illinois University, Edwardsville, Illinois 62025

E-mail address: uledzew@siue.edu

Department of Systems Science and Mathematics, Washington University, St. Louis, Missouri, 63130

E-mail address: hms@cec.wustl.edu

Proceedings of Symposia in Pure Mathematics
Volume **64**, 1999

Rolling bodies with regular surface: the holonomic case

Alessia Marigo and Antonio Bicchi

ABSTRACT. In this paper we analyse the nonholonomic nature of the constraint of pure (non–slipping) rolling between rigid bodies. Besides its theoretical interest in mechanics, the problem is of practical momentum in applications such as dexterous manipulation of objects by means of robotic hands. We study this problem by recasting it in terms of reachability analysis of a nonlinear system of ODE's. Our results include a characterization of cases where the rolling constraint is holonomic.

1. Introduction

A recent trend of research in robotics tends to meet industry needs of economy and reliability in the design of manipulation devices, by investigating minimally complex hardware systems for a given manipulation task (see e.g. [**2**]). To this end, the nonholonomic behaviour of some systems can be exploited to achieve dexterous manipulation with simple mechanical design. In fact, one of the characteristics of nonholonomic systems is that, by means of cyclic paths of some of their state variables, a controlled change in other variables can be produced.

A particularly interesting application is dexterous robotic manipulation. By "dexterity" in this context it is meant the capability of arbitrarily changing the position and orientation of a manipulated body ("object") by suitable motions of other, actuated bodies ("fingers"). The fingers and the object interact by mechanical contacts at several points. Conventional dexterous robot hands are designed based on the assumption that effects of rolling at the contact points are negligible, and typically consist of complex mechanisms with nine or more actuated joints (nine can be shown to be the theoretical minimum number of inputs to achieve dexterity if rolling is disregarded).

Exploiting the nonholonomic nature of rolling may enable substantial simplifications in the hand hardware and in the number of necessary inputs. Bicchi and Sorrentino [**4**] designed and implemented a dexterous hand using only three motors. Such hand is able to arbitrarily change the position and orientation of a manipulated object by rolling it between two planar fingertips that translate with respect to each other, provided that the object surface fulfills some conditions. The design

1991 *Mathematics Subject Classification.* Primary 93xx, 35Q72; Secondary 53Axx.
The work has been conducted with partial support of the NATO CR grant no. 960750 and ASI ARS-96-170.

of the hand was based on the conjecture ([**3**]) that *almost all* pairs of rolling objects could be driven to an arbitrary relative configuration by controlling their angular velocity.

Conditions ensuring reachability of pairs of rolling surfaces have been studied in several papers in the literature, whose results are listed below:

- Li and Canny [**5**] showed reachability for a sphere rolling on a plane, and for two spheres with different radius;
- Bicchi, Prattichizzo, and Sastry [**3**] showed reachability for objects with regular, strictly convex and axialsymmetric surfaces rolling on a plane;
- Marigo, Chitour, and Bicchi [**10**] discussed the structure of the reachable space of polyhedral surfaces rolling on a plane, showing that it is dense for generic polyhedra, while it has a lattice structure if certain geometric parameters of the surfaces are rational.

In this paper we provide a discussion of local admissibility and reachability of rolling for pairs of general bodies with regular surfaces. The main result of this paper shows that the constraint of rolling between two bodies is holonomic (hence the reachable submanifold is only two–dimensional) if and only if the bodies are specular images of each other, the mirror being the common tangent plane at the contact point.

2. Problem formulation

Consider a set of m independent Pfaffian constraints of the type

$$\mathbf{A}(\mathbf{q})\dot{\mathbf{q}} = 0, \tag{2.1}$$

where $\mathbf{q} \in M$, an n–dimensional smooth configuration manifold and $\dot{\mathbf{q}} \in T_{\mathbf{q}}M$. Constraints are said holonomic if the differential form (2.1) is integrable. In this case, there exists a family of integral submanifolds $M' \subset M$ of dimension $n-m$ that are invariant, i.e. are such that, if $\mathbf{q}(\bar{t}) \in M'$ at some $\bar{t}$, then $\mathbf{q}(t) \in M', \forall t$. If the constraint is not holonomic, there may exist invariant submanifolds of dimension $n-m+k$, $0 < k \leq m$. The number k is referred to as degree of nonholonomy. If $k = m$, the constraints are said to be maximally nonholonomic. For a maximally nonholonomic system, any two configurations $\mathbf{q}_0, \mathbf{q}_1$ in M can be linked through an admissible path $[0,1] \to M, s \to \mathbf{q}(s)$ such that $\mathbf{q}(0) = \mathbf{q}_0, \mathbf{q}(1) = \mathbf{q}_1$, and $\mathbf{A}(\mathbf{q}(s))\dot{\mathbf{q}}(s) = 0, \forall s \in [0,1]$.

To decide about the (degree of) nonholonomy of a given constraint, tools of differential–geometric nonlinear system theory are very effective. Eq.(2.1) can be recast in control form as

$$\dot{\mathbf{q}} = \mathbf{S}(\mathbf{q})\mathbf{w}, \tag{2.2}$$

where $\mathbf{S}(q) = \mathbf{A}(\mathbf{q})^{\perp}$, and $\mathbf{w} \in \mathbb{R}^{n-m}$. Chow's theorem ([**7**]), along with the fact that the distribution $\mathbf{S}(\mathbf{q})$ is smooth and nonsingular (for $\mathbf{A}(q)$ was assumed to be smooth and nonsingular), can be used to show that $k = m - n +$ LAR($\mathbf{S}(\mathbf{q})$), where LAR(Δ) denotes the rank of the Lie Algebra generated by the distribution Δ. Maximal nonholonomy is therefore equivalent to complete reachability of the n dimensional manifold by using $m < n$ control inputs.

In terms of actual construction of mechanical hands, the number of inputs is directly related to the number of actuators and joints, hence with the hardware

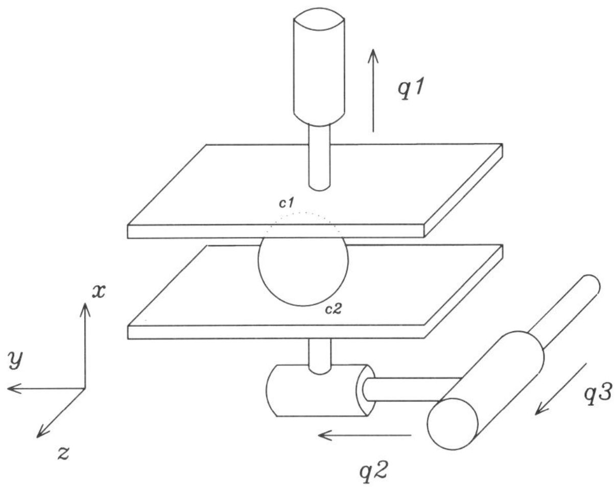

FIGURE 1. The dexterous 3–motor hand developed at the University of Pisa. The lower flat finger translates with respect to the upper finger, and makes the object roll. A suitable grasp force is maintained by the upper finger to avoid slippage (this can be done by a passive device). To achieve full dexterity, translations in the vertical direction of the fingers has to be added.

complicacy, weight, and cost. This explains the interest in exploiting the nonholonomy of rolling in manipulation.

In fact, consider an object rolling on top of a finger's surface (fixed for reference), while other finger(s) drive it by imposing a suitable two–dimensional rolling velocity. If the rolling constraint between the reference finger and the object is nonholonomic, it will be possible to obtain an arbitrary posture of the object in contact with the finger, i.e., on a 5–dimensional manifold, by just two motors for actuating rolling velocities. One more actuator will then be sufficient to achieve dexterity, i.e. reachability of $SE(3)$. An example of a device realizing this concept is schematically drawn in fig.1.

In general, it will be possible to build a dexterous hand with only three motors if it can be shown that the rolling constraint between all objects in a class of interest, and one finger is maximally nonholonomic. This motivates our study on nonholonomy of rolling pairs of bodies. In order to do so, we need some preliminaries on the differential geometry of contact.

2.1. Local geometry of contact. In this section, we report on the derivation of the mathematical model of rolling. Equations obtained are not new in the literature (compare e.g. with [**12**]), but are derived here in a different way (starting from the differential one–forms of the constraints), which is useful to report for the rest of the development.

Consider two arbitrary smooth surfaces S_1, S_2 in touch at a point $\mathbf{c}$ as represented in fig.2.

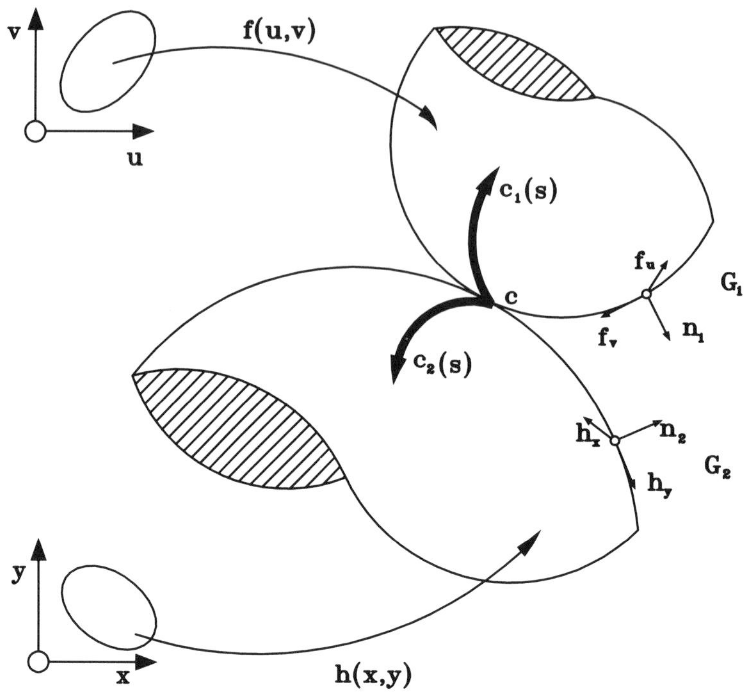

FIGURE 2. A pair of smooth rolling surfaces.

Let the two surfaces be described, in a neighborhood of the contact point, by two orthogonal parametrizations as

$$\begin{aligned} \mathbf{f}: \mathbb{R}^2 \longrightarrow \mathbb{R}^3,\ \mathbf{f}(u,v) &= (f_1(u,v), f_2(u,v), f_3(u,v)) \\ \mathbf{h}: \mathbb{R}^2 \longrightarrow \mathbb{R}^3,\ \mathbf{h}(x,y) &= (h_1(x,y), h_2(x,y), h_3(x,y)), \end{aligned}$$

respectively. These parametrization induce a Gauss frame $\mathcal{G}_1(u,v)$ and $\mathcal{G}_2(x,y)$, defined at every point in the contact neighborhood with axes $\frac{\mathbf{f}_u^T}{\|\mathbf{f}_u\|}, \frac{\mathbf{f}_v^T}{\|\mathbf{f}_v\|}, \mathbf{n}_1 = \frac{\mathbf{f}_u \times \mathbf{f}_v}{\|\mathbf{f}_u\|\|\mathbf{f}_v\|}$ and $\frac{\mathbf{h}_x^T}{\|\mathbf{h}_x\|}, \frac{\mathbf{h}_y^T}{\|\mathbf{h}_y\|}, \mathbf{n}_2 = \frac{\mathbf{h}_x \times \mathbf{h}_y}{\|\mathbf{h}_x\|\|\mathbf{h}_y\|}$, respectively (we use here and henceforth the shorthand notation $\mathbf{f}_u = \frac{\partial \mathbf{f}}{\partial u}$, etc.).

To take the rigid–body motions of our surfaces into account, we embed the above geometric parametrization in space–time by defining new maps $\mathcal{G}'_i : \mathbb{R}^2 \times \mathbb{R} \to SE(3) \times \mathbb{R}^+, i = 1,2$ as $\mathcal{G}'_i = g_i(t) \circ \mathcal{G}_i$, with $g_i(t) = (\mathbf{R}_i, \mathbf{p}_i) \in SE(3)$. By an abuse of notation, we will indicate by $\mathbf{f}(u,v,t) = g_1(t) \circ \mathbf{f}(u,v)$, and $\mathbf{h}(x,y,t) = \mathbf{g}_2(t) \circ \mathbf{h}(x,y)$. We introduce also the notation $\boldsymbol{\eta}_i = (\mathbf{t}, \boldsymbol{\omega}) \in se(3)$ for the elements of the tangent space of $SE(3)$.

At the contact point, $\mathbf{n}_2 = -\mathbf{n}_1$, while the first axis of the two Gauss frames form an angle

$$\psi = -\arccos \frac{\mathbf{f}_u^T}{\|\mathbf{f}_u\|} \frac{\mathbf{h}_x}{\|\mathbf{h}_x\|}. \tag{2.3}$$

We locally describe the configuration space of the system as the 5–dimensional manifold $M = \{\mathbf{q} = \begin{bmatrix} u & v & x & y & \psi \end{bmatrix}^T\}$. This choice of parametrization directly eliminates the obviously holonomic constraint of zero penetration–detachment velocity between the two bodies.

Let two smooth directed curves on the surfaces be $\mathbf{c}_1, \mathbf{c}_2$ (see fig.2), and let the two curves be parametrized by the arc length s. Consider a rigid motion of S_2 relative to S_1, assumed fixed in space, such that the contact point traces out the curves on the two surfaces in the given direction at unit velocity $\frac{ds}{dt} = 1$. The constraint of rolling without slipping between the two surfaces imposes that the points of the surfaces being in contact at the time of concern have zero relative velocity, and that the length of the paths covered by the image of the contact points on the two curves in a corresponding time are equal. From the first condition, taking S_1 fixed for reference (hence $g_1(t) \equiv Id$), it easily follows that also the transational velocity of S_2, $\mathbf{t}_2(t) = \frac{\partial f(u,v,t)}{\partial t}$, must be zero.

Based on this, and using the definition of the first fundamental form of the first surface, $\mathcal{I}_1$, one has $\|\dot{\mathbf{f}}\|^2 = \left[\begin{array}{cc} \dot{u} & \dot{v} \end{array}\right] \mathcal{I}_1 \left[\begin{array}{cc} \dot{u} & \dot{v} \end{array}\right]^T$. An analogous relation obviously holds for surface 2. Hence, letting $\mathbf{M}_i$ be the symmetric, positive definite square root of the first fundamental form of the i–th surface, i.e. $\mathbf{M}_i^T \mathbf{M}_i = \mathcal{I}_i$ ($\mathbf{M}_i$ is actually the Riemannian metric form), the rolling–without–slippage condition can be expressed in terms of velocities as

$$\mathbf{M}_1 \left[\begin{array}{c} \dot{u} \\ \dot{v} \end{array}\right] - \mathbf{R}_\psi \mathbf{M}_2 \left[\begin{array}{c} \dot{x} \\ \dot{y} \end{array}\right] = 0, \tag{2.4}$$

where

$$\mathbf{R}_\psi = \left[\begin{array}{cc} \cos(\psi) & -\sin(\psi) \\ -\sin(\psi) & -\cos(\psi) \end{array}\right]$$

is a matrix that takes into account the relative orientation of the axes of the Gauss frames on the tangent plane. Note that $\mathbf{R}_\psi \mathbf{R}_\psi = Id$. This constraint can be rewritten in Pfaffian form as

$$\left[\begin{array}{ccc} \mathbf{M}_1 & -\mathbf{R}_\psi \mathbf{M}_2 & 0 \end{array}\right] \dot{\mathbf{q}} = 0,$$

and the corresponding control form is

$$\dot{\mathbf{q}} = \left[\begin{array}{cc} \mathbf{M}_1^{-1} & 0 \\ \mathbf{M}_2^{-1}\mathbf{R}_\psi & 0 \\ 0 & 1 \end{array}\right] \left[\begin{array}{c} w_1 \\ w_2 \\ w_3 \end{array}\right]$$

An input $w_1 = w_2 = 0, w_3 \neq 0$ generates a rotation about the common normal direction with fixed contact point. This type of motion is actually impossible to control in most practical cases of manipulation due to friction and micro–deformations of contacting bodies. Therefore, a further constraint is to be considered on the rate of change of ψ. As the contact points move on $\mathbf{c}_1(s)$, $\mathbf{c}_2(s)$ at unit velocity, while $\mathbf{n}^T \boldsymbol{\omega} = \frac{\partial \psi}{\partial t}$ has zero component along the normal, by derivation of (2.3), noticing that, for any unit vector $\mathbf{z}$, $\left(\frac{d}{ds}\mathbf{z}\right)^T \mathbf{z} = 0$, and using the fact that

$$\begin{aligned} \frac{\mathbf{h}_x}{\|\mathbf{h}_x\|} &= \frac{\mathbf{f}_u}{\|\mathbf{f}_u\|} \cos\psi - \frac{\mathbf{f}_v}{\|\mathbf{f}_v\|} \sin\psi \\ \frac{\mathbf{f}_u}{\|\mathbf{f}_u\|} &= \frac{\mathbf{h}_x}{\|\mathbf{h}_x\|} \cos\psi - \frac{\mathbf{h}_y}{\|\mathbf{h}_y\|} \sin\psi, \end{aligned} \tag{2.5}$$

one gets

$$\dot{\psi} = \frac{\partial \psi}{\partial s}\dot{s} + \frac{\partial \psi}{\partial t} = \left((\frac{d}{ds}\frac{\mathbf{f}_u}{\|\mathbf{f}_u\|})^T(\frac{\mathbf{f}_v}{\|\mathbf{f}_v\|}) + (\frac{d}{ds}\frac{\mathbf{h}_x}{\|\mathbf{h}_x\|})^T(\frac{\mathbf{h}_y}{\|\mathbf{h}_y\|}) \right)\dot{s}. \tag{2.6}$$

Let $\mathbf{v}(s) = \dot{\mathbf{c}}_1(s)$ the tangent to the curve $\mathbf{c}_1$ at $\mathbf{c}_1(s)$. Then the geodesic curvature k_g of $\mathbf{c}_1$ at each time is defined as the algebraic value of the covariant derivative of $\mathbf{v}$ at each time, i.e.

$$k_g = \left[\frac{D\mathbf{v}}{ds}\right] = \frac{d\mathbf{v}}{ds}^T(\mathbf{n}_1 \times \mathbf{v}) \tag{2.7}$$

$\mathbf{v}(s)$ belongs to the tangent plane of surface S_1 at point $\mathbf{c}_1(s)$ then the following relations hold:

$$\begin{aligned}
\mathbf{v} &= \tfrac{\mathbf{f}_u}{\|\mathbf{f}_u\|}\cos\phi + \tfrac{\mathbf{f}_v}{\|\mathbf{f}_v\|}\sin\phi \\
\mathbf{v}\cos\phi &= \tfrac{\mathbf{f}_u}{\|\mathbf{f}_u\|}\cos^2\phi + \tfrac{\mathbf{f}_v}{\|\mathbf{f}_v\|}\sin\phi\cos\phi \\
\mathbf{v}\sin\phi &= \tfrac{\mathbf{f}_u}{\|\mathbf{f}_u\|}\cos\phi\sin\phi + \tfrac{\mathbf{f}_v}{\|\mathbf{f}_v\|}\sin^2\phi \\
(\tfrac{d\mathbf{v}}{ds})^T\tfrac{\mathbf{f}_u}{\|\mathbf{f}_u\|} + \mathbf{v}^T(\tfrac{d}{ds}\tfrac{\mathbf{f}_u}{\|\mathbf{f}_u\|}) &= -\sin\phi\tfrac{d\phi}{ds} \\
(\tfrac{d\mathbf{v}}{ds})^T\tfrac{\mathbf{f}_v}{\|\mathbf{f}_v\|} + \mathbf{v}^T(\tfrac{d}{ds}\tfrac{\mathbf{f}_v}{\|\mathbf{f}_v\|}) &= \cos\phi\tfrac{d\phi}{ds}
\end{aligned}$$

Deriving the equation for the geodesic curvature and using the relations above together with $\frac{\mathbf{f}_u}{\|\mathbf{f}_u\|}^T(\frac{d}{ds}\frac{\mathbf{f}_u}{\|\mathbf{f}_u\|}) = 0$, $\frac{\mathbf{f}_v}{\|\mathbf{f}_v\|}^T(\frac{d}{ds}\frac{\mathbf{f}_v}{\|\mathbf{f}_v\|}) = 0$ we obtain:

$$\begin{aligned}
k_g &= \left[\tfrac{D\mathbf{v}}{ds}\right] = \tfrac{d\mathbf{v}}{ds}^T(\mathbf{n}_1 \times \mathbf{v}) = \\
&= (\tfrac{d\mathbf{v}}{ds})^T(\mathbf{f}_v\cos\phi - \mathbf{f}_u\sin\phi) = \\
&= \tfrac{d\phi}{ds}\cos\phi - \mathbf{v}^T(\tfrac{d}{ds}\tfrac{\mathbf{f}_v}{\|\mathbf{f}_v\|})\cos\phi + \tfrac{d\phi}{ds}\sin\phi - \mathbf{v}^T(\tfrac{d}{ds}\tfrac{\mathbf{f}_u}{\|\mathbf{f}_u\|})\sin\phi = \\
&= \tfrac{d\phi}{ds} - \tfrac{\mathbf{f}_u}{\|\mathbf{f}_u\|}^T(\tfrac{d}{ds}\tfrac{\mathbf{f}_v}{\|\mathbf{f}_v\|})\cos^2\phi + \tfrac{\mathbf{f}_v}{\|\mathbf{f}_v\|}^T(\tfrac{d}{ds}\tfrac{\mathbf{f}_u}{\|\mathbf{f}_u\|})\sin^2\phi = \\
&= \tfrac{d\phi}{dt} + \tfrac{\mathbf{f}_v}{\|\mathbf{f}_v\|}^T(\tfrac{d}{ds}\tfrac{\mathbf{f}_u}{\|\mathbf{f}_u\|}).
\end{aligned}$$

Hence the part of $\dot{\psi}$ relative to the first surface, i.e. $\frac{\mathbf{f}_v}{\|\mathbf{f}_v\|}^T(\frac{d}{ds}\frac{\mathbf{f}_u}{\|\mathbf{f}_u\|})$, is given by the sum of the geodesic curvature and $\frac{d\phi}{ds}$ which is the derivative of the angle between $\mathbf{f}_u$ and the tangent to the curve. For the second surface it holds a similar relation but the geodesic curvature is calculated w.r.t. the coordinate frame of the tangent plane to the second surface. Then the expression of $\mathbf{v}$ in the coordinates of the tangent plane is given by: $\mathbf{v} = \frac{\mathbf{h}_x}{\|\mathbf{h}_x\|}\cos(-\phi - \psi) + \frac{\mathbf{h}_y}{\|\mathbf{h}_y\|}\sin(-\phi - \psi)$ and for the geodesic curvature we have:

$$\begin{aligned}
-k_g &= \left[\tfrac{D\mathbf{v}}{ds}\right] = (\tfrac{d\mathbf{v}}{ds})^T(\mathbf{n}_2 \times \mathbf{v}) = \\
&= -\tfrac{d\phi}{ds} - \tfrac{d\psi}{ds} + \tfrac{\mathbf{h}_y}{\|\mathbf{h}_y\|}^T(\tfrac{d}{ds}\tfrac{\mathbf{h}_x}{\|\mathbf{h}_x\|})
\end{aligned}$$

Equating the two relations obtained for the geodesic curvature we have again the expression for $\dot{\psi}$. The part of the angle ψ relative to each surface varies with the geodesic curvature and with the angle that the tangent to the curve makes with a coordinate direction. Then it is clear that along a geodesic curve the angle ψ varies only as long as the coordinate directions vary along to curve, i.e. the geodesic curve is not a coordinate one. Along a piecewise regular curve $\phi(s)$ incurs in discontinuities whereas the tangent to the curve is discontinuous. Despite this $\psi(s)$ is everywhere continuous infact it depends on the parametrization only. Also we have that along a piecewise regular closed curve on surface S_1, the integral of

$$\frac{\mathbf{f}_v}{\|\mathbf{f}_v\|}^T\left(\frac{d}{dt}\frac{\mathbf{f}_u}{\|\mathbf{f}_u\|}\right)$$

can be calculated using the Gauss-Bonnet and the "turning tangents" theorems and is equal to the integral of the gaussian curvature of surface S_1 over the region bounded by the curve. In the case where surface S_2 is a plane this result would give immediately the change of the whole of ψ, i.e. the total change of orientation of a body rolling on a plane [**8**].

By introducing Cartan's notation of *torsion forms*,

$$\mathbf{T}_1 = \begin{bmatrix} \frac{\mathbf{f}_v^T \mathbf{f}_{uu}}{\|\mathbf{f}_u\|^2\|\mathbf{f}_v\|} & \frac{\mathbf{f}_v^T \mathbf{f}_{uv}}{\|\mathbf{f}_u\|\|\mathbf{f}_v\|^2} \end{bmatrix} \mathbf{T}_2 = \begin{bmatrix} \frac{\mathbf{h}_y^T \mathbf{h}_{xx}}{\|\mathbf{h}_x\|^2\|\mathbf{h}_y\|} & \frac{\mathbf{h}_y^T \mathbf{h}_{xy}}{\|\mathbf{h}_x\|\|\mathbf{h}_y\|^2} \end{bmatrix},$$

the two constraints (2.4), (2.6) are rewritten together as

$$\begin{bmatrix} \mathbf{M}_1 & -\mathbf{R}_\psi \mathbf{M}_2 & 0 \\ \mathbf{T}_1\mathbf{M}_1 & \mathbf{T}_2\mathbf{M}_2 & -1 \end{bmatrix} \dot{\mathbf{q}} = 0, \tag{2.8}$$

and the corresponding control form is

$$\dot{\mathbf{q}} = \begin{bmatrix} \mathbf{M}_1^{-1} \\ \mathbf{M}_2^{-1}\mathbf{R}_\psi \\ \mathbf{T}_1 + \mathbf{T}_2\mathbf{R}_\psi \end{bmatrix} \begin{bmatrix} w_1 \\ w_2 \end{bmatrix} \tag{2.9}$$

The physical meaning of inputs w_1, w_2, w_3 are the velocities of the system along the allowed directions of motion. The relationship of these inputs to the physically accessible variables, i.e. to the relative angular velocity of the bodies $\boldsymbol{\omega}$, can be calculated as

$$\boldsymbol{\omega} = \mathbf{S}\left[\begin{array}{c|c} -\mathbf{M}_1^{-1}\,\mathcal{II}_1\,\mathbf{M}_1^{-1} - \mathbf{R}_\psi\mathbf{M}_2^{-1}\,\mathcal{II}_2\,\mathbf{M}_2^{-1}\mathbf{R}_\psi & 0 \\ \hline -\mathbf{T}_1 - \mathbf{T}_2\mathbf{R}_\psi & 1 \end{array}\right] \begin{bmatrix} w_1 \\ w_2 \\ \hline w_3 \end{bmatrix},$$

where $\mathcal{II}_i$ is the second fundamental form of surface S_i, and

$$\mathbf{S} = \begin{bmatrix} 0 & 1 & 0 \\ -1 & 0 & 0 \\ 0 & 0 & 1 \end{bmatrix}.$$

Observing that

$$-\mathbf{M}_1^{-1}\,\mathcal{II}_1\,\mathbf{M}_1^{-1} - \begin{bmatrix} \frac{\mathbf{f}_u^T \mathbf{n}_{1u}}{\|\mathbf{f}_u\|^2} & \frac{\mathbf{f}_u^T \mathbf{n}_{1v}}{\|\mathbf{f}_u\|\|\mathbf{f}_v\|} \\ \frac{\mathbf{f}_v^T \mathbf{n}_{1u}}{\|\mathbf{f}_u\|\|\mathbf{f}_v\|} & \frac{\mathbf{f}_v^T \mathbf{n}_{1v}}{\|\mathbf{f}_v\|} \end{bmatrix} = \mathbf{K}_1;$$

$$-\mathbf{M}_2^{-1}\,\mathcal{II}_2\,\mathbf{M}_2^{-1} = \begin{bmatrix} \frac{\mathbf{h}_x^T \mathbf{n}_{2x}}{\|\mathbf{h}_x\|^2} & \frac{\mathbf{h}_x^T \mathbf{n}_{2y}}{\|\mathbf{h}_x\|\|\mathbf{h}_y\|} \\ \frac{\mathbf{h}_y^T \mathbf{n}_{2x}}{\|\mathbf{h}_x\|\|\mathbf{h}_y\|} & \frac{\mathbf{h}_y^T \mathbf{n}_{2y}}{\|\mathbf{h}_y\|} \end{bmatrix} = \mathbf{K}_2,$$

where $\mathbf{K}_1$ and $\mathbf{K}_2$ are the *curvature forms* for the two surfaces. Introducing the *relative curvature form*

$$\mathbf{K}_R = \mathbf{K}_1 + \mathbf{R}_\psi\mathbf{K}_2\mathbf{R}_\psi,$$

a more compact equation is obtained as

$$\boldsymbol{\omega} = \mathbf{S}\begin{bmatrix} \mathbf{K}_R & 0 \\ -\mathbf{T}_1 - \mathbf{T}_2\mathbf{R}_\psi & 1 \end{bmatrix} \begin{bmatrix} w_1 \\ w_2 \\ w_3 \end{bmatrix}. \tag{2.10}$$

2.2. Admissibility of rolling contacts. Rolling contacts of solid objects imposes some further constraints than those studied in the previous section for surfaces. We will model henceforth objects as closed subsets of $\mathbb{R}^3$, whose boundary is a surface on which the normal direction is conventionally taken to be directed outwards. We will deal with conditions for admissibility of contact *per se*, and of rolling motions separately.

DEFINITION 2.1. A contact is locally admissible if the intersection of the interior parts of the objects in a neighborhood B_c of the contact point $\mathbf{f}(\bar{u}, \bar{v}) = \mathbf{h}(\bar{x}, \bar{y})$ is void.

A simple necessary condition for admissibility of contact is as follows:

LEMMA 2.2. *If the contact between two surfaces at a point $\mathbf{f}(\bar{u}, \bar{v}) = \mathbf{h}(\bar{x}, \bar{y})$, with relative orientation ψ, is locally admissible, then the relative curvature form $\mathbf{K}_R = \mathbf{K}_1 + \mathbf{R}_\psi \mathbf{K}_2 \mathbf{R}_\psi$ is semipositive definite. If $\mathbf{K}_R$ is strictly positive definite, then the contact is locally admissible.*

PROOF. The local admissibility condition can be expressed as the requirement that for any two points $\mathbf{f}(u, v), \mathbf{h}(x, y)$ on $S_1 \cap B_c$ and $S_2 \cap B_c$ such that

$$(\mathbf{f}(u,v) - \mathbf{h}(x,y)) \times \mathbf{n}_1 = 0, \tag{2.11}$$

it holds

$$(\mathbf{f}(u,v) - \mathbf{h}(x,y))^T \mathbf{n}_1 = \mathbf{f}(u,v)^T \mathbf{n}_1 + \mathbf{h}(x,y))^T \mathbf{n}_2 \leq 0. \tag{2.12}$$

Considering the second–order development of $\mathbf{f}(u, v)$ about the point of contact, the (signed) distance to the tangent plane is given by

$$\mathbf{n}_1^T \mathbf{f}(u,v) \approx \mathbf{n}_1^T \mathbf{f}(\bar{u},\bar{v}) - \frac{1}{2} \begin{bmatrix} \delta u & \delta v \end{bmatrix} \mathbf{M}_1 \mathbf{K}_1 \mathbf{M}_1 \begin{bmatrix} \delta u \\ \delta v \end{bmatrix}$$

$$= \mathbf{n}_1^T \mathbf{f}(\bar{u},\bar{v}) - \frac{1}{2} [\mathbf{n}_{1u}^T \mathbf{f}_u (\delta u)^2 + \mathbf{n}_{1u}^T \mathbf{f}_v (\delta u)(\delta v) + \mathbf{n}_{1v}^T \mathbf{f}_u (\delta u)(\delta v) + \mathbf{n}_{1v}^T \mathbf{f}_v (\delta v)^2],$$

where $\delta u = (u - \bar{u})$ and $\delta v = (v - \bar{v})$, and the relation of orthogonality between $\mathbf{n}$, $\mathbf{f}_u$ and $\mathbf{f}_v$ have been used. Similarly we obtain the expression for the second surface as

$$(\mathbf{n}_2)^T \mathbf{h}(x,y) \approx (\mathbf{n}_2)^T \mathbf{h}(\bar{x},\bar{y}) - \frac{1}{2} \begin{bmatrix} \delta x & \delta y \end{bmatrix} \mathbf{M}_2 \mathbf{K}_2 \mathbf{M}_2 \begin{bmatrix} \delta x \\ \delta y \end{bmatrix}.$$

Points that satisfy (2.11) share the same projection ζ, ξ on the tangent plane. Their coordinates relative to the contact point can be evaluated as

$$\left\{ \begin{bmatrix} \delta u \\ \delta v \end{bmatrix} = \mathbf{M}_1^{-1} \begin{bmatrix} \delta \zeta \\ \delta \xi \end{bmatrix}; \quad \begin{bmatrix} \delta x \\ \delta y \end{bmatrix} = \mathbf{M}_2^{-1} \mathbf{R}_\psi \begin{bmatrix} \delta \zeta \\ \delta \xi \end{bmatrix} \right. .$$

Hence we obtain

$$(\mathbf{n}_2)^T \mathbf{h}(x,y) = (\mathbf{n}_2)^T \mathbf{h}(\bar{x},\bar{y}) - \frac{1}{2} \begin{bmatrix} \delta u \\ \delta v \end{bmatrix} \mathbf{M}_1^T \mathbf{R}_\psi \mathbf{K}_2 \mathbf{R}_\psi \mathbf{M}_1 \begin{bmatrix} \delta u \\ \delta v \end{bmatrix}$$

The contact admissibility condition (2.12) in a neighborhood of the contact point can therefore be written

$$\frac{1}{2} \begin{bmatrix} \delta u & \delta v \end{bmatrix} \mathbf{M}_1 \mathbf{K}_R \mathbf{M}_1 \begin{bmatrix} \delta u \\ \delta v \end{bmatrix} \geq 0,$$

i.e. the relative curvature form has to be semipositive definite. If $\mathbf{K}_R$ is positive definite, the contact point is isolated in its neighborhood. We call a pair of surfaces in such configuration *relatively strictly convex.* If $\mathsf{rank}\,(\mathbf{K}_R) = 1$ and $\mathbf{K}_R$ *s.p.d.*, contact may not be admissible. If it is, there exists a one–dimensional manifold of contact points (the tangent bundle to this manifold being $\ker(\mathbf{K}_R)$). We call a pair of surfaces in such configuration *relatively weakly convex.* If $\mathsf{rank}\,(\mathbf{K}_R) = 0$, contact may not be admissible. If it is, there exists a two–dimensional manifold of contact points. We call such pair of surfaces *relatively flat*, or conformal. Note that these concepts are strictly local. If $\mathbf{K}_R$ is not definite, or seminegative definite, the contact is not admissible. □

DEFINITION 2.3. A rolling motion between two surfaces in an admissible contact configuration is admissible with respect to constraints (2.4), [resp., (2.4) and (2.6)] if there exist some nonzero relative angular velocity such that (2.4) [(2.4) and (2.6)] are satisfied at each point of contact. An admissible rolling motion is said to have p degrees-of-freedom (d.o.f.'s) if there exist a p–dimensional linear space of angular velocities satisfying (2.4) [(2.4) and (2.6)].

LEMMA 2.4. *Locally around an admissible contact configuration, the d.o.f.'s of admissible rolling motions are as reported in the table below:*

	Constraint (2.4)	*Constraint (2.4) and (2.6)*
$\mathsf{rank}\,(\mathbf{K}_R) = 2$	$p = 3$	$p = 2$
$\mathsf{rank}\,(\mathbf{K}_R) = 1$	$p \leq 1$	$p \leq 1$
$\mathsf{rank}\,(\mathbf{K}_R) = 0$	$p = 0$	$p = 0$

The cases $\mathsf{rank}\,(\mathbf{K}_R) = 1$, $p = 1$ *hold if and only if the contact manifold is linear.*

PROOF. Recall the relation (2.10) between the relative angular velocity $\boldsymbol{\omega}$ and the velocity along the unconstrained directions $\mathbf{w}$, and the fact that, under constraints (2.4) and (2.6), $w_3 = 0$. The two cases with $\mathsf{rank}\,(\mathbf{K}_R) = 2$, and the case with $\mathsf{rank}\,(\mathbf{K}_R) = 0$ under constraints (2.4) and (2.6) follow directly.

When the contact manifold is not an isolated point, condition (2.4) must hold at every point in the contact manifold. Because only one axis of fixed points may exist under a rotation $\omega_3 = w_3$, and this axis is normal to the surfaces, w_3 must be zero to avoid slippage at some points of the contact manifold, hence condition (2.4) alone is equivalent to (2.4) and (2.6). The case $\mathsf{rank}\,(\mathbf{K}_R) = 0$ under constraint (2.4) follows.

If $\mathsf{rank}\,(\mathbf{K}_R) = 1$, using the fact that the left upper 2–by–2 block of $\mathbf{S}$ is antisymmetric, and that $\mathbf{K}_R = \mathbf{K}_R^T$, one gets

$$\begin{bmatrix} 0 & -1 \\ 1 & 0 \end{bmatrix} \begin{bmatrix} \omega_1 \\ \omega_2 \end{bmatrix} \in \mathsf{range}\,(\mathbf{K}_R) \quad \Rightarrow \begin{bmatrix} \omega_1 \\ \omega_2 \end{bmatrix} \in \ker(\mathbf{K}_R).$$

From the fact that $\boldsymbol{\omega}$ is the same for all points of the rolling bodies, and that $\ker(\mathbf{K}_R)$ is the tangent bundle to the contact manifold and is one–dimensional, it follows that the contact manifold must be a line segment (rolling will be possible about an axis including the segment, such as e.g. in the case of a cylinder or a cone rolling on a plane). □

Based on the two preceding lemmas, a classification of rolling pairs is reported below for the sake of completeness. Consider the diagonal decomposition of the curvature forms

$$\mathbf{K}_1 = \mathbf{Q}_1 \begin{bmatrix} k_1 & 0 \\ 0 & k_2 \end{bmatrix} \mathbf{Q}_1^T \quad \mathbf{K}_2 = \mathbf{Q}_2 \begin{bmatrix} h_1 & 0 \\ 0 & h_2 \end{bmatrix} \mathbf{Q}_2^T$$
$$\mathbf{Q}_1^T \mathbf{R}_\psi \mathbf{Q}_2 = \begin{bmatrix} a & -b \\ -b & -a \end{bmatrix}$$

with (k_1, k_2), (h_1, h_2) the principal curvatures of the first and the second object. Assume the eigenvalues are ordered so that $k_1 \geq k_2$; $h_1 \geq h_2$; $k_1 \geq h_1$. The relative curvature form $\mathbf{K}_R$ is given by

$$\mathbf{K}_R = \mathbf{Q}_1 \left(\begin{bmatrix} k_1 + h_1 & 0 \\ 0 & k_2 + h_2 \end{bmatrix} - b(h_1 - h_2) \begin{bmatrix} b & -a \\ -a & -b \end{bmatrix} \right) \mathbf{Q}_1^T$$

The following cases apply:

A: Surfaces have an isolated contact point and have two rolling d.o.f.'s iff

$$\begin{aligned} \det \mathbf{K}_R &= (k_1 + h_1)(k_2 + h_2) + b^2(k_1 - k_2)(h_1 - h_2) > 0 \\ \mathbf{K}_R^{11} &= (k_1 + h_1) - b^2(h_1 - h_2) > 0 \\ \mathbf{K}_R^{22} &= (k_2 + h_2) + b^2(h_1 - h_2) > 0 \end{aligned}$$

where $\mathbf{K}_R^{ij}$ denotes the i, j-th element of matrix $\mathbf{K}_R$. These conditions hold $\forall \psi$ iff $-h_1 \leq -h_2 < k_2 \leq k_1$. They hold for some values of ψ iff $-h_1 < k_2 \leq -h_2 < k_1$.

B: Surfaces have a one–dimensional contact manifold iff

(i):

$$\begin{aligned} \det \mathbf{K}_R &= (k_1 + h_1)(k_2 + h_2) + b^2(k_1 - k_2)(h_1 - h_2) = 0 \\ \mathbf{K}_R^{11} &= (k_1 + h_1) - b^2(h_1 - h_2) > 0 \\ \mathbf{K}_R^{22} &= (k_2 + h_2) + b^2(h_1 - h_2) \geq 0, \end{aligned}$$

or

(ii):

$$\begin{aligned} \det \mathbf{K}_R &= (k_1 + h_1)(k_2 + h_2) + b^2(k_1 - k_2)(h_1 - h_2) = 0 \\ \mathbf{K}_R^{11} &= (k_1 + h_1) - b^2(h_1 - h_2) = 0 \\ \mathbf{K}_R^{22} &= (k_2 + h_2) + b^2(h_1 - h_2) > 0 \end{aligned}$$

Case **B(i)** holds $\forall \psi$ iff $-h_1 = -h_2 = k_2 < k_1$; $\forall \psi \neq \pm\frac{\pi}{2}$ iff $-h_1 < -h_2 = k_2 = k_1$; and for $\psi = \{0, \pi\}$ iff $-h_1 < -h_2 = k_2 < k_1$. Case **B(ii)** holds for $\psi = \pm\frac{\pi}{2}$ iff $-h_1 < k_2 \leq -h_2 = k_1$.

Moreover, if surfaces have one rolling d.o.f., then $h_1 h_2 \leq 0$ and $k_1 k_2 \leq 0$ (contact points must be hyperbolic or parabolic).

Possible cases and examples are summarized in table 1.

3. Holonomy of rolling

In this section we investigate the cases when the constraints of rolling are holonomic, or, in other words, when the maximal integral submanifold of the control system (2.9) is two–dimensional. From the application viewpoint described in the introduction, this corresponds to cases when the increase in dexterity due to nonholonomy cannot be expected to help robotic manipulation. Also, because of the application motivation, we focus our attention on the case that both rolling constraints (2.4) and (2.6) are enforced. In these conditions, (2.9) represents the

Sign of principal curvatures				Example	Rolling d.o.f.'s
h_2	h_1	k_2	k_1	$\forall\psi$	
-	-	+	+	sphere inside hollow sphere	2
-	0	+	+	sphere inside hollow cylinder	2
-	+	+	+	sphere onto saddle	2
0	+	+	+	sphere onto cylinder	2
+	+	+	+	sphere onto sphere	2
h_2	h_1	k_2	k_1	$\exists\psi$	
-	-	+	+	ellipsoid inside hollow ellipsoid	2
-	0	+	+	ellipsoid inside hollow cylinder	2
-	+	-	+	saddle onto saddle	2
-	+	0	+	cylinder onto saddle	2
-	+	+	+	ellipsoid onto saddle	2
0	+	0	+	cylinder onto cylinder	2
0	+	-	+	saddle onto cylinder	2
+	+	-	+	saddle onto ellipsoid	2
h_2	h_1	k_2	k_1	$\forall\psi$	
0	0	0	+	cylinder onto plane	1
h_2	h_1	k_2	k_1	$\exists\psi$	
-	0	0	+	cylinder inside hollow cylinder	1
-	+	-	+	ruled saddle onto ruled saddle	1
-	+	0	+	cylinder onto ruled saddle	1
0	+	-	+	ruled saddle onto cylinder	1
0	+	0	+	cylinder onto cylinder	1

TABLE 1. A classification of admissible rolling contacts

control form of the kinematics of rolling, where $\mathbf{w}$ is further subject to possible restictions. These restrictions may be entailed (because of (2.10)) by limitations on $\boldsymbol{\omega}$ as discussed in section 2.2. Consider first cases where the two surfaces are relatively weakly convex, and have instantaneously one rolling d.o.f.. From (2.10), $[\omega_1,\ \omega_2]^T \in \ker(\mathbf{K}_r) \Rightarrow [w_1 w_2]^T \in \text{ range } (\mathbf{K}_R)$.

If the set of configurations for which $\text{rank}\,(\mathbf{K}_R) = 1$ is open in the configuration manifold M, the control system (2.9) has a single nonsingular control vector field, which is obviously involutive. Therefore, the reachable submanifold is a one–dimensional submanifold of M. This is the case of a cylinder or a cone rolling on a plane, for which the allowed angular velocity is always perpendicular to the contact line.

If otherwise $\text{rank}\,(\mathbf{K}_R) = 1$ in a submanifold of M including the contact configuration $\bar{\mathbf{q}}$, the control system (2.9) has a singularity in $\bar{\mathbf{q}}$. Due to analyticity of the manifold and of the control equations, there will be maximal integral submanifolds for the system motion. However, the dimension of these submanifolds may vary from one (e.g., two cylinders rolling on each other with aligned axes), to five (e.g., a cone rolling on a cylinder, starting with aligned axes).

We consider now the case where $\mathbf{K}_R$ is full rank. Recall that the image of a surface $\mathbf{h}(x, y)$ through a rigid motion $g = (\mathbf{R}, \mathbf{z}) \in SE(3)$ is defined as $g \circ \mathbf{h}(x, y) = \mathbf{R}\mathbf{h}(x, y) + \mathbf{z}$. Let $s = (\hat{\mathbf{\Sigma}}, \hat{\mathbf{t}})$, with $\hat{\mathbf{\Sigma}} \in O(3)$ and $\det\,(\hat{\mathbf{\Sigma}}) = -1$, and $\hat{\mathbf{t}} \in \mathbb{R}^3$ denote a symmetry. In particular, by taking $\hat{\mathbf{\Sigma}} = \hat{\mathbf{R}}\mathbf{\Sigma}\hat{\mathbf{R}}^T$, with

$$\mathbf{\Sigma} = \begin{bmatrix} 1 & 0 & 0 \\ 0 & 1 & 0 \\ 0 & 0 & -1 \end{bmatrix},$$

$$\begin{aligned} \hat{\mathbf{R}} &= \left[\frac{\mathbf{f}_u(\bar{u},\bar{v})}{\|\mathbf{f}_u(\bar{u},\bar{v})\|} \quad \frac{\mathbf{f}_v(\bar{u},\bar{v})}{\|\mathbf{f}_v(\bar{u},\bar{v})\|} \quad \frac{\mathbf{f}_u(\bar{u},\bar{v}) \times \mathbf{f}_v(\bar{u},\bar{v})}{\|\mathbf{f}_u(\bar{u},\bar{v}) \times \mathbf{f}_v(\bar{u},\bar{v})\|} \right], \\ \hat{\mathbf{t}} &= -\hat{\mathbf{R}}(\mathbf{\Sigma} - Id)\hat{\mathbf{R}}^T \mathbf{f}(\bar{u}, \bar{v}) \end{aligned} \tag{3.1}$$

then $s \circ \mathbf{f}(u, v) = \hat{\mathbf{\Sigma}}\mathbf{f}(u, v) + \hat{\mathbf{t}}$ is the symmetric surface of $\mathbf{f}(u, v)$ with respect to a plane tangent to $\mathbf{f}(u, v)$ through the point $\mathbf{f}(\bar{u}, \bar{v})$.

We will also need the following definition:

DEFINITION 3.1. A surface $\mathbf{h}(x, y)$ is said to be the *specular image* of a surface $\mathbf{f}(u, v)$ at a point $\mathbf{f}(\bar{u}, \bar{v})$ if there exist a rigid motion $g = (\mathbf{R}, \mathbf{z})$, a symmetry $s = (\hat{\mathbf{R}}\mathbf{\Sigma}\hat{\mathbf{R}}^T, \hat{\mathbf{t}})$ as in (3.1), and a diffeomorphism $\phi : \mathbb{R}^2 \to \mathbb{R}^2$, $\phi : (x, y) \mapsto (u, v)$ such that $g \circ \mathbf{h}(x, y) = s \circ \mathbf{f} \circ \phi(x, y)$.

Notice explicitly that a surface and its specular image, as defined above, have a point in common, the mirror plane being the tangent plane at such contact point.

PROPOSITION 3.2. If two surfaces are specular, the change of coordinates $\phi : \mathbb{R}^2 \to \mathbb{R}^2$ in definition 3.1 is a map whose Jacobian $\mathbf{J}_\phi$ has either of these forms:

$$\mathbf{J}_\phi = \pm \begin{bmatrix} \frac{\|\mathbf{h}_x\|}{\|\mathbf{f}_u\|} & 0 \\ 0 & -\frac{\|\mathbf{h}_y\|}{\|\mathbf{f}_v\|} \end{bmatrix} \quad or \quad \mathbf{J}_\phi = \pm \begin{bmatrix} 0 & \frac{\|\mathbf{h}_x\|}{\|\mathbf{f}_v\|} \\ \frac{\|\mathbf{h}_y\|}{\|\mathbf{f}_u\|} & 0 \end{bmatrix}$$

PROOF. Suppose $g \circ \mathbf{h}(x, y) = s \circ \mathbf{f} \circ \phi(u, v)$, then

$$\begin{aligned} (g \circ \mathbf{h})_x &= (s \circ \mathbf{f})_x &\Rightarrow \quad \mathbf{R}\mathbf{h}_x &= \hat{\mathbf{\Sigma}}\mathbf{f}_u u_x + \hat{\mathbf{\Sigma}}\,\mathbf{f}_v v_x \\ (g \circ \mathbf{h})_y &= (s \circ \mathbf{f})_y &\Rightarrow \quad \mathbf{R}\mathbf{h}_y &= \hat{\mathbf{\Sigma}}\mathbf{f}_u u_y + \hat{\mathbf{\Sigma}}\,\mathbf{f}_v v_y. \end{aligned}$$

The condition of orthogonality of the local coordinates implies that the inner product $(\mathbf{R}\mathbf{h}_x)^T(\mathbf{R}\mathbf{h}_y) = \|\mathbf{f}_u\|^2 u_x u_y + \|\mathbf{f}_v\|^2 v_x v_y$ is zero. Hence the three conditions $u_x u_y = 0$, $v_x v_y = 0$ and $\text{rank}\,(J_\phi) = 2$ together gives either $u_x, v_y \neq 0$, $u_y = v_x = 0$

or $u_x = v_y = 0,\ u_y, v_x \neq 0$. Suppose the first case holds. The inner and cross products of both sides are evaluated, recalling that $\det(\hat{\mathbf{\Sigma}}) = -1$, as

$$\begin{array}{rcl} \|\mathbf{h}_x\|^2 & = & \|\mathbf{f}_u\|^2 u_x^2 \\ \|\mathbf{h}_y\|^2 & = & \|\mathbf{f}_v\|^2 v_y^2 \\ \mathbf{R}(\mathbf{h}_x \times \mathbf{h}_y) & = & -u_x v_y \hat{\mathbf{\Sigma}}\,(\mathbf{f}_u \times \mathbf{f}_v) \end{array},$$

hence

$$|u_x| = \frac{\|\mathbf{h}_x\|}{\|\mathbf{f}_u\|}; \quad |v_y| = \frac{\|\mathbf{h}_y\|}{\|\mathbf{f}_v\|}. \tag{3.2}$$

The sign of the components of $\mathbf{J}_\phi$ is derived considering the convention of outwardly pointing normals. This implies $-1 = \frac{(\mathbf{f}_u \times \mathbf{f}_v)^T}{\|\mathbf{f}_u \times \mathbf{f}_v\|}\mathbf{R}\frac{(\mathbf{h}_x \times \mathbf{h}_y)}{\|\mathbf{h}_x \times \mathbf{h}_y\|} = -\frac{(\mathbf{f}_u \times \mathbf{f}_v)^T}{\|\mathbf{f}_u\|\|\mathbf{f}_v\|}\hat{\mathbf{\Sigma}}\frac{(\mathbf{f}_u \times \mathbf{f}_v)}{\|\mathbf{f}_u\|\|\mathbf{f}_v\|}\frac{u_x v_y}{|u_x||v_y|}$. Using the fact that $\frac{(\mathbf{f}_u \times \mathbf{f}_v)^T}{\|\mathbf{f}_u\|\|\mathbf{f}_v\|}\hat{\mathbf{\Sigma}}\frac{(\mathbf{f}_u \times \mathbf{f}_v)}{\|\mathbf{f}_u\|\|\mathbf{f}_v\|} = -1$, we get $\frac{u_x v_y}{|u_x||v_y|} = -1$, hence u_x, v_y have opposite signs and $\mathbf{J}_\phi$ has the first of the two forms in the proposition statement.

If instead $u_x = v_y = 0,\ u_y, v_x \neq 0$, similar arguments can be used to prove that $\mathbf{J}_\phi$ has the second form. □

Our main result is stated as follows:

THEOREM 3.3. *The rolling contact constraint between two relatively strictly convex surfaces is holonomic if and only if they are specular images of each other.*

PROOF. Part 1: Holonomy $\Leftarrow$ Specularity.
The control vector fields in (2.9) are written explicitly as

$$\mathbf{g}_1 = \begin{bmatrix} \frac{1}{\|\mathbf{f}_u\|} \\ 0 \\ \frac{\cos\psi}{\|\mathbf{h}_x\|} \\ \frac{-\sin\psi}{\|\mathbf{h}_y\|} \\ \hat{t}_1 \end{bmatrix} \quad \text{and} \quad \mathbf{g}_2 = \begin{bmatrix} 0 \\ \frac{1}{\|\mathbf{f}_v\|} \\ \frac{-\sin\psi}{\|\mathbf{h}_x\|} \\ \frac{-\cos\psi}{\|\mathbf{h}_y\|} \\ \hat{t}_2 \end{bmatrix},$$

where

$$\begin{array}{rcl} \hat{t}_1 & = & t_1 + \cos\psi\tilde{t}_1 - \sin\psi\tilde{t}_2 \\ \hat{t}_2 & = & t_2 - \sin\psi\tilde{t}_1 - \cos\psi\tilde{t}_2 \\ t_1 & = & \frac{\mathbf{f}_v^T\mathbf{f}_{uu}}{\|\mathbf{f}_u\|^2\|\mathbf{f}_v\|} \\ t_2 & = & \frac{\mathbf{f}_v^T\mathbf{f}_{uv}}{\|\mathbf{f}_u\|\|\mathbf{f}_v\|^2} \\ \tilde{t}_1 & = & \frac{\mathbf{h}_y^T\mathbf{h}_{xx}}{\|\mathbf{h}_x\|^2\|\mathbf{h}_y\|} \\ \tilde{t}_2 & = & \frac{\mathbf{h}_y^T\mathbf{h}_{xy}}{\|\mathbf{h}_x\|\|\mathbf{h}_y\|^2}. \end{array}$$

The Lie bracket $\mathbf{g}_3 = [\mathbf{g}_1, \mathbf{g}_2]$ is evaluated as

$$\mathbf{g}_3 = \begin{bmatrix} -\frac{t_1}{\|\mathbf{f}_u\|} \\ -\frac{t_2}{\|\mathbf{f}_v\|} \\ \frac{1}{\|\mathbf{h}_x\|}(\tilde{t}_1 - \cos\psi\hat{t}_1 + \sin\psi\hat{t}_2) \\ \frac{1}{\|\mathbf{h}_y\|}(\tilde{t}_2 + \sin\psi\hat{t}_1 + \cos\psi\hat{t}_2) \\ \frac{t_{2u}}{\|\mathbf{f}_u\|} - \frac{t_{1v}}{\|\mathbf{f}_v\|} - \frac{\tilde{t}_{2x}}{\|\mathbf{h}_x\|} + \frac{\tilde{t}_{1y}}{\|\mathbf{h}_y\|} - \hat{t}_1(\hat{t}_1 - t_1) - \hat{t}_2(\hat{t}_2 - t_2) \end{bmatrix}, \tag{3.3}$$

where by t_{iz} we denote the derivative of t_i with respect to the variable z. It will be shown below that $\mathbf{g}_3 = -t_1\mathbf{g}_1 - t_2\mathbf{g}_2$.

In fact, the first two components trivially satisfy this condition. The third and fourth components do also, by simply observing that

$$\tilde{t}_1 - \hat{t}_1 \cos\psi + \hat{t}_2 \sin\psi = -t_1 \cos\psi + t_2 \sin\psi$$

and

$$\tilde{t}_2 + \hat{t}_1 \sin\psi + \hat{t}_2 \cos\psi = +t_1 \sin\psi + t_2 \cos\psi.$$

The last component of $\mathbf{g}_3$ satisfies the condition if the surfaces are specular. In this case, we know that there exists a change of coordinates ϕ such that $g\circ\mathbf{h}(x,y) = s\circ\mathbf{f}(\phi(x,y))$ with $\mathbf{J}_\phi$ as in Proposition 3.2. If

$$\mathbf{J}_\phi = \pm\begin{bmatrix} \frac{\|\mathbf{h}_x\|}{\|\mathbf{f}_u\|} & 0 \\ 0 & -\frac{\|\mathbf{h}_y\|}{\|\mathbf{f}_v\|} \end{bmatrix}$$

then

$$\begin{aligned} \mathbf{R}\mathbf{h}_x &= \frac{\|\mathbf{h}_x\|}{\|\mathbf{f}_u\|}\hat{\Sigma}\mathbf{f}_u \\ \mathbf{R}\mathbf{h}_y &= -\frac{\|\mathbf{h}_y\|}{\|\mathbf{f}_v\|}\hat{\Sigma}\mathbf{f}_v \end{aligned}$$

and

$$\begin{aligned} (g\circ\mathbf{h})_{xx} &= \hat{\Sigma}\frac{\mathbf{f}_u}{\|\mathbf{f}_u\|}\frac{\partial\|\mathbf{h}_x\|}{\partial x} + \|\mathbf{h}_x\|\hat{\Sigma}\frac{\partial}{\partial u}\frac{\mathbf{f}_u}{\|\mathbf{f}_u\|} \\ (g\circ\mathbf{h})_{xy} &= \hat{\Sigma}\frac{\mathbf{f}_u}{\|\mathbf{f}_u\|}\frac{\partial\|\mathbf{h}_x\|}{\partial y} + \|\mathbf{h}_x\|\hat{\Sigma}\frac{\partial}{\partial v}\frac{\mathbf{f}_u}{\|\mathbf{f}_u\|}, \end{aligned}$$

hence

$$\tilde{t}_1 = \frac{\mathbf{h}_y^T\mathbf{h}_{xx}}{\|\mathbf{h}_y\|\|\mathbf{h}_x\|^2} = -\frac{\mathbf{f}_v^T\mathbf{f}_{uu}}{\|\mathbf{f}_v\|\|\mathbf{f}_u\|^2} = -t_1 \quad \text{and} \quad \tilde{t}_2 = \frac{\mathbf{h}_y^T\mathbf{h}_{xy}}{\|\mathbf{h}_y\|^2\|\mathbf{h}_x\|} = \frac{\mathbf{f}_v^T\mathbf{f}_{uv}}{\|\mathbf{f}_v\|^2\|\mathbf{f}_u\|} = t_2,$$

and

$$\begin{aligned} \frac{t_{1v}}{\|\mathbf{f}_v\|} &= \frac{\tilde{t}_{1y}}{\|\mathbf{h}_y\|} \\ \frac{t_{2u}}{\|\mathbf{f}_u\|} &= \frac{\tilde{t}_{2x}}{\|\mathbf{h}_x\|}. \end{aligned}$$

If otherwise $\mathbf{J}_\phi$ has the second form in the statement of Proposition 1, then, by similar calculations and taking into account that, because of the orthogonality of the parametrization, it holds $\mathbf{f}_u^T\mathbf{f}_{vv} = -\mathbf{f}_v^T\mathbf{f}_{uv}$ and $\mathbf{f}_u^T\mathbf{f}_{uv} = -\mathbf{f}_v^T\mathbf{f}_{uu}$, one gets

$$\begin{aligned} \tilde{t}_1 = \frac{\mathbf{h}_y^T\mathbf{h}_{xx}}{\|\mathbf{h}_y\|\|\mathbf{h}_x\|^2} &= \frac{-\mathbf{f}_v^T\mathbf{f}_{uv}}{\|\mathbf{f}_u\|\|\mathbf{f}_v\|^2} = -t_2 \\ \tilde{t}_2 = \frac{\mathbf{h}_y^T\mathbf{h}_{xy}}{\|\mathbf{h}_y\|^2\|\mathbf{h}_x\|} &= \frac{-\mathbf{f}_v^T\mathbf{f}_{uu}}{\|\mathbf{f}_u\|^2\|\mathbf{f}_v\|} = -t_1 \end{aligned}$$

and

$$\begin{aligned} \frac{t_{1v}}{\|\mathbf{f}_v\|} &= -\frac{\tilde{t}_{2x}}{\|\mathbf{h}_x\|} \\ \frac{t_{2u}}{\|\mathbf{f}_u\|} &= -\frac{\tilde{t}_{1y}}{\|\mathbf{h}_y\|} \end{aligned}$$

To verify that the last component of $\mathbf{g}_3$ in (3.3) is equal to $-t_1\hat{t}_1 - t_2\hat{t}_2$, it suffices to notice that

$$-\hat{t}_1(\hat{t}_1 - t_1) - \hat{t}_2(\hat{t}_2 - t_2) = -\hat{t}_1 t_1 - \hat{t}_2 t_2.$$

(end of Proof of the $\Leftarrow$ part).

Part 2: Holonomy $\Rightarrow$ Specularity.
By hypothesis, the constraint equation (2.8) are integrable, i.e. there exists a two

dimensional submanifold described by $\{\mathbf{q} \in M : \mathbf{F}(\mathbf{q}) = 0\}$, with $\mathbf{F}$ a C^∞ map $\mathbf{F} : \mathbb{R}^2 \times \mathbb{R}^2 \times S^1 \to \mathbb{R}^2 \times S^1$ such that its rank–3 Jacobian is

$$D\mathbf{F} = \begin{bmatrix} \mathbf{M}_1 & -\mathbf{R}_\psi \mathbf{M}_2 & 0 \\ \mathbf{T}_1\mathbf{M}_1 & \mathbf{T}_2\mathbf{M}_2 & -1 \end{bmatrix}.$$

By the Implicit Function Theorem, there exists locally a C^∞ map $\hat{\phi} : \mathbb{R}^2 \to \mathbb{R}^3; \hat{\phi}(x,y) = (u,v,\psi)$ such that $\mathbf{F}\left(x,y,\hat{\phi}(x,y)\right) = 0$. Let ϕ denote the trivial restriction of $\hat{\phi}$, $\phi : \mathbb{R}^2 \to \mathbb{R}^2 : \phi(x,y) = (u,v)$.

Consider the surface $\tilde{\mathbf{f}}(x,y) = s \circ \mathbf{f} \circ \phi$, which is the specular image of $\mathbf{f}(u,v)$ at point $(\bar{u},\bar{v}) = \phi(x,y)$. The theorem will be proved by showing that the surface $\tilde{\mathbf{f}}(x,y)$ and $\mathbf{h}(x,y)$ are indeed the same.

The elements of the first fundamental form of $\tilde{\mathbf{f}}$ are evaluated as

$$\begin{aligned} \tilde{\mathbf{f}}_x &= \hat{\boldsymbol{\Sigma}}\mathbf{f}_u \frac{\|\mathbf{h}_x\|\cos\psi}{\|\mathbf{f}_u\|} - \hat{\boldsymbol{\Sigma}}\mathbf{f}_v \frac{\|\mathbf{h}_x\|\sin\psi}{\|\mathbf{f}_v\|} \\ \tilde{\mathbf{f}}_y &= -\hat{\boldsymbol{\Sigma}}\mathbf{f}_u \frac{\|\mathbf{h}_y\|\sin\psi}{\|\mathbf{f}_u\|} - \hat{\boldsymbol{\Sigma}}\mathbf{f}_v \frac{\|\mathbf{h}_y\|\cos\psi}{\|\mathbf{f}_v\|} \\ \tilde{\mathbf{n}} &= \hat{\boldsymbol{\Sigma}} \frac{\mathbf{f}_u \times \mathbf{f}_v}{\|\mathbf{f}_u\|\|\mathbf{f}_v\|} \end{aligned}$$

which implies

$$\begin{aligned} \|\tilde{\mathbf{f}}_x\| &= \|\mathbf{h}_x\| \\ \|\tilde{\mathbf{f}}_y\| &= \|\mathbf{h}_y\|, \end{aligned}$$

hence the first fundamental forms of $\tilde{\mathbf{f}}(x,y)$ and $\mathbf{h}(x,y)$ coincide,

$$\tilde{\mathcal{I}} = \mathcal{I}_2. \tag{3.4}$$

Further, it holds that

$$\begin{aligned} \tilde{\mathbf{n}}_x &= \hat{\boldsymbol{\Sigma}}\mathbf{n}_u \frac{\|\mathbf{h}_x\|\cos\psi}{\|\mathbf{f}_u\|} - \hat{\boldsymbol{\Sigma}}\mathbf{n}_v \frac{\|\mathbf{h}_x\|\sin\psi}{\|\mathbf{f}_v\|} \\ \tilde{\mathbf{n}}_y &= -\hat{\boldsymbol{\Sigma}}\mathbf{n}_u \frac{\|\mathbf{h}_y\|\sin\psi}{\|\mathbf{f}_u\|} - \hat{\boldsymbol{\Sigma}}\mathbf{n}_v \frac{\|\mathbf{h}_y\|\cos\psi}{\|\mathbf{f}_v\|}. \end{aligned}$$

The elements of the second fundamental form of $\tilde{\mathbf{f}}$ are evaluated as

$$\begin{aligned} -\tilde{\mathbf{n}}^T\tilde{\mathbf{f}}_{xx} &= \tilde{\mathbf{n}}_x^T\tilde{\mathbf{f}}_x \\ -\tilde{\mathbf{n}}^T\tilde{\mathbf{f}}_{yx} &= \tilde{\mathbf{n}}_x^T\tilde{\mathbf{f}}_y \\ -\tilde{\mathbf{n}}^T\tilde{\mathbf{f}}_{xy} &= \tilde{\mathbf{n}}_y^T\tilde{\mathbf{f}}_x \\ -\tilde{\mathbf{n}}^T\tilde{\mathbf{f}}_{yy} &= \tilde{\mathbf{n}}_y^T\tilde{\mathbf{f}}_y \end{aligned}$$

and, by some calculations, the relation between the second fundamental forms of $\tilde{\mathbf{f}}(x,y)$ and $\mathbf{f}(u,v)$ is found as

$$\widetilde{\mathbf{M}}^{-1}\widetilde{\mathcal{II}}\widetilde{\mathbf{M}}^{-1} = \mathbf{R}_\psi \mathbf{M}_1^{-1}\mathcal{II}_1\mathbf{M}_1^{-1}\mathbf{R}_\psi \tag{3.5}$$

where $\widetilde{\mathbf{M}} = \sqrt{\tilde{\mathcal{I}}}$.

The first and second fundamental forms $\mathcal{I}_2, \mathcal{II}_2$ for $\mathbf{h}(x,y)$ holonomically rolling on $\mathbf{f}(u,v)$ are derived considering once again the no–slippage constraint (2.5), rewritten as

$$\begin{aligned} \frac{\mathbf{f}_u}{\|\mathbf{f}_u\|}\cos\psi - \frac{\mathbf{f}_v}{\|\mathbf{f}_v\|}\sin\psi &= \frac{\mathbf{h}_x}{\|\mathbf{h}_x\|} \\ -\frac{\mathbf{f}_u}{\|\mathbf{f}_u\|}\sin\psi - \frac{\mathbf{f}_v}{\|\mathbf{f}_v\|}\cos\psi &= \frac{\mathbf{h}_y}{\|\mathbf{h}_y\|} \end{aligned}$$

It directly follows that $\mathcal{I}_2 = \mathcal{I}_1$. Further, observing that

$$\begin{aligned} \dot{\mathbf{n}}_1^T(\frac{\mathbf{f}_u}{\|\mathbf{f}_u\|}\cos\psi - \frac{\mathbf{f}_v}{\|\mathbf{f}_v\|}\sin\psi) &= \dot{\mathbf{n}}_2^T\frac{\mathbf{h}_x}{\|\mathbf{h}_x\|} \\ \dot{\mathbf{n}}_1^T(-\frac{\mathbf{f}_u}{\|\mathbf{f}_u\|}\sin\psi - \frac{\mathbf{f}_v}{\|\mathbf{f}_v\|}\cos\psi) &= \dot{\mathbf{n}}_2^T\frac{\mathbf{h}_y}{\|\mathbf{h}_y\|}, \end{aligned}$$

by some calculations one gets

$$\mathbf{M}_2^{-1}\mathcal{II}_2\mathbf{M}_2^{-1} = \mathbf{R}_\psi\mathbf{M}_1^{-1}\mathcal{II}_1\mathbf{M}_1^{-1}\mathbf{R}_\psi$$

where $\mathbf{M}_2 = \sqrt{\mathcal{I}_2}$. Therefore, as (3.4) implies $\widetilde{\mathbf{M}} = \mathbf{M}_2$, we have

$$\widetilde{\mathcal{II}} = \mathcal{II}_2. \tag{3.6}$$

Recalling from Bonnet's theorem ([**6**]) that two surfaces with equal fundamental forms coincide up to a rigid motion in $SE(3)$, it is proved that a surface S_2 rolling holonomically on S_1 must be specular of S_1 at the contact points. □

Acknowledgements

The authors wish to acknowledge Prof. Roger Brockett's early stimulus of their interest in the problem of rolling bodies, and Prof. Shankar S. Sastry for the thrust he contributed to this work in its initial phase. Prof. Yacine Chitour is gratefully acknowledged for the helpful discussions and inspiration to parts of the work.

References

[1] Y. Aiyama, M. Inaba and H. Inoue, *Pivoting: A New Method of Graspless Manipulation of Object by Robot Fingers*, Proc. IEEE/RSJ Int. Conf. on Int. Robots and Systems, IROS'93, 1993.

[2] A. Bicchi and K. Goldberg, eds., *Minimalism in Robotic Manipulation*, Proc. of the IEEE 1996 Workshop on Minimalism in Robotic Manipulation, Minneapolis, MA, April 22–28, 1996.

[3] A. Bicchi, D. Prattichizzo, and S. S. Sastry, *Planning Motions of Rolling Surfaces*, IEEE Conf. on Decision and Control, 1995.

[4] A. Bicchi, and R. Sorrentino, *Dexterous Manipulation through Rolling*, Proc. IEEE Int. Conf. on Robotics and Automation, 1995.

[5] Z. Li and J. Canny, *Motion of Two Rigid Bodies with Rolling Constraint*, IEEE Trans. on Robotics and Automation, volume 6, number 1, pp. 62–72, 1990.

[6] M. P. do Carmo, *Differential Geometry of Curves and Surfaces*, Prentice-Hall, Inc., Englewood Cliffs, New Jersey, 1976.

[7] R. Hermann and A. Krener, *Nonlinear Controllability and Observability*, IEEE Trans. on Automatic Control, volume AC-22, number 5, 1977.

[8] M. Levi, *Geometric Phases in the Motion of Rigid Bodies*, Arch. Rational Mech. Anal. 122(1993), pp. 213-229.

[9] Y. Liu, and R. Popplestone, *A Group Theoretic Formalization of Surface Contact*, Int. J. of Robotics Research, volume 13, number 2, pp. 148–161, 1994.

[10] A.Marigo, Y.Chitour, A.Bicchi. *Manipulation of Polyhedral Parts by Rolling* Proc. IEEE Int. Conf. on Robotics and Automation, volume 4, pp.2992–2997 , 1997

[11] R. M. Murray, Z. Li, S. S. Sastry, *Robotic Manipulation.* CRC Press, 1994.

[12] D.J. Montana, *The Kinematics of Contact and Grasp*, Int. J. of Robotics Research, 7(3), pp. 17–32, 1988.

Centro "E. Piaggio", Università di Pisa, 56100 Pisa, Italy.
E-mail address: `marigo@piaggio.ccii.unipi.it`

Centro "E. Piaggio" and Dept. of Electrical Systems and Automation, Università di Pisa, 56100 Pisa, Italy.
E-mail address: `bicchi@piaggio.ccii.unipi.it`

Proceedings of Symposia in Pure Mathematics
Volume **64**, 1999

On Degenerate Monge - Ampere Equations

Viswanath Ramakrishna

Abstract

This paper shows how the method of the intermediate integral, due to E. Goursat, in the theory of parabolic Monge-Ampere equations can be extended in the presence of certain singularities in the data. More precisely, one concrete situation wherein the method of intermediate integrals can be used to solve a degenerate Monge-Ampere equation is provided. Other mechanisms via which the same could be achieved are also discussed.

1 Introduction

Much of the geometric theory of partial differential equations is carried out in the presence of regularity conditions (see, for instance, [3, 14]). To the best of our knowledge the only exception to this has been the work of Kossowski, [9], on equations which change type. On the other hand there is a substantial amount of literature on various versions of the Frobenius theorem in the presence of singularities, [6, 11, 12, 18, 19]. Since the Frobenius theorem intervenes at various points in the theory of exterior differential systems, it would seem that some of the regularity conditions could be weakened. Part of the reason maybe (as was suggested to the author by Professor R. Gardner) that many of the "Frobenius with singularities" type results are in the analytic category whereas much of the theory of exterior differential systems is carried out in the smooth category, which has perhaps not been conducive to interaction between the two communities. However, there are situations which naturally lend themselves to the possibility of examining the effect of singularities. In particular, if some problem can be reduced to the study of an ideal of forms, whose generators are decomposable, then the various versions of the Frobenius theorem (not all are equally effective in a given situation) become immediately relevant. Of course, it is impossible to encompass the bewildering variety of singular phenomena into a single theorem. Nevertheless, it is worthwhile to examine which constructions go through in the presence of singularities. The

---

1991 Mathematics Subject Classifications Primary 35 A30, Secondary 58 A15
Work supported in part by a grant from the ARO and the NSA.

purpose of this paper is to point out that one such concrete situation, namely that of the method of the intermediate integral in the geometric theory of parabolic Monge - Ampere equations, does admit extensions in the presence of singularities.

2 The Method of Intermediate Integrals

The method of intermediate integrals rarely appears in any modern discussion of Monge-Ampere equations To the best of our knowledge the sole exception is the text of R. Hermann, [7], of which the author became aware after he began the work relevant to this paper. We will discuss this text briefly later in this section. After the initial submission of this paper a referee pointed out a recent paper, [5], (posted on the web after this current paper was first submitted) which corrects an error in [7]. This error persisted in the initial version of this paper also. The author is grateful to the anonymous referee for pointing out both the error and its correction in [5]. The correction of [5] will also be discussed later in this section. It is worthwhile pointing out that both [5] and [7] consider the method of intermediate integrals in the non-singular case alone.

Consider a Monge-Ampere equation:

$$A + Bz_{xx} + 2Cz_{xy} + Dz_{yy} + E(z_{xx}z_{yy} - z_{xy}^2) = 0 \qquad (2.1)$$

As usual the coordinate system (x, y, z, p, q) on (an open set of) $M = R^5$ will be used to study (2.1) geometrically. So, A through E are functions of (x, y, z, p, q). As is well known, a solution of (2.1) can be thought of as a 2 dimensional submanifold $\phi : N \to M$ such that $\phi^*(dx \wedge dy) \neq 0$ and which annihilates the ideal generated I on M generated by the contact one - form $\theta = dz - pdx - qdy$, $d\theta$ and the two form

$$\Omega = Adx \wedge dy + Bdp \wedge dy + C(dx \wedge dp + dq \wedge dy) + Ddx \wedge dq + Edp \wedge dq \quad (2.2)$$

and conversely, every such submanifold N gives rise to a solution of the equation (2.1).

Assume that Ω is a decomposable two form, say $\Omega = \beta_1 \wedge \beta_2$ for some one-forms, β_1 and β_2 on M (so that I is simply generated). Suppose furthermore that there is a function $F : M \to R$ which is not locally constant and is a weak first integral for $\hat{I} = \text{span}\{\beta_1, \beta_2\}$, i.e.,

$$dF \wedge \Omega = 0 \qquad (2.3)$$

Assume for the moment that Ω has no zeroes on M (if need be we will choose the origin as a reference point and some open set U containing the origin such that Ω never vanishes on U and then restrict attention to U). If we can now find a solution to the *first order partial differential equation* $F(x,y,z,p,q) = 0$, then indeed $\phi(x,y)$ solves (2.1). Indeed, since β_1 and β_2 are linearly independent, we also have:

$$dF = a\beta_1 + b\beta_2 \tag{2.4}$$

for some germs of functions a, b which do not vanish simulatneously in a neighbourhood of the reference point. In other words, F is, in the terminology of [12], a strong first integral for I. This is adequate to show that ϕ solves Equation (2.1). We provide a proof this statement assuming that Ω may vanish at some points, since we will need this fact later. In the statement of this proposition the manifold N contains the projection of the origin in U. The origin in U is the reference point and is allowed to be a singular point.

Proposition 2.1 *Suppose F is a strong first integral for $\hat{I}$ on an open $U \subseteq M$ containing the origin and $\phi(x,y)$, $(x,y) \in N$, satisfies the first order partial differential equation $F(x,y,z,p,q) = 0$. Then ϕ solves Equation (2.1).*

Proof: On U we have $dF = a\beta_1 + b\beta_2$ for some functions a, b on U (of the appropriate degree of smoothness) such that a and b do not vanish simultaneously on U. Since $\phi^*(\theta) = 0$ trivially and $\phi^*(dF) = 0$ as ϕ solves $F = 0$. Now ϕ cannot solve (2.1) only if $\phi^*(\beta_1 \wedge \beta_2) \neq 0$. This can happen only if $\phi^*(a) = 0 = \phi^*(b)$, i.e., the functions a and b vanish simultancously on all points in U of thc form $(x, y, \phi(x,y), \phi_x(x,y), \phi_y(x,y))$, $(x,y) \in N$. This is a contradiction to the fact that F is a strong first integral for $\hat{I}$ on U. This finishes the proof.

Remark 2.1 i) Regarding the above proposition it is useful to note that in the non-singular case the fact that a and b both vanish at $(x, y, \phi(x,y), \phi_x(x,y), \phi_y(x,y))$ would preclude F being even locally constant. This is because (by tacitly shrinking the size of U) one can ensure that dF never vanishes at any point in U. However, in the singular case this cannot be achieved. In particular, dF could vanish at the reference point. Therefore, we need the stronger hypothesis that F

is a strong first integral. In other words the argument above need not work for an F which is only a weak first integral.

ii) Note that it is possible that $F = 0$ may not have global solutions, i.e., in the above result the size of N has to be shrunk tacitly to reflect this possibility.

Definition 2.1 *A strong first integral for $\hat{I}$ is called an intermediate integral for Equation (2.1).*

The author became aware that this method is due to Goursat and that it has the appellation "the method of intermediate integrals" upon reading the text [7]. If there are more intermediate integrals such that the corresponding first order PDEs admit solutions then we have more solutions to the equation (2.1).

Remark 2.2 The definition of intermediate integral adopted here is less general than that in [5, 7]. In these references an intermediate integral is a strong first integral for $I = \text{span}\{\beta_1, \beta_2, \theta\}$. However, the proof of the validity of Goursat's technique in [7] has an error. In effect, the proof in [7] is the same as the proof of Proposition (2.1) above in the non-singular case but with the more general definition of intermediate integral. However, now it is perfectly possible that both a and b vanish at certain points without there being any contradiction to the fact that F is a strong first integral for I (because c - the coefficient of θ in the expression of dF as a linear combination of the generators of I - need not vanish at such points). The elegant analysis of [5] not only corrects this error by providing sufficient conditions for salvaging the proof in [7] but also provides (in the non-singular case, of course) a canonical counterexample upto contact diffeomorphisms. We have chosen a more restrictive definition of intermediate integrals because the main aim of this paper is to provide an illustration of how E. Goursat's method can be extended. However, under some real-analyticity hypotheses it is possible to extend the analysis of [5]. For instance, it is shown in [5] that if I is not integrable but F is an intermediate integral and $d\theta = 0 \text{mod} I$, then indeed the argument of [7] goes through. Assume that the set of points where θ, β_1 and β_2 are linearly independent forms a dense subset. If in addition, the projection onto the (x, y) space of this set is also dense then once again the conclusions of [5] are valid. Indeed, the argument of [5] is that in the absence of singularities one can complete $\{\beta_1, \beta_2, \theta\}$ to a basis for the cotangent space. Once this is done then (after possibly a basis change for $\hat{I}$) the hypotheses yield that the intermediate integral F must satisfy $dF = a(\beta_1 - s\theta)$ for some functions s and a, with a non-vanishing. If now this happened only on the dense subset of regular points, then by density we would still have that the solution ϕ to $F(x, y, z, p, q) = 0$ satisfies (2.1) provided that F and ϕ are defined in a neighbourhood of the singular point of interest. Thus the stumbling blocks are the existence of a strong first integral for codistributions of varying rank and a solution to the possibly singular partial differential equation that this intermediate integral gives rise to. In this paper the primary interest lies in the existence of classical solutions to (2.1) in relation to the coefficients A through E. Thus, a position intermediate to that of geometers and that of analysts will be taken in that the ideal $\hat{I}$ and its first integrals will be studied as opposed to the behaviour of the ideal I. Hopefully, it will be clear how to extend (or, for that matter, identify the stumbling blocks to) these techniques to the study of geometric solutions.

Remark 2.3 It is worthwhile pointing out that, whilst it is F's being a strong first integral that enables one to find a solution (2.1), in practice it is more convenient to use (2.3) to find a weak first integral and then check if it is a strong integral. Indeed (2.3) is cast in the form of an equality whereas in practice (2.4) is at best an inclusion condition.

Thus, the method of intermediate integral requires us to solve two first order partial differential equations to solve (2.1). The first, (2.3), is an overdetermined system and the second, the equation $F(x, y, z, p, q) = 0$ is a nonlinear scalar equation.

The crucial ingredients to this technique are: i) The decomposability of the form Ω; ii) the use of the *division property* which allows the passage from (2.3) to (2.4); and then iii) solving the corresponding first order equation.

The first of these requires the condition $\Omega \wedge \Omega = 0$ which, in the non-singular case, (since we are dealing with a two-form) is both a necessary and sufficient condition for decomposability. Since a direct calculation shows that

$$\Omega \wedge \Omega = (AE - BD + C^2) dx \wedge dy \wedge dp \wedge dq$$

we see that the method of the intermediate integral requires that the given Monge - Ampere equation be *parabolic*.

Remark 2.4 i) As observed in [5] (in the non-singular case) the decomposability requirement does not force us to restrict ourselves to parabolic equations. For instance, we can replace Ω by $\Omega + \lambda d\theta$ for a suitable function λ so that the resultant 2 - form is decomposable. However, if the form Ω vanishes at some point then λ will also vanish at this point. Thus, one cannot circumvent the problem of singularity via this artifice. Furthermore, without additional conditions on the coefficients of the form Ω even the smoothness of $\Omega + \lambda d\theta$ cannot be guaranteed in the event that Ω has zeroes. Therefore, in this paper we will work only with Ω. This is also in keeping with the modest aim of this paper and the choice of working with the ideal $\hat{I}$ instead of I, mentioned above.

ii) In his interesting book, [7], R. Hermann discusses the method of the intermediate integral in the nonsingular case. This book and [5] are, to the best of our knowledge, the only modern treatments of this method. However, since [7] starts out with a pair of forms β_1, β_2 and poses the question of which partial differential equations arise when ϕ^* annihilates $\beta_1 \wedge \beta_2$ and θ, the crucial role played by the decomposability requirement is occluded. Thus, when it is mentioned that Monge-Ampere equations arise as a special case (Page 247, [7]), no mention of the parabolicity condition on the equation is made. On the other hand, the exposition of [7] makes it clear that much of the next section is equally valid for other singular second order partial differential equations.

iii) It may seem that the insistence on the importance of the division property as being crucial is unwarranted, since only F's being a strong integral is being used. However, in the singular case every result that we are aware of which produces strong first integrals makes use of some division property (see, for instance, the theorem of Saito in the next section). Thus, it seems worth emphasizing it.

In the next section, we will examine which of these steps survive when singularities are introduced. As mentioned above singularities come in various forms and thus it is impossible to present a single (or even a few) theorem which covers all cases. Theorem (3.4) below comes, in our opinion, closest to proving a positive result in the singular situation wherein the conclusions of the result can be closely related to the coefficients of the equation. However, there are other mechanisms via which the method of intermediate integrals provides a classical solution to a degenerate Monge-Ampere equation. Two such instances are provided by the contents of remarks (2.1) and (3.7). Both impose strong density requirements. In particular, they preclude cases where the coefficients of (2.1) do not depend on x and y. Nevertheless, they are of value since the contents of these remarks identify mechanisms via which one could obtain solutions to degenerate Monge-Ampere equations. In other words, the mechanisms identified in these remarks could work even when the sufficient conditions for the assumptions going into these remarks do not hold (precisely because they are only sufficient conditions). Thus they are not only of interest, but also suggest questions (e.g., the study of derived systems in the singular case).

3 Degenerate Monge-Ampere Equations

Usually, by a point of degeneracy of a differential equation, one means a point at which the order of the equation drops. In the case of the Monge- Ampere equation this occurs when B, C, D and E vanish at some point, and thus it drops order by one. However, this is not a point of degeneracy from the point of view of the method of the intermediate integral if A also does not vanish at this point. Indeed by choosing $X = \partial_x$ and $Y = \partial_y$ (where X and Y are as in Equation (3.5) below), we can ensure that the one-forms into which Ω decomposes are linearly independent if A does not vanish. One may object that this choice of X, Y may lead to a violation of the Frobenius integrability condition which may be needed for the method of intermediate integral to work . However, it will be shown below that the integrability condition depends only on Ω. Thus, we will now assume that all the coefficients A, B, C, D and E of the equation (2.1) vanish at some operating point, which is taken to be the origin (note : no claim about the origin being an isolated zero is being made at this stage). A non-afficionado of geometric methods for partial differential equations may further object that a point of degeneracy ought to be defined as a point in N (coordinatized by (x, y)) at which all the coefficients of (2.1) vanish. Whilst qualitatively this situation can be subsumed in the general case, it may affect the actual results, since most of the sufficient conditions for the division property and the existence of first integrals for I depend on the "dimension" of the zero locus of A, B, C, D and E. This will be addressed later.

So now assume that the coefficients of (2.1) vanish at the origin. Thus the form Ω also vanishes at the origin. Suppose that the equation is parabolic. The question is whether the decomposability of Ω still persists. An explicit expression for the pair of one-forms into which Ω decomposes into is available. Indeed, define the forms

$$\theta_1 = i_X\Omega, \; \theta_2 = i_Y\Omega \tag{3.5}$$

where X and Y form a pair of vector fields which are linearly independent. such that $i_X\Omega$ and $i_Y\Omega$ are not identically zero. Next define,

$$\omega_1 = \frac{\theta_1}{i_Y i_X\Omega}, \; \omega_2 = \frac{\theta_2}{i_Y i_X\Omega} \tag{3.6}$$

Then we have

$$\Omega = \theta_1 \wedge \omega_2 = \omega_1 \wedge \theta_2 \tag{3.7}$$

For purposes which will become clear later define also

$$(i_Y i_X\Omega)\Omega = \tilde{\Omega} = \theta_1 \wedge \theta_2 \tag{3.8}$$

We note that the decomposition of Ω described by by (3.7) is unique (modulo the arbitrariness in the choice of X and Y) since such a decomposition has to hold pointwise. So the question to be addressed is whether these one-forms are smooth at the zero of Ω, and if not what is the effect on the method of the intermediate integral. An additional question is whether by making different choices for X and Y one can ameliorate any of the problems due to the singulartities. Clearly, θ_1 and θ_2 are well defined at the zero of Ω, whereas ω_1 and ω_2 need not due to the $i_Y i_X\Omega$ factor in the denominator.

It is appropriate at this point to introduce the result of Saito alluded to earlier. Saito's result is stated in the general context of division properties of the exterior complex over a Noetherian ring. Translating this to the case of the ring of germs of C^ω real-valued functions, we have:

Theorem 3.1 *(K. Saito) Suppose $\omega_1, \ldots, \omega_p$ are germs of C^ω one-forms (not necessarily satisfying the Frobenius integrability condition and not necessarily pointwise independent). Suppose that profondeur of the ideal generated by the components of the wedge product of the ω_a's in the ring of germs of C^∞ (C^ω) functions is bigger than q. Let ω be the germ of a q-form (smooth/real anal). Then the following conditions are equivalent: a) $\omega \wedge \omega_1 \wedge \ldots \wedge \omega_p = 0$; and b) $\omega = \sum_{i=1}^{p} \theta_i \wedge \omega_i$ for some germs of C^ω $q-1$ forms θ_i.*

We refer the reader to [17] for the definition of profondeur. We will just be content with saying that it is essentially an algebraic notion. However, in the holomorphic category to say that an ideal of germs of holomorphic functions has profondeur q is to say that the codimension of the associated variety is also q. There is an extension of the above result to the case of C^∞ germs due to Tougeron (see [4]). Global decomposition theorems, i.e., ones where the $\omega_i, i = 1, \ldots, p$ and ω are defined on an open set U and yield the condition that the θ_i's in the above theorem are also defined on all of U, are known only in the C^∞ category (since the proof uses partitions of unity), see e.g., [12]. Moussu's theorem, [12], requires the zeroes of $\omega_1 \wedge \ldots \wedge \omega_p$ to be algebraically isolated, which is stronger than the profondeur condition. However, it seems to the author that the same proof works with the more general profondeur condition, since the algebraically isolated condition is just used to establish local division.

We can now address the question of whether decomposability still persists when Ω has zeroes.

Proposition 3.1 *Fix an open set U in M containing the origin which is assumed to be a zero point of Ω. Suppose that the complement of the zero set of Ω is dense in U. Assume that $\Omega \wedge \Omega = 0$. Suppose there is a vector field X on U such that the ideal generated by the germ of the one-form $i_X\Omega$ at the origin has profondeur at least 3, then Ω decomposes into the wedge product of the germs of 2 one-forms at the origin.*

Proof: Let V denote the complement in U of the zero set of Ω. Since $\Omega \wedge \Omega = 0$ and $\Omega \neq 0$ on V there are one-forms α_1 and α_2 on V such that Ω is their wedge product on V. We may assume that X does not vanish at the reference point (see remark (3.2) below for this). By shrinking the size of U one can hence assume that X does not vanish anywhere on U. Now choose α_1 to be $i_X\Omega$ (where the X here is the restriction to V of X) Consider the one-form on U given by $\beta_1 = i_X\Omega$. Since V is dense in U and $\Omega \wedge \beta_1 = 0$ on V the same holds on U. Now consider the germs of the forms Ω and β_1 at the origin. The profondeur assumption and the result of Saito with $q = 2$ imply that there is a germ of a one-form β_2 such that $\Omega = \beta_1 \wedge \beta_2$. This yields the proof. With appropriate assumptions one can also obtain a global decomposability result.

Remark 3.1 i) This decomposability proposition has applications beyond the problem at hand. Saito's result is, in our opinion, the key step to extending various results in the geometric theory of partial differential equations to the situation where singularities are present. This and other related results also play a role in control theory as well, [16].

ii) Note that the assumption on the profondeur of the ideal generated by $i_X\Omega$ implies that the profondeur of the ideal generated by Ω is also at least 3. Indeed, the above proposition shows that the zero set of β_1 is contained in the zero set of Ω. However, we cannot replace the assumption on $i_X\Omega$ by the this latter more elegant looking assumption on Ω, since it is not known beforehand that $\Omega = \beta_1 \wedge \beta_2$.

iii) The last result viewed out of the context of the problem at hand implies some curious divisability results. These will not be discussed as they are not relevant here.

We now go back to the notation of (3.7) In the same vein as the second part of the previous Remark, let us try to get a feel for what the hypotheses on $i_X\Omega$ means in terms of Ω itself. Assume that X is some coordinate vector field on U (we tacitly shrink U so that X may be expressed as a coordinate vector field on all of U). For the purposes of the discussion immediately below, we will just concentrate on ω_2 (since the argument is symmetric in ω_1 and ω_2). Choose X to be a coordinate vector field except $\frac{\partial}{\partial z}$ (since then $i_X\Omega$ will not be everywhere 0.). The question as to whether this choice of X affects either the analogue of the passage from (2.3) to (2.4) or the existence of an intermediate integral will be addressed later. If, for example, we choose $X = \partial_x$ and Y is also a coordinate vector field, say ∂_y, we get

$$\theta_1 = Ady + Cdp + Ddq; \omega_2 = \frac{1}{A}(-Adx - Bdp - Cdq)$$

whereas if we have $X = \partial_p$ and $Y = \partial_q$ we get

$$\theta_1 = Bdy - Cdx + Edq; \omega_2 = \frac{1}{E}(Cdy - Ddx - Edp)$$

If both X and Y are coordinate vector fields then the denominator appearing in ω_2 will be one of the coefficients of (2.1). For concreteness, let us illustrate all issues via the choice $X = \partial_x$ and $Y = \partial_y$. The hypothesis on $i_X\Omega$ is thus a hypothesis on A, C and D. *If it also turns out* that the second factor Ω decomposes into is of the form ω_2 with $Y = \partial_y$, then the conclusion is that that A divides B and C

in the ring of germs of functions. This may seem curious. To put things in perspective, we first emphasize that an assumption on the expression for the second factor that Ω decomposes into is being made (which is not a conclusion of any result). This assumption is in force only to simplify subsequent discussions and is not really central to any result. With this assumption's role clarified, a plausibility argument for the fact that a conclusion about B may be drawn from only a knowledge of A, C and D runs as follows. First note that we already have $AE = BD - C^2$. So A obviously divides $BD - C^2$. Now the profondeur assumption (which can loosely be interpreted as a statement regarding the variety cut out by A, C and D) says for this to happen A must also divide B.

Going back to former discussion it can thus be asserted that Ω is decomposable (and with both factors smooth or C^ω). Of course, if A divided C and D, we would work with ω_1 and θ_2. At this point the following remark is relevant:

Remark 3.2 It is tempting to make the above discussion coordinate free by appealing to the flowbox theorem, i.e., by working with a rectifying chart for some vector field X which does not vanish at the origin (it should be clear, in light of the profondeur connection, that if X also vanished at the origin then the profondeur hypothesis is not likely to be met. Thus there is no great loss of generality in assuming that the X which satisfies the required hypothesis is non-vanishing at the origin.). However, even for the purposes of obtaining geometric solutions one cannot use arbitrary coordinate changes. The condition that the ideal generated by the germs of A, C and D has profondeur at least 3 is itself coordinate free. But, to give the whole result a flavour which is both coordinate free and is explicit in the coefficients of Ω itself requires the assumption that $X = \partial_x$ and this is not coordinate free. However, in practice this would be the way to proceed.

If it turns out that there is no X such that the hypothesis on $i_X\Omega$ is met, then we could work with the generators $\tilde{\Omega} = (i_X i_Y \Omega)\Omega$ and θ. When $X = \partial_x$ and $Y = \partial_y$, these forms generate I away from the zero-locus of A. Let us call the situation wherein at least one of ω_1 or ω_2 is smooth at the zero of Ω as *Scenario 1* and the situation wherein neither is (and thus we have to work with $\tilde{\Omega}$) as *Scenario 2*.

Notice, however, that the consideration of *Scenario 2* would be irrelevant in situations where Remark (3.7) (which appears later in the paper) is applicable. To avoid circumlocution only *Scenario 1* will be considered in the balance of the paper, unless explicit reference to the contrary is made. We then have the following, easily proved, theorem.

Theorem 3.2 *Suppose that we have been able to find an F which is not locally constant and which satisfies the following assumptions:*

Assumption 3.1 $dF \wedge \theta_1 \wedge \omega_2 = 0$ *on all of* U

Assumption 3.2 $dF \wedge \theta_1 \wedge \omega_2 = 0$ *on* U *implies that* $dF = a\theta_1 + b\omega_2$ *for some smooth (or* C^ω*) functions* a *and* b *defined on* U *such that neither vanishes simultaneously on* U

Assumption 3.3 *The first order equation* $F(x, y, z, p, q) = 0$ *has a solution,* $\phi(x, y)$.

Then ϕ *solves (2.1). Once again if* $F = 0$ *has only a locally defined solution about* $(x, y) = (0, 0)$, *then the size of the open set of definition of the solution to (2.1) is assumed to be appropriately shrunk.*

Remark 3.3 Note, however that if the $U - \{A = 0\}$ was dense in U and so was its projection onto the (x, y) space, then we could have just demanded that Assumption (3.2) hold only on $U - \{A = 0\}$. This is because we have assumed that F is defined in a neighbourhood of the origin (the singular point) and that the solution ϕ is defined in a neighbourhood $(x, y) = (0, 0)$. We can ensure Assumption (3.2) does indeed hold on all of $U - \{A = 0\}$ because all results are local and because of the non-singularity we can tacitly shrink U to achieve this. However, if we are interested in demonstrating a global result, i.e., one which shows that $dF \wedge \Omega \wedge \theta = 0$ on U implies $dF \in \text{span}\{\omega_2, \theta_1, \theta\}$ on all of U without resorting to shrinking the size of U then we need a global division result. Such a result is available in the smooth category. This is the result of Moussu mentioned above.

Let us next address the questions of when the assumptions (3.1), (3.2), and (3.3) hold. Sufficient conditions for Assumption (3.2) are also provided by the result of Saito and analogues. Of course, we could dispense with the use of Saito's result when $U - \{A = 0\}$ is dense in U by using the non-singular division results.

Thus we see that Assumption (3.2) requires that the ideal generated by the coefficients of Ω (i.e., the ideal generated by A, B, C, D and E) has profondeur at least 2,

Next let us ask when there is an F satisfying Assumptions (3.1) exists. *The results below also show that such an F is automatically a strong first integral.* Now if θ_1 (resp. ω_2) is to have a first integral then it must satisfy $d\theta_1 \wedge \theta_1 = 0$ (resp. $d\omega_2 \wedge \omega_2 = 0$). *Since ω_2 is non-vanishing at the origin one can use the classical Frobenius theorem to conclude the existence of a first integral for ω_2 and hence for $\hat{I}$.* However, for θ_1 we need the full force of the singular versions of the Frobenius theorem to be discussed below.

Now if none of these forms are individually integrable then we can still inquire if the codistribution $\hat{I}$ has first integrals. Again we need integrability conditions $d\theta_1 \wedge \Omega = d\omega_2 \wedge \Omega = 0$. So the first question that arises is whether the choice of X and Y in forming θ_1 and ω_2 affects the integrability condition. Before addressing this issue let us point out (the obvious but useful) fact that $\hat{I}$ could have first integrals even if Frobenius type integrability conditions are not satisfied by the generators of $\hat{I}$. For instance some linear combination of these generators could satisfy the Frobenius integrability condition even if these individual generators do not. However, we cannot do anything beyond turning this possibility into an assumption (pending further research at least). This is especially true in the presence of singularities wherein most known results, on the existence of first integrals, impose a Frobenius type integrability condition. Therefore, this interesting line of research will not be followed in this paper.

Proposition 3.2 *Suppose $\Omega = \beta_1 \wedge \beta_2 = \beta_3 \wedge \beta_4$ for some smooth (or real analytic) one-forms on an open set U. Assume that the pair $\{\beta_1, \beta_2\}$ satisfies the division property for one-forms on U, i.e., if a one-form γ is such that $\gamma \wedge \beta_1 \wedge \beta_2 = 0$ on U then $\gamma = a\beta_1 + b\beta_2$ on a possibly smaller open set $V \subseteq U$ then (a, b are smooth or real analytic functions on V). Then $d\beta_i \wedge \Omega = 0, i = 1, 2$ implies that $d\beta_i \wedge \Omega = 0, i = 3, 4$.*

Proof: This is obvious from the fact that $\beta_i \wedge \Omega = 0 = \beta_i \wedge \beta_1 \wedge \beta_2, i = 3, 4$. Notice that necessarily the corresponding a, b are non-vanishing on V.

Thus the integrability condition depends only on the singularities of Ω, i.e., if the singularities allow the division property to hold then integrability does not depend on the choice of the factors Ω decomposes into. Furthermore, requiring the the division property to hold on a dense subset would have been adequate, since $d\beta_i \wedge \Omega = 0, i = 3, 4$ on a dense subset implies the same everywhere.

Let us now consider versions of the Frobenius theorem with singularities. Since our interest is in finding first integrals for codistributions only such versions will be discussed. Malgrange [11] proves the following theorem:

Theorem 3.3 *(Malgrange) Consider a collection of germs at* 0 *of* p *holomorphic* 1*-forms* $\omega_i, i = 1, \ldots, p$ *which also satisfy the integrability condition* $d\omega_l \wedge \omega_1 \wedge \ldots \omega_p = 0, l = 1, \ldots, p$. *Then, if hypothesis that the codimension of the set where the codistribuion spanned by* $\omega_i, i = 1, \ldots, p$ *has rank less than* p *is at least* 3*, there exist germs of holomorphic functions* $f_i, g_{ik}, i = 1, \ldots, p; k = 1, \ldots, p$ *so that* $\omega_i = \sum_k g_{ik} df_k i = 1, \ldots, p$. *Furthermore* $det(g_{ik}(0)) \neq 0$.

The key idea (besides considerable technical difficulties) in this proof is to use the codimension condition to show that each $d\omega_i$ can be "divided" by the $\omega_i, i = 1, \ldots, p$. One could directly use his result for real-analytic data via the usual procedure of complexification. However, similar results are available in the smooth and real-analytic category, [12, 4]. The results in the smooth category require (if a strong first integral is desired) an additional topological requirement on the foliation generated by the one-forms away from their singular points, [12]. Once again the main step is repeated applications of the division property.

Remark 3.4 Since all the above results actually use the division property to achieve their proofs we see that whenever a sufficient condition for a stong integral holds sufficient conditions for the division property are also in effect.

Remark 3.5 Applying the above results to Scenario 2, we see that (by choosing $X = \partial_x, Y = \partial_y$ for instance) that if either θ_1 or θ_2 satisfy the Frobenius integrability condition **individually** then it suffices it either zero-locus of A, C and D or the zero-locus of $A, BandC$ is of adequately high codimension then there will be a strong first integral for span$\{\theta_1, \theta_2\}$. This is reasonably likely since we are requiring that the zero-locus of 3 functions be, for instance, of codimension 3. However, if individually these forms do not satisfy the Frobenius condition but they do jointly, then the sufficient conditions need to be satisfied by the zero-locus of A, which is unlikely.

Remark 3.6 A result of a different nature is provided by M. Freeman in [6]. Under a finite-generation hypothesis (which is always satisfied in the C^ω category) he exhibits first integrals (see Equation (3.3$'$ of [6]) for a Pfaffian system whose rank varies, but is such that its dimension at the origin is positive. His results do not require the Frobenius integrability condition, but rather require a "pointwise involutivity" condition. His results are relevant to our problem when we need θ to find first integrals, i.e., when the more general definition of intermediate integral, as in [5, 7], is used. This is because I has dimension 1 at the origin.

The above discussion leads to the following result which includes Theorem (3.2) as a special case by assuming that β_1 and β_2 turn out to be $i_X\Omega$ and $\frac{i_Y\Omega}{i_Y i_X\Omega}$ respectively in Proposition (3.1). *A short sketch of the proof is included after the statement in the interest of clarity.*

Theorem 3.4 *Suppose that there is a vector field X on U such that the germs of the coefficients of the the one-form $i_X\Omega$ at the origin generate an ideal of profondeur at least 3. Suppose that the complement of the singular set of Ω in U is dense in U. Suppose that $\Omega\wedge\Omega = 0$ and that the one-forms it decomposes into satisfy the Frobenius integrability condition (recall that this does not depend on the one-forms into which it decomposes into). If some intermediate integral F of $\hat{I}$ thereby obtained is such that the equation $F = 0$ has a solution ϕ in a neighbourhood of $(x, y) = (0, 0)$. Then ϕ is a solution to (2.1).*

Proof: Since $\Omega\wedge\Omega = 0$ the form Ω is decomposable away from the singular set. So on the complement of the singuar set $\Omega\wedge i_X\Omega = 0$, since one of the factors into which Ω decomposes may always be taken as $\alpha_1 = i_X\Omega$. Thus by density we have $\Omega\wedge i_X\Omega = 0$ everywhere on U. The profondeur assumption and Saito's result imply now that there is (a germ of) another one form α_2 such that Ω decomposes into the wedge product of this form and $i_X\Omega$. Now note that the ideal generated by the coefficients of $i_X\Omega$ is contained in the ideal generated by the coefficients of Ω. So the profondeur condition on $i_X\Omega$ implies that Ω also has profondeur at least 3. If the one-forms into which Ω was decomposed also satisfy the Frobenius integrability condition (this depends only on Ω as explained before) then there is a strong first integral for these one-forms, i.e., for $\hat{I}$ (of course, for smooth data the additional topological hypothesis of [12] has to be made. We will ignore this technicality here since it is not central to the paper. At any rate the statement assumes there is a first integral). So, we have $dF = a\alpha_1 + b\alpha_2$, where a and b do not simultaneously vanish and α_1, α_2 is the pair of one-forms into which Ω decomposes. *We emphasize that F is defined in a neighbourhood of the singular point.* Since ϕ solves $F = 0$, the same argument as in Proposition (2.1) yields the conclusions. *Note that the density assumption was used only to show that Ω decomposes in a neighbourhood of the singular point, i.e., the usage of density is different from the manner in which it is used in remarks (2.1) and (3.7).* This finishes the sketch of the proof.

Next the issue of the solvability of the equation $F(x, y, z, p, q) = 0$ will be considered. Since we are primarily interested in *classical* solutions $z(x, y)$ it is difficult to make general observations. If we allowed multivalued (i.e., geometric) solutions then one could use the various normal forms provided in the classic paper of Lychagin ([10]). This direction is being pursued by the author. Let us focus attention on the existence of classical solutions. Now it is possible that dF is 0 when Ω is 0, and it is this situation that is of interest (for otherwise the local solvability of $F = 0$ is well understood). So one may expect that the associated first order partial differential equation will also be singular, in the sense that the corresponding first order PDE drops order to a zeroth order equation at some points. This need not be the case. One can easily concoct examples where stationary Hamilton-Jacobi equations arise, and thus local solvability (in classical solutions) is well understood. However, the more interesting case is when the equation $F = 0$ does indeed become singular, i.e., it drops order at $(x, y) = (0, 0)$. The main technique to address the existence of classical solutions is to establish that formal power series solutions exist and then show that this solution either converges to give a C^ω solution or to modify the formal solution to establish the existence of C^∞ or C^k solutions. For the former the main idea is to use non small-divisor conditions to establish convergence of the formal power series (see e.g., [1]). For the latter typically essential use of E. Borel's lemma is made (see e.g., [8]).

As a prototype of the results about the solvability of degenerate first order partial differential equations, consider the following result of Bengel and Gerard, [1]: Suppose the equation $Xz = f(x, y, z)$ is such that the vector field $X(x, y)$ is real analytic and a Poincaré vector field and vanishes at the origin (see [15] for the definition of a Poincaré vector field) and we also have $f(0,0,0) = 0$ and $\frac{\partial f}{\partial z}(0,0,0) = 0$ then there is a C^ω solution to the equation $Xz = f(x,y,z)$. Their techniques can be adapted to certain fully non-linear equations. On the other hand, their method provides only the zero solution for the special case of a linear equation $Xz = c(x,y)z$ (see [15] for obtaining non-zero solutions to $Xz = c(x,y)z$).

In light of the above, a preliminary question is the existence of first integrals quadratic in the variables x, y, z, p, q for I, since then the equation $F = 0$ assumes a simple form. Furthermore, there are frequently (though not always) affirmative answers to the sovability of the equation $F = 0$ in such a situation. For instance, if F is of the form $xp+yq+x^2+y^2$ then techniques provided by [1] address the solvability of $F = 0$. If F is of the form $p^2 + q^2 + x^2 + y^2$ then we are dealing with a Hamilton-Jacobi equation etc., Unfortunately there is no easy answer to the above question if classical solutions are the focus. On the other hand if we allow geometric solutions then there indeed exist some answers to this question. In fact, it is possible to modify Nelson's proof of the Sternberg linearization theorem, [13], to the contact case. This in turn provides criteria for the existence of a coordinate change (which preserves the contact one-form) in which I has quadratic first integrals.

Remark 3.7 Throughout this paper we have only considered smooth or real analytic first integrals for $\hat{I}$. However, if $\hat{I}$ does not have such first integrals, it may still have meromorphic first integrals. Indeed since the forms $\omega_i, i = 1, 2$ have meromorphic quotients (they are what are, in the holomorphic category, called forms with logarithmic singularities) it is likely that they will have meromorphic first integrals. In fact even for smooth but vanishing forms such integrals are the norm, e.g., the one-form $ydx - xdy$ has the meromorphic first integral $F = \frac{x}{y}$. Sufficient criteria for their existence are known, see e.g., the bibliography in [2]. Such meromorphic first integrals may be useful for solving (2.1). Indeed given such an F, we consider a related first integral $F - c$, where c is a suitable constant. Then the first order equation $F - c = 0$ can be written as a first order equation with smooth coefficients upon clearing denominators. If the $U - \{A = 0\}$ is dense and so is its projection onto (x, y) space, and the resulting first order equation has a smooth solution defined in a neighbourhood of $(x, y) = (0, 0)$ then once again we may conclude that this solution also solves (2.1). Thus in this situation the only step which may cause problems is the unsolvability of $F - c = 0$ in a neighbourhood of $(x, y) = (0, 0)$.

4 Examples

Here is a simple minded example which illustrates the mechanics of remark (3.7).

Example 4.1 Consider the equation:

$$-(z_y)^2 + 2xz_y z_{xy} + -yz_y z_{yy} + x^2(z_{xx}z_{yy} - z_{xy}^2) = 0 \qquad (4.9)$$

This equation degenerates to a first order equation whenever $(x, y) = (0, 0)$ (from the point of view of a non-afficionado of geometric methods). So here $A = -q^2, B = 0, C = -qx; D = -qy$ and $E = x^2$. Clearly $AE - BD + C^2 = 0$ and thus the equation is parabolic. Choosing $X = \partial_p$ and $Y = \partial_q$ we get: $\theta_1 = -qxdx + x^2 dq$ and $\omega_2 = \frac{1}{x^2}(-qxdy + qydy - x^2 dp) = \frac{-q}{x}dy + \frac{qy}{x^2}dx - 1dp$. As one can check directly $\omega_2 = d(F)$, where $F = -\frac{px+qy}{x} - 1$. So the equation $F(x, y, z, p, q) = 0$ becomes

$$xz_x + yz_y = -x.$$

This equation has many solutions, one of which is $z(x, y) = -\frac{x^2}{2}$. As was mentioned in Remark (3.7), the choice of the -1 factor in defining F was made precisely so that the resulting first order PDE has smooth coefficients.

Example 4.2 Finally consider an example which shows how a one-dimensional stationary Hamilton-Jacobi equation arises as the equation $F(x, y, z, p, q) = 0$:

$$-x^3 z_{yy} + z_x(z_{xx}z_{yy} - z_{xy}^2) = 0 \qquad (4.10)$$

So, here $A = 0; B = 0, C = 0, D = -x^3, E = p$ and we have $AE - BD + C^2 = 0$ and so we have a parabolic equation. Choosing $X = \partial_q$ and $Y = \partial_x$ we get $\theta_1 = x^3 dx - pdp$ $\omega_2 = -\frac{1}{x^3}x^3 dq = -dq$ As $\theta_1 = d(\frac{x^4}{2} - p^2)$, the equation $F(x, y, z, p, q) = 0$ is in this case $(\frac{\partial z}{\partial x})^2 = \frac{x^4}{2}$ which has the solution $\phi(x, y) = \frac{x^3}{3\sqrt{2}} + \psi(y)$ for any smooth function ψ. Choosing a first integral, $F = -q$ for ω_2, we get $\phi(x, y) = \lambda(x)$ for any smooth function $\lambda(x)$.

5 Concluding Remarks

There are several problems which deserve study. One obvious one is to pursue the existence of geometric solutions so that present paper can be made coordinate free. Another problem is to study singularities of derived systems so that a cleaner theory applicable to the general definition of intermediate integrals (as defined in [5, 7]) is obtained. More generally, it is worthwhile examining if the sufficient conditions for the assumptions going into remarks (2.1) and (3.5) can be packaged into assumptions which can be verified directly from the equation (2.1).

The circle of ideas above should be applicable in other situations in the geometric theory of partial differential equations and we hope this line of research will be pursued by researchers in the theory of exterior differential systems. Whilst the method of intermediate integrals is somewhat orthogonal to the general theory of exterior differential systems, it seems that the techniques here (especially the division results) could be applied to the general theory. For this, it seems to the author, a preliminary question is to provide a definition of the Cartan system, $C(I)$, associated to a exterior differential ideal I which does not depend on the characteristic distribution, $A(I)$. Since $C(I)$ plays an important role in practically every normal form for I this question needs to be addressed first.

A different extension of the ideas presented in this paper is an approximate solution technique for equations which change type. Indeed, close to a point where the equation changes type the function $AE - BD + C^2$ is close to 0. Thus the wedge product $\theta_1 \wedge \omega_2$ is nearly equal to Ω. Having sacrificed exact decomposability for approximate decomposability there is no need to insist on exact integrability of the codistribution generated by θ_1 and ω_2. Approximate integration of distributions has been well studied in the context of control theory. This question and the question of providing coordinate free criteria for the existence of geometric solutions via the method of intermediate integrals for nonlinear second order partial differential equations is being currently investigated.

6 Acknowledgement

This paper greatly benefitted from the comments of an alert referee. The author is grateful for this assistance. It is also a pleasure to thank late Professor R. Gardner and Professor G. Wilkens for several email correspondences on the general subject of exterior differential systems. This manuscript is dedicated to the memory of late Professor R. Gardner. The author remembers him as a person extremely generous to others with his prodgious expertise.

References

[1] G. Bengel and R. Gérard, *"Formal and Convergent Solutions to Singular Partial Differential Equations*, Manuscripta Mathematica **38** 1982, 343-373.

[2] M.Berthier and D. Cerveau, *Quelques Calculs de Cohomologie Relative*, preprint, IRMAR, Dec 1992.

[3] R. Bryant, S. Chern, R. Gardner, H. Goldscmidt and P. Griffiths, *Exterior Differential Systems*, MSRI Publications, Springer-Verlag, New York, 1991.

[4] D. Cerveau, *Thèse de 3eme Cycle*, Université de Dijon, 1978.

[5] J. N. Clelland, *On the Intermediate Integral for Monge -Ampére Equations*, Preprint available as math.AP/9804029, 6 April 1998 on http://xxx.lanl.gov/ archive/math under Analysis of PDEs.

[6] M. Freeman, *Fully Integrable Pfaffian Systems*, Annals of Mathematics, **119**, 1984, 465-510.

[7] R. Hermann, *Geometry, Physics and Systems*, Marcel- Dekker, New York, 1977.

[8] L. Hörmander, *"Linear Partial Differential Operators, Vol III"*, Springer-Verlag, Heidelberg, 1985.

[9] M. Kossowski, *Local Existence of Analytic Multivalued Solutions to Symplectic Monge-Ampere Equations (the nondegenerate and type changing cases)*, Indiana Univ J of Math, **40**, 1991, 123 - 148.

[10] V. Lychagin, *Local Classification of Non-linear First Order Partial Differential Equations*, Russian Math Surveys, **30**, 1975, 105-175.

[11] B. Malgrange, *Frobenius Avec Singularités - Le Cas Général*, Inventiones Math, **39**, 1977, 67-89.

[12] R. Moussu, *Sur La Existence D'Integrales Premiérés Pour Un Germe De Forme De Pfaff*, Ann Instt Fourier, **26**, 1976, 171-220.

[13] E. Nelson, *Topics in Dynamics I : Flows*, Princeton University Press, Princeton, NJ, 1969.

[14] P. Olver, *Equivalence, Invariants and Symmetry*, Cambridge University Press, 1995.

[15] V. Ramakrishna, *"Local Solvability of Degenerate, Overdetermined Systems - A Control Theoretic Perspective"*, V. Ramakrishna, The Journal of Differential Equations, **123**, 1995, 260-304.

[16] V. Ramakrishna, *Summary of Disturbance Decoupling Via Differential Forms*, J of Mathematical Systems, Estimation and Control, **3**, 1996, 367-369 - full version will appear later, but has been published electronically on July 1 1996 and is available from the journal.

[17] K. Saito, *On a Generalization of De-Rham Lemma*, Ann Instt Fourier, **26**, 1976, 165-170.

[18] P. Stefan, *Accessible sets, Orbits and Foliations with Singularities*, Proc London Math Soc, **29** 1974, 699-713.

[19] H. Sussmann, *Orbits of families of vector fields and the integrability of distributions*, Transactions of the American Mathematical Society **180**, 1973, 171-188.

MATH SCIENCES, UNIVERSITY OF TEXAS AT DALLAS, RICHARDSON, TX 75083.
E-mail addrcss:vish@utdallas.edu

Proceedings of Symposia in Pure Mathematics
Volume **64**, 1999

Lie brackets and impulsive controls: an unavoidable connection

Franco Rampazzo

ABSTRACT. In this note we aim to give a brief account of some results concerning control problems where, generally speaking, trajectories are allowed to be discontinuous. The interest in this extension of the ordinary notion of trajectory is motivated e.g. by optimal control problems *with slow growth*, where minimizing sequences possibly converge to discontinuous maps. We stress two main facts. First, the genuine nonlinear nature of the problem implies that a naive distributional approach does not work, even if the control is scalar valued. In other words the dynamical equation governing the motion cannot be interpreted as an equality between distributions. Secondly, as soon as the control is vector-valued, the fact that the Lie brackets of the non-drift terms do not vanish identically makes the problem of determining a discontinuity of the trajectory much more involved. For instance, assume that the control [appears linearly in the dynamics and] consists of a [vector-valued] measure concentrated at time $\hat{t}$. It turns out that the mere knowledge of this measure is not sufficient for determining the corresponding jump of the trajectory. In fact, in order to compute this jump one needs a description a the path bridging (instantaneously at $t = \hat{t}$) the gap of the control's primitive. This leads to a notion of *space–time control*, which can be regarded as the limit of the graphs of primitives of ordinary —i.e. measurable— controls. Let us point out that two distinct space–time controls may have that same spatial projection. And, unless all involved Lie brackets vanish identically, two such space–time controls possibly produce distinct (space–time) trajectories and, hence, distinct costs. Let us remark that the set of [the graphs of] ordinary trajectories is dense in the set of space–time trajectories. Moreover, for a large class of minimum problems an optimal space-time trajectory does exist. Hence,the space-time embedding can be considered as a natural *extension* of the original problem with slow growth. Here, besides presenting some basic elements of the theory, we recall some results concerning dynamic programming. Nonlinear control problems with slow growth arise e.g. in classical mechanics, when moving holonomic constraints are regarded as controls [**7**][**31**][**9**]. Further applications can be found in mathematical modeling of optimal advertising [**15**][**19**] and in the study of driven stochastic equations [**16**].

1991 *Mathematics Subject Classification.* 49J15, 49N25, 93B29.

The participation of the author to the 1997 Summer Research Institute on Differential Geometry and Control was supported by NSF.

1. Introduction

While there is no difficulty to give a robust notion of solution to

$$\begin{cases} \dot{x} = g_0(x) + G\xi \\ x(\bar{t}\,) = \overline{x} \end{cases} \qquad t \in [\bar{t}, T] \tag{1.1}$$

when ξ is a first order distribution — i.e., $\xi = \dot{u}$, $u \in L^1_{\mathrm{loc}}$ — a *distributional approach* is not adequate for a nonlinear system

$$\begin{cases} \dot{x} = g_0(x) + G(x)\xi = g_0(x) + g_1(x)\xi_1 + \cdots + g_m(x)\xi_m \\ x(\bar{t}\,) = \overline{x}\,. \end{cases} \tag{1.2}$$

In fact, for the *linear* system (1.1) — where the term linear refers to the essential fact that G is an $n \times m$ matrix *independent of* x — we observe that x is a solution corresponding to a control $\xi \in L^1$ if and only if the map

$$z = x - G \cdot u, \tag{1.3}$$

with $u(t) \doteq \int_{\bar{t}}^t \xi(s)\, ds$, is a solution of

$$\begin{cases} \dot{z} = \widehat{g}_0(z, u) \doteq g_0(z + Gu) \\ z(\bar{t}\,) = \overline{x}. \end{cases} \tag{1.4}$$

Notice that since ξ belongs to L^1, the map u belongs to AC, where AC denotes the set of absolutely continuous maps. Yet both (1.3) and (1.4) are meaningful for a control $u \in L^1$ as well. Moreover in view of the linearity of (1.3) and of well-known properties of the input-output map of (1.4) we may *pass to the limit* when a sequence of controls $u_n \in AC$ converges (e.g., in the L^1-norm) to a control $u \in L^1$. Hence, given a map $u \in L^1$, one can define *the* solution x of (1.1) corresponding to (the distribution) $\xi = \dot{u}$ as

$$x \doteq z + Gu,$$

where z is the solution to (1.4).

In other words the *input-output map* $\Phi : AC \to L^1$, which to a control u associates the solution $x = \Phi(u)$ of (1.1) corresponding to $\xi = \dot{u}$ [is continuous when AC is endowed with the L^1-topology and] can be *continuously extended* to a map $\widetilde{\Phi} : L^1 \to L^1$. That is, the continuous map $\widetilde{\Phi}$ renders the diagram

$$\begin{array}{ccc} AC & \xrightarrow{\ \Phi\ } & L^1 \\ {\scriptstyle e}\big\downarrow & \nearrow_{\tilde{\Phi}} & \\ L^1 & & \end{array} \tag{1.5}$$

commutative, where e denotes the (dense) embedding of AC in L^1. And the above-given notion of solution x corresponding to a control $u \in L^1$ is nothing but $\widetilde{\Phi}(u)$. Notice that the solution $\widetilde{\Phi}(u)$ is defined up to a set of null measure. (Yet, a notion of solution pointwise defined of a subset $I \subset [\bar{t}, T]$ can be trivially given by considering sequences u_n converging to u in L^1 and pointwise on I (see [**4**] and [**5**])).

We remark that as soon as u has bounded variation — which implies that $\xi = \dot{u}$ is a Radon-measure — the solution $x = \widetilde{\Phi}(u)$ is a solution "in measure" of (1.1), that is, (1.1) is verified as an identity between the measure $\dot{x}$ and the measure on the right-hand side.

Summing up the above considerations we can say that for equations like (1.1) no difficulty arises when one wishes to extend the notion of solution in order to include the case where the control ξ is the (distributional) derivative of a L^1 function (see also [**35**] for the case when G depend on t). This can be done by continuous extension of the input-output map. And, as soon as u has bounded variation, this is equivalent to give a notion of solution "in measure".

However, there are essentially two crucial drawbacks in the attempt of extending the arguments of the linear case to the nonlinear equation (1.2):

(i) for $m = 1$, i.e., when (1.2) reduces to

$$\begin{cases} \dot{x} = g_0(x) + g_1(x)\xi \\ x(\bar{t}\,) = \bar{x} \end{cases} \tag{1.6}$$

the approach based on the continuous extension of the input-output map still works. On the contrary, the approach "in measure" lacks natural features of well-posedness (see below);

(ii) the "continuous extension" approach, valid for the scalar control-case (i.e., $m = 1$), does not hold any longer when $m \geq 2$.

While the problem described in (i) is a genuine analytical question (relying essentially on the fact that a naive distributional approach does not work for nonlinear systems), the difficulty pointed out in (ii) is directly related to a more differential geometric issue, namely the fact that (in general)

$$[g_i, g_j](x) \neq 0 \ ,$$

where $[g_i, g_j]$ is the Lie bracket of g_i and g_j:

$$[g_i, g_j] = \nabla g_j g_i - \nabla g_i g_j \ .$$

Actually, when

$$[g_i, g_j](x) = 0 \qquad \forall x \in \mathbb{R}^n, \quad i, j = 1, \ldots, m, \tag{1.7}$$

the "extension approach" *does* work [**5**], just as in the case when $m = 1$ (which trivially verifies the commutativity hypothesis (1.6)).

To conclude this introduction, let us remark that there exists a vast literature on impulsive control problems, see for example [**2**] and [**3**]. However these references include a dynamics where the fields multiplied by unbounded controls are constant, as in (1.1), so that the distributional approach is sufficient to deal with those problems, and no noncommutativity questions arise. Moreover the methods described in these references requires that the cost of a jump be given *a priori*. This fact makes those problems substancially different from ours. In particular such problems *are not* extensions of standard (slow growth) problems.

2. The scalar case and the commutative case

As mentioned above, when $m = 1$, we can address the problem of giving a robust notion of solution to (1.2) by exploiting the same extension argument which turned out to be successful in the linear case. At various levels of generality this approach has been pursued in [**38**], [**4**] (see also [**33**], [**39**], [**18**],[**34**] and the references therein) More precisely, under standard regularity assumptions one proves that the input-output map of (1.6)

$$\Phi : AC \longrightarrow L^1$$

— where $\Phi(u)$ is the classical (i.e., Caratheodory) solution of (1.6) corresponding to $u(t) = \int_{\bar t}^t \xi(s)\,ds$ — is uniformly continuous when AC is endowed with the L^1 topology: hence Φ can be continuously extended to (1.6).

Therefore, just as in the linear case, there is a continuous mapping $\widetilde{\Phi}$ which renders the diagram

$$\begin{array}{ccc} AC & \xrightarrow{\Phi} & L^1 \\ {\scriptstyle e}\big\downarrow & \nearrow_{\tilde\Phi} & \\ L^1 & & \end{array} \tag{2.1}$$

commutative. The reason why this happens relies on the fact that (1.6) can be transformed into a linear system. Indeed, let us consider the system

$$\begin{cases} \dot\alpha = \xi \\ \dot x = g_0(x) + g_1(x)\xi \\ (\alpha, x)(\bar t\,) = (0, \bar x), \end{cases} \tag{2.2}$$

which is formally obtained from (1.6) by adding the variable α and the equation $\dot\alpha = \xi$.

It is straightforward to show that the change of coordinates

$$(\alpha, y) = \Gamma(\alpha, x) \doteq (\alpha, \exp[-\alpha g_1]x)$$

transforms the vector field $(1, g_1)$ into the constant vector $(1, 0, \dots, 0)$.

Hence $(\alpha, x)(\cdot) = (u, x)(\cdot)$ is the solution of (2.2) corresponding to ξ if and only if

$$(\alpha, y) \doteq \Gamma \circ (\alpha, x)$$

is the solution to

$$\begin{cases} \dot\alpha = \xi \\ \dot y = \widehat{g}_0(\alpha, y) \\ (\alpha, y)(\bar t\,) = (0, \bar x), \end{cases} \tag{2.3}$$

where, for every $(\alpha, y) = \Gamma(\alpha, x)$,

$$(0, \widehat{g}_0)(\alpha, y) \doteq d\Gamma(\alpha, x) \cdot (0, g_0(x)),$$

$d\Gamma$ denoting the derivative of Γ.

Let us set $u \doteq \int_{\bar t}^t \xi(s)\,ds$ and let us denote the solution to

$$\begin{cases} \dot y = \widehat{g}_0(u, y) \\ y(\bar t\,) = \bar x \end{cases}$$

by $\widehat{\Phi}(u)$. Hence

$$(u, \Phi(u)) = \Gamma^{-1} \cdot (u, \widehat{\Phi}(u))\,. \tag{2.4}$$

It is easy to show that $\widehat{\Phi}(u)$ can be extended from AC to L^1, continuously with respect to the L^1-topology (see [**4**], [**5**]). Hence by (2.4), the same can be stated for Φ.

In other words the solution x of (1.6) corresponding to $\xi = \dot u$, with $u \in L^1$, coincides with the L^1-limit of (classical) solutions x_n corresponding to a sequence $u_n \in AC$ converging to u in L^1. The above arguments show that x is independent of the particular sequence u_n approximating u and is defined up to a set of measure zero. (Yet the notion of solution can be refined in order to get a trajectory pointwise defined e.g., at the final time).

We conclude this section by remarking that a direct approach "in measure" does not work although it works for the transformed system (2.3). Moreover, it is easy to check that the above given notion of solution x corresponding to a distribution $\xi = \dot{u}$ is nothing but the Γ^{-1}-transform of the (measure) solution (y_0, y) of (2.3), namely $(u, x) = \Gamma^{-1}(y_0, y))$.

A quite critical (and convincing) argument against a direct approach in measure can be found in [**21**]. An heuristic, non-rigorous explanation of the reason why a "solution in measure" cannot verify elementary properties of well-posedness can be argued by considering, the control $\xi = \dot{u}$ where $u(t) = 0$ for all $t \in [\bar{t}, \widehat{t}\,]$ and $u(t) = 1$ for $t \in]\widehat{t}, T]$. It is clear that a *solution* x to (1.6) corresponding to ξ *has* to jump at $t = \widehat{t}$ and that the jump should satisfy a relation like

$$x(\widehat{t}_+) - x(\widehat{t}_-) = g_1(x(\widehat{t}\,))(u(\widehat{t}_+) - u(\widehat{t}_-)) \,.$$

Now the question is: which is the "value of $x(\widehat{t}\,)$" where g has to be evaluated? It obvious that an answer to this question is crucial in the attempt of giving to (1.6) a distributional meaning. Yet, any such answer (for example one could decide to take the right-limit of x at $\widehat{t}$, or the left-limit, the intermediate point, ect.) turns out to be unsatisfactory in that i does not agree with elementary well-posedness requirements (again, see [**21**]).

When $g_1, \dots, g_m$ verify the commutativity conditions

$$[g_i, g_j] = 0 \qquad \forall i, j = 1, \dots, m \tag{2.5}$$

the approach pursued in the scalar case can be easily extended to system (1.2) for any $m \geq 1$. The details can be found e.g. in [**5**]. We only remark that a transformation of coordinates analogous to the one performed in the linear-case plays a crucial role in the proof of the extendability of the input-output map. More precisely one considers the transformation $\Gamma : \mathbb{R}^m \times \mathbb{R}^n \to \mathbb{R}^m \times \mathbb{R}^n$ defined by

$$\begin{aligned}\Gamma(\alpha, x) \doteq (\alpha, \exp[-\alpha_1 g_1] \circ \cdots \circ \exp[-\alpha_m g_m] x),& \\ \text{for every } (\alpha, x) = (\alpha_1, \dots, \alpha_m, x_1, \dots, x_n) \in \mathbb{R}^m \times \mathbb{R}^n \,.&\end{aligned} \tag{2.6}$$

It is straightforward to check that Γ changes the system

$$\begin{cases} \dot{\alpha} = \xi \\ \dot{x} = g_0(x) + \displaystyle\sum_{i=1}^{m} g_i(x)\xi_i \\ (\alpha, x)(\bar{t}\,) = (0, \bar{x}) \end{cases} \tag{2.7}$$

into the simpler system

$$\begin{cases} \dot{\alpha} = \xi \\ \dot{y} = \widehat{g}_0(\alpha, y) \\ (\alpha, y)(\bar{t}\,) = (0, \bar{x}), \end{cases} \tag{2.8}$$

where, for every $(\alpha, y) = \Gamma(\alpha, y)$,

$$(0, \widehat{g}_0)(\alpha, y) \doteq d\Gamma(\alpha, x) \cdot (0, g_0(x)) \,.$$

Therefore one can proceed as in the scalar case and prove the extendability of the input-output map of (2.8), which in turn implies the extendability of the input-output map of (2.7).

3. Non commutative systems

For the purpose of continuously extending the input-output map from the set of L^1 controls to the set of controls which are the distributional derivatives of L^1-maps the commutativity condition (2.5) is crucial, as is illustrated by the following example.

EXAMPLE 3.1. Let g_1 and g_2 be smooth vector fields in $\mathbb{R}^n$ such that $[g_1, g_2] \neq 0$ and let us consider the system

$$\begin{cases} \dot{x} = g_1(x)\xi_1 + g_2(x)\xi_2 \\ x(0) = 0, \end{cases} \qquad t \in [0,2] \,.$$

Let us fix the control $(\xi_1, \xi_2) = \dot{u}$, $u = (\chi_{[0,1[}, \chi_{[0,1[})$, where χ_E denotes the characteristic function of E.

Both sequences

$$u_n^1(t) = \begin{cases} (1,1) & t \in [0, 1-1/n] \\ (1, 1-n(t-1+1/n)) & t \in [1-1/n, 1] \\ (1-n(t-1), 0) & t \in [1, 1+1/n] \\ (0,0) & t \in [1+1/n, 2] \end{cases}$$

and

$$u_n^2(t) = \begin{cases} (1,1) & t \in [0, 1-1/n] \\ (1-n(t-1+1/n), 1) & t \in [1-1/n, 1] \\ (0, 1-n(t-1)) & t \in [1, 1+1/n] \\ (0,0) & t \in [1+1/n, 2] \,. \end{cases}$$

(are in AC and) converge to u in the L^1-topology.

Yet the solutions x_n^1 corresponding to the controls u_n^1 converge to the map

$$x^1(t) = a^1 \chi_{[1,2]}, \qquad a^1 \doteq \exp(g_2) \circ \exp(g_2)(0)$$

while the solutions x_n^2 corresponding to the controls u_n^2 converge to the map

$$x^2(t) = a^2 \chi_{[1,2]} \qquad a^2 \doteq \exp(g_1) \circ \exp(g_1)(0) \,.$$

Since $[g_1, g_2] \neq 0$, in general one has $a^1 \neq a^2$.

This example might suggest that, whereas the input-output of (1.6) fails to be continuously extendable to a set of controls that includes measures, one could try to extend the input-output map corresponding to the graphs of the controls (and of the trajectories). Actually this is the idea underlying the approach pursued in **[4]**, **[22]**-**[25]**,**[26]**-**[30]**, **[36]**-**[37]**. Let us briefly illustrate the main idea of this approach.

To begin with, let us consider the set

$$\mathcal{U}_K \doteq \{ u \in AC([\bar{t}, T], \mathbb{R}^m), V_{\bar{t}}^T(u) \leq K \},$$

where $AC([\bar{t},T],\mathbb{R}^m)$ denotes the set of absolutely continuous maps from $[\bar{t},T]$ into $\mathbb{R}^m$ and $V_{\bar{t}}^T(u)$ denotes the variation of u in $[\bar{t},T]$. Since u is absolutely continuous, one has

$$V_{\bar{t}}^T(u) = \int_{\bar{t}}^T |\dot{u}|\,dt\,.$$

Now let $\varphi_0 : [0,1] \to [\bar{t},T]$ be a differentiable, surjective map, with $\varphi_0'(s) > 0$ $\forall s \in [0,1]$, where the sign "$'$" denotes differentiation with respect to the variable s. It is trivial to verify that a map x is the trajectory of (1.6) corresponding to a control $\xi = \dot{u}$, $u \in \mathcal{U}_K$, if and only if the *space-time trajectory* $(y_0,y) : [0,1] \to \mathbb{R}^{1+n}$ defined by

$$(y_0,y)(s) \doteq (\varphi_0(s), x \circ \varphi_0(s)) \qquad \forall s \in [0,1]$$

is the solution of

$$\begin{cases} y_0' = \varphi_0' \\ y' = g_0(y)\varphi_0' + \sum g_i(y)\varphi_i' \\ (y_0,y)(0) = (0,\bar{x}) \end{cases} \tag{3.1}$$

corresponding to the *space-time control*

$$(\varphi_0,\varphi)(s) \doteq (\varphi_0(s), u \circ \varphi_0(s))\,.$$

In other words (3.1) is the system resulting from (1.6) — more precisely, from (1.6) supplemented with the equation $\dot{t} = 1$ — after reparameterizing time by means of $t = \varphi_0(s)$. And the *equivalence* between (1.6) and (3.1) relies on the injectivity of φ_0. Yet, in principle there is no problem in considering space-time controls (φ_0,φ) with a φ_0 merely non decreasing. Of course, unless φ_0 is injective, the space-time control (φ_0,φ) is not the reparametrization of (the graph of a) control $u \in \mathcal{U}_K$.

In fact a "spatial projection" of (φ_0,φ), say a selection of $\varphi \circ \varphi_0^{-1}(t)$, is in general discontinuous. Hence it cannot belong to $\mathcal{U}_K$. In a sense, (φ_0,φ) contains the information of a "spatial control" $u(t) \in \varphi \circ \varphi_0^{-1}(t)$ *plus* the extra-information represented by the restriction of φ to those intervals where φ_0 is constant. Actually, let $[s_1,s_2] \subseteq [0,1]$ be such that $\varphi_0'(s) = 0$ $\forall s \subset [s_1,s_2]$. One has $\varphi_0(s) = \varphi_0(s_1) = \hat{t}$ $\forall s \in [s_1,s_2]$, while in general $\varphi(s_1) \neq \varphi(s_2)$. Hence $(\varphi_0,\varphi)(s)$ cannot be the reparametrization of the graph of a continuous control $u(t)$: actually one has $u(\hat{t}_-) = \varphi(s_1)$, $u(\hat{t}_+) = \varphi(s_2)$. The extra-information is here provided by the restriction $\varphi_{|[s_1,s_2]}$. Indeed, in order to compute the "jump" of the trajectory x at t_1, namely the vector

$$x(\hat{t}_+) - x(\hat{t}_-) = y(s_2) - y(s_1),$$

it is not sufficient to know the gap $u(\hat{t}_+) - u(\hat{t}_-) = \varphi(s_2) - \varphi(s_1)$, because "during" the s-interval $[s_1,s_2]$ the state y evolves according to the controlled dynamics

$$y' = \sum_{i=1}^m g_i(y)\varphi_i'\,. \tag{3.2}$$

Therefore in order to calculate the "jump" $y(s_2) - y(s_1)$ one has to know the whole restriction $\varphi_{|[s_1,s_2]}$.

REMARK 3.2. Of course in the commutative case the mere knowledge of the gap $\varphi(s_2) - \varphi(s_1)$ is sufficient for the computation of $y(s_2) - y(s_1)$. This very fact, together with the density theorem below, explain why there are no problems in extending the input-output map when the vector fields g_i commute.

Let us make the above considerations more precise. We shall consider the set Φ_K of *space-time controls with variation bounded by* K, defined by

$$\Phi_K \doteq \Big\{ (\varphi_0, \varphi) \in \mathrm{Lip}([0,1], [\bar{t}, T] \times \mathbb{R}^m) : \\ |(\varphi_0', \varphi')(s)| \leq K + T,\ \varphi_0'(s) \geq 0 \text{ a.e. in } [0,1], \\ \varphi_0([0,1]) = [\bar{t}, T],\ V_0^1(\varphi) \leq K \Big\},$$

where $0 \leq \bar{t} < T$ and $\mathrm{Lip}(E, F)$ denotes the set of Lipschitz continuous maps from E into F. Let us denote the set of corresponding solution to (3.1) by Λ_K. In the space-time setting, the set $\mathcal{U}_K$ has to be identified with the subset

$$\Phi_K^+ \doteq \Big\{ (\varphi_0, \varphi) \in \Phi_K, \varphi_0' > 0 \text{ a.e. in } [0,1] \Big\} \subset \Phi_K$$

in a obvious way (see e.g., [**26**], [**28**], [**32**]). Let us denote the corresponding set of solutions to (3.1) by Λ_K^+. The following can be trivially proved:

LEMMA 3.3. *Let* $\overline{\Phi}_K^+$ *denote the closure of* Φ_K^+ *in* $C^0([0,1], \mathbb{R}^{1+n})$. *Then* $\overline{\Phi}_K^+ = \Phi_K$.

REMARK 3.4. In Lemma 3.3 and in the results below concerning the density of Λ_K^+ and the continuity of the input-output map a topology could be considered on both Φ_K and Λ_K that is (weaker and) more appropriate than the C^0-topology (see e.g., [**4**] and [**26**]). We only point out that this topology keeps track of the free-parameter character of system (3.1), namely of the fact that if $(y_0, y)(s)$ is the solution corresponding to $(\varphi_0, \varphi)(s)$ and $s(\sigma)$ is a reparametrization of $[0,1]$ than $(y_0, y) \circ s(\sigma)$ is the solution corresponding to $(\varphi_0, \varphi) \circ s(\sigma)$. Yet we use the C^0 topology, for this choice is not too restrictive and, at the same time, allows one to simplify the presentation of several subjects.

THEOREM 3.5 ([**4**]). *The input-output map* $\mathcal{I} : \Phi_K :\longrightarrow \Lambda_K$, *which to every* $(\varphi_0, \varphi) \in \Phi_K$ *associates the corresponding solution to* (3.1), *is continuous w.r. to the* C^0 *norm on both* Φ_K *and* Λ_K.

¿From the relations $\Lambda_K^+ = \mathcal{I}(\Phi_K^+)$, $\Lambda_K = \mathcal{I}(\Phi_K)$ and the continuity of $\mathcal{I}$ we obtain:

COROLLARY 3.6. *Let* $\overline{\Lambda}_K^+$ *denote the closure of* Λ_K^+ *in the* C^0 *norm. Then* $\overline{\Lambda}_K^+ = \Lambda_K$.

Now let us come back to our original question, that is: *can one give a notion of solution to* (1.2) *corresponding to a control* $\xi = \dot{u}$, *with* u *an* L^1*-map?* Let us restrict our attention (see Remark 3.8 below) to the set of controls

$$\mathcal{U}_K^* \doteq \Big\{ u \in BV([\bar{t}, T], \mathbb{R}^m), V_{\bar{t}}^T(u) \leq K \Big\}.$$

Observe that the original set $\mathcal{U}_K$ is dense in $\mathcal{U}_K^*$ w.r. to the L^1-topology.

Fix $u \in \mathcal{U}_K^*$. On one hand, the previous example show that we cannot extend continuously the notion of solution corresponding to a control u from $\mathcal{U}_K$ to $\mathcal{U}_K^*$. In fact, distinct sequences (u_n), $(\widetilde{u}_n)$ in $\mathcal{U}_K$ approximating u give rise to sequences of solutions (x_n) and $(\widetilde{x}_n)$, respectively, that in general converge to distinct limits.

On the other hand one can consider a *graph-completion* of u.

DEFINITION 3.7. A *graph-completion* of u is a space-time control $(\varphi_0, \varphi) \in \Phi_K$ such that $u([\bar{t}, T]) \subseteq \varphi([0,1])$.

Notice that a graph-completion always exists: it is sufficient to bridge the gaps of u with straight lines and to parameterize the so-obtained (rectifiable) $(m+1)$-path with an abscissa that is proportional to the arc-length (in this case the total variation of φ equals the total variation of u). Of course, there is no special reason to prefer *this* graph-completion to another one where gaps are filled with curves which are not rectilinear (see the next section).

Now, whereas we are not able to define a solution of (1.2) corresponding to a control $u \in \mathcal{U}_K^*$, we can consider the solution $(y_0, y)(s)$ of (3.1) corresponding to a graph completion (φ_0, φ) of u. If one wishes to come back to the parameter t, any selection of $y_0 \circ y_0^{-1}(t)$ can be considered as a solution of (1.2) corresponding to the graph completion (φ_0, φ).

Notice that in view of Corollary 3.6 the space-time trajectory (y_0, y) can be uniformly approximated by means of trajectories $(y_0^n, y^n) \in \Lambda_K^+$ corresponding to space-time controls (φ_0^n, φ^n) uniformly converging to (φ_0, φ). In practice this means that the solution (y_0, y) corresponding to a graph -completion (φ_0, φ) of u can be approached by the graphs of the solutions x^n corresponding to controls u^n whose (reparameterized) graphs converge to (φ_0, φ).

We have constructed a class of trajectories, namely the set Λ_K, in which the class of original trajectories, here identified with Λ_K^+, is dense. The set Λ_K coincides with the set of outputs of system (3.1) corresponding to the inputs in Φ_K, the set of graph-completions (it is obvious that *each* space–time control (φ_0, φ) is the graph-completion of a suitable control $u \in \mathcal{U}_K^*$). This gives an answer to the question of defining a notion of output of (1.2) corresponding to a control $\xi = \dot{u}$, with $u \in \mathcal{U}_K^*$. In fact, in order to define such an output we have to *choose* —in principle, arbitrarily — a graph-completion (φ_0, φ) of u and to consider the corresponding trajectory. Let us remark that for a wide class of optimal control problems having (1.2) as dynamics the optimal controls *exists* only in the set Φ_K (see next section). In this case the optimal strategy consists of a control $u \in \mathcal{U}_K^*$ *and* a *suitable* graph-completion of u.

REMARK 3.8. When the variation of u is not bounded the situation is much more involved. For example, the input-output map is not continuous *even when it is restricted to continuous controls*, as is shown in [**38**]. The question is addressed in [**6**] where a concept of *looping control* is introduced: roughly speaking a looping control is a representation of the *limit* of a sequence of controls with larger and larger variation.

4. Optimal control problems

In order to give an idea of the questions involved in optimal control problems subject to a dynamics of the form (1.2), let us consider the following Mayer problem: let $\psi : \mathbb{R}^n \to \mathbb{R}$ be a continuous map and let us define the final cost functional as

$$\mathcal{C}(u) \doteq \psi(x[u](T)),$$

where $x[u](\cdot)$ denotes the trajectory of

$$\left\{\begin{aligned} &\dot{x}=g_0(x)+\sum_{i=1}^{m} g_i(x)\dot{u}_i \\ &x(\bar{t})=\overline{x}\end{aligned}\right. \tag{4.1}$$

corresponding to the control $\xi=\dot{u}$. The problem

$$\underset{u\in\mathcal{U}_K}{\text{minimize}}\, \mathcal{C}(u) \tag{$\mathcal{P}$}$$

consists in looking for a control $\widehat{u}\in\mathcal{U}_K$ such that $\mathcal{C}(\widehat{u})=\min_{u\in\mathcal{U}_K}\mathcal{C}(u)$.

We begin by observing that an optimal control $\widehat{u}\in\mathcal{U}_K$ in general does not exist. This is due essentially to the lack of compactness of $\mathcal{U}_K$. In order to frame this phenomenon of non-existence in a more theoretical picture we begin by observing that it is not related to any lack of convexity: no chattering controls are needed here. Instead, the non-existence of an optimal control $\widehat{u}\in\mathcal{U}_K$ is linked to a fact that, in analogy with a similar pathology occurring in the Calculus of Variations, could be called *lack of coerciveness.* As a matter of fact, in the classical problem

$$\underset{\substack{x\in AC\\ x(a)=A\\ x(b)=B}}{\text{minimize}} \int_a^b L(x,\dot{x})\,dt, \tag{4.2}$$

the term *coerciveness* refers to the fact that the Lagrangian L is *superlinear* in $\dot{x}$. Roughly speaking, this avoids the occurrence of minimizing sequences of trajectories that converge to a discontinuous map. In Control Theory, on one hand several problems involve a *bounded* dynamics: this is sufficient to avoid jumps of the optimal trajectories. On the other hand it often happens that, although the dynamics is unbounded, a sort of coerciveness condition penalizes the use of too large velocities. For instance, this is the case of the so-called linear-quadratic problems.

On the contrary, problem $(\mathcal{P})$ involves a unbounded dynamics — $\dot{u}$ has no bounds— and no condition makes the use of larger and larger controls unfavorable for the pursuit of the infimum of $\mathcal{C}$.

Example 4.1 below provides a typical situation of non-existence of an optimal control. Successively we shall present an extension of the general problem $(\mathcal{P})$ and test this extension with the special problem of Example 4.1.

EXAMPLE 4.1. Let us consider the system

$$\left\{\begin{aligned} &\dot{x}_1=\dot{u}_1\\ &\dot{x}_2=\dot{u}_2\\ &\dot{x}_3=\arctan(x_1^2+x_2^2)-\frac{x_2\dot{u}_1}{2}+\frac{x_1\dot{u}_2}{2}\\ &(x_1,x_2,x_3)=(0,0,0)\end{aligned}\right. \qquad t\in[0,1] \tag{4.3}$$

and let us address the problem of minimizing the Mayer functional

$$\mathcal{C}(u)=x_3[u](1)$$

over the set $\mathcal{U}_\pi$, where $x[u](\cdot)$ denotes the solution to (4.3) and $x_i[u]$, $i=1,2,3$, is its i-th component. Since the integral

$$\int_0^\alpha \frac{x_1(\sigma)\dot{x}_2(\sigma)-\dot{x}_2(\sigma)x_1(\sigma)}{2}\,d\sigma$$

coincides with the *area spanned by the vector* $(x_1, x_2)(\sigma)$ *in the interval* $[0, \alpha]$, it is not difficult to check that the maps

$$\widehat{u}_n(t) \doteq [\cos(\pi n(1 - 1/n - t)) - 1, \sin(\pi n(1 - 1/n - t))]\chi_{[1-1/n,1]}$$

define a minimizing sequence. Moreover, one has

$$\inf_{u \in \mathcal{U}_\pi} \mathcal{C}(u) = \lim_{n\to\infty} \mathcal{C}(\widehat{u}_n) = -\pi/2 .$$

Now, one can easily verify that every minimizing sequence $\widetilde{u}_n$ verifies

$$\lim_{n\to\infty} x_3(\widetilde{u}_n)(t) = 0 \qquad \forall t \in [0,1[,$$
$$\lim_{n\to\infty} x_3(\widetilde{u}_n)(t) = -\pi/2,$$

which, in particular, implies that no optimal control exists in $\mathcal{U}_\pi$. Also, observe that although the sequence

$$\check{u}_n \doteq [\cos(\pi n(t - 1 + 1/n)) - 1, \sin(\pi n(t - 1 + 1/n))]\chi_{[1-1/n,1]}$$

is not minimizing (in that $\lim_{n\to\infty} \mathcal{C}(\check{u}_n) = \pi/2$) one has

$$\lim |\widehat{u}_n(t) - \check{u}_n(t)| = 0 \qquad \forall t \in [0,1] .$$

Extension of problem $(\mathcal{P})$**.** We now introduce an *extension* of the problem $(\mathcal{P})$, that is, a new minimum problem $(\mathcal{P}_e)$ having the following features:

(i) the infimum value of $(\mathcal{P})$ coincides with the infimum value of $(\mathcal{P}_e)$;
(ii) the class of elements where one seeks the infimum of $(\mathcal{P})$ is dense in the corresponding class for $(\mathcal{P}_e)$, with respect to a topology for which the functional of $(\mathcal{P}_e)$ is continuous.

We shall consider the extended cost functional

$$\mathcal{C}_e(\varphi_0, \varphi) \doteq \psi(y[\varphi_0, \varphi](1))$$

where $(y_0, y)[\varphi_0, \varphi](\,\cdot\,)$ denotes the solution of

$$\begin{cases} y_0' = \varphi_0' \\ y' = g_0(y)\varphi_0' + \sum_{i=1}^{m} g_i(y)\varphi_i' \\ (y_0, y)(0) = (0, \overline{x}) \end{cases} \tag{4.4}$$

corresponding to the space-time control $(\varphi_0, \varphi) \in \Phi_K$. We define the extended problem $(\mathcal{P}_e)$ as

$$(\mathcal{P}_e) \qquad \operatorname*{minimize}_{(\varphi_0,\varphi)\in\Phi_K} \mathcal{C}_e(\varphi_0, \varphi) .$$

THEOREM 4.2. *Problem* $(\mathcal{P}_e)$ *admits an optimal control* $(\widehat{\varphi}, \widehat{\varphi}) \in \Phi_K$*, that is*

$$\mathcal{C}_e(\widehat{\varphi}_0, \widehat{\varphi}) = \inf_{(\varphi_0,\varphi)\in\Phi_K} \mathcal{C}(\varphi_0, \varphi) . \tag{4.5}$$

Moreover, for every sequence $(\varphi_0^n, \varphi^n)(\,\cdot\,)$ in Φ_K^+ which converges to (φ_0, φ) one has that the sequence (u^n) in $\mathcal{U}_K$ defined by

$$u^n(t) = \varphi^n \circ (\varphi_0^n)^{-1}(t) \qquad \forall t \in [\bar{t}, T]$$

verifies

$$\lim_{n\to\infty} \mathcal{C}(u^n) = \inf_{u\in\mathcal{U}_K} \mathcal{C}(u) = \mathcal{C}_e(\widehat{\varphi}_0, \varphi) . \tag{4.6}$$

REMARK 4.3. In particular (4.5)-(4.6) yield

$$\inf_{u\in\mathcal{U}_K} \mathcal{C}(u) = \inf_{(\varphi_0,\varphi)\in\Phi_K} \mathcal{C}_e(\varphi_0,\varphi)\,.$$

PROOF OF THEOREM 4.2. By Theorem 3.5 the map

$$(\varphi_0,\varphi) \longmapsto \psi(y[\varphi](1))$$

is continuous with respect to the C^0-norm. Moreover by Ascoli-Arzelà's theorem the set Φ_K is compact, from which the existence of an optimal control follows. The remaining part of the thesis is a straightforward consequence of the density of Φ_K^+ in Φ_K and of the continuity of the input- output map $\mathcal{I}$. □

Application to Example 4.1. The space-time versions of the system and the cost-functional in Example 4.1 are given by

$$\begin{cases} y_0' = \varphi_0' \\ y_1' = \varphi_1' \\ y_2' = \varphi_2' \\ y_3' = \arctan(y_1^2 + y_2^2) - \dfrac{y_2\varphi_1'}{2} + \dfrac{y_1\varphi_2'}{2} \\ (y_0, y_1, y_2, y_3)(0) = (0,0,0,0), \end{cases}$$

and

$$\mathcal{C}_e(\varphi_0,\varphi) = y_3[\varphi_0,\varphi_1](1)\,,$$

respectively. By Theorem 4.2 the extended problem

$$\operatorname*{minimize}_{(\varphi_0,\varphi)\in\Phi_\pi} \mathcal{C}_e(\varphi_0,\varphi)$$

has one solution. By the considerations made in Example 4.1 it is clear that the space- time control

$$(\widehat{\varphi}_0, \widehat{\varphi}(s)) \doteq (2s,0,0)\chi_{[0,1/2[}(s) + (0, \cos(\pi - 2\pi s) - 1, \sin(\pi - 2\pi s))\chi_{[1/2,1]}(s)$$

is optimal for the extended problem.

It is straightforward to check that the sequence $(\widehat{\varphi}_0^n, \widehat{\varphi}^n)$ defined by

$$(\widehat{\varphi}_0^n, \widehat{\varphi}^n(s)) \doteq \Big((s - s/n)\chi_{[0,1/2[}(s) + (1 + s/n - 1/n)\chi_{[1/2,1]}(s), \widehat{\varphi}(s)\Big)$$

lies in Φ_π, converges to $(\widehat{\varphi}_0, \widehat{\varphi})$, and verifies

$$\widehat{u}_n(t) = \widehat{\varphi}^n \circ (\widehat{\varphi}_0^n)^{-1}(t)$$

for every n and every $t \in [0,1]$, where $(\widehat{u}_n)$ is the sequence of maps defined in Example 4.1. This explains why $(\widehat{u}_n)$ is a minimizing sequence, while $(\check{u}_n)$ *is not* a minimizing sequence. In fact, the space-time controls $(\check{\varphi}_0^n, \check{\varphi}^n)$ defined by

$$(\check{\varphi}_0^n, \check{\varphi}^n)(s) \doteq (\widehat{\varphi}_0^n(s), \check{\varphi}(s))$$
$$\check{\varphi}(s) \doteq (0, \cos(2\pi s - \pi) - 1, \sin(2\pi s - \pi))$$

are space-time representations of the $\check{u}_n$, that is

$$\check{u}_n(t) = \check{\varphi}^n \circ (\check{\varphi}_0^n)^{-1}(t),$$

and they converge uniformly to the space-time control $(\widehat{\varphi}_0, \check{\varphi})$, which is not optimal, for

$$\mathcal{C}_e(\widehat{\varphi}_0, \check{\varphi}) = \frac{\pi}{2} \cdot$$

We conclude by remarking that both $(\widehat{u}_n)$ and $(\check{u}_n)$ converge pointwise to the control $\widehat{u} = (-1,0)\chi_{\{1\}}$. Yet the information that $(\widehat{u}_n)$ [resp. $\check{u}_n$] is [resp. is not] a minimizing sequence cannot be deduced by the mere knowledge of $\widehat{u}$ (or, equivalently, of $\xi = (-1,0)\dot{\hat{u}} = (-1,0)\delta_{\{1\}}$). On the contrary, this information is contained in the specification of the graph-completion $(\widehat{\varphi}_0, \widehat{\varphi})$ of $\widehat{u}$.

5. Some investigated problems

The presentation of the previous sections is intentionally simplified and confined to a model-problem which displays a number of advantages from the topological viewpoint.Yet, on one hand more general systems than (1.2) have been investigated in the last years. Several minimum problems connected with these systems share the fact that too large velocities are not penalized (*slow growth problems*), so that optimal trajectories possibly are not found within continuous curves. On the other hand, for this more general class of systems, other questions have been addressed besides existence. Let us just give a list of some investigated problems (other directions of research can be found in the bibliographies of the quoted papers).

More general systems. Instead of taking values in the whole $\mathbb{R}^m$, the control ξ could be allowed to range in a cone $C \subset \mathbb{R}^m$, e.g., the cone of vectors having non-negative components. Moreover one could consider an *ordinary* control v besides the unbounded control ξ. By saying that v is ordinary we mean that it takes values in a bounded set V.

Finally, the dynamics (1.2) could be generalized into a dynamics which is merely *sublinear* in ξ (see e.g.[**28**],[**25**]). More precisely, one can consider systems of the form

$$\begin{cases} \dot{x} = f(t,x,v,\xi) \\ x(\bar{t}) = \bar{x} \end{cases} \tag{5.1}$$

where the ordinary control $v(\cdot)$ takes value in a compact set $V \subseteq \mathbb{R}^q$, the unbounded control ξ ranges in a closed cone $C \subseteq \mathbb{R}^m$ and the dynamics f is sublinear in ξ, that is f verifies

$$|f(t,x,v,\xi)| \leq M(t,x,v) + N(t,x,v)|\xi|$$

with M and N suitable continuous maps.

Under some regularity assumptions on f the hypothesis of sublinearity in ξ allows one to consider the *recession map* f^∞ of f,

$$f^\infty(t,x,v,w) \doteq \lim_{r\to 0} rf(t,x,v,w/r) \, ,$$

and the space-time extension of f

$$\overline{f}(t,x,v,w_0,w) \doteq \begin{cases} w_0 f\Big(t,x,v,\dfrac{w}{w_0}\Big) & \text{if } w_0 \neq 0 \\ f^\infty(t,x,v,w) & \text{if } w_0 = 0 \, . \end{cases}$$

The space-time system

$$\begin{cases} y_0' = \varphi_0' \\ y' = \overline{f}(y_0, y, v, \varphi_0', \varphi') \\ (y_0, y)(0) = (0, \overline{x}) \end{cases} \tag{5.2}$$

plays the role of (3.1) when the original system (1.2) is replaced by the more general system (5.1). One can prove density results analogous to the one valid for the special case (3.1). Yet, we remark that even in the space-time setting an optimal control in general does not exist: indeed both the presence of the ordinary control v and the fact that f is no longer affine in ξ cause a loss of the arguments of compactness and continuity that in the special case treated in the previous sections yielded the existence of optimal (space-time) controls. Actually the closure of the set of (space-time) trajectories can be recovered by imposing suitable convexity-assumptions. Instead there are no means for recovering the continuity of the input-output map. This contunuity can be replaced only by the lower continuity of the map which maps a (space-time) control into the corresponding cost.

State-space constraints. The density result (Corollary (3.6)), which justifies the introduction of the space-time system, is no longer valid as soon as the state x is constrained in the closure $\overline{\mathcal{O}}$ of a open set $\mathcal{O} \subseteq \mathbb{R}^n$. Then a question arises of looking for sufficient conditions which guarantee a dense embedding of the set of solutions to (5.1) in the set of solutions to (5.2). Conditions of this type can be found in [**28**] in the form of hypotheses on the directions of the field $\overline{f}$ on the boundary $\partial\mathcal{O}$.

Dynamic Programming. We have seen that a question of *slow growth* in the variable $\dot{u}$ lies at the basis of the fact that a problem like

$$\underset{(u,v)\in(\mathcal{U}\mathcal{V})_K}{\text{minimize}} \; \mathcal{C}(u,v)$$

$$\begin{cases} \dot{x} = f(t, x, v, \dot{u}) \\ x(\bar{t}\,) = \overline{x} \end{cases}$$

[where $(\mathcal{U}\mathcal{V})_K$ is the natural generalization of $\mathcal{U}_K$ when an ordinary control v is included in the dynamics, i.e.,

$$(\mathcal{U}\mathcal{V})_K = \mathcal{U}_K \times L^1([\bar{t}, T], V),$$

and $\mathcal{C}(u,v)$ is the obvious generalization of $\mathcal{C}(u)$] possibly does not have a solution — even in the presence of convexity conditions, as in the case when (5.1) reduces to (1.2).

This same lack of growth impedes a standard application of methods of Dynamic Programming. In fact — after observing that there is a *hidden* state variable in the problem, namely the variation k of u — let us consider the value function

$$\mathcal{V}(\bar{t}, \overline{x}, \overline{k}\,) = \inf_{(u,v)\in(\mathcal{U},\mathcal{V})_{K-\overline{k}}} \mathcal{C}_{(\bar{t},\overline{x})}(u,v),$$

where we have written $\mathcal{C}_{(\bar{t},\overline{x})}$ instead of $\mathcal{C}$ in order to stress the fact that the cost function (and hence the value function) depend on the initial data. The hidden variable k verifies the trivial initial value problem

$$\begin{cases} \dot{k} = |\dot{u}| \\ k(\bar{t}) = \bar{k} \,. \end{cases}$$

If we assume that the maximal available amount of *energy* $\int_{\bar{t}}^{T} |\dot{u}(t)|\, dt$ is equal to $K - \bar{k}$, we can regard to our problem as an optimal control problem subject to the state constraint $k(t) \leq K$.

The *formal* Bellman equation for the value function $\mathcal{V}$ is given by

$$H\Big(t, x, \frac{\partial W}{\partial t}, \frac{\partial W}{\partial x}, \frac{\partial W}{\partial k}\Big) = 0$$

where $\partial W/\partial x$ denotes the gradient of W with respect to x and H is defined by

$$H(t, x, p_t, p_x, p_k) \doteq p_t + \inf_{\substack{v \in V \\ \xi \in C}} \Big\{ p_x \cdot \overline{f}(t, x, v, \xi) + p_k |\xi| \Big\} \,.$$

Now it is clear why the sublinearity of f in ξ represents a drawback for the standard application of Dynamic Programming methods (though, for a quite special case, a direct approach has been pursued in [**1**]) . Indeed — unlike the linear-quadratic problem, where the superlinear growth of the lagrangian guarantees that the infimum in the definition of the hamiltonian is in fact a minimum— the hamiltonian H may happen to be equal to $-\infty$ for a lot of values of (t, x, p_t, p_x, p_k). In [**27**], [**29**] the question is approached by means of the introduction of a new, *continuous* hamiltonian such that $\mathcal{V}$ is the unique (viscosity) solution of the associated Bellman equation. The main tool for obtaining this result is the combination of ad hoc reparametrization techniques and recent comparison results in the theory of viscosity solutions.

Let us recall one of the main result in this direction. Let us define the hamiltonian

$$\mathcal{H}(t, x, p_t, p_x, p_k) \doteq \min_{\substack{v \in V \\ w \in C \\ |w_0, w| = 1}} \Big\{ p_t w_0 + p_x \overline{f}(t, x, v, w_0, w) + p_k |w| \Big\}$$

which is continuous, for the minimization domain on the right-hand side is compact.

Under mild assumptions on f and for a continuous, bounded ψ one can prove the following result (see e.g.[**8**] for the definition of viscosity solution on a non open set).

THEOREM 5.1. [**29**] *The value function $\mathcal{V}$ is continuous and bounded. Moreover it is the unique continuous, bounded map solving the boundary value problem* (BVP) *defined below.*

DEFINITION 5.2. A continuous map $\nu : [0, T] \times \mathbb{R}^n \times [0, K] \to \mathbb{R}$ solves the boundary value problem (BVP) if:

(i) ν is a viscosity subsolution of

$$\text{(BE)} \qquad -\mathcal{H}\Big(t, x, \frac{\partial \nu}{\partial t}, \frac{\partial \nu}{\partial x}, \frac{\partial \nu}{\partial k}\Big) = 0 \quad \text{on } [0, T[\, \times \mathbb{R}^n \times [0, K[$$

(ii) ν is a viscosity supersolution of (BE) on $[0, T[\, \times \mathbb{R}^n \times [0, K]$

(iii) for all $(x, k) \in \mathbb{R}^n \times [0, K]$, either $\nu(T, x, k) = \psi(x)$ or ν is a viscosity supersolution of (BE) at (T, x, k).

Theorem 5.1 has been generalized to the case where x is constrained in the closure $\overline{\mathcal{O}}$ of an open subset $\mathcal{O} \subset \mathbb{R}^n$ (see [**29**]). Moreover the case where the variation of u (i.e., the L^1-norm of ξ) is not bounded has been addressed in [**30**].

A minimum time problem with unbounded strategies and the possibility of impulses has been investigated in [**32**].

Free boundary problems. Some free boundary problems where impulsive controls are involved have been investigated in [**20**] and [**17**].

6. Some open problems

Necessary Conditions. Problem $(\mathcal{P})$, introduced in section 4, in general does not have a solution, while an optimal (space-time) control $(\widehat{\varphi}_0, \widehat{\varphi})$ always exists for the extended problem $(\mathcal{P}_e)$. Hence it is natural to look for necessary conditions which have to be verified by $(\widehat{\varphi}_0, \widehat{\varphi})$. Since $(\mathcal{P}_e)$ is a standard optimal control problem subject to an inequality —namely $\int_0^1 |\varphi'(s)|ds \leq K$— which can be regarded as a state–space constraint, one can apply some version of the maximum principle for this kind of problems (see e.g. [**12**]). However the state constraint here considered has a particular form: if $\alpha'(s) = |\varphi'(s)|$, $\alpha(0) = 0$, it reduces to

$$\alpha - K \leq 0.$$

Therefore one could try to construct some ad hoc *variation* in order to obtain necessary conditions which are independent of the ones provided by the standard maximum principle. Of course the interest in such question extends to the more general systems considered in section 5.

Existence of optimal space–time controls. An optimal space-time controls always exists for problem $(\mathcal{P}_e)$(see section 4). As soon as some convexity is assumed, the same can be stated for more general problems involving dynamics like (5.1). Yet these existence results rely on the essential fact that the variation of φ, namely the integral $\int_0^1 |\varphi'(s)|ds$, is constrained to be less than or equal to a constant K. Indeed, via a reparametrization technique and a suitable application of Ascoli-Arzelà's theorem, this constraint can be interpreted as a compactness argument which is sufficient for the existence of a minimum. Of course, the lack of this constraint makes the problem of finding an optimal space-time control much harder. One obvious way to approach the problem consists in proving that the cost functional is greater than a function of α which tends to infinity as k goes to infinity (see e.g. [**11**] for an analogous approach when the problem reduces to a problem of Calculus of Variations). This would be always in the spirit of the *direct method.* Alternatively one could try to show that each extremal trajectory implies an expenditure of *energy* α which is bounded by a quantity K. Of course this (indirect) method involve the exploitation of necessary conditions.

Existence of non-impulsive optimal control. Though problem $(\mathcal{P}_e)$ has an optimal control in Φ_K, in general an optimal control for problem $(\mathcal{P})$ does not exist in $\mathcal{U}_K$. This is equivalent to state that in general no optimal space-time controls exist *in* Φ_K^+ for problem $(\mathcal{P}_e)$. The situation is akin to the one occurring in Calculus of Variations when problems without coercivity assumptions (*slow growth problems*) are addressed. Some recent advancements in the study of these problems are provided in [**10**],[**13**]. From our point of view the analogous question in Control

Theory reduces to a problem of *regularity* (i.e.: does the optimal space-time control $(\widehat{\varphi}_0, \widehat{\varphi}) \in \Phi_K$ belong to Φ_K^+?).

References

[1] E.N.BARRON, R.JENSEN and J.L. MENALDI, *Optimal control and differential games with measures,* Nonlinear Analysis T.M.A., **21**, 241–268 (1993)

[2] A. BENSOUSSAN and J.-L. LIONS, *Impulsive Control and Quasivariational Inequalities*, Bordas, Paris, (1984)

[3] A. BLAQUIÈRE, *Impulsive optimal control with finite or infinite time horizon*, J. Optimization Theory and App., **46**, 431-439 (1985)

[4] ALBERTO BRESSAN and F. RAMPAZZO, *On differential systems with vector–valued impulsive controls,* Boll. Un. Mat.Ital., (7) **2–B**, 641–656 (1988)

[5] ALBERTO BRESSAN and F. RAMPAZZO, *Impulsive control systems with commutative vector fields,* Journal of Optimization Theory and Applications, **71**, 67–83 (1991)

[6] ALBERTO BRESSAN and F. RAMPAZZO, *Impulsive control systems without commutativity assumption,* Journal of Optimization Theory and Applications, **81**, 435–457 (1994)

[7] ALDO BRESSAN, *Hyperimpulsive motions and controllizable coordinates for Lagrangian systems,* Atti Accad. Naz. Lincei, Mem.Cl. Sc. Fis. Mat. Natur., **19**, (1991)

[8] I. CAPUZZO–DOLCETTA and P.L. LIONS, *Hamilton–Jacobi equations and state constrained problems,* Trans. Amer. Math. Soc., **318**, 643–668 (1990)

[9] F. CARDIN and M.FAVRETTI, *Hyperimpulsive motion on manifolds* , to appear on J. of Dynamics of Continuous, Discrete and Impulsive Systems.

[10] A. CELLINA, G. TREU and S. ZAGATTI, *On the minimum problem for a class of noncoercive functionals,* J. of Diff. Eq. **127**, 225-262 (1996)

[11] L. CESARI, *Optimization-Theory and Applications*, Springer-Verlag, New York, Heidelberg, Berlin (1984)

[12] F.H.CLARKE, *Methods of dynamic and nonsmooth optimization*, SIAM, Philadelphia (1989)

[13] F.H. CLARKE, *An indirect method in the calculus of variations* Trans. Amer. Math. Soc., **336**, 655–673 (1993)

[14] G. DAL MASO and F. RAMPAZZO, *On systems of ordinary differential equations with measures as controls,* Differential and Integral Equations, **4–4**, 739–765 (1991)

[15] J.R. DORROH and G. FERREYRA, *A multi–state, multi–control problem with unbounded controls,* SIAM J. Control and Optimization, **32**, 1322–1331 (1994)

[16] J.R.DORROH and G. FERREYRA *A free boundary problem in R^d with both smooth and nonsmooth fit*, SIAM Journal on Control and Optimization, **36**, No. 2, (1998) pp. 579-589.

[17] J. R. DORROH G. FERREYRA and P. SUNDAR, *A technique for stochastic control problems with unbounded control set*, to appear in J. Theoretical Probability.

[18] V.A. DYKHTA, *Impulse trajectory extension of degenerated optimal control problems,* The Liapunov functions methods and applications, P. Borne and V. Matrosov eds., J.C. Baltzer AG, Scientific Publishing Co., 103–109 (1990)

[19] G. FERREYRA *The optimal control problem for the Vidale-Wolfe advertising model revisited*, Optimal Control Appl. and Methods **11** , 363-368 (1990).

[20] G. FERREYRA and O. HIJAB, *A simple free boundary problem in R^d*, SIAM Journal on Control and Optimization **32**, 501-515 (1994)

[21] O. HÁJEK, Book review on Bull. Am. Math Soc.**12-2**, 272-279 (1975)

[22] B.M. MILLER, *Optimization of dynamic systems with a generalized control,* Automation and Remote Control, **50**, (1989)

[23] B.M. MILLER, *Conditions for the optimality in problems of generalized control. I. Necessary conditions for optimality,* Automation and Remote Control, **53**, (1992)

[24] B.M. MILLER, *The generalized solutions of ordinary differential equations in the impulse control problems* Journal of Mathematical Systems, Estimation and Control, **4**, 385–388 (1994)

[25] B.M. MILLER, *The generalized solutions of nonlinear optimization problems with impulse control,* SIAM J. Control and Optimization, **34**, 1420–1440 (1996)

[26] M. MOTTA and F.RAMPAZZO, *Space–time trajectories of nonlinear systems driven by ordinary and impulsive controls,* Differential and Integral Equations, **8–2**, 269–288 (1995)

[27] M. MOTTA and F.RAMPAZZO, *Dynamic programming for nonlinear systems driven by ordinary and impulsive controls,* SIAM J. Control and Optimization, (1996)

[28] M. MOTTA and F.RAMPAZZO, *Nonlinear systems with unbounded controls and state constraints: a problem of proper extension,* NoDEA – Nonlinear Differential Equations and Applications,**3**, 191–216 (1996)

[29] M. MOTTA and F.RAMPAZZO, *The value function of a slow growth control problem with state constraints*, J. Math. Syst. Est. and Control, to appear (available electronic manuscript, short version in J. Math. Syst. Est. and Control **7**, 375-378 (1997)

[30] M. MOTTA and F.RAMPAZZO, , *State constrained control problems with neither coercivity nor L^1 bounds on the controls* Appl. Math. and Optim., to appear

[31] F. RAMPAZZO, *On the Riemannian structure of a Lagrangian system and the problem of adding time–dependent constraints as controls,* Eur. J. Mech., A/Solids, Gauthier–Villars, **10**, 405–431 (1991)

[32] F.RAMPAZZO and C. SARTORI *The minimum time function with unbounded controls*, J. Math Syst. Est. and Control, to appear

[33] R.W.RISHEL, *An extended Pontryagin principle for control systems whose control laws contain measures,* SIAM J. Control, **3**, 191–205 (1965)

[34] A.V. SARYCHEV, *Nonlinear systems with impulsive and generalized function controls,* Proc. Conf. on Nonlinear Synthesis, Sopron, Hungary, (1989)

[35] W.W. SCHMAEDEKE, *Optimal control theory for nonlinear differential equations containing measures,* SIAM J. Control,**3**, 231–279 (1965)

[36] A.N. SESEKIN, *Nonlinear differential equations in the class of functions of bounded variation,* Automation and Remote Control, (1990)

[37] A.N. SESEKIN, *Impulse extension in the problem of the optimization of the energy functional,* Automation and Remote Control, (1993)

[38] H.J. SUSSMANN, *On the gap between deterministic and stochastic ordinary differential equations,* Ann. of Probability, **6**, 17–41 (1978)

[39] H. J. SUSSMANN, *Semigroup representations, bilinear approximation of input-output maps, and generalized inputs*, Math. Sys. Theory, Lecture Notes in Econ. and Math Sys., Proc. Int. Symp., Udine, **131**, 172-191 (1976)

Dipartimento di Matematica Pura e Applicata, Università di Padova, Via Belzoni 7, 35131 Padova, Italy

Current address: same

E-mail address: rampazzo@math.unipd.it

Proceedings of Symposia in Pure Mathematics
Volume **64**, 1999

Survey on Controllability of Invariant Systems on Solvable Lie Groups

Yuri L. Sachkov

ABSTRACT. Known and new results on controllability of right-invariant systems on solvable Lie groups are presented and discussed. The main ideas and technique used are outlined, illustrating examples are given. Some open questions are suggested.

1. Introduction

Controllability properties of right-invariant systems on Lie groups have been a subject of active research in the mathematical control theory during the last 25 years. This question is motivated both by important applications (mechanics, geometry, bilinear systems) and by essential links with various branches of mathematics outside control theory (Lie groups and Lie algebras, differential geometry, Lie semigroups, dynamical systems).

The problem was stated and the basic properties of right-invariant systems and their reachable sets were established by R. W. Brockett [**8**], and V. Jurdjevic and H. J. Sussmann [**20**]. Controllability theory for various classes of Lie groups different from solvable ones was developed by V. Jurdjevic and H. J. Sussmann [**20**] (compact Lie groups), V. Jurdjevic and I. Kupka [**21, 22**], J. P. Gauthier and G. Bornard [**11**], J. P. Gauthier, I. Kupka and G. Sallet [**12**], R. El Assoudi and J. P. Gauthier [**2, 3**], R. El Assoudi, J. P. Gauthier, and I. Kupka [**4**], F. Silva Leite and P. E. Crouch [**35**], L. A. B. San Martin [**33**], L. A. B. San Martin and P. A. Tonelli [**34**] (semisimple Lie groups), B. Bonnard, V. Jurdjevic, I. Kupka and G. Sallet [**7**] (semidirect products of vector Lie groups with compact Lie groups), J. Hilgert [**14**] (reductive Lie groups).

The aim of this paper is to describe the state of the art in controllability on solvable Lie groups: to cover all results published, outline the main ideas and techniques used, and give illustrating examples. Notice that the case of solvable Lie groups is incompletely covered by surveys and textbooks on control theory and Lie semigroup theory (R. W. Brockett [**9**], G. Sallet [**30, 32**], A. A. Agrachev, S. V. Vakhrameev

1991 *Mathematics Subject Classification.* 93B05, 17B20.

This work was partially supported by the Russian Foundation for Fundamental Research, projects No. 98-01-01028 and No. 97-1-1a/22. The author gratefully acknowledges the support from the American Mathematical Society for participation in the Summer Research Institute on Differential Geometry and Control, June 30–July 18, 1997, Boulder CO, USA. The author is a recipient of the Russian State Scientific Stipend for 1997.

and R.V. Gamkrelidze [**1**], I. Kupka [**24**], J. Hilgert and K.H. Neeb [**15**], V. Jurdjevic [**23**]).

The structure of the paper is as follows. In Section 2 we introduce the general notation and definitions. In Section 3 we consider an important special class of solvable Lie groups — nilpotent ones. Subsection 3.1 is devoted to the controllability criterion for arbitrary right-invariant systems on a nilpotent Lie group by J. Hilgert, K.H. Hofmann and J.D. Lawson [**13**]. In Subsection 3.2 we present and discuss the controllability test for affine in control systems on simply connected nilpotent Lie groups due to V.`Ayala Bravo [**6**]. In Section 4 we present controllability conditions for Lie groups with cocompact radical; this class of Lie groups includes all solvable Lie groups. This basic result is due to J.D. Lawson [**25**]. In Section 5 we consider a subclass of solvable Lie groups called completely solvable ones and a generalization of the controllability test of Subsection 3.2 from nilpotent to completely solvable Lie groups. Section 6 is devoted to single-input systems on Lie groups that differ from their derived subgroups; this class contains all solvable Lie groups. In Subsection 6.1 we give necessary and sufficient controllability conditions for such systems. These results yield the characterization of controllability on metabelian Lie groups presented in Subsection 6.2. In Subsection 6.3 we study right-invariant systems on a particular matrix metabelian Lie subgroup of the group of affine transformations of the n-space, and in Subsection 6.4 we treat bilinear systems in $\mathbb{R}^n$ that are projections of right-invariant systems on this matrix group. Finally, in Section 7 we classify single-input controllable systems on simply connected solvable Lie groups up to dimension 6, inclusive. The details on original sources are collected in Section 8; no references to authors are made immediately in the text. Some open questions related to the results presented are discussed in Section 9.

2. Definitions and basic facts

Let G be a Lie group and L its Lie algebra (i.e., the Lie algebra of right-invariant vector fields on G).

A *right-invariant control system* on G is an arbitrary subset $\Gamma \subset L$. An important particular class of systems is formed by systems *affine in control* $\Gamma = \{\, A + \sum_{i=1}^{m} u_i B_i \mid u \in \mathbb{R}\,\}$, where A, $B_1, \dots, B_m$ are some elements of L.

The *attainable set* $\mathbb{A}$ of a system Γ is the subsemigroup of G generated by the set

$$\exp(\mathbb{R}_+\Gamma) = \{\, \exp(tX) \mid X \in \Gamma,\, t \in \mathbb{R}_+ \,\}.$$

A system Γ is called *controllable* if $\mathbb{A} = G$.

The basic conditions necessary for controllability of a right-invariant system Γ on a Lie group G were given by V. Jurdjevic and H.J. Sussmann [**20**]: the Lie group G should be connected and the system Γ should satisfy the *rank controllability condition*, i.e., the Lie algebra $\mathrm{Lie}(\Gamma)$ generated by the system Γ must coincide with the whole Lie algebra L. That is why these conditions may be assumed without loss of generality. In particular, in the sequel all Lie groups are supposed connected.

The *derived series* of a Lie algebra L is constructed as follows:

$$L^{(1)} = [L, L],\ L^{(2)} = [L^{(1)}, L^{(1)}],\ \dots,\ L^{(i)} = [L^{(i-1)}, L^{(i-1)}],\ \dots,\quad i \in \mathbb{N}.$$

$L^{(1)}$ is the *derived subalgebra* of L, and the corresponding subgroup denoted by $G^{(1)}$ is the *derived subgroup* of G. The algebra L is called *solvable* if its derived series stabilizes at zero:

$$L \supset L^{(1)} \supset L^{(2)} \supset \ldots \supset L^{(N)} = \{0\}$$

for some $N \in \mathbb{N}$. Finally, a Lie group G is *solvable* if its Lie algebra L is solvable.

We denote topological closure and interior of a set M by $\operatorname{cl} M$ and $\operatorname{int} M$ respectively.

3. Nilpotent Lie groups

In this section we present controllability conditions for right-invariant systems on nilpotent Lie groups. Recall that a Lie algebra L is called *nilpotent* if its descending central series

$$L_{(1)} = [L, L],\ L_{(2)} = [L, L_{(1)}],\ \ldots,\ L_{(i)} = [L, L_{(i-1)}],\ \ldots,\qquad i \in \mathbb{N},$$

stabilizes at zero:

$$L \supset L_{(1)} \supset L_{(2)} \supset \ldots \supset L_{(N)} = \{0\}$$

for some $N \in \mathbb{N}$. Any nilpotent Lie algebra is solvable since $L_{(i)} \supset L^{(i)}$, $i \in \mathbb{N}$. Another equivalent characterization of nilpotency of L is that all adjoint operators $\operatorname{ad} x$, $x \in L$, are nilpotent and thus have zero spectrum (this is important in Section 5).

3.1. Arbitrary systems. Controllability of a right-invariant system $\Gamma \subset L$ on a nilpotent Lie group G is completely characterized in terms of the wedge, i.e., the topologically closed convex cone $W(\Gamma) \subset L$ generated by Γ. It is standard that Γ and $W(\Gamma)$ are controllable or noncontrollable simultaneously.

THEOREM 3.1. *Let G be a nilpotent Lie group with Lie algebra L and let W be a wedge in L which generates L as a Lie algebra. Then W is controllable on G iff one of the following conditions holds:*

1. $\operatorname{int}_{W-W} W \cap L^{(1)} \neq \emptyset$,
2. $\operatorname{int} \operatorname{cl}(L^{(1)} + W) \cap \exp^{-1}(e) \neq \emptyset$.

REMARK. Here $\operatorname{int}_{W-W} W$ is the interior of the wedge W relative to the vector space $W - W$ generated by W, and e is the identity of the Lie group G.

The sufficiency in Theorem 3.1 is based on the description of maximal open subsemigroups S of nilpotent Lie groups in terms of their tangent objects

$$L(S) = \{\, x \in L \mid \exp(\mathbb{R}_+ x) \subset \operatorname{cl}(S) \,\}.$$

An open subsemigroup S of a Lie group G is proper, i.e., $S \neq G$ if and only if $e \notin S$. Hence the set of open subsemigroups in G is inductive, and any proper subsemigroup is contained in a *maximal* one.

THEOREM 3.2. *Let G be a nilpotent Lie group and let S be a maximal open proper subsemigroup of G. Then $L(S)$ is a halfspace bounded by a hyperplane algebra in $L(G)$.*

For necessity in Theorem 3.1, the Hahn-Banach theorem gives that $W + L^{(1)}$ is contained in a halfspace in L; then $\exp\left(\operatorname{int}\left(W + L^{(1)}\right)\right)$ is a proper open semigroup of G, which implies that $\exp(W)$ is contained in a proper subsemigroup of G.

The controllability test of Theorem 3.1 is essentially nilpotent. This result is false for the group $\mathrm{SL}(2, \mathbb{R})$; it also fails in the following solvable non-nilpotent example.

EXAMPLE. Let G be the (unique) two-dimensional connected simply connected non-abelian Lie group, which is represented by the matrices

$$\left\{ \begin{pmatrix} x & y \\ 0 & 1 \end{pmatrix} \mid x > 0,\ y \in \mathbb{R} \right\}.$$

The Lie group G is solvable but not nilpotent. Its Lie algebra has the form

$$L = \left\{ \begin{pmatrix} a & b \\ 0 & 0 \end{pmatrix} \mid a, b \in \mathbb{R} \right\}.$$

Consider the following wedge in L:

$$W = \left\{ \begin{pmatrix} a & b \\ 0 & 0 \end{pmatrix} \mid a \in \mathbb{R},\ b \geq 0 \right\}.$$

Direct computations show that

$$\exp(\mathbb{R}_+ W) = \left\{ \begin{pmatrix} x & y \\ 0 & 1 \end{pmatrix} \mid x > 0,\ y \geq 0 \right\},$$

which is a proper subsemigroup of G, thus W is not controllable on G. On the other hand, it is easy to see that both conditions 1, 2 of Theorem 3.1 hold for the wedge W in this example.

3.2. Affine systems. For right-invariant systems affine in control

$$\Gamma = \left\{ A + \sum_{i=1}^{m} u_i B_i \mid u_i \in \mathbb{R} \right\} \subset L, \tag{3.1}$$

on simply connected nilpotent Lie groups there is a simple controllability criterion in terms of the Lie subalgebra L_0 of L generated by the non-drift vector fields:

$$L_0 = \operatorname{Lie}(B_1, \dots, B_m),$$

and the connected subgroup G_0 of G with Lie algebra L_0.

THEOREM 3.3. *Let G be a simply connected nilpotent Lie group. Then system* (3.1) *is controllable on G iff $L_0 = L$.*

Sufficiency of the condition $L_0 = L$ for controllability of Γ is a well-known fact valid for arbitrary Lie groups G. So the essential part is necessity. Here the key role is played by the necessary controllability conditions in terms of the notion of a *symplectic vector.*

Consider the co-adjoint representation ρ^* of the group G in the dual space L^* of L. For any covector $\lambda \in L^*$, the co-adjoint orbit θ_λ of λ by the ρ^* action $\theta_\lambda = \rho^*_G(\lambda)$ is a smooth submanifold of L^* diffeomorphic to the homogeneous space G/E_λ, where E_λ is the isotropy subgroup of λ, $E_\lambda = \{ g \in G \mid \rho^*_g(\lambda) = \lambda \}$. Further, the system Γ can be projected from G onto the homogeneous space $G/E_\lambda \simeq \theta_\lambda$, and controllability of Γ on G obviously implies controllability of its projection Γ_λ on G/E_λ. This leads to necessary controllability conditions in terms of the co-adjoint representation.

DEFINITION. $\lambda \in L^*$ is called a *symplectic vector* for $w \in L$ if the co-adjoint orbit θ_λ is not trivial and $\langle w, \beta \rangle > 0$ for all $\beta \in \theta_\lambda$.

(We denote by $\langle \cdot, \cdot \rangle$ the pairing of a vector and covector.)

THEOREM 3.4. *If there is a vector field $\xi \in L$ belonging to the centralizer of the subalgebra L_0 such that the nonzero vector field $[A, \xi]$ has a symplectic vector, then system* (3.1) *cannot be controllable on G.*

In fact, the existence of such vector field $\xi \in L$ yields that the function

$$f_\xi : \theta_\lambda \to \mathbb{R}, \qquad \beta \mapsto f_\xi(\beta) = -\langle \xi, \beta \rangle$$

is strictly increasing on the trajectories of projection of Γ onto the co-adjoint orbit θ_λ. Indeed, the solution of the Cauchy problem $\dot{g}(t) = A(g(t))$, $g(0) = g_0$ is given by $g(t) = \exp(tA)g_0$. Further, the function

$$h(t) = \operatorname{Ad}\left(g(t)^{-1}\right) = \operatorname{Ad}\left(g_0^{-1}\right) \circ \exp(-t \operatorname{ad} A)$$

has the derivative

$$\dot{h}(t) = \operatorname{Ad}\left(g_0^{-1}\right) \circ \exp(-t \operatorname{ad} A) \circ (-\operatorname{ad} A) = -\operatorname{Ad}\left(g(t)^{-1}\right) \circ \operatorname{ad} A.$$

Now for $\lambda \in L^*$ the co-adjoint action ρ^* of the element $g \in G$ is determined by

$$\rho^*_g(\lambda) = \operatorname{Ad}^*\left(g^{-1}\right)\lambda,$$

consequently, for any $\xi \in L$, $\lambda \in L^*$

$$\begin{aligned} \frac{d}{dt} f_\xi(\rho^*_{g(t)}(\lambda)) &= -\frac{d}{dt}\langle \xi, \rho^*_{g(t)}(\lambda)\rangle = -\frac{d}{dt}\langle \xi, \operatorname{Ad}^*\left(g(t)^{-1}\right)\lambda\rangle \\ &= -\frac{d}{dt}\langle \operatorname{Ad}\left(g(t)^{-1}\right)\xi, \lambda\rangle = \langle \operatorname{Ad}\left(g(t)^{-1}\right) \circ \operatorname{ad} A(\xi), \lambda\rangle \\ &= \langle [A, \xi], \rho^*_{g(t)}(\lambda)\rangle. \end{aligned}$$

Thus if λ is a symplectic vector for $[A, \xi]$, then f_ξ is increasing along co-adjoint orbits of trajectories of the field A. If in addition $\operatorname{ad}\xi$ vanishes on the subalgebra L_0, then the same holds for trajectories of the whole system Γ, which is impossible for a controllable system.

Another important fact for necessity in Theorem 3.3 is the following proposition related to *hypersurface* systems, i.e., affine systems (3.1) with L_0 a codimension one subalgebra of L.

THEOREM 3.5. *Let $\Gamma \subset L$ be an affine system* (3.1) *on G such that L_0 is a codimension one ideal of L.*

1. *If G_0 is closed in G, then Γ is controllable iff $A \notin L_0$ and $G/G_0 \simeq S^1$.*
2. *If G_0 is not closed in G, then Γ is controllable iff $A \notin L_0$.*

REMARK. In fact, the previous theorem holds without the assumption that L_0 is an ideal; this is important for a generalization of Theorem 3.3 to a subclass of solvable Lie groups including nilpotent ones (see Section 5 below).

Now we outline the scheme of proof of necessity in Theorem 3.3. Suppose the system Γ is controllable on the group G. Then the theory of symplectic vectors implies that the subalgebra L_0 is an ideal of L. The rank condition for Γ is satisfied: $\operatorname{Lie}(\Gamma) = \operatorname{Lie}(A, L_0) = L$, hence L_0 has codimension 0 or 1 in L. But the codimension one case is impossible since then Theorem 3.5 yields $G/G_0 \simeq S^1$, which contradicts simple connectedness of G. Thus $L_0 = L$, and necessity in Theorem 3.3 follows.

EXAMPLE. Let G be the Heisenberg group of dimension $2p+1$. It may be represented as a subgroup of $\mathrm{GL}(p+2,\mathbb{R})$ generated by the matrices

$$\mathrm{Id}+X_i,\ \mathrm{Id}+Y_i,\ Z,\quad i=1,\dots,p,$$

where

$$X_i = E_{1,i+1},\ Y_i = E_{i+1,p+2},\quad i=1,\dots,p.$$

(Id denotes the identity matrix, and E_{ij} stands for the square matrix with all zero entries except one unit in the i-th row and the j-th column.)

The Lie algebra L of G is spanned by the matrices

$$X_i,\ Y_i,\ Z,\quad i=1,\dots,p,$$

with the nonzero brackets

$$[X_i,Y_i]=Z,\quad i=1,\dots,p.$$

The Heisenberg group G is simply connected and nilpotent, hence Theorem 3.3 describes all controllable systems on G.

4. Lie groups with cocompact radical

Denote by $\operatorname{Rad} G$ the radical of a Lie group G, i.e., the maximal solvable normal subgroup of G. In this section we suppose that a Lie group G has cocompact radical, that is, the quotient group $K = G/\operatorname{Rad} G$ is compact. This wide class of Lie groups contains:

- solvable Lie groups ($K=\{e\}$),
- compact Lie groups,
- semidirect products of a vector space V with a compact Lie group ($V \subset \operatorname{Rad} G$).

The next theorem gives a Lie-algebraic description of controllability on such Lie groups, complete in the simply connected case.

THEOREM 4.1. *Suppose that $G/\operatorname{Rad} G$ is compact, and let $\Gamma \subset L$ be a system satisfying the rank condition* $\mathrm{Lie}(\Gamma)=L$. *If Γ is not contained in any half-space of L with boundary a subalgebra, then Γ is controllable on G. The converse holds if G is simply connected.*

This result is a consequence of the following classification of maximal subsemigroups of Lie groups with cocompact radical.

DEFINITION. A subsemigroup M of G is called a *maximal subsemigroup of* G if the only subsemigroups containing M are M and G, and M is not a subgroup.

THEOREM 4.2. *The maximal subsemigroups M with non-empty interior of a simply connected Lie group G with $G/\operatorname{Rad} G$ compact are in one-to-one correspondence with their tangent objects $L(M)$, and the latter are precisely the closed half-spaces with boundary a subalgebra. Further, M is the semigroup generated by* $\exp(L(M))$.

Theorem 4.1 follows from Theorem 4.2 since the attainable set of any noncontrollable right-invariant system $\Gamma \subset L$, $\mathrm{Lie}(\Gamma)=L$, is a proper subsemigroup of G contained in some maximal subsemigroup with non-empty interior.

5. Completely solvable Lie groups

In this section we suppose that Γ is a system affine in control as in (3.1).

DEFINITION. A solvable Lie algebra L is called *completely solvable* if all adjoint operators $\operatorname{ad} x$, $x \in L$, have real spectra. A Lie group is *completely solvable* if it has completely solvable Lie algebra.

(Such Lie algebras and Lie groups are also called *triangular over* $\mathbb{R}$ or *algebras*, resp. *groups, of type* (R) [**36**].)

The triangular group $\mathrm{T}(n)$ (see the example below) is completely solvable, as well as any of its subgroups. Nilpotent Lie groups are completely solvable since adjoint operators in nilpotent Lie algebras have zero spectrum. On the other hand, e.g., the group of motions of the plane $\mathrm{E}(2)$ is solvable but not completely solvable (the group $\mathrm{E}(2)$ and its simply connected covering $\widetilde{\mathrm{E}(2)}$ are treated in Section 6.3).

It turns out that the controllability criterion for systems affine in control on nilpotent Lie groups (Theorem 3.3) is valid for completely solvable Lie groups as well.

THEOREM 5.1. *Let G be a simply connected completely solvable Lie group. Then system* (3.1) *is controllable on G iff $L_0 = L$.*

Theorem 5.1 is based on the following general characterization of controllability for hypersurface systems.

THEOREM 5.2. *Let $\Gamma \subset L$ be an affine system* (3.1) *on G such that L_0 is a codimension one subalgebra of L.*

1. *If G_0 is closed in G, then Γ is controllable iff $A \notin L_0$ and $G/G_0 \simeq S^1$.*
2. *If G_0 is not closed in G, then Γ is controllable iff $A \notin L_0$.*

REMARK. Theorem 5.2 generalizes the analogous criterion of Theorem 3.5 with the additional assumption that L_0 is an ideal of L.

Theorem 5.2 implies the following *hypersurface principle* — a general necessary controllability condition for simply connected Lie groups.

THEOREM 5.3. *Let G be simply connected. Suppose that there exists a codimension one subalgebra l of L containing L_0. Then system* (3.1) *is not controllable.*

The sense of this proposition is that if the codimension one subalgebra $l \supset L_0$ exists, then attainable set of Γ lies "to one side" of the connected codimension one subgroup of G corresponding to l: by simple connectedness of G, this codimension one subgroup separates G into two disjoint parts.

Now we outline the proof of the controllability test of Theorem 5.1. Sufficiency of $L_0 = L$ for controllability is obvious, while the necessity follows from Theorem 5.3 and the observation that any proper subalgebra of a completely solvable Lie algebra L is contained in a codimension one subalgebra of L.

EXAMPLE. Let $G = \mathrm{T}(n)$ be the Lie group of all $n \times n$ upper triangular matrices with positive diagonal entries. $\mathrm{T}(n)$ is connected, simply connected and completely solvable. Its Lie algebra L consists of all $n \times n$ upper triangular matrices. By Theorem 5.1, an affine system Γ is controllable on G if and only if $L_0 = L$. The analogous result holds for any Lie subgroup of the triangular group $\mathrm{T}(n)$.

6. Lie groups differing from their derived subgroups

Lie groups G which satisfy the condition $G \neq G^{(1)}$ make up a wide class containing solvable Lie groups. On the other hand, no semisimple Lie groups belong to this class.

In this section we present controllability conditions for single-input systems

$$\Gamma = \{ A + uB \mid u \in \mathbb{R} \} = A + \mathbb{R}B \tag{6.1}$$

on such Lie groups.

Throughout this section we make the following assumptions:

1. $L \neq L^{(1)}$,
2. the adjoint operator $\operatorname{ad} B|_{L^{(1)}}$ has simple spectrum.

Notice that while the first assumption is obviously imposed by the class of Lie groups under consideration, the second one is added for simplicity of exposition and can be removed.

To formulate the results we need the following notation. Spectra of the adjoint operators in the first and second derived subalgebras $\operatorname{ad} B|_{L^{(1)}}$ and $\operatorname{ad} B|_{L^{(2)}}$ are denoted by $\operatorname{Sp}^{(1)}$ and $\operatorname{Sp}^{(2)}$ respectively. For any eigenvalue $a \in \operatorname{Sp}^{(1)}$ the eigenspace of $\operatorname{ad} B|_{L^{(1)}}$ corresponding to a is denoted by $L(a)$. In view of simplicity of the spectrum of $\operatorname{ad} B|_{L^{(1)}}$, there is the following decomposition of the derived subalgebra of L into a direct sum of one- and two-dimensional eigenspaces:

$$L^{(1)} = \sum\nolimits^{\oplus} \{ L(a) \mid a \in \operatorname{Sp}^{(1)}, \operatorname{Im} a \geq 0 \}.$$

If $B \notin L^{(1)}$ and $\dim L^{(1)} = \dim L - 1$, then we have

$$L = \mathbb{R}B \oplus L^{(1)} = \mathbb{R}B \oplus \sum\nolimits^{\oplus} \{ L(a) \mid a \in \operatorname{Sp}^{(1)}, \operatorname{Im} a \geq 0 \}$$

and the corresponding decomposition for a vector $A \in L$:

$$A = A_B + \sum \{ A(a) \in L(a) \mid a \in \operatorname{Sp}^{(1)}, \operatorname{Im} a \geq 0 \}.$$

(The signs $\oplus$ and $\sum^{\oplus}$ above denote direct sums of vector spaces.)

6.1. Controllability conditions. There are restrictive necessary controllability conditions for single-input systems on simply connected Lie groups $G \neq G^{(1)}$.

THEOREM 6.1. *Let a Lie group G be simply connected and $G \neq G^{(1)}$. If system* (6.1) *is controllable, then:*

1. $\dim L^{(1)} = \dim L - 1$,
2. $B \notin L^{(1)}$,
3. $\operatorname{Sp}^{(1)} \cap \mathbb{R} \subset \operatorname{Sp}^{(2)}$,
4. $A(a) \neq 0$ *for all* $a \in \operatorname{Sp}^{(1)} \setminus \operatorname{Sp}^{(2)}$.

Notice that controllability of (at least one) single-input system on a simply connected Lie group $G \neq G^{(1)}$ implies restriction 1, $\dim G^{(1)} = \dim G - 1$, on the Lie group, not on the system. In fact, for solvable Lie algebras L restriction 3 of Theorem 6.1 depends on L but not on the system (provided that conditions 1 and 2 hold). This restrictive force of necessary conditions of Theorem 6.1 is crucial for classification of controllable systems on small-dimensional simply connected solvable Lie groups in Section 7.

Theorem 6.1 is based on the rank controllability condition and the necessary controllability condition of Theorem 5.3. The justification of the fact that these two results should be enough to give necessary controllability conditions on simply connected solvable Lie groups is provided by the general controllability test for simply connected Lie groups with cocompact radical of Theorem 4.1.

There are sufficient controllability conditions close to the previous necessary ones in the case of simply connected Lie groups.

THEOREM 6.2. *Suppose that the following conditions are satisfied for Lie algebra L of a Lie group G and system* (6.1)*:*

1. $\dim L^{(1)} = \dim L - 1$,
2. $B \notin L^{(1)}$,
3. $\mathrm{Sp}^{(1)} \cap \mathbb{R} \subset \mathrm{Sp}^{(2)}$,
4. $A(a) \neq 0$ *for all* $a \in \mathrm{Sp}^{(1)} \setminus \mathbb{R}$.
5. $\mathrm{Sp}^{(1)} \cap \mathbb{R} = \emptyset$ *or* $\mathrm{Sp}^{(1)} \subset \{\mathrm{Re}\, z > 0\}$ *or* $\mathrm{Sp}^{(1)} \subset \{\mathrm{Re}\, z < 0\}$.

Then system (6.1) *is controllable on G.*

This theorem is obtained with the help of the Lie saturation technique introduced by V. Jurdjevic and I. Kupka [**22**] (see also G. Sallet [**31**] and V. Jurdjevic [**23**]): a sequence of increasing lower bounds of the tangent cone to the attainability set at the identity is shown to stabilize at the whole Lie algebra L.

The above controllability conditions for Lie groups $G \neq G^{(1)}$ yield complete description of controllability for several particular classes of Lie groups: metabelian ones, some subgroups of the group of affine transformations of the n-space, and small-dimensional simply connected solvable Lie groups. These results are presented in the following subsections and in Section 7.

6.2. Metabelian groups. Lie algebras L having derived series of length 2:

$$L \supset L^{(1)} \supset L^{(2)} = \{0\},$$

are called *metabelian.* A Lie group with a metabelian Lie algebra is also called *metabelian.*

A metabelian Lie algebra is obviously solvable. Thus results of the previous subsection yield controllability conditions for metabelian Lie groups.

THEOREM 6.3. *Let G be a metabelian Lie group. Then the following conditions are sufficient for controllability of system* (6.1) *on G:*

1. $\dim L^{(1)} = \dim L - 1$,
2. $B \notin L^{(1)}$,
3. $\mathrm{Sp}^{(1)} \cap \mathbb{R} = \emptyset$,
4. $A(a) \neq 0$ *for any eigenvalue* $a \in \mathrm{Sp}^{(1)}$.

If the group G is simply connected, then conditions 1 – 4 *are also necessary for controllability of system* (6.1) *on G.*

6.3. Semidirect products. Let V be a real finite-dimensional vector space, $\dim V = n$, and M a nonzero linear operator in V. Consider the metabelian Lie algebra $L(M)$ which is the semidirect product of the abelian Lie algebra V with the one-dimensional Lie algebra $\mathbb{R}M$. This Lie algebra can be represented by $(n+1)\times(n+1)$ matrices:

$$L(M) = \left\{ \begin{pmatrix} Mt & b \\ 0 & 0 \end{pmatrix} \mid t \in \mathbb{R},\ b \in \mathbb{R}^n \right\} \subset \mathrm{gl}(n+1, \mathbb{R}). \tag{6.2}$$

The connected Lie subgroup of $\mathrm{GL}(n+1,\mathbb{R})$ corresponding to $L(M)$ is denoted by $G(M)$. It is the semidirect product of the vector Lie group $\mathbb{R}^n$ with the one-dimensional Lie group $G_1 = \{\exp(Mt) \mid t \in \mathbb{R}\}$. Elements of the group $G(M)$ are the matrices

$$\begin{pmatrix} \exp(Mt) & p \\ 0 & 1 \end{pmatrix}, \quad t \in \mathbb{R},\ p \in \mathbb{R}^n,$$

thus $G(M)$ may be viewed as a subgroup of the group of affine transformations of the n-space generated by the one-parameter group of automorphisms G_1 and all translations $p \in \mathbb{R}^n$. The group $G(M)$ is not simply connected iff the one-parameter subgroup G_1 is periodic, which obviously occurs iff

$$\left.\begin{array}{l} \text{the matrix } M \text{ is semisimple,} \\ \mathrm{Sp}(M) = ir \cdot (k_1, \dots, k_n) \quad \text{for some } r \in \mathbb{R},\ (k_1, \dots, k_n) \in \mathbb{Z}^n. \end{array}\right\} \tag{6.3}$$

If conditions (6.3) hold, then a controllability test on $G(M)$ is given by the result of B. Bonnard, V. Jurdjevic, I. Kupka and G. Sallet [**7**] for Lie groups that are semidirect products of vector spaces with compact Lie groups. Otherwise the controllability test for simply connected metabelian Lie groups (Theorem 6.3) implies the following.

THEOREM 6.4. *Suppose that conditions* (6.3) *are violated. System* (6.1) *is controllable on the group* $G(M)$ *if and only if the following conditions hold:*

1. $\mathrm{Sp}(M) \cap \mathbb{R} = \emptyset$,
2. $B \notin L^{(1)}$,
3. $\mathrm{span}(B, A, (\mathrm{ad}\, B)A, \dots, (\mathrm{ad}\, B)^{n-1}A) = L$.

REMARK. Suppose that conditions (6.3) hold, i.e., the group $G(M)$ is not simply connected. Then conditions 1–3 of the previous theorem are necessary and sufficient for controllability of system (6.1) on $\widetilde{G(M)}$ — the simply connected covering of $G(M)$. And for the group $G(M)$ itself conditions 1–3 are then sufficient for controllability.

EXAMPLE. Let $G = \mathrm{E}(2)$ be the Euclidean group of motions of the plane $\mathbb{R}^2$. It can be represented as the group of 3×3 matrices of the form

$$\begin{pmatrix} \cos t & -\sin t & s_1 \\ \sin t & \cos t & s_2 \\ 0 & 0 & 1 \end{pmatrix},$$

with the rotation matrix and translation vector respectively

$$\begin{pmatrix} \cos t & -\sin t \\ \sin t & \cos t \end{pmatrix} \in \mathrm{SO}(2) \text{ for } t \in \mathbb{R}, \quad \begin{pmatrix} s_1 \\ s_2 \end{pmatrix} \in \mathbb{R}^2.$$

The corresponding matrix Lie algebra L is spanned by the matrices

$$x = \begin{pmatrix} 0 & -1 & 0 \\ 1 & 0 & 0 \\ 0 & 0 & 0 \end{pmatrix}, \quad y = \begin{pmatrix} 0 & 0 & 0 \\ 0 & 0 & 1 \\ 0 & 0 & 0 \end{pmatrix}, \quad z = \begin{pmatrix} 0 & 0 & 1 \\ 0 & 0 & 0 \\ 0 & 0 & 0 \end{pmatrix},$$

and has form (6.2):

$$L = L(M), \quad M = \begin{pmatrix} 0 & -1 \\ 1 & 0 \end{pmatrix}.$$

It is solvable (in fact, metabelian):

$$L^{(1)} = \mathrm{span}(y, z) \supset L^{(2)} = \{0\},$$

but not completely solvable:

$$\operatorname{Sp}(\operatorname{ad} x) = \{\pm i, 0\}.$$

The Lie group $\mathrm{E}(2) = G(M)$ is connected but not simply connected, compare with (6.3).

Consider the system $\Gamma = A + \mathbb{R}B$ on $\widetilde{\mathrm{E}(2)}$ — the simply connected covering of $\mathrm{E}(2)$. A complete characterization of controllability of Γ on $\widetilde{\mathrm{E}(2)}$ is derived from the remark after Theorem 6.4:

THEOREM 6.5. *System* (6.1) *is controllable on* $\widetilde{\mathrm{E}(2)}$ *if and only if the vectors* A, B *are linearly independent and* $B \notin \operatorname{span}(y, z)$.

Compare the controllability conditions for $\widetilde{\mathrm{E}(2)}$ with the following conditions for $\mathrm{E}(2)$ (derived from Theorem 1 [**7**]):

THEOREM 6.6. *System* (6.1) *is controllable on* $\mathrm{E}(2)$ *if and only if the vectors* A, B *are linearly independent and* $\{A, B\} \not\subset \operatorname{span}(y, z)$.

6.4. Bilinear systems. Global controllability conditions for bilinear systems of the form

$$\dot{x} = uAx + b, \qquad x \in \mathbb{R}^n, \quad u \in \mathbb{R}, \tag{Σ}$$

where A is a constant real $n \times n$ matrix and $b \in \mathbb{R}^n$, are obtained with the help of results for the matrix group of the previous subsection.

The system Σ may be rewritten as the following bilinear system in the hyperplane $\{x_{n+1} = 1\} \subset \mathbb{R}^{n+1}$:

$$\frac{d}{dt}\begin{pmatrix} x \\ 1 \end{pmatrix} = \begin{pmatrix} 0 & b \\ 0 & 0 \end{pmatrix}\begin{pmatrix} x \\ 1 \end{pmatrix} + u\begin{pmatrix} A & 0 \\ 0 & 0 \end{pmatrix}\begin{pmatrix} x \\ 1 \end{pmatrix}, \quad \begin{pmatrix} x \\ 1 \end{pmatrix} \in \mathbb{R}^{n+1} \tag{Σ'}$$

with the matrices

$$\overline{A} = \begin{pmatrix} 0 & b \\ 0 & 0 \end{pmatrix}, \ \overline{B} = \begin{pmatrix} A & 0 \\ 0 & 0 \end{pmatrix} \in L(A).$$

Moreover, Σ' is projection of the single-input right-invariant system $\overline{\Gamma} = \overline{A} + \mathbb{R}\overline{B} \subset L(A)$ on the matrix group $G(A)$ onto the hyperplane $\{x_{n+1} = 1\}$. Hence controllability of $\overline{\Gamma}$ on $G(A)$ implies global controllability of Σ' in the hyperplane $\{x_{n+1} = 1\}$ and thus global controllability of Σ in $\mathbb{R}^n$.

On the other hand, necessary controllability conditions for Σ can easily be obtained by studying codimension 1 and 2 invariant spaces of this system in $\mathbb{R}^n$.

This gives the following controllability test for Σ:

THEOREM 6.7. *The system* Σ *is globally controllable on* $\mathbb{R}^n$ *if and only if the following conditions hold:*

1. *the matrix* A *has a purely complex spectrum,*
2. $\operatorname{span}(b, Ab, \dots, A^{n-1}b) = \mathbb{R}^n$.

7. Small-dimensional solvable Lie groups

Given a Lie algebra L, there is the "largest" connected Lie group G having Lie algebra L — the simply connected one. All other connected Lie groups with Lie algebra L are "smaller" than G in the sense that they are quotients G/C, where C is a discrete subgroup of center of G. A right-invariant system $\Gamma \subset L$ may thus be considered on any of these groups, and the simply connected group G is the hardest among them to control. Hence given a right-invariant system Γ on a Lie group (or a homogeneous space of a Lie group) H, it is natural first to study its controllability on the simply connected covering $\widetilde{H}$ of H. If Γ is controllable on $\widetilde{H}$, then it is obviously controllable on H (and on all its homogeneous spaces); in the opposite case one should use particular geometric properties of H (e.g., existence of periodic one-parameter subgroups) to verify controllability of Γ on H. It is obvious and remarkable that controllability conditions on a simply connected Lie group G should have a completely Lie-algebraic form: they are fully determined by the Lie algebra L and its subset Γ (see, e.g., Theorems 3.3, 5.1, 6.1, 6.3, 6.4).

This motivates the following definition.

DEFINITION. A system $\Gamma \subset L$ is called *controllable* if it is controllable on the (unique) connected simply connected Lie group with Lie algebra L.

And the next definition makes sense at least for solvable Lie algebras in small dimensions.

DEFINITION. A Lie algebra L is called *controllable* if there are $A, B \in L$ such that the system $\Gamma = A + \mathbb{R}B$ is controllable.

Indeed, it turns out that controllability conditions on solvable Lie groups (Sections 4 and 6) imply that for solvable small-dimensional Lie algebras L:

- existence of a controllable single-input system $\Gamma \subset L$, i.e., controllability of L, is a strong restriction on L;
- if L is controllable, then almost all pairs $(A, B) \in L \times L$ give rise to controllable systems $\Gamma = A + \mathbb{R}B$;
- controllability of a system $\Gamma \subset L$ depends primarily on L but not on Γ.

Moreover, these results yield a complete description of controllability in small-dimensional solvable Lie algebras presented in the following subsections.

Up to dimension 6 inclusive we describe all solvable Lie algebras L that are controllable, and give controllability tests for single-input systems $\Gamma = A + \mathbb{R}B \subset L$ (the only gap in this picture is the class $L_{6,IV}$ of six-dimensional Lie algebras not completely studied).

The general "bird's-eye view" of controllable small-dimensional solvable Lie algebras is as follows:

$\dim L = 1$: the (unique) Lie algebra is controllable;
$\dim L = 2$: the two Lie algebras are noncontrollable;
$\dim L = 3$: there is one family of controllable Lie algebras $L_3(\lambda)$, $\lambda \in \mathbb{C} \setminus \mathbb{R}$;
$\dim L = 4$: there is one family of controllable Lie algebras $L_4(\lambda)$, $\lambda \in \mathbb{C} \setminus \mathbb{R}$;
$\dim L = 5$: there are two families of controllable Lie algebras:

1. $L_{5,I}(\lambda, \mu)$, $\lambda, \mu \in \mathbb{C} \setminus \mathbb{R}$, $\lambda \neq \mu, \bar{\mu}$,
2. $L_{5,II}(\lambda)$, $\lambda \in \mathbb{C} \setminus \mathbb{R}$;

$\dim L = 6$: there are five families of controllable Lie algebras:

1. $L_{6,I}(\lambda, \mu)$, $\lambda, \mu \in \mathbb{C} \setminus \mathbb{R}$, $\lambda \neq \mu, \bar{\mu}$,
2. $L_{6,II}(\lambda, \mu, k)$, $\lambda, \mu \in \mathbb{C} \setminus \mathbb{R}$, $\operatorname{Re}\lambda = \operatorname{Re}\mu$, $\lambda \neq \mu, \bar{\mu}$, $k \in \mathbb{R} \setminus \{0\}$,
3. $L_{6,III}(\lambda, k, l)$, $\lambda \in \mathbb{C} \setminus (\mathbb{R} \cup i\mathbb{R})$, $k, l \in \mathbb{R}$, $k^2 + l^2 \neq 0$,
4. $L_{6,IV}(\lambda, k, l)$, $\lambda \in \mathbb{C} \setminus (\mathbb{R} \cup i\mathbb{R})$, $k, l \in \mathbb{R}$, $k^2 + l^2 \neq 0$,
5. $L_{6,V}(\lambda, k, l)$, $\lambda \in \mathbb{C} \setminus (\mathbb{R} \cup i\mathbb{R})$, $k, l \in \mathbb{R}$, $k^2 + l^2 \neq 0$,

and one exceptional class $L_{6,IV}(bi)$, $b \in \mathbb{R} \setminus \{0\}$, containing both controllable and noncontrollable Lie algebras.

All controllable Lie algebras L are presented by a scheme in the complex plane $\mathbb{C}$ containing eigenvalues of the adjoint operator $\operatorname{ad} B|_{L^{(1)}}$, $B \in L \setminus L^{(1)}$, and arrows between these eigenvalues describing Lie brackets between eigenvectors of the operator $\operatorname{ad} B|_{L^{(1)}}$ (these schemes are given at the very end of this section). Notice that for solvable Lie algebras L with codimension one subalgebras $L^{(1)}$ (and only such solvable Lie algebras may be controllable, see condition 1 of Theorem 6.1), spectra of all adjoint operators $\operatorname{ad} B|_{L^{(1)}}$, $B \in L \setminus L^{(1)}$, are homothetic with respect to the origin $0 \in \mathbb{C}$, and the homothety equivalence class of spectra of $\operatorname{ad} B|_{L^{(1)}}$, $B \in L \setminus L^{(1)}$, is determined not by $B \in L \setminus L^{(1)}$ but by L itself (in fact, by the isomorphism class of L).

Now we present the classification of controllability in small-dimensional solvable Lie algebras. These results are obtained by virtue of controllability conditions of Sections 4 and 6.

7.1. One-dimensional Lie algebras. The unique one-dimensional Lie algebra is abelian and isomorphic to $\mathbb{R}$.

THEOREM 7.1. *The one-dimensional Lie algebra $\mathbb{R}$ is controllable.*
A system $\Gamma = A + \mathbb{R}B \subset \mathbb{R}$ is controllable iff $B \neq 0$.

7.2. Two-dimensional Lie algebras. There are two nonisomorphic two-dimensional Lie algebras: abelian $\mathbb{R}^2$, and solvable non-abelian $S_2 = \operatorname{span}(x, y)$, $[x, y] = y$.

THEOREM 7.2. *Both two-dimensional Lie algebras $\mathbb{R}^2$ and S_2 are not controllable.*

7.3. Three-dimensional Lie algebras.

CONSTRUCTION. The Lie algebra $L_3(\lambda)$, $\lambda \in \mathbb{C} \setminus \mathbb{R}$, Fig. 1.

$$L_3(\lambda) = \operatorname{span}(x, y, z),$$
$$\operatorname{ad} x|_{\operatorname{span}(y,z)} = \begin{pmatrix} a & -b \\ b & a \end{pmatrix}, \qquad \lambda = a + bi.$$

The Lie algebra $L_3(\lambda)$ is schematically represented in Fig. 1 by the eigenvalues $\lambda, \bar{\lambda} \in \mathbb{C}$ and realifications of the eigenvectors $y, z \in L_3(\lambda)$ of the adjoint operator $\operatorname{ad} x|_{\operatorname{span}(y,z)}$.

THEOREM 7.3. *A three-dimensional solvable Lie algebra is controllable iff it is isomorphic to $L_3(\lambda)$, $\lambda \in \mathbb{C} \setminus \mathbb{R}$.*

THEOREM 7.4. *Let $L = L_3(\lambda)$, $\lambda \in \mathbb{C} \setminus \mathbb{R}$, and let $A, B \in L$. The system $\Gamma = A + \mathbb{R}B \subset L$ is controllable iff the following conditions are satisfied:*
1. *$B \notin L^{(1)}$,*
2. *the vectors A and B are linearly independent.*

7.4. Four-dimensional Lie algebras.

CONSTRUCTION. The Lie algebra $L_4(\lambda)$, $\lambda \in \mathbb{C} \setminus \mathbb{R}$, Fig. 2.

$$L_4(\lambda) = \operatorname{span}(x, y, z, w),$$
$$\operatorname{ad} x|_{\operatorname{span}(y,z,w)} = \begin{pmatrix} a & -b & 0 \\ b & a & 0 \\ 0 & 0 & 2a \end{pmatrix}, \qquad \lambda = a + bi,$$
$$[y, z] = w.$$

The arrows in the schematic representation of the Lie algebra $L_4(\lambda)$ in Fig. 2 mean that Lie bracket of the vectors y and z gives the vector w.

THEOREM 7.5. *A four-dimensional solvable Lie algebra is controllable iff it is isomorphic to* $L_4(\lambda)$, $\lambda \in \mathbb{C} \setminus \mathbb{R}$.

THEOREM 7.6. *Let* $L = L_4(\lambda)$, $\lambda \in \mathbb{C} \setminus \mathbb{R}$, *and let* $A, B \in L$. *The system* $\Gamma = A + \mathbb{R}B \subset L$ *is controllable iff the following conditions are satisfied:*

1. $B \notin L^{(1)}$,
2. $A(\lambda) \neq 0$.

7.5. Five-dimensional Lie algebras.

CONSTRUCTION. The Lie algebra $L_{5,I}(\lambda, \mu)$, $\lambda, \mu \in \mathbb{C} \setminus \mathbb{R}$, Fig. 3.

$$L_{5,I}(\lambda, \mu) = \operatorname{span}(x, y, z, u, v),$$

$$\operatorname{ad} x|_{\operatorname{span}(y,z,u,v)} = \begin{pmatrix} a & -b & 0 & 0 \\ b & a & 0 & 0 \\ 0 & 0 & c & -d \\ 0 & 0 & d & c \end{pmatrix}, \quad \lambda = a + bi, \ \mu = c + di.$$

CONSTRUCTION. The Lie algebra $L_{5,II}(\lambda)$, $\lambda \in \mathbb{C} \setminus \mathbb{R}$, Fig. 4.

$$L_{5,II}(\lambda) = \operatorname{span}(x, y, z, u, v),$$

$$\operatorname{ad} x|_{\operatorname{span}(y,z,u,v)} = \begin{pmatrix} a & -b & 0 & 0 \\ b & a & 0 & 0 \\ 1 & 0 & a & -b \\ 0 & 1 & b & a \end{pmatrix}, \quad \lambda = a + bi.$$

The circles around the eigenvalues λ, $\bar{\lambda}$ in Fig. 4 mean that they have double algebraic multiplicity. (Notice that according to the previous matrix their geometric multiplicity is single.)

THEOREM 7.7. *A five-dimensional solvable Lie algebra is controllable iff it is isomorphic to* $L_{5,I}(\lambda, \mu)$, $\lambda, \mu \in \mathbb{C} \setminus \mathbb{R}$, $\lambda \neq \mu, \bar{\mu}$, *or* $L_{5,II}(\lambda)$, $\lambda \in \mathbb{C} \setminus \mathbb{R}$.

THEOREM 7.8. *Let* $L = L_{5,I}(\lambda, \mu)$, $\lambda, \mu \in \mathbb{C} \setminus \mathbb{R}$, $\lambda \neq \mu, \bar{\mu}$, *and let* $A, B \in L$. *The system* $\Gamma = A + \mathbb{R}B \subset L$ *is controllable iff the following conditions are satisfied:*

1. $B \notin L^{(1)}$,
2. $A(\lambda) \neq 0$ *and* $A(\mu) \neq 0$.

THEOREM 7.9. *Let* $L = L_{5,II}(\lambda)$, $\lambda \in \mathbb{C} \setminus \mathbb{R}$, *and let* $A, B \in L$. *The system* $\Gamma = A + \mathbb{R}B \subset L$ *is controllable iff the following conditions are satisfied:*

1. $B \notin L^{(1)}$,
2. $\operatorname{top}(A, \lambda) \neq 0$.

REMARK. The notation $\operatorname{top}(A, \lambda) \neq 0$ in Theorem 7.9 (and in Theorem 7.15 below) means that the vector A has a nonzero component corresponding to the higher order root space of the operator $\operatorname{ad} B|_{L^{(1)}}$ corresponding to its eigenvalue λ.

7.6. Six-dimensional Lie algebras.

CONSTRUCTION. The Lie algebra $L_{6,I}(\lambda, \mu)$, $\lambda, \mu \in \mathbb{C} \setminus \mathbb{R}$, Fig. 5.

$$L_{6,I}(\lambda, \mu) = \operatorname{span}(x, y, z, u, v, w),$$

$$\operatorname{ad} x|_{\operatorname{span}(y,z,u,v,w)} = \begin{pmatrix} a & -b & 0 & 0 & 0 \\ b & a & 0 & 0 & 0 \\ 0 & 0 & c & -d & 0 \\ 0 & 0 & d & c & 0 \\ 0 & 0 & 0 & 0 & 2a \end{pmatrix}, \quad \lambda = a + bi, \ \mu = c + di,$$

$$[y, z] = w.$$

CONSTRUCTION. The Lie algebra $L_{6,II}(\lambda, \mu, k)$, $\lambda, \mu \in \mathbb{C} \setminus \mathbb{R}$, $\operatorname{Re} \lambda = \operatorname{Re} \mu$, $k \in \mathbb{R} \setminus \{0\}$, Fig. 6.

$$L_{6,II}(\lambda, \mu, k) = \operatorname{span}(x, y, z, u, v, w),$$

$$\operatorname{ad} x|_{\operatorname{span}(y,z,u,v,w)} = \begin{pmatrix} a & -b & 0 & 0 & 0 \\ b & a & 0 & 0 & 0 \\ 0 & 0 & a & -d & 0 \\ 0 & 0 & d & a & 0 \\ 0 & 0 & 0 & 0 & 2a \end{pmatrix}, \quad \lambda = a + bi, \ \mu = a + di,$$

$$[y, z] = w, \quad [u, v] = kw.$$

CONSTRUCTION. The Lie algebra $L_{6,III}(\lambda, k, l)$, $\lambda \in \mathbb{C} \setminus (\mathbb{R} \cup i\,\mathbb{R})$, $k, l \in \mathbb{R}$, $k^2 + l^2 \neq 0$, Fig. 7.

$$L_{6,III}(\lambda, k, l) = \operatorname{span}(x, y, z, u, v, w),$$

$$\operatorname{ad} x|_{\operatorname{span}(y,z,u,v,w)} = \begin{pmatrix} a & -b & 0 & 0 & 0 \\ b & a & 0 & 0 & 0 \\ 0 & 0 & 3a & -b & 0 \\ 0 & 0 & b & 3a & 0 \\ 0 & 0 & 0 & 0 & 2a \end{pmatrix}, \quad \lambda = a + bi,$$

$$[w, y] = ku + lv, \quad [w, z] = -lu + kz.$$

CONSTRUCTION. The Lie algebra $L_{6,IV}(\lambda, k, l)$, $\lambda \in \mathbb{C} \setminus (\mathbb{R} \cup i\,\mathbb{R})$, $k, l \in \mathbb{R}$, $k^2 + l^2 \neq 0$, Fig. 8.

$$L_{6,IV}(\lambda, k, l) = \operatorname{span}(x, y, z, u, v, w),$$

$$\operatorname{ad} x|_{\operatorname{span}(y,z,u,v,w)} = \begin{pmatrix} a & -b & 0 & 0 & 0 \\ b & a & 0 & 0 & 0 \\ 0 & 0 & -a & -b & 0 \\ 0 & 0 & b & -a & 0 \\ 0 & 0 & 0 & 0 & 0 \end{pmatrix}, \quad \lambda = a + bi,$$

$$[y, v] = -[z, u] = kw, \quad [y, u] = [z, v] = lw.$$

CONSTRUCTION. The Lie algebra $L_{6,V}(\lambda,k,l)$, $\lambda \in \mathbb{C} \setminus (\mathbb{R} \cup i\,\mathbb{R})$, $k,l \in \mathbb{R}$, $k^2+l^2 \neq 0$, Fig. 9.

$$L_{6,V}(\lambda,k,l) = \operatorname{span}(x,y,z,u,v,w),$$

$$\operatorname{ad} x|_{\operatorname{span}(y,z,u,v,w)} = \begin{pmatrix} a & -b & 0 & 0 & 0 \\ b & a & 0 & 0 & 0 \\ 1 & 0 & a & -b & 0 \\ 0 & 1 & b & a & 0 \\ 0 & 0 & 0 & 0 & 2a \end{pmatrix}, \qquad \lambda = a+bi,$$

$$[y,z] = kw, \qquad [y,u] = [z,v] = lw.$$

CONSTRUCTION. The class of Lie algebras $L_{6,VI}(bi)$, $b \in \mathbb{R} \setminus \{0\}$, Fig. 10.

A Lie algebra L belongs to the class $L_{6,VI}(bi)$ if:

$L = \operatorname{span}(x,y,z,u,v,w)$,

$L^{(1)} = \operatorname{span}(y,z,u,v,w)$,

$\operatorname{Sp}(\operatorname{ad} x|_{L^{(1)}}) = \{\pm bi, 0\}$,

both eigenvalues $\pm bi$ have double algebraic multiplicity,

$w \in L^{(2)}$.

The class $L_{6,VI}$ contains a lot of nonisomorphic Lie algebras in which multiplication can not be described in detail as in Lie algebras $L_{6,I}$–$L_{6,V}$.

THEOREM 7.10. *Let a six-dimensional solvable Lie algebra L not belong to the class $L_{6,VI}(bi)$, $b \in \mathbb{R} \setminus \{0\}$. Then L is controllable iff it is isomorphic to one of the following Lie algebras:*

1. $L_{6,I}(\lambda,\mu)$, $\lambda,\mu \in \mathbb{C} \setminus \mathbb{R}$, $\lambda \neq \mu, \bar{\mu}$;
2. $L_{6,II}(\lambda,\mu,k)$, $\lambda,\mu \in \mathbb{C} \setminus \mathbb{R}$, $\operatorname{Re}\lambda = \operatorname{Re}\mu$, $\lambda \neq \mu,\bar{\mu}$, $k \in \mathbb{R} \setminus \{0\}$;
3. $L_{6,III}(\lambda,k,l)$, $\lambda \in \mathbb{C} \setminus (\mathbb{R} \cup i\,\mathbb{R})$, $k,l \in \mathbb{R}$, $k^2+l^2 \neq 0$;
4. $L_{6,IV}(\lambda,k,l)$, $\lambda \in \mathbb{C} \setminus (\mathbb{R} \cup i\,\mathbb{R})$, $k,l \in \mathbb{R}$, $k^2+l^2 \neq 0$;
5. $L_{6,V}(\lambda,k,l)$, $\lambda \in \mathbb{C} \setminus (\mathbb{R} \cup i\,\mathbb{R})$, $k,l \in \mathbb{R}$, $k^2+l^2 \neq 0$.

THEOREM 7.11. *Let $L = L_{6,I}(\lambda,\mu)$, $\lambda,\mu \in \mathbb{C} \setminus \mathbb{R}$, $\lambda \neq \mu,\bar{\mu}$, and let $A, B \in L$. The system $\Gamma = A + \mathbb{R}B \subset L$ is controllable iff the following conditions are satisfied:*

1. $B \notin L^{(1)}$,
2. $A(\lambda) \neq 0$ *and* $A(\mu) \neq 0$.

THEOREM 7.12. *Let $L = L_{6,II}(\lambda,\mu,k)$, $\lambda,\mu \in \mathbb{C} \setminus \mathbb{R}$, $\operatorname{Re}\lambda = \operatorname{Re}\mu$, $\lambda \neq \mu,\bar{\mu}$, $k \in \mathbb{R} \setminus \{0\}$, and let $A, B \in L$. The system $\Gamma = A + \mathbb{R}B \subset L$ is controllable iff the following conditions are satisfied:*

1. $B \notin L^{(1)}$,
2. $A(\lambda) \neq 0$ *and* $A(\mu) \neq 0$.

THEOREM 7.13. *Let $L = L_{6,III}(\lambda,k,l)$, $\lambda \in \mathbb{C} \setminus (\mathbb{R} \cup i\,\mathbb{R})$, $k,l \in \mathbb{R}$, $k^2+l^2 \neq 0$, and let $A, B \in L$. The system $\Gamma = A + \mathbb{R}B \subset L$ is controllable iff the following conditions are satisfied:*

1. $B \notin L^{(1)}$,
2. $A(\lambda) \neq 0$.

THEOREM 7.14. *Let $L = L_{6,IV}(\lambda,k,l)$, $\lambda \in \mathbb{C} \setminus (\mathbb{R} \cup i\,\mathbb{R})$, $k,l \in \mathbb{R}$, $k^2+l^2 \neq 0$, and let $A, B \in L$. The system $\Gamma = A + \mathbb{R}B \subset L$ is controllable iff the following conditions are satisfied:*

1. $B \notin L^{(1)}$,
2. $A(\lambda) \neq 0$ *and* $A(-\lambda) \neq 0$.

THEOREM 7.15. *Let* $L = L_{6,V}(\lambda, k, l)$, $\lambda \in \mathbb{C} \setminus (\mathbb{R} \cup i\,\mathbb{R})$, $k, l \in \mathbb{R}$, $k^2 + l^2 \neq 0$, *and let* $A, B \in L$. *The system* $\Gamma = A + \mathbb{R}B \subset L$ *is controllable iff the following conditions are satisfied:*

1. $B \notin L^{(1)}$,
2. $\operatorname{top}(A, \lambda) \neq 0$.

REMARK. The class $L_{6,VI}(bi)$, $b \in \mathbb{R} \setminus \{0\}$, contains both controllable and noncontrollable Lie algebras.

Controllable solvable Lie algebras up to dimension 6:

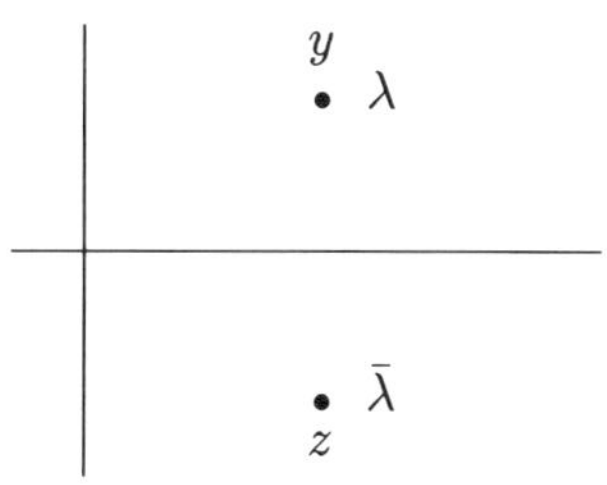

Fig. 1. $L_3(\lambda)$.

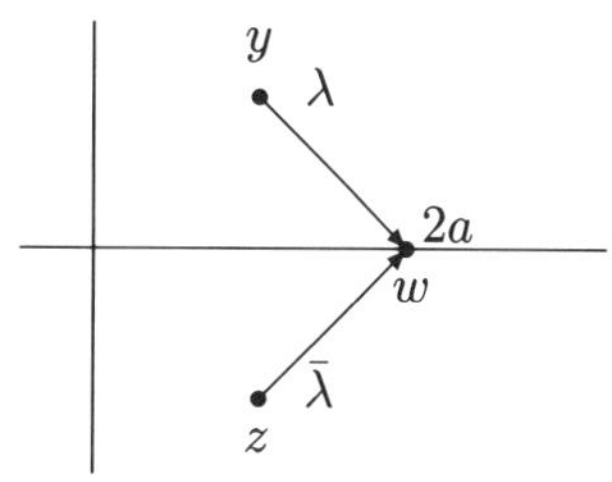

Fig. 2. $L_4(\lambda)$, $\operatorname{Re}\lambda = a$.

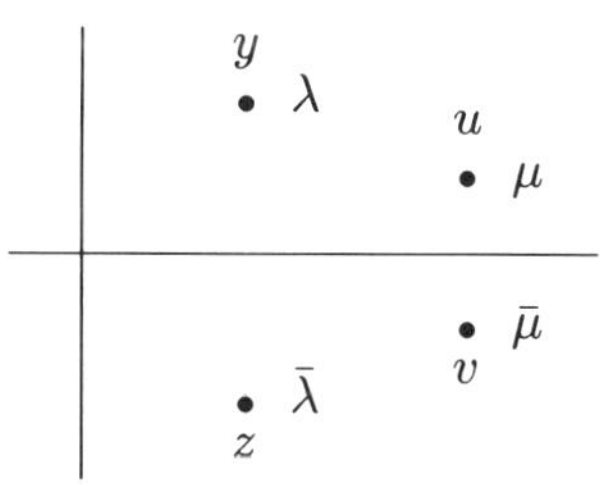

Fig. 3. $L_{5,I}(\lambda, \mu)$.

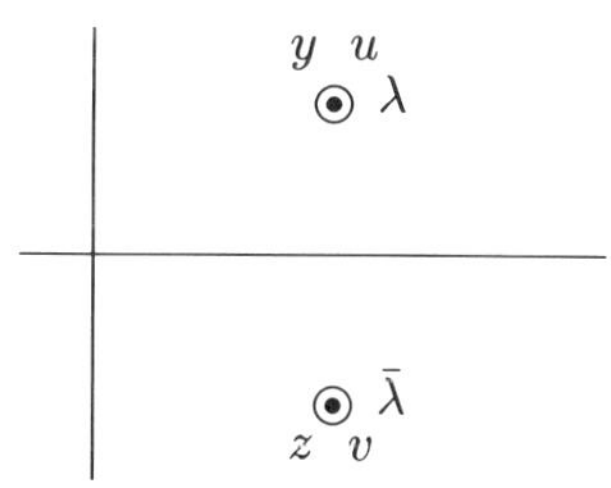

Fig. 4. $L_{5,II}(\lambda)$.

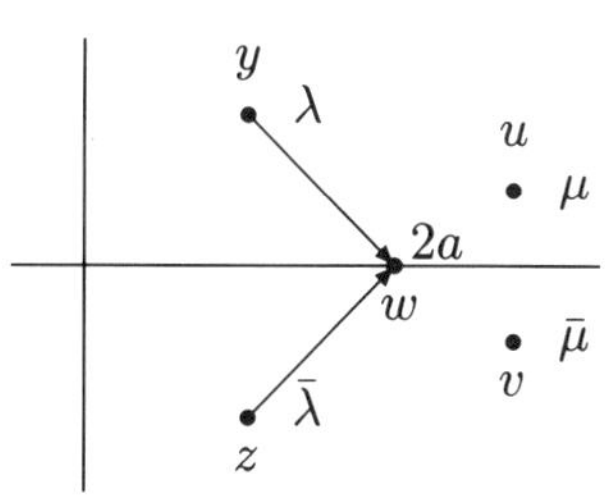

Fig. 5. $L_{6,I}(\lambda, \mu)$, $\operatorname{Re}\lambda = a$.

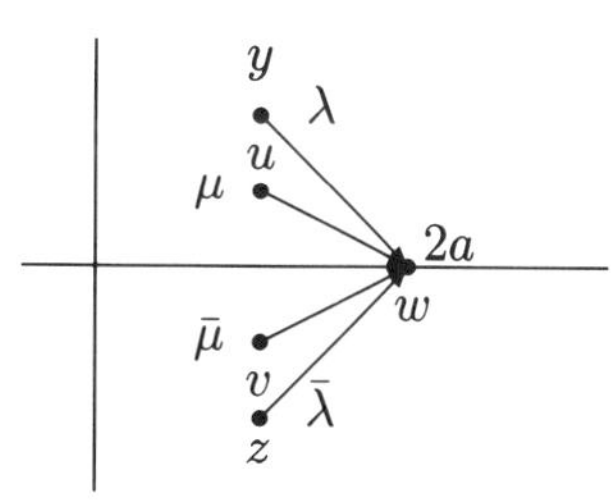

Fig. 6. $L_{6,II}(\lambda, \mu, k)$, $\operatorname{Re}\lambda = \operatorname{Re}\mu = a$.

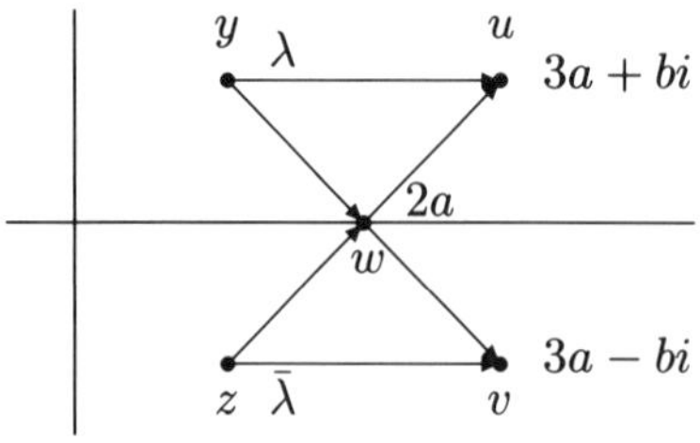

Fig. 7. $L_{6,III}(\lambda, k, l)$, $\lambda = a + bi$.

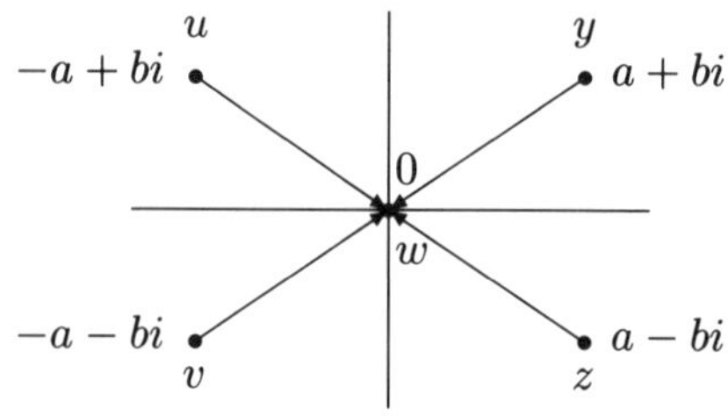

Fig. 8. $L_{6,IV}(\lambda, k, l)$, $\lambda = a + bi$.

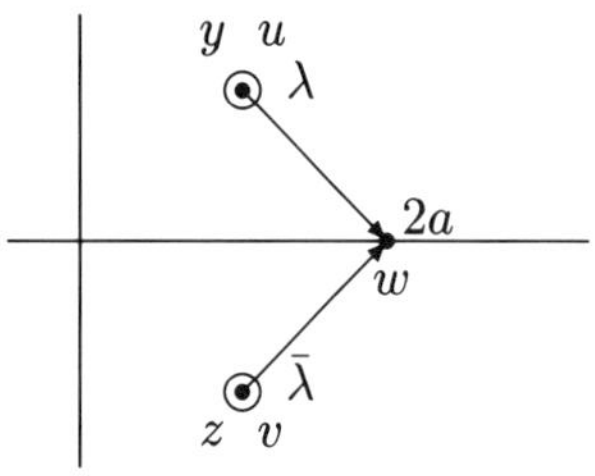

Fig. 9. $L_{6,V}(\lambda, \mu)$, $\operatorname{Re} \lambda = a$.

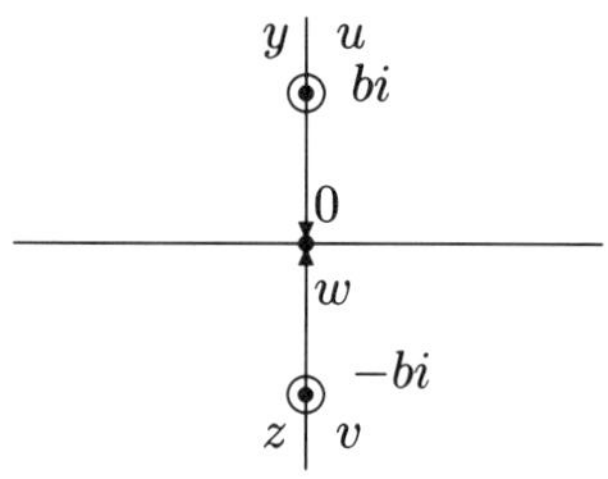

Fig. 10. $L_{6,VI}(bi)$.

8. Bibliographical notes

The results presented in this paper were obtained by the following authors:
Subsection 3.1: J. Hilgert, K. H. Hofmann, and J. D. Lawson [**13**].
Subsection 3.2: V. Ayala Bravo [**6**].
Section 4: J. D. Lawson [**25**].
Sections 5–7: Yu. L. Sachkov [**26, 27, 28**].

9. Questions and suggestions

In this section we present and discuss several challenging open questions related to the results considered above.

9.1. The hypersurface principle. The hypersurface principle given by Theorem 5.3, Section 5, is a necessary controllability condition for an arbitrary simply connected Lie group. If a simply connected Lie group has cocompact radical, then this principle is also sufficient for controllability (Theorem 4.1, Section 4). Is it possible to extend the class of simply connected Lie groups with cocompact radical so that the hypersurface principle remain a criterion of controllability?

9.2. Lie algebras hard to control. For any Lie group G and any system affine in control $\Gamma = \{\, A + \sum_{i=1}^m u_i B_i \mid u_i \in \mathbb{R} \,\}$ on G controllability of the homogeneous part $\Gamma_0 = \{\, \sum_{i=1}^m u_i B_i \mid u_i \in \mathbb{R} \,\}$ is sufficient for controllability of Γ on G. We call a Lie algebra L *hard to control* if any affine in control system $\Gamma \subset L$ and its homogeneous part Γ_0 are simultaneously controllable or noncontrollable (on the connected simply connected Lie group G corresponding to L). In Lie algebras L

hard to control the drift term A in an affine system $\Gamma \subset L$ does not help in control, which is not the case for general Lie algebras.

There is the expanding chain of classes of Lie algebras hard to control:

$$\text{abelian} \subset \text{nilpotent} \subset \text{completely solvable.} \qquad (*)$$

The abelian case is obvious, the nilpotent one is Theorem 3.3, Subsection 3.2, and the completely solvable one is Theorem 5.1, Section 5.

On the other hand, the Lie algebra of the group $\mathrm{E}(2)$ of motions of the plane is solvable, not completely solvable and not hard to control (see the example in Subsection 6.3).

Corollary 3.3 [**26**] states that all Lie algebras satisfying the following property:

$$\left.\begin{array}{l}\text{any subalgebra } l \subset L,\ l \neq L, \text{ is contained}\\ \text{in a codimension one subalgebra of } L\end{array}\right\} \qquad (**)$$

are included in the set of Lie algebras hard to control. The author does not know, whether this inclusion is strict. (By Lemma 4.2 [**26**], completely solvable Lie algebras satisfy property (**).)

Are there any Lie algebras hard to control not contained in chain (*)? If yes, can this chain be continued by any reasonable class of Lie algebras?

The theory of K. H. Hofmann on hyperplane subalgebras of Lie algebras [**16, 17, 19**] may be important for this question.

9.3. Small-dimensional groups. A complete and visual classification of controllable systems might be obtained for small-dimensional groups with the help of the known results for the following classes of groups: compact [**20**], semi-simple [**21**], [**22**], [**11**], reductive [**14**], nilpotent [**13**], [**6**], and solvable [**25**], [**26**], [**27**], [**28**]. An attempt in this direction was made in [**29**].

9.4. Solvable not simply connected Lie groups. The results of K. H. Hofmann on compact elements in solvable Lie algebras [**18**] might be applied in order to understand controllability for solvable Lie groups without the assumption of simple connectedness essential in Sections 5–7.

9.5. General groups. On the basis of the results listed in Subsection 9.3, controllability theory for general Lie groups can be started synthesizing the "semi-simple" and "solvable" theory via Levi decomposition (I. Kupka [**24**]).

9.6. Nilpotent and solvable manifolds. Controllability of projections of right-invariant systems onto nilpotent and solvable manifolds can be studied via application of the theory of flows on these manifolds [**5**]. This may be important for studying local controllability of nonlinear systems via nilpotent approximations (P. E. Crouch and C. I. Byrnes [**10**]).

9.7. Codimension one and two subalgebras. The solution of the controllability problem for completely solvable Lie groups (see Section 5) is based upon the following fact: any proper subalgebra of a real *completely solvable* Lie algebra is contained in a codimension one subalgebra. On the other hand, any proper subalgebra of a real *solvable* Lie algebra is included in some subalgebra of codimension one or two.

This suggests the following approach to controllability on solvable Lie groups. Project a system along the connected subgroup corresponding to the indicated codimension one or two subalgebra. Then: 1) if this group is closed and normal, we obtain a right-invariant system on a one- or two-dimensional Lie group (such systems are transparent); 2) if this subgroup is closed, we obtain a nonlinear system on a one- or two-dimensional smooth manifold (such systems are tractable by the nonlinear controllability theory); 3) and if this subgroup is not closed, then try to apply and develop the theory of control systems on foliations.

The progress in this direction may be useful for the controllability theory on general Lie groups (Subsection 9.5) as well.

9.8. Rank condition and hypersurface principle. The customary procedure to show noncontrollability is either to show the violation of the rank controllability principle [**20**] or to construct a (not necessarily smooth) hypersurface in the state space of a system intersected by all trajectories of the system in one direction only. For right-invariant systems on simply connected Lie groups with cocompact radical such hypersurface can always be found among codimension one subgroups (Section 4).

Does every full-rank noncontrollable right-invariant system have such codimension one subgroup? A positive answer will give a new method of obtaining sufficient controllability conditions, and a negative one will give an example of a complex obstacle to controllability.

Acknowledgments

The author wishes to thank Professors K. H. Hofmann, J .D. Lawson, and V. Ayala Bravo for supplying their papers and valuable discussions of the subject of this work. The author is grateful to the anonymous referee for several important linguistic and mathematical corrections.

References

[1] A. A. Agrachev, S. V. Vakhrameev, and R. V. Gamkrelidze, *Differential-Geometric and Group-Theoretic Methods in Optimal Control Theory*, In: Problemy geometrii (Itogi nauki i tekhniki, VINITI AN SSSR) (in Russian) **14** (1983), 3–56.

[2] R. El Assoudi and J. P. Gauthier, *Controllability of Right Invariant Systems on Real Simple Lie Groups of Type F_4, G_2, C_n, and B_n*, Math. Control Signals Systems **1** (1988), 293–301.

[3] R. El Assoudi and J. P. Gauthier, *Controllability of Right-Invariant Systems on Semi-simple Lie Groups*, In: New trends in nonlinear control theory, Springer-Verlag **122** (1989), 54–64.

[4] R. El Assoudi, J. P. Gauthier, and I. Kupka, *On Subsemigroups of Semisimple Lie Groups*, Ann. Inst. Henri Poincaré **13** (1996), 1: 117–133.

[5] L. Auslander, L. Green, and F. Hahn, *Flows on Homogeneous Spaces*, Annals of Mathematics Studies, No. 53, Princeton Univ. Press, Princeton, New Jersey, 1963.

[6] V. Ayala Bravo, *Controllability of nilpotent systems*, In: Geometry in nonlinear control and differential inclusions, Banach Center Publications, Warszawa, **32** (1995), 35–46.

[7] B. Bonnard, V. Jurdjevic, I. Kupka and G. Sallet, *Transitivity of Families of Invariant Vector Fields on the Semidirect Products of Lie Groups*, Trans. Amer. Math. Soc. **271** (1982), 2, 525–535.

[8] R. W. Brockett, *System Theory on Group Manifolds and Coset Spaces*, SIAM J. Control **10** (1972), 265–284.

[9] R. W. Brockett, *Lie Algebras and Lie Groups in Control Theory*, In: Geometric methods in system theory, D. Q. Mayne and R. W. Brockett, Eds., Proceedings of the NATO Advanced study institute held at London, August 27 – September 7, 1983, Dordrecht – Boston, D. Reidel Publishing Company, 1973, 43–82.

[10] P. E. Crouch and C. I. Byrnes, *Symmetries and Local Controllability*, In: Algebraic and Geometric Methods in Nonlinear Control Theory, M. Fliess and M. Hazewinkel eds., Reidel Publishing Company, Holland, 1986.

[11] J. P. Gauthier and G. Bornard, *Contrôlabilité des systèmes bilinèaires*, SIAM J. Control Optim. **20** (1982), 3, 377–384.

[12] J. P. Gauthier, I. Kupka and G. Sallet, *Controllability of Right Invariant Systems on Real Simple Lie Groups*, Systems & Control Letters **5** (1984), 187–190.

[13] J. Hilgert, K. H. Hofmann and J. D. Lawson, *Controllability of Systems on a Nilpotent Lie Group*, Beiträge zur Algebra und Geometrie **20** (1985), 185–190.

[14] J. Hilgert, *Controllability on Real Reductive Lie Groups*, Mathematische Zeitschrift **209** (1992), 463–466.

[15] J. Hilgert and K. H. Neeb, *Lie Semigroups and their Applications*, Lecture Notes in Math. **1552** (1993).

[16] K. H. Hofmann, *Lie algebras with subalgebras of codimension one*, Illinois J. Math. **9** (1965), 636–643.

[17] K. H. Hofmann, *Hyperplane Subalgebras of Real Lie Algebras*, Geometriae Dedicata **36** (1990), 207–224.

[18] K. H. Hofmann, *Compact Elements in Solvable Real Lie Algebras*, Seminar Sophus Lie (now: Journal of Lie theory) **2** (1992), 41–55.

[19] K. H. Hofmann, *Memo to Yurii Sachkov on Hyperplane Subalgebras of Lie Algebras*, E-mail message, March 1996.

[20] V. Jurdjevic and H. Sussmann, *Control Systems on Lie Groups*, Journ. Differ. Equat. **12** (1972), 313–329.

[21] V. Jurdjevic and I. Kupka, *Control Systems Subordinated to a Group Action: Accessibility*, J. Differ. Equat. **39** (1981), 186–211.

[22] V. Jurdjevic and I. Kupka, *Control Systems on Semi-simple Lie Groups and their Homogeneous Spaces*, Ann. Inst. Fourier, Grenoble **31** (1981), 4, 151–179.

[23] V. Jurdjevic, *Geometric Control Theory*, Cambridge University Press, 1997.

[24] I. Kupka, *Applications of Semigroups to Geometric Control Theory*, In: The analytical and topological theory of semigroups — Trends and developments, K. H. Hofmann, J. D. Lawson and J. S. Pym, Eds., de Gruyter Expositions in Mathematics **1** (1990), 337–345.

[25] J. D. Lawson, *Maximal Subsemigroups of Lie Groups that are Total*, Proc. Edinburgh Math. Soc. **30** (1985), 479–501.

[26] Yu. L. Sachkov, *Controllability of Hypersurface and Solvable Invariant Systems*, Journal of Dynamical and Control Systems **2** (1996), 55–67.

[27] Yu. L. Sachkov, *Controllability of Right-Invariant Systems on Solvable Lie Groups*, Journal of Dynamical and Control Systems **3** (1997), 531–564.

[28] Yu. L. Sachkov, *Classification of Controllability in Small-Dimensional Solvable Lie Algebras*, in preparation.

[29] Yu. L. Sachkov, *Controllability on Lie groups and Singularities*, submitted.

[30] G. Sallet, *Complete Controllabilite sur les Groupes Lineaires*, In: Qutils et modeles math. autom. Anal. syst. et trait signal., vol. 1, Paris, 1981, 215–227.

[31] G. Sallet, *Extension Techniques*, In: Systems and Control Encyclopedia II, (1987), 1581–1583.

[32] G. Sallet, *Lie Groups: Controllability*, In: Systems and Control Encyclopedia, (1987), 2756–2759.

[33] L. A. B. San Martin, *Invariant Control Sets on Flag Manifolds*, Math. Control Signals Systems **6** (1993), 41–61.

[34] L. A. B. San Martin and P. A. Tonelli, *Semigroup Actions on Homogeneous Spaces*, Semigroup Forum **14** (1994), 1–30.

[35] F. Silva Leite and P. C. Crouch, *Controllability on Classical Lie Groups*, Math. Control Signals Systems **1** (1988), 31–42.

[36] E. B. Vinberg, V. V. Gorbatcevich, and A. L. Onishchik, *Construction of Lie Groups and Lie Algeras*, Itogi Nauki i Tekhniki, Sovremennyje Problemy Matematiki, Fundamental'nyje Napravlenija (in Russian) **41** (1989).

PROGRAM SYSTEMS INSTITUTE, RUSSIAN ACADEMY OF SCIENCES, 152140 PERESLAVL-ZALESSKY, RUSSIA

E-mail address: `sachkov@sys.botik.ru`

Proceedings of Symposia in Pure Mathematics
Volume **64**, 1999

Centro-affine geometry in the plane and feedback invariants of two-state scalar control systems

George R. Wilkens

ABSTRACT. The goal of this paper is to establish the precise connection between the centro-affine invariants of plane curves and the feedback invariants of nonlinear scalar control systems in the plane. We will also show how the centro-affine structure provides a shortcut to the structure equations for feedback equivalence one obtains by applying Cartan's equivalence method.

0. Introduction

A direct observation of the way that control systems transform under the action of feedback transformations leads us to consider a classical problem in the geometry of submanifolds of $\mathbf{R}^n$. Specifically, let $\dot{\mathbf{x}} = \mathbf{f}(\mathbf{x}, u)$ represent a control system, where $\mathbf{x} \in \mathbf{R}^n$ and $u \in \mathbf{R}^m$, with $m < n$. Let $\mathbf{x}$ be any point in state space and let $T_{\mathbf{x}}$ denote the tangent space at $\mathbf{x}$. The control system determines the mapping $u \mapsto \mathbf{f}(\mathbf{x}, u)$ from the control space, $\mathbf{R}^m$, to the tangent space $T_{\mathbf{x}}$. If we make the reasonable assumption that the rank of $\partial\mathbf{f}/\partial u$ is maximal, then we see that this mapping determines an m-dimensional parameterized submanifold of $T_{\mathbf{x}}$. As we vary $\mathbf{x}$, the control system $\mathbf{f}(\mathbf{x}, u)$ determines an n-parameter family of submanifolds, and we are interested in the action of feedback transformations on this family.

Let $\bar{\mathbf{x}} = \boldsymbol{\phi}(\mathbf{x})$, $\bar{u} = \psi(\mathbf{x}, u)$ be a feedback transformation taking $\dot{\mathbf{x}} = \mathbf{f}(\mathbf{x}, u)$ to $\dot{\mathbf{x}} = \bar{\mathbf{f}}(\mathbf{x}, \bar{u})$. To see how the submanifolds $\bar{u} \mapsto \bar{\mathbf{f}}(\bar{\mathbf{x}}, \bar{u})$ are related to the original submanifolds, we need only apply the chain rule. Since

$$\bar{\mathbf{f}} = \frac{d\bar{\mathbf{x}}}{dt} = \frac{\partial\boldsymbol{\phi}}{\partial\mathbf{x}} \cdot \frac{d\mathbf{x}}{dt} = \frac{\partial\boldsymbol{\phi}}{\partial\mathbf{x}} \cdot \mathbf{f},$$

we see that applying the nonsingular matrix $\partial\boldsymbol{\phi}/\partial\mathbf{x}$ to the image of $u \mapsto \mathbf{f}(\mathbf{x}, u)$ yields the image of $\bar{u} \mapsto \bar{\mathbf{f}}(\bar{\mathbf{x}}, \bar{u})$; moreover we have also reparametrized the image with the new parameter $\bar{u} = \psi(\mathbf{x}, u)$. Since $\boldsymbol{\phi}(\mathbf{x})$ can be any diffeomorphism, $\partial\boldsymbol{\phi}/\partial\mathbf{x}$ can be any GL(n,$\mathbf{R}$) matrix at $\mathbf{x}$. Together, these observations show that

1991 *Mathematics Subject Classification.* 49,53.

Key words and phrases. Feedback equivalence, control systems, centro-affine geometry, equivalence method, moving frames.

Supported by the Ministry of Colleges and Universities of Ontario and the Natural Sciences and Engineering Research Council of Canada the feedback transformations induce an arbitrary reparametrization and an arbitrary general linear action on each submanifold $u \mapsto \mathbf{f}(\mathbf{x}, u)$, which introduces the problem of classifying the general linear orbits of submanifolds of $\mathbf{R}^n$.

For this classical submanifold problem, we view GL(n,**R**) as a group of motions on $\mathbf{R}^n$ and try to classify submanifolds of $\mathbf{R}^n$ up to a GL(n,**R**)-motion. This problem is exactly analogous to the problem of classifying submanifolds of $\mathbf{R}^n$ up to a motion of the Euclidean group in $\mathbf{R}^n$. The term *centro-affine geometry* is used to describe the geometry on $\mathbf{R}^n$ induced by GL(n,**R**), just as Euclidean geometry is used to describe the geometry on $\mathbf{R}^n$ induced by the Euclidean group. Once we know the centro-affine invariants of submanifolds of $\mathbf{R}^n$, our expectation is that these invariants should also appear in some form as feedback invariants of a control system. When we know how to recognize them, we will be able to assign geometric meaning to them and use our centro-affine insight to interpret them. We will show exactly how this occurs in the special case of a scalar control system in the plane.

In n-dimensional Euclidean geometry, we often make use of frame fields to study submanifolds. For example, the Frenet frame on a regular curve is a Euclidean invariant of the curve, and this frame determines the arc-length and $n-1$ curvatures of the curve. These functions are also Euclidean invariants, and they classify the curve up to a Euclidean motion. In an entirely analogous manner, we can associate a GL(n,**R**)-invariant frame field to a curve, and this frame field determines an "arc-length" and $n-1$ "curvatures" which classify the curve up to GL(n,**R**)-motions [**GW97**].

In section 1 we use Cartan's method of moving frames [**Ca51**] to construct a centro-affine invariant frame field for a curve in the plane. For a regular class of curves we are able to find a GL(2,**R**)-invariant curve parameter, which we call the centro-affine arc-length, and an invariant function we call the centro-affine curvature. The arc-length and curvature determine the curve up to a centro-affine motion. In section 2, we show how a general nonlinear feedback transformation induces centro-affine actions on a family of curves naturally associated to a scalar control system in the plane. Then we show exactly how the centro-affine invariants of plane curves appear as feedback invariants of the control system. In effect, we solve the problem of finding a complete set of invariants for feedback equivalence. Finally, in section 3, we consider some examples and exhibit some normal forms.

1. Centro-affine geometry in the punctured plane.

Moving frames on the punctured plane. The method of moving frames is a technique that is well suited to the study of submanifolds of a homogeneous space. In [**Ca51**], Cartan shows that when a Lie group acts transitively and effectively on a manifold, one can construct a bundle of frames over the manifold. Cartan develops this theory to study two submanifold problems: the problem of contact and the problem of equivalence. The first problem involves determining the order of contact two submanifolds have at a point. The second problem involves determining when there exists an element of the given Lie group that translates one submanifold onto another. Cartan devotes most of his book to specific geometric examples of these problems. More recent descriptions of the method of moving frames can be found in [**Gn78**], [**Gr74**] and an excellent recent book [**Sh96**]. Green's paper [**Gn78**] and Sharpe's book [**Sh96**] carefully point out subtle features in the theory and include many examples. Our goal in this section will be to classify planar curves up to general linear actions.

We define the *punctured plane* to be the plane of column vectors, $\mathbf{R}^2$, with the origin removed; and we will denote it by $\mathbf{R}^2_0$. Since the group GL(2,$\mathbf{R}$) acts transitively and effectively on $\mathbf{R}^2_0$, we can use the method of moving frames to study the GL(2,$\mathbf{R}$)-invariants of curves in the punctured plane. A frame on $\mathbf{R}^2_0$ will consist of a pair of linearly independent vectors in $\mathbf{R}^2_0$, $(\mathbf{e}_0, \mathbf{e}_1)$. Let $\mathcal{F}$ be the set of all frames on $\mathbf{R}^2_0$. We may also think of the frame $(\mathbf{e}_0, \mathbf{e}_1)$ as the GL(2,$\mathbf{R}$) matrix $({e^i}_j)_{0\le i,j\le 1}$, where $\mathbf{e}_j$ is the column vector $\mathbf{e}_j = {}^t({e^0}_j, {e^1}_j)$ (we use ^{t} on the left to denote transpose). Now GL(2,$\mathbf{R}$) is an open subset of the four dimensional vector space L(2,$\mathbf{R}$) of all two by two matrices. The projections

$$\begin{array}{ccc} & \mathrm{L}(2,\mathbf{R}) & \\ \mathbf{e}_0 \swarrow & & \searrow \mathbf{e}_1 \\ \mathbf{R}^2 & & \mathbf{R}^2 \end{array}$$

from L(2,$\mathbf{R}$) onto the left and right columns of the matrices are linear maps, and therefore their restrictions to any open subset are differentiable. In particular, we have two differentiable mappings $\mathbf{e}_0, \mathbf{e}_1 \colon \mathcal{F} \to \mathbf{R}^2_0$. Expressing the derivatives of these mappings in terms of themselves gives the structure equations on $\mathcal{F}$:

$$\begin{aligned} d\mathbf{e}_0 &= \mathbf{e}_0\, {\omega^0}_0 + \mathbf{e}_1\, {\omega^1}_0 \\ d\mathbf{e}_1 &= \mathbf{e}_0\, {\omega^0}_1 + \mathbf{e}_1\, {\omega^1}_1 \\ d{\omega^i}_j &= -{\omega^i}_0 \wedge {\omega^0}_j - {\omega^i}_1 \wedge {\omega^1}_j \qquad 0 \le i,j \le 1. \end{aligned} \tag{1}$$

In these equations we view $\mathbf{e}_0$ and $\mathbf{e}_1$ as functions from $\mathcal{F}$ to $\mathbf{R}^2$. The equations for $d{\omega^i}_j$ are consequences of $d^2\mathbf{e}_0 = d^2\mathbf{e}_1 = 0$. The 1-forms ${\omega^0}_0, {\omega^0}_1, {\omega^1}_0, {\omega^1}_1$ form a basis for the 1-forms on $\mathcal{F}$.

Zeroth order frames. Let $I \subset \mathbf{R}$ be an open interval and let $\mathbf{x} \colon I \to \mathbf{R}^2_0$ be a smooth immersed curve. We call

$$\{(u; \mathbf{e}_0, \mathbf{e}_1) \in I \times \mathcal{F} \mid \mathbf{e}_0 = \mathbf{x}(u)\}$$

the set of *zeroth order frames* for the curve $\mathbf{x}$, and we denote this set by $\mathcal{F}^{(0)}_{\mathbf{x}}$. We think of the zeroth order frames as the restriction of the frames on $\mathbf{R}^2_0$ to the curve $\mathbf{x}$ by requiring that the first leg of the frame, $\mathbf{e}_0$, be the position vector of the curve; that is we let $\mathbf{e}_0(u) = \mathbf{x}(u)$ for all $u \in I$. If we differentiate this equation and use the structure equations (1), we get that

$$\mathbf{x}'(u)\, du = d\mathbf{x} = d\mathbf{e}_0 = \mathbf{e}_0\, {\omega^0}_0 + \mathbf{e}_1\, {\omega^1}_0. \tag{2}$$

This implies that ${\omega^0}_0$ and ${\omega^1}_0$ are multiples of du. In Cartan's language, the curve parameter u is called a *principal parameter*, and the components of $d\mathbf{e}_j$ that are multiples of du are called *principal components*. We want to choose the second leg of the frame, $\mathbf{e}_1$, in a way that is adapted to the geometry of the curve. The natural thing to do is to choose $\mathbf{e}_1$ to be tangent to the curve, however this is not always possible. Since $\mathbf{e}_0$ and $\mathbf{e}_1$ must be linearly independent, we can require $\mathbf{e}_1$ to be tangent to the curve if and only if $\mathbf{x}(u)$ and $\mathbf{x}'(u)$ are linearly independent. Since $\mathbf{x}(u) \wedge \mathbf{x}'(u)\, du = \mathbf{e}_0 \wedge d\mathbf{e}_0 = \mathbf{e}_0 \wedge \mathbf{e}_1\, {\omega^1}_0$, we see that $\mathbf{x}(u)$ and $\mathbf{x}'(u)$ are linearly independent if and only if ${\omega^1}_0 \neq 0$. Since $\mathbf{x}(u)$ and $\mathbf{x}'(u)$ are linearly dependent if and only if the curve is tangent to the ray through $\mathbf{x}(u)$ emanating from the origin, we see that ${\omega^1}_0$ vanishes exactly when the curve is tangent to radial directions. We will consider the two extreme cases for this condition: ${\omega^1}_0$ either vanishes or is nonzero for every point on the curve. We will encounter several similar conditions in the rest of this paper. We will use the term *regular* to mean that a condition is satisfied for every point in the domain, and we will always restrict our consideration to regular cases. If we consider the regular case satisfying ${\omega^1}_0 = 0$ for every point on the curve, then $d\mathbf{e}_0 \wedge \mathbf{e}_0 = 0$ for all $u \in I$. This implies that the ray through $\mathbf{e}_0$ is constant, and we see that the curve $\mathbf{x}(u)$ lies on a ray.

First order frames. We now restrict to the regular case where ${\omega^1}_0 \neq 0$ for all $u \in I$. In this case we can define the first order frames $\mathcal{F}_{\mathbf{x}}^{(1)}$ to be the set

$$\mathcal{F}_{\mathbf{x}}^{(1)} = \{(u; \mathbf{e}_0, \mathbf{e}_1) \in \mathcal{F}_{\mathbf{x}}^{(0)} \mid \mathbf{e}_1 \text{ points in the direction of } \mathbf{x}'(u)\}.$$

REMARK. This choice of first order frames is adapted to *oriented* curves. If we allow the curve to reverse orientation, then we can only require that $\mathbf{e}_1$ be tangent to the curve, but not that it point in a particular direction.

Since $\mathbf{x}'$ is parallel to $\mathbf{e}_1$, equation (2) implies that $d\mathbf{e}_0 = \mathbf{e}_1 {\omega^1}_0$; equivalently ${\omega^0}_0 = 0$. This shows that ${\omega^1}_0$ is a non-zero multiple of du, so a 1-form on $\mathcal{F}_{\mathbf{x}}^{(1)}$ is a multiple of du if and only if it is a multiple of ${\omega^1}_0$. In particular, principal components must be multiples of ${\omega^1}_0$. Differentiating the last relation and using equations (1) gives $0 = d{\omega^0}_0 = -{\omega^0}_1 \wedge {\omega^1}_0$. Therefore ${\omega^0}_1$ is a principal component and we may write ${\omega^0}_1 = a\,{\omega^1}_0$, for some function a. A subgroup of GL(2,$\mathbf{R}$) acts on the function a, and we can compute the infinitesimal action by differentiating both sides of the last equation. (See pages 40–43 of [**Ga89**] for a discussion of group actions and how to identify them from their infinitesimal actions.) Differentiating both sides, we get

$$\begin{aligned} d{\omega^0}_1 &= -{\omega^0}_1 \wedge {\omega^1}_1 \\ &= -a\,{\omega^1}_0 \wedge {\omega^1}_1 \qquad \text{and} \\ d(a\,{\omega^1}_0) &= da \wedge {\omega^1}_0 - a\,{\omega^1}_1 \wedge {\omega^1}_0. \end{aligned}$$

Subtracting the second equation from the third equation we arrive at the equation $0 = (da - 2a\,{\omega^1}_1) \wedge {\omega^1}_0$. This shows that the function a is acted on by a square. We can in fact compute the function a explicitly. The first order frames are parametrized by $I \times \mathbf{R}_+$, where (u, ν) determines the frame $(\mathbf{e}_0, \mathbf{e}_1) = (\mathbf{x}(u), \mathbf{x}'(u)\,\nu^{-1})$, $\nu > 0$. Hence, $d\mathbf{e}_0 = \mathbf{x}'(u)\,du = (\mathbf{x}'(u)\,\nu^{-1})(\nu\,du) = \mathbf{e}_1\,\nu\,du = \mathbf{e}_1\,{\omega^1}_0$, and we see that ${\omega^1}_0 = \nu\,du$. By taking determinants we can isolate ${\omega^0}_1$ in the structure equation for $d\mathbf{e}_1$. Explicitly,

$$\det(d\mathbf{e}_1, \mathbf{e}_1) \overset{\text{by (1)}}{=} \det(\mathbf{e}_0, \mathbf{e}_1)\,{\omega^0}_1 = \det(\mathbf{x}(u), \mathbf{x}'(u))\,\nu^{-1}\,{\omega^0}_1.$$

We also have that

$$\det(d\mathbf{e}_1, \mathbf{e}_1) = \det(d(\mathbf{x}'(u)\,\nu^{-1}), \mathbf{x}'(u)\,\nu^{-1}) = \det(\mathbf{x}''(u), \mathbf{x}'(u))\,\nu^{-2}\,du,$$

so

$${\omega^0}_1 = \frac{\det(\mathbf{x}''(u), \mathbf{x}'(u))}{\det(\mathbf{x}(u), \mathbf{x}'(u))}\,\nu^{-1}\,du = \frac{\det(\mathbf{x}''(u), \mathbf{x}'(u))}{\det(\mathbf{x}(u), \mathbf{x}'(u))}\,\nu^{-2}\,{\omega^1}_0 = a\,{\omega^1}_0.$$

We see that the function $a = [\det(\mathbf{x}''(u), \mathbf{x}'(u))/\det(\mathbf{x}(u), \mathbf{x}'(u))]\,\nu^{-2}$, and the square action of ν^{-1} is evident. From this formula it is clear that a vanishes if and only if $\mathbf{x}''(u)$ and $\mathbf{x}'(u)$ are linearly dependent, which occurs exactly at the inflection points of the curve. If we consider the regular case where a is identically zero, we see that this happens if and only if every point on the curve is an inflection point. This means that the curve must lie on a line, and since ${\omega^1}_0 \neq 0$, this line can not pass through the origin. Thus a identically zero classifies the affine lines.

REMARK. We do not need the explicit formula for a to see that the vanishing of a determines affine lines. This is because the vanishing of a is equivalent to the condition $d\mathbf{e}_1 \wedge \mathbf{e}_1 = 0$. Since $\mathbf{e}_1$ determines the direction of the tangent line, this equation says that the tangent line is constant and therefore $\mathbf{x}(u)$ lies on a line.

Since a is acted on by a square, the sign of a is an invariant. The formula for a makes it clear that a will be positive when the position vector $\mathbf{x}(u)$ and the acceleration vector $\mathbf{x}''(u)$ point to the same side of the tangent line, and a will be negative when the acceleration vector and the position vector point to opposite sides of the tangent line. In other words, that a is negative when the curve is convex, and a is positive when the curve is concave.

Second order frames. We now consider the last regular case, namely the case where $a \neq 0$ for every point on the curve. In this case we can determine a unique set of second order frames, $\mathcal{F}^{(2)}_{\mathbf{x}}$, by choosing the first order frames for which $a = -\epsilon$, where $\epsilon = 1$ if the curve is convex and $\epsilon = -1$ if the curve is concave. With $a = -\epsilon$, the parameter ν takes on the value $\nu = \sqrt{|\det(\mathbf{x}''(u), \mathbf{x}'(u))/\det(\mathbf{x}(u), \mathbf{x}'(u))|}$, and the frame $(\mathbf{e}_0, \mathbf{e}_1) = (\mathbf{x}(u), \mathbf{x}'(u)\,\nu^{-1})$ is uniquely defined along the curve. The 1-form ${\omega^1}_0 = \nu\, du = \sqrt{|\det(\mathbf{x}''(u), \mathbf{x}'(u))/\det(\mathbf{x}(u), \mathbf{x}'(u))|}\, du$ is well defined on the curve, and determines the *centro-affine arc-length* of the curve. We will denote the centro-affine arc-length parameter by s, which implies that ${\omega^1}_0 = ds$. Notice that the centro-affine arc-length is second order in $\mathbf{x}(u)$, as is the vector $\mathbf{e}_1$. The 1-form ${\omega^1}_1$, which is obtained by differentiating $\mathbf{e}_1$, must be a multiple of ds. We will write ${\omega^1}_1 = \kappa(s)\, ds$, and we will call κ the *centro-affine curvature* of the curve. Notice that κ is third order.

The structure equations (1) reduce to the Frenet type equations

$$
\begin{aligned}
d\mathbf{e}_0 &= \qquad\qquad \mathbf{e}_1\, ds \\
d\mathbf{e}_1 &= -\epsilon\, \mathbf{e}_0\, ds + \mathbf{e}_1\, \kappa\, ds.
\end{aligned}
\tag{3}
$$

A standard argument shows that the centro-affine arc-length and the centro-affine curvature determine the curve $\mathbf{x}(u)$ up to a GL(2,**R**)-motion. It is also easy to check that if you reverse the orientation of the curve, then κ changes sign. Thus the centro-affine arc-length and the absolute value of κ determine unoriented curves up to a GL(2,**R**)-motion.

Structure theory of centro-affine curves. We can collect the previous results into a classification theorem for centro-affine curves. Essentially the same results appear in the Shirokows' book on affine geometry [**SS62,pp. 49–51, pp. 57–61**]. They attribute these results to Mayer and Myller [**MM33**].

THEOREM (MAYER AND MYLLER). *Let $I \subset \mathbf{R}$ be an open interval and let $\mathbf{x}\colon I \to \mathbf{R}^2_0$ be a smooth immersed curve. We have the following three regular classes of curves.*

- *If $\mathbf{x}(u)$ and $\mathbf{x}'(u)$ are linearly dependent for all u in I, then the curve lies on a ray emanating from the origin.*

- *If* $\mathbf{x}(u)$ *and* $\mathbf{x}'(u)$ *are linearly independent and* $\mathbf{x}'(u)$ *and* $\mathbf{x}''(u)$ *are linearly dependent for all* u *in* I*, then the curve lies on an affine line.*
- *If* $\mathbf{x}(u)$ *and* $\mathbf{x}'(u)$ *are linearly independent and* $\mathbf{x}'(u)$ *and* $\mathbf{x}''(u)$ *are linearly independent for all* u *in* I*, then the curve possesses a* $GL(2,\mathbf{R})$*-invariant arc-length* s *and a* $GL(2,\mathbf{R})$*-invariant frame* $(\mathbf{e}_0(s), \mathbf{e}_1(s))$ *satisfying structure equations (3). The arc-length* s *and the curvature* $\kappa(s)$ *determine the curve up to a* $GL(2,\mathbf{R})$*-motion.*

The following observation provides some insight to the meaning of the arc-length parameter s. If the position vector and the velocity vector of the curve are linearly independent, then we can use $\mathbf{x}(u)$ and $\mathbf{x}'(u)$ as a basis for $\mathbf{R}^2_0$. Therefore, we can write the acceleration vector in terms of this basis, giving

$$\mathbf{x}''(u) = \alpha(u)\,\mathbf{x}(u) + \beta(u)\,\mathbf{x}'(u).$$

If we reparametrize the curve, letting $u = u(t)$, then we get

$$\mathbf{x}''(t) = \hat{\alpha}(t)\,\mathbf{x}(t) + \hat{\beta}(t)\,\mathbf{x}'(t), \quad \text{where } \hat{\alpha}(t) = u'(t)^2\,\alpha(u(t)).$$

We can always reparametrize the curve so that $\hat{\alpha}(t) = -\epsilon$, $\epsilon = \pm 1$. With this parametrization we have $\mathbf{x}''(t) = -\epsilon\,\mathbf{x}(t) + \hat{\beta}(t)\,\mathbf{x}'(t)$. If we write this as a system of equation, $\mathbf{x}'(t) = \mathbf{e}_1(t)$, $\mathbf{e}_1'(t) = -\epsilon\,\mathbf{x}(t) + \hat{\beta}(t)\,\mathbf{e}_1(t)$, we see that this system exactly matches the structure equations (3) (recall that $\mathbf{e}_0$ is the position vector, $\mathbf{x}$). Thus t must be the arc-length parameter s and $\hat{\beta}$ must be the curvature κ. We see that the arc-length parameter is the one which normalizes the coefficient of $\mathbf{x}$ when we express the acceleration vector in terms of the position vector and the velocity vector. The coefficient of $\mathbf{x}'$ is then the curvature. Another interpretation of the arc-length parameter is that it is the parameter for which $\det(\mathbf{x}''(s), \mathbf{x}'(s)) = -\epsilon\,\det(\mathbf{x}(s), \mathbf{x}'(s))$, since in a general parameter we would have $\det(\mathbf{x}''(u), \mathbf{x}'(u)) = \alpha(u)\,\det(\mathbf{x}(u), \mathbf{x}'(u))$.

The previous paragraph gives us one interpretation of the curvature κ, it is the coefficient of the velocity vector when we express the acceleration in terms of the position vector and the velocity vector (all derivatives are with respect to arc-length). However, this is not a very satisfying description of the curvature. To get a better description of the curvature, notice that the frame $(\mathbf{e}_0(s), \mathbf{e}_1(s))$ determines the volume element $\mathbf{e}_0 \wedge \mathbf{e}_1$ at each point of the curve. If we differentiate this volume element with respect to arc-length and use the structure equations (3), we get

$$d(\mathbf{e}_0 \wedge \mathbf{e}_1) = d\mathbf{e}_0 \wedge \mathbf{e}_1 + \mathbf{e}_0 \wedge d\mathbf{e}_1 = \mathbf{e}_0 \wedge \mathbf{e}_1\,\kappa(s)\,ds.$$

We see that the curvature is the relative rate of change of the volume element $\mathbf{e}_0 \wedge \mathbf{e}_1$ with respect to arc-length. In terms of determinants, this is equivalent to $\kappa(s) = d[\ln|\det(\mathbf{x}(s), \mathbf{x}'(s))|]/ds$. Those curves for which κ is identically zero satisfy Kepler's second law of sweeping out equal area in equal time, although for centro-affine curves time is replaced by arc-length.

We can give a good description of the curves with constant centro-affine curvature. Structure equations (3) imply that a centro-affine curve must satisfy the second order differential equation $\mathbf{x}''(s) = \kappa\,\mathbf{x}'(s) - \epsilon\,\mathbf{x}(s)$. The characteristic equation for this differential equation is $r^2 - \kappa\,r + \epsilon = 0$. The two roots λ_1 and λ_2 are such that $\lambda_1 + \lambda_2 = \kappa$ and $\lambda_1\,\lambda_2 = \epsilon$.

In the concave case, $\epsilon = -1$ which means that the roots have opposite sign. Let λ_1 be the positive root. There are two vectors $\mathbf{w}_1$ and $\mathbf{w}_2$ such that $\mathbf{x}(s) =$

$\mathbf{w}_1\, e^{\lambda_1\, s} + \mathbf{w}_2\, e^{\lambda_2\, s}$. Since $\mathbf{x}(0)$ and $\mathbf{x}'(0)$ must be linearly independent, $\mathbf{w}_1$ and $\mathbf{w}_2$ must also be linearly independent. Thus, we can apply a GL(2,$\mathbf{R}$) matrix to the curve and put it in the standard form $\mathbf{x}(s) = {}^t(1,0)\, e^{\lambda_1\, s} + {}^t(0,1)\, e^{\lambda_2\, s}$. We see that the curve is a saddle solution to a linear ordinary differential equation, and up to a general linear action lies on the curve $y^{\lambda_1} = x^{\lambda_2}$, $\lambda_1\, \lambda_2 = -1$, $\lambda_1 + \lambda_2 = \kappa$.

The convex case branches into three subcases depending on the sign of $\kappa^2 - 4$. If $|\kappa| > 2$, then we have distinct real roots λ_1 and λ_2 and the solution curve can be put in the standard form $\mathbf{x}(s) = {}^t(1,0)\, e^{\lambda_1\, s} + {}^t(0,1)\, e^{\lambda_2\, s}$, $\lambda_1\, \lambda_2 = 1$, $\lambda_1 + \lambda_2 = \kappa$. We see that the curve is a nodal solution to a linear ordinary differential equation, and up to a general linear action lies on the curve $y^{\lambda_1} = x^{\lambda_2}$.

If $|\kappa| = 2$, then we have two equal real roots λ. Since $2\lambda = \kappa$, λ is either 1 or -1. The solution curve can be put in the standard form $\mathbf{x}(s) = {}^t(1,0)\, e^{\lambda\, s} + {}^t(0,1)\, s\, e^{\lambda\, s}$. We see that the curve is an improper node, and lies on the curve $\lambda\, y = x\, \ln(x)$. If $\kappa = 2$, then $\lambda = 1$ and $\mathbf{x}(s)$ moves away from the origin, while if $\kappa = -2$, then $\lambda = -1$ and $\mathbf{x}(s)$ moves toward from the origin.

If $|\kappa| < 2$, then we have a non-real conjugate pair of roots, λ and $\bar{\lambda}$, with κ being two times the real part of λ. If we let λ be the root with positive imaginary part, then $\lambda = \kappa/2 + \sqrt{-1}\, b$, where b is the positive root of the equation $\kappa^2/4 + b^2 = 1$. We can put the solution curve into the standard form $\mathbf{x}(s) = e^{\kappa\, s/2}[{}^t(1,0)\, \cos(b\, s) + {}^t(0,1)\, \sin(b\, s)]$. If $|\kappa| > 0$ we have a spiral solution which moves away from the origin if κ is positive and moves towards the origin if κ is negative. When $\kappa = 0$ we have a circle centered at the origin.

Notice that when $\kappa = 0$, our curve is a hyperbola in the concave case and a circle in the convex case. Thus $\kappa = 0$ exactly picks out the non-degenerate quadric curves centered at the origin. This fits in with the centro-affine theory of hypersurfaces. In this lowest dimensional case, the symmetric form $\omega^0{}_1\, \omega^1{}_0 = -\epsilon\, (ds)^2$ is the Blaschke form, and $\kappa\, (ds)^3$ is the Pick form [**Bl23**]. It is exactly the vanishing of the Pick form that characterizes the non-degenerate quadric hypersurfaces.

2. Two state scalar control

Our goal in this section is to use our new intuition for centro-affine curves to compute the feedback invariants of a two state, single input control system. We will also have the benefit of a geometric interpretation of these invariants. The motivation for considering centro-affine geometry comes from viewing a control system as a submanifold of the tangent bundle, and observing the effect that feedback has on this submanifold.

Consider the control system

$$\dot{\mathbf{x}} = \mathbf{f}(\mathbf{x}, u), \tag{4}$$

where $\mathbf{x} = {}^t(x^1, x^2) \in \mathbf{R}^2$, $u \in \mathbf{R}$ and $\mathbf{f}(x^1, x^2, u)$ is a smooth function from $\mathbf{R}^2 \times \mathbf{R}$ to $\mathbf{R}^2$. For each $(\mathbf{x}, u)$, $\mathbf{f}(\mathbf{x}, u)$ represents a tangent to vector to $\mathbf{R}^2$ at $\mathbf{x}$. Thus the mapping from $\mathbf{R}$ to $T_{\mathbf{x}}\mathbf{R}^2$ defined by $u \mapsto \mathbf{f}(\mathbf{x}, u)$ is a curve in the tangent space to $\mathbf{x}$. Therefore the mapping $V\colon \mathbf{R}^2 \times \mathbf{R} \to T\mathbf{R}^2$ defined by $(\mathbf{x}, u) \mapsto (\mathbf{x}, \mathbf{f}(\mathbf{x}, u))$ gives a submanifold of the tangent bundle whose intersection with each tangent space is a curve parametrized by u.

Let $(\bar{\mathbf{x}}, \bar{u}) = \mathbf{\Phi}(\mathbf{x}, u) = (\phi(\mathbf{x}), \psi(\mathbf{x}, u))$ be a feedback transformation and apply it to the control system (4). Using the chain rule we compute the new control system

$$\bar{\mathbf{f}}(\bar{\mathbf{x}},\bar{u}) = \frac{d\bar{\mathbf{x}}}{dt} = \frac{\partial\phi}{\partial\mathbf{x}}\cdot\frac{d\mathbf{x}}{dt} = \frac{\partial\phi}{\partial\mathbf{x}}\cdot\mathbf{f}(\mathbf{x},u),$$

and since $\phi(\mathbf{x})$ can be an arbitrary diffeomorphism, we see that the curve $\mathbf{f}(\mathbf{x},u)$ can be acted on by an arbitrary GL(2,$\mathbf{R}$) matrix. Further, since $\bar{u} = \psi(\mathbf{x},u)$, we see that feedback allows the curve $\mathbf{f}(\mathbf{x},u)$ to be reparametrized. Thus the action on the curve is a centro-affine action, and we should expect to see centro-affine invariants in the feedback invariants of the control system (4).

To make effective use of the previous section, we introduce the frame bundle on $\mathbf{R}^2$. Let $\mathcal{F}$ denote the bundle of frames on $\mathbf{R}^2$. A point $e \in \mathcal{F}$ is a linear isomorphism from a tangent space $T_{\mathbf{x}}\mathbf{R}^2$ to $\mathbf{R}^2$. The mapping $e \mapsto \mathbf{x}$ is the projection from $\mathcal{F}$ to $\mathbf{R}^2$. The point e determines a triple $(\mathbf{x},\mathbf{e}_0,\mathbf{e}_1)$, where $(\mathbf{e}_0,\mathbf{e}_1)$ is the basis of $T_{\mathbf{x}}\mathbf{R}^2$ that e maps to the standard basis of $\mathbf{R}^2$. In other words, $e(\mathbf{e}_0) = {}^t(1,0)$ and $e(\mathbf{e}_1) = {}^t(0,1)$. We may view $\mathbf{x}, \mathbf{e}_0$ and $\mathbf{e}_1$ as functions from $\mathcal{F}$ to $\mathbf{R}^2$. The derivatives of these maps determine the *structure equations* of $\mathcal{F}$,

$$\begin{aligned}
d\mathbf{x} &= \mathbf{e}_0\sigma^0 + \mathbf{e}_1\sigma^1\\
d\mathbf{e}_0 &= \mathbf{e}_0\omega^0{}_0 + \mathbf{e}_1\omega^1{}_0\\
d\mathbf{e}_1 &= \mathbf{e}_0\omega^0{}_1 + \mathbf{e}_1\omega^1{}_1\\
d\sigma^0 &= -\omega^0{}_0\wedge\sigma^0 - \omega^0{}_1\wedge\sigma^1\\
d\sigma^1 &= -\omega^1{}_0\wedge\sigma^0 - \omega^1{}_1\wedge\sigma^1\\
d\omega^i{}_j &= -\omega^i{}_0\wedge\omega^0{}_j - \omega^i{}_1\wedge\omega^1{}_j \quad 0\le i,j\le 1.
\end{aligned} \tag{5}$$

The six 1-forms $\sigma^0,\sigma^1,\omega^0{}_0,\omega^0{}_1,\omega^1{}_0,\omega^1{}_1$ form a basis for the 1-forms on $\mathcal{F}$. Notice that the structure equations for $\mathbf{e}_0$, $\mathbf{e}_1$ and $\omega^0{}_0,\omega^0{}_1,\omega^1{}_0,\omega^1{}_1$ agree with (1) and reflect the GL(2,$\mathbf{R}$), or centro-affine, structure on each tangent space.

Zeroth order frames. The first step for centro-affine curves is to let $\mathbf{e}_0$ be the position vector of the curve. We apply the same technique here and define the zeroth order frame bundle $\mathcal{F}^{(0)}_{\mathbf{f}}$ to be the set of all frames with $\mathbf{e}_0 = \mathbf{f}(\mathbf{x},u)$. Thus a zeroth order frame can be viewed as a triple $(\mathbf{x},\mathbf{f}(\mathbf{x},u),\mathbf{e}_1)$, where $\mathbf{f}$ and $\mathbf{e}_1$ are linearly independent. Notice that the formula for $d\mathbf{x}$ in (5) becomes $d\mathbf{x} = \mathbf{f}\,\sigma^0 + \mathbf{e}_1\sigma^1$, which implies, on using $\langle,\rangle$ to denoting the pairing of vectors with forms, that $\langle\mathbf{f},\sigma^0\rangle = 1$ and $\langle\mathbf{f},\sigma^1\rangle = 0$.

We wish now to compute the centro-affine invariants of the curves in each tangent space. Since a tangent space is determined by setting $\mathbf{x}$ equal to a constant, all we have to do is compute the derivatives modulo dx^1 and dx^2.

Since $\mathbf{e}_0 = \mathbf{f}$, the equation for $d\mathbf{e}_0$ in (5) shows that $\omega^0{}_0$ and $\omega^1{}_0$ are multiples of du, dx^1 and dx^2. We therefore know that $\omega^0{}_0 \equiv m^0 du$ and $\omega^1{}_0 \equiv m^1 du$ modulo dx^1, dx^2. We would like to choose $\mathbf{e}_1$ so that it points in the same direction as $\partial\mathbf{f}/\partial u$, but since $\mathbf{e}_0$ and $\mathbf{e}_1$ must be linearly independent, we can only do that if $\mathbf{e}_0 = \mathbf{f}$ and $\partial\mathbf{f}/\partial u$ are linearly independent. Being independent is equivalent to $\omega^1{}_0$ being nonzero modulo dx^1, dx^2, which is the same as the function m^1 being nonzero. If m^1 is identically zero, then $\mathbf{f}(\mathbf{x},u)$ lies on a ray emanating from the origin of $T_{\mathbf{x}}\mathbf{R}^2$, for each $\mathbf{x}\in\mathbf{R}^2$. Thus there is a feedback transformation taking $\mathbf{f}(\mathbf{x},u)$ to a control linear normal form $\mathbf{g}(\mathbf{y})\,v$, where $(\mathbf{y},v) = \mathbf{\Phi}(\mathbf{x},u)$.

First order frames. We will assume that m^1 is always nonzero. Then we may define the first order frames $\mathcal{F}^{(1)}_{\mathbf{f}}$ by requiring $\mathbf{e}_1$ to point in the same direction as

$\partial \mathbf{f}/\partial u$, which implies that $\omega^0{}_0 \equiv 0 \pmod{dx^1, dx^2}$. As in the curve case, this condition implies that $\omega^0{}_1 \equiv a\,\omega^1{}_0 \pmod{dx^1, dx^2}$. Differentiation shows that a is acted on by a square, and we have three cases. If a is identically zero, then $\mathbf{f}(\mathbf{x}, u)$ lies on an affine line in each tangent space and thus there is a feedback transformation taking $\mathbf{f}(\mathbf{x}, u)$ to a control linear normal form $\mathbf{g}_0(\mathbf{y}) + \mathbf{g}_1(\mathbf{y})\,v$. We will now consider the final case.

Second order frames. If a is never zero, then we may define the second order frames $\mathcal{F}^{(2)}_{\mathbf{f}}$ by choosing $\mathbf{e}_1$ so that $a = -\epsilon$, where $\epsilon = \pm 1$. This determines the centro-affine arc-length in each tangent space. In fact, using the arc-length formula derived in the previous section, we see that

$$\omega^1{}_0 \equiv \sqrt{\left| \frac{\det(\frac{\partial^2 \mathbf{f}}{\partial u^2}, \frac{\partial \mathbf{f}}{\partial u})}{\det(\mathbf{f}, \frac{\partial \mathbf{f}}{\partial u})} \right|}\, du \qquad \pmod{dx^1, dx^2}.$$

Comparing again with centro-affine curves, we see that $\omega^1{}_1$ must be congruent to a multiple of $\omega^1{}_0$, say $\omega^1{}_1 \equiv \kappa(\mathbf{x}, u)\,\omega^1{}_0 \pmod{dx^1, dx^2}$, and for each tangent space $T_{\mathbf{x}}\mathbf{R}^2$, $\kappa(\mathbf{x}, u)$ is the centro-affine curvature of $\mathbf{f}(\mathbf{x}, u)$.

Taken together, we see that we may choose frames adapted to the control system $\mathbf{f}$ so that in each tangent space, the frames satisfy the structure equations for centro-affine plane curves. We can summarize the above congruences with the matrix equation

$$\begin{pmatrix} \omega^0{}_0 & \omega^0{}_1 \\ \omega^1{}_0 & \omega^1{}_1 \end{pmatrix} \equiv \begin{pmatrix} 0 & -\epsilon \\ 1 & \kappa(\mathbf{x}, u) \end{pmatrix} \omega^1{}_0 \qquad \pmod{dx^1, dx^2}.$$

Comparing the expression for $d\mathbf{x}$ in equations (5) with the expression

$$d\mathbf{x} = \frac{\partial}{\partial x^1}\, dx^1 + \frac{\partial}{\partial x^2}\, dx^2$$

we see that σ^0, σ^1 must be a nonsingular linear combination of dx^1, dx^2. Thus any congruence modulo dx^1, dx^2 is equivalent to a congruence modulo σ^0, σ^1. We can therefore rewrite the last equation as

$$\text{(6)} \qquad \begin{pmatrix} \omega^0{}_0 & \omega^0{}_1 \\ \omega^1{}_0 & \omega^1{}_1 \end{pmatrix} \equiv \begin{pmatrix} 0 & -\epsilon \\ 1 & \kappa(\mathbf{x}, u) \end{pmatrix} \omega^1{}_0 \qquad \pmod{\sigma^0, \sigma^1}.$$

Equation (6) shows how $\mathbf{f}$ varies in a given tangent space, but we also need to know how $\mathbf{f}$ varies as we move from tangent space to tangent space, that is as we vary the state variables $\mathbf{x}$. We do so by putting equation (6) back into the derivative equations for $d\sigma^0, d\sigma^1$ in (5). Writing these equations in matrix form, we have

$$\begin{pmatrix} d\sigma^0 \\ d\sigma^1 \end{pmatrix} = - \begin{pmatrix} 0 & -\epsilon \\ 1 & \kappa(\mathbf{x}, u) \end{pmatrix} \omega^1{}_0 \wedge \begin{pmatrix} \sigma^0 \\ \sigma^1 \end{pmatrix} + \begin{pmatrix} A\,\sigma^0 \wedge \sigma^1 \\ B\,\sigma^0 \wedge \sigma^1 \end{pmatrix}.$$

A short calculation shows that by adding terms to $\omega^1{}_0$ we can rewrite this as

$$\text{(7)} \quad \begin{pmatrix} d\sigma^0 \\ d\sigma^1 \end{pmatrix} = - \begin{pmatrix} 0 & -\epsilon \\ 1 & \kappa(\mathbf{x}, u) \end{pmatrix} [\omega^1{}_0 + \epsilon\, A\,\sigma^0 + (B + \epsilon\,\kappa\, A)\,\sigma^1] \wedge \begin{pmatrix} \sigma^0 \\ \sigma^1 \end{pmatrix}.$$

Notice that equation (6), which was arrived at by considering the centro-affine structure of $\mathbf{f}$ on each tangent space, only defines $\omega^1{}_0$ up to linear combinations of

σ^0 and σ^1. Thus we may add such linear terms to $\omega^1{}_0$ and preserve equation (6), equivalently we preserve the centro-affine structure equations for $\mathbf{f}$. We therefore define a new 1-form, $\mu = \omega^1{}_0 + \epsilon\, A\,\sigma^0 + (B + \epsilon\,\kappa\,A)\,\sigma^1$. Substituting into (7) gives

$$\begin{pmatrix} d\sigma^0 \\ d\sigma^1 \end{pmatrix} = -\begin{pmatrix} 0 & -\epsilon \\ 1 & \kappa(\mathbf{x}, u) \end{pmatrix} \mu \wedge \begin{pmatrix} \sigma^0 \\ \sigma^1 \end{pmatrix}. \tag{8}$$

It is easy to see that equation (8) determines μ uniquely. Suppose $\bar{\mu}$ also satisfied equation (8). Then setting the two expressions equal and subtracting gives two equations, $\epsilon(\bar{\mu} - \mu) \wedge \sigma^1 = 0$ and $(\bar{\mu} - \mu) \wedge \sigma^0 + \kappa(\bar{\mu} - \mu) \wedge \sigma^1 = 0$. The first equations implies that there is a function p such that $\bar{\mu} - \mu = p\,\sigma^1$. Substituting into the second equation gives $p\,\sigma^1 \wedge \sigma^0 = 0$, thus $p = 0$ and $\bar{\mu} = \mu$.

We are now at an important point. The three independent 1-forms, σ^0, σ^1 and μ, form a basis for the 1-forms on state-control space, $\mathbf{R}^2 \times \mathbf{R}$. Moreover, this basis remains invariant under the action of feedback transformations. We see this in the following way. To each control system, we assign the centro-affine "Frenet frame" $(\mathbf{e}_0, \mathbf{e}_1)$ in each tangent space. Since a feedback transformation induces a centro-affine action in each tangent space, this frame is feedback invariant. The 1-forms σ^0, σ^1 are dual to the frame $(\mathbf{e}_0, \mathbf{e}_1)$, and are defined by the equation $d\mathbf{x} = \mathbf{e}_0\,\sigma^0 + \mathbf{e}_1\,\sigma^1$. Since the frame is feedback invariant, so are the 1-forms σ^0 and σ^1. Equation (8) determines the 1-form μ. Since feedback preserves σ^0 and σ^1, feedback must also preserve equation (8). Therefore, since equation (8) determines μ uniquely, feedback must also preserve μ.

Using (8) we compute that $0 = d^2\sigma^0 = \epsilon\, d\mu \wedge \sigma^1$, thus $d\mu$ has the form

$$d\mu = -K\,\sigma^0 \wedge \sigma^1 - J\,\mu \wedge \sigma^1.$$

For any function $h(x^1, x^2, u)$, the functions h_{σ^0}, h_{σ^1} and h_μ are uniquely defined by the equation $dh = h_{\sigma^0}\,\sigma^0 + h_{\sigma^1}\,\sigma^1 + h_\mu\,\mu$. With this notation and with equation (8) we compute

$$0 = d^2\sigma^1 = J\,\mu \wedge \sigma^1 \wedge \sigma^0 - d\kappa\, \wedge \mu \wedge \sigma^1 = (J - \kappa_{\sigma^0})\mu \wedge \sigma^1 \wedge \sigma^0.$$

From this we see that J is a derivative of κ. Finally, computing $d^2\mu = 0$ gives the relation

$$J_{\sigma^0} + \kappa\,K - K_\mu = 0.$$

The above discussion leads to the following structure theorem for two state, single input systems.

THEOREM. *There are three regular classes for the two state, single input control system (4).*

- *If $\mathbf{f}(\mathbf{x}, u)$ and $\frac{\partial \mathbf{f}}{\partial u}(\mathbf{x}, u)$ are linearly dependent for all $(\mathbf{x}, u)$, then in each tangent space $\mathbf{f}(\mathbf{x}, u)$ lies on a ray emanating from the origin and is therefore feedback equivalent to the form $\dot{\mathbf{y}} = \mathbf{g}(\mathbf{y})\,v$.*
- *If $\mathbf{f}(\mathbf{x}, u)$ and $\frac{\partial \mathbf{f}}{\partial u}(\mathbf{x}, u)$ are linearly independent and $\frac{\partial \mathbf{f}}{\partial u}(\mathbf{x}, u)$ and $\frac{\partial^2 \mathbf{f}}{\partial u^2}(\mathbf{x}, u)$ are linearly dependent for all $(\mathbf{x}, u)$, then in each tangent space $\mathbf{f}(\mathbf{x}, u)$ lies on an affine line and is therefore feedback equivalent to the form $\dot{\mathbf{y}} = \mathbf{g}_0(\mathbf{y}) + \mathbf{g}_1(\mathbf{y})\,v$.*
- *If $\mathbf{f}(\mathbf{x}, u)$ and $\frac{\partial \mathbf{f}}{\partial u}(\mathbf{x}, u)$ are linearly independent and $\frac{\partial \mathbf{f}}{\partial u}(\mathbf{x}, u)$ and $\frac{\partial^2 \mathbf{f}}{\partial u^2}(\mathbf{x}, u)$ are linearly independent for all $(\mathbf{x}, u)$, then in each tangent space we may assign a centro-affine frame $(\mathbf{e}_0, \mathbf{e}_1)$ to $\mathbf{f}$. This frame uniquely determines dual 1-forms σ^0 and σ^1 satisfying the equation $d\mathbf{x} = \mathbf{e}_0\,\sigma^0 + \mathbf{e}_1\,\sigma^1$. The*

1-form μ together with $\epsilon = \pm 1$ and the function κ are uniquely determined by equation (8). They additionally satisfy the equations

$$\left.\begin{aligned} d\mathbf{e}_0 &\equiv \mathbf{e}_1\,\mu \\ d\mathbf{e}_1 &\equiv -\mathbf{e}_0\,\epsilon\,\mu + \mathbf{e}_1\,\kappa\,\mu \end{aligned}\right\} \quad (\text{mod } \sigma^0, \sigma^1), \tag{9}$$

which imply, when we restrict to a fixed tangent space on the states, that μ is the centro-affine arc-length and κ is the centro-affine curvature of $\mathbf{f}$. *Finally, the derivative of μ has the form*

$$d\mu = -K\,\sigma^0 \wedge \sigma^1 - J\,\mu \wedge \sigma^1, \tag{10}$$

and the functions κ, K and J are a fundamental set of feedback invariant functions on state-control space. These functions must further satisfy the integrability relations

$$\begin{aligned} 0 &= J - \kappa_{\sigma^0} \\ 0 &= J_{\sigma^0} + \kappa\,K - K_{\mu}. \end{aligned} \tag{11}$$

Writing equations (8) and (10) as

$$\begin{aligned} d\sigma^0 &= \epsilon\,\mu \wedge \sigma^1 \\ d\sigma^1 &= -\mu \wedge \sigma^0 - \kappa\,\mu \wedge \sigma^1 \\ d\mu &= -K\,\sigma^0 \wedge \sigma^1 - J\,\mu \wedge \sigma^1, \end{aligned}$$

we see that they agree (up to sign) with the structure equations that R. Gardner derives using Cartan's equivalence method [**Ga89**]. (Comparing with Gardner's equation, $\eta^1 = \sigma^0$, $\eta^2 = \sigma^1$, and $I = -\kappa$.) We arrived at the *same* set of structure equations, and *entirely bypassed* the equivalence method. Moreover, we have a *geometric interpretation* of the quantities. The 1-form μ represents the centro-affine arc-length of $\mathbf{f}$ in each tangent space, and $\kappa(\mathbf{x}, u)$ represents the centro-affine curvature in each tangent space. The sign of ϵ determines the convexity of $\mathbf{f}$. The functions $K(\mathbf{x}, u)$ and $J(\mathbf{x}, u)$ describe the variation of $\mathbf{f}$ from one tangent space to another.

3. Examples

The fundamental invariants are constant. Suppose now that κ, K and J are constant. We see from (11) that J must be 0 and $\kappa\,K = 0$. We will consider the case $\kappa \neq 0$ first. Thus $K = 0$ and $\epsilon = \pm 1$. We will examine the case $\epsilon = -1$, the other case being similar. We begin by recalling that a curve in the centro-affine plane with structure equations (3), $\epsilon = -1$, and κ equal to a constant is equivalent to the standard form $\mathbf{x}(s) = {}^t(1,0)\,e^{\lambda_1 s} + {}^t(0,1)\,e^{\lambda_2 s}$, where λ_1 and λ_2 are the roots of $r^2 - \kappa\,r - 1 = 0$. We arrived at this form by using the fact that equations (3) imply that $\mathbf{x}(s)$ satisfies the differential equation $\mathbf{x}''(s) - \kappa\,\mathbf{x}'(s) - \mathbf{x}(s) = 0$. In this form we see that $\kappa = \lambda_1 + \lambda_2$. The equivalent normal form

$$\mathbf{x}(s) = \left[\begin{pmatrix} \cosh(\sqrt{\kappa^2 + 4}\,s/2) \\ 0 \end{pmatrix} + \begin{pmatrix} 0 \\ \sinh(\sqrt{\kappa^2 + 4}\,s/2) \end{pmatrix}\right] e^{\kappa\,s/2}$$

has the advantage that the constant κ is easier to read off. What we shall show is that, under the current assumptions, the control system $\mathbf{f}(\mathbf{x}, u)$ is feedback equivalent to the nonlinear system

(12)
$$\begin{aligned}\mathbf{g}(\mathbf{y}, v) = \mathbf{g}(v) &= \left[\begin{pmatrix}\cosh(\sqrt{\kappa^2+4}\,v/2)\\ 0\end{pmatrix} + \begin{pmatrix}0\\ \sinh(\sqrt{\kappa^2+4}\,v/2)\end{pmatrix}\right] e^{\kappa\, v/2}\\ &= \begin{pmatrix}\cosh(\sqrt{\kappa^2+4}\,v/2)\\ \sinh(\sqrt{\kappa^2+4}\,v/2)\end{pmatrix} e^{\kappa\, v/2}.\end{aligned}$$

Since J and K vanish, the structure equations (8) and (10) are

$$\begin{aligned} d\sigma^0 &= -\mu\wedge\sigma^1\\ d\sigma^1 &= -\mu\wedge\sigma^0 - \kappa\,\mu\wedge\sigma^1\\ d\mu &= 0.\end{aligned} \tag{13}$$

We see that there is a function $v = v(\mathbf{x}, u)$, defined up to a constant, such that $\mu = dv$. Choose $v(\mathbf{x}, u)$ so that 0 is in its range. Since σ^0, σ^1 and μ are independent, and since σ^0 and σ^1 are linear combinations of dx^1 and dx^2 only, we see that $\partial v/\partial u \neq 0$. Recalling that $\mathbf{e}_0 = \mathbf{f}(\mathbf{x}, u)$, equations (9) imply that $\mathbf{e}_1 = \partial\mathbf{f}/\partial v$ and $\partial^2\mathbf{f}/\partial v^2 - \kappa\,\partial\mathbf{f}/\partial v - \mathbf{f} = 0$. We will now restrict to the two dimensional submanifold defined by $v = 0$. Equations (13) show that, along $v = 0$, the Lie bracket $[\mathbf{e}_0, \mathbf{e}_1] = 0$. Thus, we may apply a change of state variable, $\mathbf{y} = \boldsymbol{\phi}(\mathbf{x})$ such that

$$\begin{aligned} \mathbf{e}_0\big|_{v=0} &= \frac{\partial}{\partial y^1}\\ \mathbf{e}_1\big|_{v=0} &= \frac{\kappa}{2}\frac{\partial}{\partial y^1} + \frac{\sqrt{\kappa^2+4}}{2}\frac{\partial}{\partial y^2}.\end{aligned} \tag{14}$$

Together, $\mathbf{y}$ and v determine a feedback transformation that transforms $\mathbf{f}(\mathbf{x}, u)$ to $\mathbf{g}(\mathbf{y}, v)$, where $\mathbf{g}(\mathbf{y}, 0) = \partial/\partial y^1$ and $\partial\mathbf{g}/\partial v(\mathbf{y}, 0) = \kappa/2\,\partial/\partial y^1 + \sqrt{\kappa^2+4}/2\,\partial/\partial y^2$. Moreover, since all of our centro-affine structure is preserved under feedback, we must also have that $\mathbf{g}$ satisfies the same differential equation as $\mathbf{f}$, namely for all $\mathbf{y}$, $\partial^2\mathbf{g}/\partial v^2 - \kappa\,\partial\mathbf{g}/\partial v - \mathbf{g} = 0$, with the initial conditions at $v = 0$ given above. The function given in (12) satisfies this differential equation and these initial conditions, so $\mathbf{g}(\mathbf{y}, v)$ must be equal to (12). Thus we have a normal form for control systems of this type.

If $\epsilon = +1$ and $\kappa \neq 0$, a similar argument gives one of three normal forms. The particular form depends on whether $\kappa^2 - 4$ is positive, zero or negative. We find that $\mathbf{f}(\mathbf{x}, u)$ is respectively equivalent to either

- $\mathbf{g}(\mathbf{y}, v) = {}^t(\cosh[\sqrt{\kappa^2-4}\,v/2], \sinh[\sqrt{\kappa^2-4}\,v/2])e^{\kappa\, v/2}$,
- $\mathbf{g}(\mathbf{y}, v) = {}^t(1, v)e^{\pm v}$, or
- $\mathbf{g}(\mathbf{y}, v) = {}^t(\cos[\sqrt{4-\kappa^2}\,v/2], \sin[\sqrt{4-\kappa^2}\,v/2])e^{\kappa\, v/2}$.

The case $\kappa = 0$. With $\kappa = 0$, we still must have that $J = 0$, but K can be nonzero. Equations (8) and (10) become

$$\begin{aligned} d\sigma^0 &= \epsilon\,\mu\wedge\sigma^1\\ d\sigma^1 &= -\mu\wedge\sigma^0\\ d\mu &= -K\,\sigma^0\wedge\sigma^1.\end{aligned} \tag{15}$$

We recall that the centro-affine curvature of a plane curve vanishes if and only if the curve lies on a quadric (convex when $\epsilon = 1$, concave when $\epsilon = -1$). Since we are assuming that $\kappa = 0$, we see that the control system $\mathbf{f}(\mathbf{x}, u)$ determines a quadric curve in each tangent space. We can use this curve to define the length of a tangent vector. We do this by declaring $\|\mathbf{f}(\mathbf{x}, u)\|$ to be 1 for all $(\mathbf{x}, u)$. This suggests that we have either a Riemannian or a pseudo-Riemannian metric on the state space. In fact, we can show that this metric is given by $\|\mathbf{w}\|^2 = \sigma^0(\mathbf{w})^2 + \epsilon\,\sigma^1(\mathbf{w})^2$, for any $\mathbf{w} \in T_{\mathbf{x}}\mathbf{R}^2$. We have to be a bit careful here, because the 1-forms σ^0 and σ^1 are defined on $(\mathbf{x}, u)$-space, and we are claiming that the quadratic combination $(\sigma^0)^2 + \epsilon\,(\sigma^1)^2$ is well defined on $\mathbf{x}$-space. To show this, let Z be any vector field tangent to the fiber of $(\mathbf{x}, u)$-space over $\mathbf{x}$ space, that is, tangent to the fiber of state-control space over state space. Since σ^0, σ^1 are nonsingular linear combinations of dx^1, dx^2, we see that every such vector Z is defined by the equations $\langle Z, \sigma^0\rangle = \langle Z, \sigma^1\rangle = 0$. Using Cartan's formula for Lie derivative, $\mathcal{L}_Z\,\theta = Z \lrcorner\, d\theta + d(Z \lrcorner\, \theta)$, and using equations (15) we can compute the Lie derivative $\mathcal{L}_Z\,(\sigma^0)^2 + \epsilon\,(\sigma^1)^2 = 2(\sigma^0\mathcal{L}_Z\,\sigma^0 + \epsilon\,\sigma^1\mathcal{L}_Z\,\sigma^1) = 2\langle Z, \mu\rangle(\epsilon\,\sigma^0\,\sigma^1 - \epsilon\,\sigma^1\,\sigma^0) = 0$. Since Z is an arbitrary vertical vector field, this shows that $(\sigma^0)^2 + \epsilon\,(\sigma^1)^2$ is a well defined metric on state space. Moreover, we can interpret our control system $\mathbf{f}(\mathbf{x}, u)$ as simply saying that we move along the state space at unit speed.

Since the 1-forms σ^0 and σ^1 diagonalize the Riemannian metric, we recognize that equations (15) are simply the structure equations of a surface, with μ representing the Levi-Civita connection 1-form and K representing the Gaussian curvature of the surface. We see instantly that if K is constant, then our control system must be equivalent to a control system that parametrizes unit speed curves on constant curvature surfaces. For example, in the $\epsilon = 1$ case, if $(\sigma^0)^2 + (\sigma^1)^2$ is isometricly equivalent to the constant curvature metric

$$\frac{(dy^1)^2 + (dy^2)^2}{\left(1 + \frac{K}{4}\,[(y^1)^2 + (y^2)^2]\right)^2}$$

then the control system $\dot{\mathbf{x}} = \mathbf{f}(\mathbf{x}, u)$ must be feedback equivalent to the control system

$$\dot{\mathbf{y}} = \mathbf{g}(\mathbf{y}, v) = \begin{pmatrix} \cos v \\ \sin v \end{pmatrix} \left(1 + \frac{K}{4}\,[(y^1)^2 + (y^2)^2]\right),$$

since this is equivalent to the metric norm being equal to one, $\|\dot{\mathbf{y}}\|^2 = 1$.

Constant κ in each fiber. Suppose now that in each fiber, $\mathbf{f}(\mathbf{x}, u)$ defines a constant curvature centro-affine curve, but that the constant can vary as we move to different fibers. In other words, assume that κ is a nonconstant function depending only on $\mathbf{x}$. We can construct examples of this kind of system by taking a normal form such as (12) and allow κ to be a function of $\mathbf{x}$. For this kind of system, we can see that J is arrived at in a very natural way. Since κ only depends on $\mathbf{x}$, $\kappa_\mu = 0$ and therefore $d\kappa = \kappa_{\sigma^0}\,\sigma^0 + \kappa_{\sigma^1}\,\sigma^1$. Since $\langle \mathbf{f}, \sigma^0\rangle = 1$ and $\langle \mathbf{f}, \sigma^1\rangle = 0$, we see that $\langle \mathbf{f}, d\kappa\rangle = \kappa_{\sigma^0}$, but $\kappa_{\sigma^0} = J$ by (11), so we see that J is simply the derivative of κ by the control vector field $\mathbf{f}$. So, in this case at least, we have a nice geometric interpretation of J. It is giving us the rate at which the centro-affine curvature changes as we move from fiber to fiber along the state space in the direction of our control.

A Gauss-Bonnet theorem. Suppose that in each fiber $\mathbf{f}(\mathbf{x}, u)$ defines a strictly convex simple closed curve surrounding the origin. Then we may view these curves as the unit vectors, or indicatrix, for a Finsler metric on state space. Equations (8) and (10) are the structure equations for a Finsler surface. We have already seen that the vanishing of κ means that the Finsler metric is in fact a Riemannian metric, K is the Gaussian curvature and μ is the connection 1-form.

Let $S_\mathbf{x} \subset T_\mathbf{x}\mathbf{R}^2$ denote the indicatrix at $\mathbf{x}$. The centro-affine perimeter of $S_\mathbf{x}$ is given by

$$P_\mathbf{x} = \int_{S_\mathbf{x}} \mu.$$

Computing the variational derivative $\delta \int \mu = 0$ shows that $P_\mathbf{x}$ is constant exactly when the invariant J vanishes. These are the Landsberg surfaces. Assuming that J vanishes, equation (10) gives

$$d\mu = -K\,\sigma^0 \wedge \sigma^1.$$

From this equation we see that the 2-form $K\,\sigma^0 \wedge \sigma^1$ is well defined on state space. We also see that μ is a transgression for $K\,\sigma^0 \wedge \sigma^1$ on state-control space. These observations show how we may directly translate the proofs of the standard Gauss-Bonnet theorems into this setting, with the constant perimeter $P_\mathbf{x}$ replacing 2π.

4. Closing

When we view a control system as a family of submanifolds of the linear tangent spaces to the state space, we see that the action of the feedback group on each unparametrized submanifold is exactly a centro-affine action. This observation suggests that we take a look at centro-affine geometry, that is, the geometry of submanifolds of $\mathbf{R}^n$ under the action of GL(n,$\mathbf{R}$), in the hope that we will gain geometric insight into the overall geometry of the control system.

We studied the very simplest case in this paper, that is the case of curves in the plane. These dimensions correspond exactly to the case of a single input control system on the plane. More generally, centro-affine curves in $\mathbf{R}^n$ correspond to single input control systems on $\mathbf{R}^n$. We gave a complete description of the invariants of centro-affine plane curves, along with a collection of normal forms. We carried this understanding of centro-affine curves over to the two state, single input control problem. The strength of our approach is that it allows us to analyze very nonlinear systems on the plane. In fact, the system must be so nonlinear that is cannot even lie on an affine line in each tangent space. Another advantage to this approach is that we find the same fundamental invariants and structure equations we are led to using Cartan's method of equivalence. Moreover, we gain a geometric interpretation of the invariants and of the structure equations that is difficult to discern from the equivalence method.

We found normal forms for several nonlinear systems, as well as a class of systems possessing a Riemannian metric. In fact, the presence of this metric, which corresponds to the vanishing of κ, is easily predicted from centro-affine considerations. We know that for centro-affine curves, the vanishing of κ is equivalent to having a quadric curve centered at the origin. Knowing this, the vanishing of the feedback invariant κ gives a quadric curve in each tangent space. We may use these curves as a set of unit vectors in each tangent space, which defines the metric on state space. The control system itself may then be interpreted as requiring that a particle move through state space at unit speed.

To extend this study to higher dimensions, we need to study higher dimensional centro-affine geometries. This has been done in part. We have computed the structure of centro-affine hypersurfaces in $\mathbf{R}^n$ and applied these results to feedback control [**GW96**], [**Wi96**]. We have also computed the structure of centro-affine curves in $\mathbf{R}^n$ [**GW97**] and applications to feedback control are under way.

References

Bl23. W. Blaschke, *Vorlesungen über Differentialgeometrie. Affine Differentialgeometrie*, vol. 2, Springer-Verlag, Berlin, 1923.

Ca51. É. Cartan, *La théorie des groupes finis et continus et la géométrie différentielle*, Gauthier-Villars, Paris, 1951.

Ga89. R.B. Gardner, *The Method of Equivalence and Its Applications*, vol. 58, SIAM, Philadelphia, PA, 1989.

GW96. R.B. Gardner and G.R. Wilkens, *A pseudo-group isomorphism between control systems and certain generalized Finsler structures*, Finsler Geometry (Seattle, WA, July 1995) (D. Bao, S.S. Chern and Z. Shen, eds.), vol. Contemporary Math. 196, 1996, pp. 231–243.

GW97. R.B. Gardner and G.R. Wilkens, *The fundamental theorems of curves and hypersurfaces in centro-affine geometry*, Bull. Belgian Math. Society **4** (1997), no. 3, 379–401.

Gn78. M.L. Green, *The moving frame, differential invariants and rigidity theorems for curves in homogeneous spaces*, Duke Mathematical Journal **45** (1978), 735–779.

Gr74. P.A. Griffiths, *On Cartan's method of Lie groups and moving frames as applied to uniqueness and existence questions in differential geometry*, Duke Mathematical Journal **41** (1974), 775–814.

MM33. O. Mayer and A. Myller, *La géométrie centroaffine des courbes planes*, Annales Scíentifíques de l'Université de Jassy **18** (1933), 234–280.

SS62. P. Schirokow and A. Schirokow, *Affine Differentialgeometrie*, B.G. Teubner Verlagsgesellschaft, Leipzig, 1962.

Sh96. R.W. Sharpe, *Differential Geometry: Cartan's Generalization of Klein's Erlangen Programme*, vol. 166, Springer-Verlag, 1996.

Wi96. G.R. Wilkens, *Finsler geometry in low dimensional control theory*, Finsler Geometry (Seattle, WA, July 1995) (D. Bao, S.S. Chern and Z. Shen, eds.), vol. Contemporary Math. 196, 1996, pp. 245–257.

Fields Institute for Research in Mathematical Sciences, 185 Columbia Street West, Waterloo, Ontario N2L 5Z5 CANADA

Current address: Department of Mathematics, University of Hawaii at Manoa, Honolulu, HI 96822

E-mail address: grw@math.hawaii.edu

Proceedings of Symposia in Pure Mathematics
Volume **64**, 1999

On hypersurfaces satisfying a basic equality

Jie Yang

ABSTRACT. In the paper, we describe a recent joint work with B. Y. Chen on hypersurfaces in real space forms related to a basic inequality proved by Chen in 1993. We classify CMC hypersurfaces in a Riemannian space form which satisfy the equality case of the inequality, and by utilizing Jacobi's elliptic functions and Theta function we obtain the complete classification of conformally flat hypersurfaces in Riemannian space forms which satisfy the equality.

For a submanifold M^n in an m-dimensional Riemannian space form $R^m(\epsilon)$ of constant sectional curvature ϵ, B. Y. Chen proves in [**4**] a sharp inequality involving the sectional curvature K of M^n, the scalar curvature n τ of M^n and the squared mean curvature H^2. It is convenient to introduce a Riemannian invariant δ_M of M^n by $\delta_M(p) = \tau(p) - \inf K(p)$, where $\inf K$ is the function assigning to each $p \in M^n$ the infimum of $K(\pi)$, where π runs over all 2-planes in T_pM and τ is defined by $\tau = \sum_{i<j} K(e_i \wedge e_j)$, and $\{e_1, \dots, e_n\}$ is an orthonormal basis of T_pM^n. For $n = 2$, this invariant vanishes trivially. The inequality proved in [4] can be written as follows:

$$\delta_M \leq \frac{n^2(n-2)}{2(n-1)} H^2 + \tfrac{1}{2}(n+1)(n-2)\epsilon. \tag{1}$$

Inequality (1) has some important applications, for examples, it gives rise to the second Riemannian obstruction for a Riemannian manifold to admit a minimal isometric immersion into a Euclidean space. It also gives rise to an obstruction to Lagrangian isometric immersions from compact Riemannian manifolds with finite fundamental group π_1 into complex space forms (see [6] for details).

Since (1) is a very general and sharp inequality, it is natural and interesting to investigate and to understand submanifolds in a Riemannian space form which verify the equality of this inequality, i.e. the equality:

$$\delta_M = \frac{n^2(n-2)}{2(n-1)} H^2 + \tfrac{1}{2}(n+1)(n-2)\epsilon. \tag{2}$$

1991 *Mathematics Subject Classification.* 53B40. 53C42.

Key words and phrases. Jacobi's elliptic functions, Theta function, basic equality, conformal flat hypersurfaces, CMC hypersurfaces.

The complete version of this paper will be submitted to elsewhere.

Submanifolds satisfying this basic equality were studied recently in several papers (cf. [4,5,6,8,9,12,15,16]). In this respect, in particular we would like to point out that 3-dimensional totally real submanifolds satisfying this basic equality in the nearly Kähler 6-sphere $S^6(1)$ have been completely classified in [16] by F. Dillen and L. Vrancken.

We investigate the most fundamental case; namely hypersurfaces satisfying the basic equality.

THEOREM 1. *A hypersurface M^n $(n > 2)$ of a Euclidean $(n+1)$-space $\mathbb{E}^{n+1}$ with constant mean curvature satisfies equality* (2) *if and only if either M^n is minimal or M^n is an open portion of a spherical hypercylinder $\mathbb{R} \times S^{n-1}(r)$.*

THEOREM 2. *Let M^n $(n > 2)$ be a hypersurface with constant mean curvature in the sphere $S^{n+1}(1)$. Then M^n satisfies equality* (2) *if and only if one of the following two cases occurs.*

1. *M^n is a totally geodesic hypersurface.*
2. *There is an open dense subset U of M^n and a non-totally geodesic, isometric, minimal immersion $\phi : B^2 \to S^{n+1}(1)$ from a surface B^2 into $S^{n+1}(1)$ such that U is an open subset of the unit normal bundle NB^2 defined by*

$$N_pB^2 = \left\{\xi \in T_{\phi(p)}S^{n+1}(1) | \langle \xi, \xi\rangle = 1 \text{ and } \left\langle \xi, \phi_\star(T_pB^2)\right\rangle = 0\right\}.$$

Let $\mathbb{E}_1^{n+2}$ denote the $(n+2)$-dimensional Minkowski space-time with the Lorenzian metric $g = -dx_1^2 + dx_2^2 + \cdots + dx_{n+2}^2$. Recall that the unit hyperbolic space $H^{n+1}(-1)$ and the unit de Sitter space-time $S_1^{n+1}(1)$ are isometrically imbedded in $\mathbb{E}_1^{n+2}$ respectively in the following standard ways:

$$H^{n+1}(-1) = \{\mathbf{x} = (x_1, \dots, x_{n+2}) \in \mathbb{E}_1^{n+2} \mid \langle \mathbf{x}, \mathbf{x}\rangle = -1\},$$
$$S_1^{n+1}(1) = \{x \in \mathbb{E}_1^{n+2} | \langle x, x\rangle = 1\}.$$

THEOREM 3. *Let M^n $(n > 2)$ be a hypersurface with constant mean curvature in the hyperbolic space $H^{n+1}(-1)$. Then M^n satisfies equality* (2) *if and only if one of the following three cases occurs.*

1. *M^n is a totally geodesic hypersurface.*
2. *M^n is a tubular hypersurface with radius $r = \coth^{-1}(\sqrt{2})$ about a 2-dimensional totally geodesic surface of $H^{n+1}(-1)$.*
3. *There is an open dense subset U of M^n and a non-totally geodesic, isometric, minimal immersion $\phi : B^2 \to S_1^{n+1}$ from a surface B^2 into the de Sitter space-time $S_1^{n+1}(1)$ such that U is an open subset of the unit normal bundle NB^2 of B^2 defined by*

$$N_pB^2 = \left\{\xi \in T_{\phi(p)}S_1^{n+1}(1) | \langle \xi, \xi\rangle = -1 \text{ and } \left\langle \xi, \phi_\star(T_pB^2)\right\rangle = 0\right\}.$$

In order to state our next results, we recall the three families of Riemannian manifolds, P_a^n $(a > 1), C_a^n$ $(a > 1), D_a^n$ $(0 < a < 1)$ and the two exceptional spaces, F^n, L^n, introduced in [7].

Let $\mathrm{cn}(u,k), \mathrm{dn}(u,k)$ and $\mathrm{sn}(u,k)$ denote the three main Jacobi's elliptic functions with modulus k. The nine other elliptic functions $\mathrm{nd}(u,k), \mathrm{nc}(u,k), \mathrm{ns}(u,k)$, $\mathrm{sc}(u,k), \mathrm{cd}(u,k), \mathrm{ds}(u,k), \mathrm{cs}(u,k), \mathrm{dc}(u,k), \mathrm{sd}(u,k)$ are defined by taking reciprocals and quotients; for examples, $\mathrm{sd}(u,k) = \mathrm{sn}(u,k)/\mathrm{dn}(u,k)$, $\mathrm{nd}(u,k) = 1/\mathrm{dn}(u,k)$ (cf. [17,19] for details).

We put

$$\mu_a = ak\,\mathrm{cn}(ax,k), \quad k = \frac{\sqrt{a^2-1}}{\sqrt{2}a}, \quad a>1, \tag{3}$$

$$\eta_a = \frac{a}{k}\,\mathrm{dn}(\frac{a}{k}x,k), \quad k = \frac{\sqrt{2}a}{\sqrt{a^2+1}}, \quad 0<a<1. \tag{4}$$

$$\rho_a = ak\,\mathrm{cn}(ax,k), \quad k = \frac{\sqrt{a^2+1}}{\sqrt{2}a}, \quad a>1, \tag{5}$$

Let $S^n(c)$ and $H^n(-c)$ denote the n-sphere with constant sectional curvature c and the hyperbolic n-space with constant sectional curvature $-c$, respectively. For $n>2$, P^n_a, D^n_a and C^n_a are the Riemannian n-manifolds given by the warped product manifolds $I\times_{\mu_a} S^{n-1}(\frac{a^4-1}{4})$, $\mathbb{R}\times_{\eta_a} H^{n-1}(\frac{a^4-1}{4})$ and $I\times_{\rho_a} S^{n-1}(\frac{a^4-1}{4})$ with warped functions μ_a, η_a and ρ_a, respectively, where I are the open intervals on which the corresponding warped functions are positive. The two exceptional spaces F^n and L^n are the warped product manifolds $\mathbb{R}\times_{1/\sqrt{2}} H^{n-1}(-\frac{1}{4})$ and $\mathbb{R}\times_{\mathrm{sech}(x)}\mathbb{R}^{n-1}$, respectively.

D^n_a, F^n, L^n are complete Riemannian n-manifolds, but P^n_a and C^n_a are not complete. Topologically, S^n is the two point compactification of both P^n_a and C^n_a. From [10] we know that the Riemannian metrics defined on P^n_a and C^n_a can be extended smoothly to their two point compactifications S^n. We denote by $\hat{P}^n_a$ and $\hat{C}^n_a$ the S^n together with the Riemannian metrics given by the (smooth) extensions of the metrics on P^n_a and C^n_a to S^n, respectively. We remark that P^n_a, D^n_a, C^n_a are indeed isometric to the warped products n-manifolds $I\times_{f_1} S^{n-1}(1), I\times_{f_2} H^{n-1}(-1), I\times_{f_3} S^{n-1}(1)$ with warped functions $2\rho_a/(a^2+1), 2\mu_a/(a^2-1), 2\eta_a/(a^2+1)$, respectively.

Let A^n_a $(a>1)$, B^n_a $(0<a<1)$, G^n, H^n_a $(a>0)$ and Y^n_a $(0<a<1)$ denote the warped product manifolds:

$$\mathbb{R}\times_{\sqrt{a^2-1}\cosh x} S^{n-1}(1), \quad (-\frac{\pi}{2},\frac{\pi}{2})\times_{\sqrt{1-a^2}\cos x} S^{n-1}(1),$$
$$\mathbb{R}\times_{\cosh x}\mathbb{E}^{n-1}, \quad \mathbb{R}\times_{\sqrt{a^2+1}\cosh x} H^{n-1}(-1), \quad (0,\infty)\times_{\sqrt{1-a^2}\sinh x} S^{n-1}(1),$$

respectively. When $n-2$, the second factor S^{n-1} or H^{n-1} in each of the warped product manifolds shall be replaced either by $S^1(1)$ or by $\mathbb{R}$. The geometry of A^2_a, G^2 and H^2_a are similar in the sense that one can be obtained from the others by applying some suitable scalings on the first factor $\mathbb{R}$.

Clearly, A^n_a, G^n, H^n_a are complete Riemannian manifolds. Topologically, S^n is the two point compactification of B^n_a. As for P^n_a and C^n_a, the warped metric on B^a_n can be extended smoothly to its two point compactification; a fact follows from (1.8). We denote by $\hat{B}^n_a$ the S^n together with the Riemannian metrics on S^n extended from the metric on B^n_a.

For $n\geq 2$ and any real number $a>0$, there is a well-known Lagrangian immersion from the unit n-sphere S^n into a complex Euclidean n-space $\mathbb{C}^n$ defined by

$$w_a(y_0,y_1,\dots,y_n) = \frac{a}{1+y_0^2}(y_1,\dots,y_n,y_0y_1,\dots,y_0y_n), \tag{6}$$

where $y_0^2 + y_1^2 + \cdots + y_n^2 = 1$. This Whitney's immersion w_a has a unique self-intersection point $w_a(-1,0,\ldots,0) = w_a(1,0,\ldots,0)$. The S^n together with the metric induced from Whitney's immersion w_a, denoted by W_a^n, is called a *Whitney n-sphere.*

We completely classify conformally flat hypersurfaces satisfying the basic equality. In this respect, first we sharpen Proposition 2 of [15] to the following

THEOREM 4. *Let $\mathbf{x} : M^n \to \mathbb{E}^{n+1}$ $(n > 2)$ be an isometric immersion of a conformally flat n-manifold into a Euclidean $(n+1)$-space. Then it satisfies equality (2) if and only if one of the following four cases occurs.*

1. *M^n is totally geodesic.*
2. *M^n is an open portion of a spherical hypercylinder $S^{n-1} \times \mathbb{R}$.*
3. *M^n is an open portion of a round hypercone.*
4. *$n = 3$ and M^3 is an open portion of a Whitney 3-sphere W_a^3 for some $a > 0$ and, up to rigid motions, the immersion $\mathbf{x} : W_a^3 \to \mathbb{E}^4$ is given by*

$$\mathbf{x}(x, y_1, y_2, y_3) = \left(\frac{1}{2}\int_0^x sd^2\left(\frac{\sqrt{2}}{a}x\right)dx, ay_1\, sd\left(\frac{\sqrt{2}}{a}x\right), ay_2\, sd\left(\frac{\sqrt{2}}{a}x\right), ay_3\, sd\left(\frac{\sqrt{2}}{a}x\right)\right), \tag{7}$$

where $y_1^2 + y_2^2 + y_3^2 = \frac{1}{2}$ and $k = \frac{1}{\sqrt{2}}$ is the modulus of the Jacobi's elliptic functions.

Our next main results are the following.

THEOREM 5. *Let $\mathbf{x} : M^n \to S^{n+1}(1) \subset \mathbb{E}^{n+2}$ $(n > 2)$ be an isometric immersion of a conformally flat n-manifold. Then it satisfies equality (1.2) if and only if one of the following three cases occurs.*

(i) *M^n is an open portion of $S^n(1)$ and the immersion $\mathbf{x} : M^n \to S^{n+1}(1)$ is totally geodesic.*

(ii) *M^n is an open portion of $\hat{B}_a^n$ for some $a \in (0,1)$ and, up to rigid motions, the immersion $\mathbf{x} : \hat{B}_a^n \to S^{n+1}(1) \subset \mathbb{E}^{n+2}$ is given by*

$$\mathbf{x}(x, y_1, \ldots, y_n) = (\sin x, a\cos x, \sqrt{1-a^2}\, y_1 \cos x, \ldots, \sqrt{1-a^2}\, y_n \cos x) \tag{8}$$

with $y_1^2 + y_2^2 + \cdots + y_n^2 = 1$.

(iii) *$n = 3$, M^3 is an open portion of $\hat{P}_a^3$ for some $a > 1$ and, up to rigid motions, the immersion $\mathbf{x} : \hat{P}_a^3 \to S^4(1) \subset \mathbb{E}^5$ is given by*

$$\begin{gathered}\mathbf{x}(x, y_1, y_2, y_3) = \frac{1}{ak'}\,(y_1\, cn(ax), y_2\, cn(ax), y_3\, cn(ax), \\ \sqrt{a^2k'^2 - cn^2(ax)}\cos\left(\frac{k'}{k}x + \frac{i}{2}\left(\ln\frac{\Theta(ax-\gamma)}{\Theta(ax+\gamma)} + 2aZ(\gamma)x\right)\right) \\ \sqrt{a^2k'^2 - cn^2(ax)}\sin\left(\frac{k'}{k}x + \frac{i}{2}\left(\ln\frac{\Theta(ax-\gamma)}{\Theta(ax+\gamma)} + 2aZ(\gamma)x\right)\right)\Bigg), \\ y_1^2 + y_2^2 + y_3^2 = 1, \quad \gamma = sn^{-1}\left(\frac{i}{ak^2}\right), \quad i = \sqrt{-1},\end{gathered} \tag{9}$$

where $k = \sqrt{a^2-1}/(\sqrt{2}a)$, $k' = \sqrt{a^2+1}/(\sqrt{2}a)$ are the modulus and the complementary modulus of Jacobi's elliptic functions respectively, $\Theta(u) = \Theta(u,k)$ is the Theta function and $Z(u) = Z(u,k)$ the Zeta function.

THEOREM 6. *Let* $\mathbf{x} : M^n \to H^{n+1}(-1) \subset \mathbb{E}_1^{n+2}$ $(n > 2)$ *be an isometric immersion of a conformally flat n-manifold. Then it satisfies equality* (1.2) *if and only if one of the following nine cases occurs.*

(i) M^n *is an open portion of* $H^n(-1)$ *and the immersion* $\mathbf{x} : M^n \to H^{n+1}(-1)$ *is totally geodesic.*

(ii) M^n *is an open portion of* A_a^n *for some* $a > 1$ *and, up to rigid motions, the immersion* $\mathbf{x} : A_a^n \to H^{n+1}(-1) \subset \mathbb{E}_1^{n+2}$ *is given by*

$$\mathbf{x}(x, y_1, \dots, y_n) = (a\cosh x, \sinh x, \sqrt{a^2-1}\, y_1 \cosh x, \dots \\ \dots, \sqrt{a^2-1}\, y_n \cosh x) \tag{10}$$

with $y_1^2 + y_2^2 + \cdots + y_n^2 = 1$.

(iii) M^n *is an open portion of* G^n *and, up to rigid motions, the immersion* $\mathbf{x} : G^n \to H^{n+1}(-1) \subset \mathbb{E}_1^{n+2}$ *is given by*

$$\mathbf{x}(x, u_2, \dots, u_n) = \left(\left(1 + \frac{u_2^2 + \cdots + u_n^2}{2}\right)\cosh x, \\ \frac{u_2^2 + \cdots + u_n^2}{2}\cosh x, \sinh x, u_2 \cosh x, \dots, u_n \cosh x\right). \tag{11}$$

(iv) M^n *is an open portion of* H_a^n *for some* $a > 0$, *and, up to rigid motions, the immersion* $\mathbf{x} : H_a^n \to H^{n+1}(-1) \subset \mathbb{E}_1^{n+2}$ *is given by*

$$\mathbf{x}(x, y_1, \dots, y_n) = (\sqrt{a^2+1}\, y_1 \cosh x, \dots, \sqrt{a^2+1}\, y_n \cosh x, \\ a\cosh x, \sinh x) \tag{12}$$

with $y_1^2 - y_2^2 + \cdots - y_n^2 = 1$.

(v) M^n *is an open portion of* Y_a^n *for some* a, $0 < a < 1$ *and, up to rigid motions, the immersion* $\mathbf{x} : Y_a^n \to H^{n+1}(-1) \subset \mathbb{E}_1^{n+2}$ *is given by*

$$\mathbf{x}(x, y_1, \dots, y_n) = (\cosh x, a \sinh x, \sqrt{1-a^2}\, y_1 \sinh x, \dots \\ \dots, \sqrt{1-a^2}\, y_n \sinh x) \tag{13}$$

with $y_1^2 + y_2^2 + \cdots + y_n^2 = 1$.

(vi) $n = 3$, M^3 *is an open portion of* F^3 *and, up to rigid motions, the immersion* $\mathbf{x} : M^3 \to H^4(-1) \subset \mathbb{E}_1^5$ *is given by*

$$\mathbf{x}(x, u, v) = \left(\sqrt{2}\cosh u \cosh v, \sqrt{2}\cosh u \sinh v, \sqrt{2}\sinh u, \cos x, \sin x\right). \tag{14}$$

(vii) $n = 3$, M^3 *is an open portion of* L^3 *and, up to rigid motions, the immersion* $\mathbf{x} : L^3 \to H^4(-1) \subset \mathbb{E}_1^5$ *is given by*

$$\mathbf{x}(x, u, v) = \operatorname{sech} x \left(x^2 + u^2 + v^2 + \cosh^2 x + \frac{1}{4}, \\ x^2 + u^2 + v^2 + \cosh^2 x - \frac{1}{4}, x, u, v\right). \tag{15}$$

(viii) $n = 3$, *M^3 is an open portion of* $\hat{C}^3_a$ *for some* $a > 1$ *and, up to rigid motions, the immersion* $\mathbf{x} : \hat{C}^3_a \to H^4(-1) \subset \mathbb{E}^5_1$ *is given by*

$$\begin{aligned}
&\mathbf{x}(x,y_1,y_2,y_3) = \\
&\quad \frac{1}{ak'}\left(\sqrt{a^2k'^2 + cn^2(ax)}\cosh\left(\frac{k'}{k}x - \frac{1}{2}\ln\frac{\Theta(ax-\gamma)}{\Theta(ax+\gamma)} - aZ(\gamma)x\right),\right. \\
&\quad \sqrt{a^2k'^2 + cn^2(ax)}\sinh\left(\frac{k'}{k}x - \frac{1}{2}\ln\frac{\Theta(ax-\gamma)}{\Theta(ax+\gamma)} - aZ(\gamma)x\right), \\
&\quad \left. y_1\, cn(ax), y_2\, cn(ax), y_3\, cn(ax)\right), \qquad y_1^2 + y_2^2 + y_3^2 = 1,
\end{aligned} \tag{16}$$

where $\gamma = sn^{-1}\left(1/(ak^2)\right)$, $k = \sqrt{a^2+1}/(\sqrt{2}a)$ *and* $k' = \sqrt{a^2-1}/(\sqrt{2}a)$.

(ix) $n = 3$, *M^3 is an open portion of* D^3_a *for some* $0 < a < 1$ *and, up to rigid motions, the immersion* $\mathbf{x} : D^3_a \to H^4(-1) \subset \mathbb{E}^5_1$ *is given by*

$$\begin{aligned}
&\mathbf{x}(x,u,v) = \frac{1}{ak'}\left(k\, dn(\frac{a}{k}x)\cosh u \cosh v,\right. \\
&\quad k\, dn(\frac{a}{k}x)\cosh u \sinh v, k\, dn(\frac{a}{k}x)\sinh u, \\
&\quad \sqrt{k^2 dn^2(\frac{a}{k}x) - a^2k'^2}\cos\left(k'x - \frac{i}{2}\ln\frac{\Theta(\frac{a}{k}x-\gamma)}{\Theta(\frac{a}{k}x+\gamma)} - i\frac{a}{k}Z(\gamma)x\right) \\
&\quad \left.\sqrt{k^2 dn^2(\frac{a}{k}x) - a^2k'^2}\sin\left(k'x - \frac{i}{2}\ln\frac{\Theta(\frac{a}{k}x-\gamma)}{\Theta(\frac{a}{k}x+\gamma)} - i\frac{a}{k}Z(\gamma)x\right)\right)
\end{aligned} \tag{17}$$

where $\gamma = sn^{-1}(k/a)$, $k = \sqrt{2}a/\sqrt{1+a^2}$ *and* $k' = \sqrt{1-a^2}/\sqrt{1+a^2}$.

References

[1] R. Bellman, *A Brief Introduction to Theta Function*, Holt, Rinehart and Winston, New York, 1961.

[2] E. Cartan, *Familles de surfaces isoparamětriques dans le espacex à courbure constante*, Ann. Mat. Pura Appl. **17** (1938), 177–191.

[3] B. Y. Chen, *Geometry of Sumbmanifolds*, M. Dekker, New York, 1973.

[4] B. Y. Chen, *Some pinching and classification theorems for minimal submanifolds*, Archiv der Math. **60** (1993), 568–578.

[5] B. Y. Chen, *Tubular hypersurfaces satisfying a basic equality*, Soochow J. Math **20** (1994), 569–586 (A special issue in memory of Prof. T.-H. Teng).

[6] B. Y. Chen, *A Riemannian invariant and its applications to submanifold theory*, Results in Math. **27** (1995), 17–26 (A special issue dedicated to 70th birthday of Prof. K. Nomizu).

[7] B. Y. Chen, *Jacobi's elliptic functions and Lagrangian immersions*, Proc. Royal Soc. Edinburgh Ser. A **126** (1996), 687–704.

[8] B. Y. Chen, F. Dillen, L. Verstraelen and L. Vrancken, *Two equivariant totally real immersions into the nearly Kähler 6-sphere and their characterization*, Japan. J. Math. **21** (1995), 207–222.

[9] B. Y. Chen, F. Dillen, L. Verstraelen and L. Vrancken, *Characterization a class of totally real submanifolds of S^6 by their sectional curvatures*, Tôhoku Math. J. **47** (1995), 185–198.

[10] B. Y. Chen and L. Vrancken, *Lagrangian submanifolds satisfying a basic equality*, Math. Proc. Cambridge Phil. Soc. **120** (1996), 291–307.

[11] M. Dajczer and D. Gromoll, *Gauss parametrizations and rigidity aspects of submanifolds*, J. Differential Geometry **22** (1985), 1–12.

[12] R. Deszez, L. Verstraelen and S. Yaprak, *Hypersurfaces with pseudosymmetric Weyl tensor in conformally flat manifolds*, Dept. Math. Agricultural Univ., Wroclaw, Preprint Ser. No. 23 (1995).

[13] F. Dillen and S. Nölker, *Semi-parallelity, multi-rotation surfaces and the helix-property*, J. Reine Angew. Math. **435** (1993), 33-63.

[14] S. Hiepko, *Eine innere Kennzeichung der verzerrten Produkte*, Math. Ann. **241** (1979), 209–215.
[15] F. Dillen, M. Petrovic, L. Verstraelen, *Einstein, conformally flat and semi-symmetric submanifolds satisfying Chen's equality*, Israel J. Math. (to appear).
[16] F. Dillen, L. Vrancken, *Totally real submanifolds in $S^6(1)$ satisfying Chen's equality*, Trans. Amer. Math. Soc. **348** (1996), 1633-1646.
[17] D. F. Lawden, *Elliptic functions and applications*, Springer-Verlag, New York, 1989.
[18] E. Picard, *Sur une application de la théorie des fonctions elliptiques*, C. R. Acad. Sc. Paris **89** (1879), 74–76.
[19] H. E. Rauch and A. Lebowitz, *Elliptic functions, Theta functions and Riemann surfaces*, Williams & Wilkins, 1973.

DEPARTMENT OF MATHEMATICS, MICHIGAN STATE UNIVERSITY, EAST LANSING, MI 48824, U. S. A.

Current address: Department of Computer Science and Systems, McMaster University, Hamilton, ON L8S 4K1, Canada

E-mail address: jyang@dcss.mcmaster.ca

Selected Titles in This Series

(*Continued from the front of this publication*)

42 **Anil Nerode and Richard A. Shore, Editors,** Recursion theory (Cornell University, Ithaca, New York, June/July 1982)

41 **Yum-Tong Siu, Editor,** Complex analysis of several variables (Madison, Wisconsin, April 1982)

40 **Peter Orlik, Editor,** Singularities (Humboldt State University, Arcata, California, July/August 1981)

39 **Felix E. Browder, Editor,** The mathematical heritage of Henri Poincaré (Indiana University, Bloomington, April 1980)

38 **Richard V. Kadison, Editor,** Operator algebras and applications (Queens University, Kingston, Ontario, July/August 1980)

37 **Bruce Cooperstein and Geoffrey Mason, Editors,** The Santa Cruz conference on finite groups (University of California, Santa Cruz, June/July 1979)

36 **Robert Osserman and Alan Weinstein, Editors,** Geometry of the Laplace operator (University of Hawaii, Honolulu, March 1979)

35 **Guido Weiss and Stephen Wainger, Editors,** Harmonic analysis in Euclidean spaces (Williams College, Williamstown, Massachusetts, July 1978)

34 **D. K. Ray-Chaudhuri, Editor,** Relations between combinatorics and other parts of mathematics (Ohio State University, Columbus, March 1978)

33 **A. Borel and W. Casselman, Editors,** Automorphic forms, representations and L-functions (Oregon State University, Corvallis, July/August 1977)

32 **R. James Milgram, Editor,** Algebraic and geometric topology (Stanford University, Stanford, California, August 1976)

31 **Joseph L. Doob, Editor,** Probability (University of Illinois at Urbana-Champaign, Urbana, March 1976)

30 **R. O. Wells, Jr., Editor,** Several complex variables (Williams College, Williamstown, Massachusetts, July/August 1975)

29 **Robin Hartshorne, Editor,** Algebraic geometry – Arcata 1974 (Humboldt State University, Arcata, California, July/August 1974)

28 **Felix E. Browder, Editor,** Mathematical developments arising from Hilbert problems (Northern Illinois University, Dekalb, May 1974)

27 **S. S. Chern and R. Osserman, Editors,** Differential geometry (Stanford University, Stanford, California, July/August 1973)

26 **Calvin C. Moore, Editor,** Harmonic analysis on homogeneous spaces (Williams College, Williamstown, Massachusetts, July/August 1972)

25 **Leon Henkin, John Addison, C. C. Chang, William Craig, Dana Scott, and Robert Vaught, Editors,** Proceedings of the Tarski symposium (University of California, Berkeley, June 1971)

24 **Harold G. Diamond, Editor,** Analytic number theory (St. Louis University, St. Louis, Missouri, March 1972)

23 **D. C. Spencer, Editor,** Partial differential equations (University of California, Berkeley, August 1971)

22 **Arunas Liulevicius, Editor,** Algebraic topology (University of Wisconsin, Madison, June/July 1970)

21 **Irving Reiner, Editor,** Representation theory of finite groups and related topics (University of Wisconsin, Madison, April 1970)

20 **Donald J. Lewis, Editor,** 1969 Number theory institute (State University of New York at Stony Brook, Stony Brook, July 1969)

19 **Theodore S. Motzkin, Editor,** Combinatorics (University of California, Los Angeles, March 1968)

(See the AMS catalog for earlier titles)

ISBN 0-8218-0887-7